XV International Conference on Electrical Bio-Impedance (ICEBI) & XIV Conference on Electrical Impedance Tomography (EIT)

Journal of Physics: Conference Series Volume 434

Heilbad Heiligenstadt, Germany
22-25 April 2013

ISBN: 978-1-62748-311-7
ISSN: 1742-6588

Printed from e-media with permission by:

Curran Associates, Inc.
57 Morehouse Lane
Red Hook, NY 12571

Some format issues inherent in the e-media version may also appear in this print version.

Printed by Curran Associates, Inc. (2013)

Institute of Physics
Dirac House, Temple Back
Bristol BS1 6BE UK

Phone: 44 1 17 929 7481
Fax: 44 1 17 920 0979

techtracking@iop.org

Additional copies of this publication are available from:

Curran Associates, Inc.
57 Morehouse Lane
Red Hook, NY 12571 USA
Phone: 845-758-0400
Fax: 845-758-2634
Email: curran@proceedings.com
Web: www.proceedings.com

TABLE OF CONTENTS

THEORY

INSTRUMENTATION

ELECTRODES

SIGNAL PROCESSING

ULTRA WIDE BAND TECHNIQUE

CLINICAL APPLICATIONS

EIS-APPLICATION IN CARDIOLOGY

GENERAL APPLICATIONS

BODY IMPEDANCE ANALYSIS (BIA)

SPECIAL APPLICATIONS

ELECTRICAL IMPEDANCE TOMOGRAPHY (EIT)

MAGNETIC INTERACTION

NEW APPROACH

PREFACE

Over recent years advanced measurement methods have facilitated outstanding achievements not only in medical instrumentation but also in biotechnology. Impedance measurement is a simple and innocuous way to characterize materials. For more than 40 years biological materials, most of them based on cells, have been characterized by means of electrical impedance for quality control of agricultural products, monitoring of biotechnological or food processes or in health care. Although the list of possible applications is long, very few applications successfully entered the market before the turn of the century. This was, on the one hand, due to the low specificity of electrical impedance with respect to other material properties because it is influenced by multiple factors. On the other hand, equipment and methods for many potential applications were not available.

With the appearance of microcontrollers that could be easily integrated in applications at the beginning of the 1980s, impedance measurement advanced as a valuable tool in process optimization and lab automation. However, established methods and data processing were mostly used in a new environment. This has changed significantly during the last 10 years with a dramatic growth of the market for medical instrumentation and also for biotechnological applications. Today, advanced process monitoring and control require fast and highly parallel electrical characterization which in turn yields incredible data volumes that must be handled in real time.

Many newer developments require miniaturized but precise sensing methods which is one of the main parts of Lab-on-Chip technology. Moreover, biosensors increasingly use impedometric transducers, which are not compatible with the large expensive measurement devices that are common in the laboratory environment.

Following the achievements in the field of bioimpedance measurement, we will now witness a dramatic development of new electrode structures and electronics. Structures down to sub-micrometer range and complex impedance measurements tools integrated at single chips are now affordable.

Moreover, the introduction of alternative signals and data processing algorithms focuses on very fast and parallel electrical characterization which in turn pushes this technique to new applications and markets. Electrical impedance tomography today yields pictures in real time with a resolution that was impossible 10 years ago.

The XVth International Conference on Electrical Bio-Impedance in conjunction with the XIVth Electrical Impedance Tomography ICEBI/EIT 2013 organized by the Institute for Bioprocessing and Analytical Measurement Techniques, Heilbad Heiligenstadt, Germany, together with the EIT-group at the University of Göttingen, Germany, brings world leading scientists in these fields together. It is a platform to present the latest developments in instrumentation and signal processing but also points to new applications, especially in the field of biosensors and non-linear phenomena.

Two Keynote lectures will extend the view of the participants above the mainstream of bio-impedance measurement. Friederich Kremer (University of Leipzig) delivers the plenary lecture on broad bandwidth dielectric spectroscopy. New achievements in the research of ligand gated ionic channels will be presented by Klaus Benndorf (University of Jena).

Leading scientists in the field of bio-impedance measurement, such as, Sverre Grimnes, Orjan Martinsen, Andrea Robitzki, Richard Bayford, Jan Gimsa and Mart Min will give lectures for students but also more experienced scientists in a pre-conference tutorial which is a good opportunity to learn or refresh the basics.

List of committees

Conference Chair
Dr Uwe Pliquett
Professor Dieter Beckmann
Institut für Bioprozess- und Analysenmesstechnik eV, Rosenhof, Heilbad Heiligenstadt, Germany

Technical Program Chair
Maik Hiller
Conventus Congressmanagement & Marketing GmbH, Carl-Pulfrich-Str. 1 – 07745 Jena

Pre-Conference Tutorial Coordinator
Uwe Pliquett

International Advisory committee
Kenneth R Foster, USA
Sverre Grimnes, Norway
David Holder, UK
Alexander V Korjenewski, Russia
Ørjan G Martinsen, Norway
Mart Min, Estonia
Stig Ollmar, Sweden
Tadeusz Palko, Poland
Pere J Riu, Spain
Andrea Robitzki, Germany
Hermann Scharfetter, Austria
Leigh C Ward, Australia

Conference logo

Sponsor logos

Journal of Physics: Conference Series **434** (2013) 011002 doi:10.1088/1742-6596/434/1/011002

Pre-conference Tutorial XV ICEBI / XIV EIT 2013
21.April 2013 in Heilbad Heiligenstadt, Lecture Hall at Institut für Bioprozeß- und Analysenmeßtechnik

10:00 – 10:45	Impedance basics	Orjan Martinsen
10:45 - 11:30	Modeling	Sverre Grimnes
Coffee break		
11:45 – 12:30	Electrode Polarization	Friedrich Kremer
12:30 - 13:15	EIT basics	Richard Bayford
lunch		
14:00 – 14:45	AC-electrokinetic phenomena	Jan Gimsa
14:45 - 15:30	Lab-on-Chip-application	Andrea Robitzki
Coffee break		
15:45 – 16:30	Time Domain Spectroscopy	Uwe Pliquett
16:30 – 17:15	Signal processing	Mart Min

Orjan Martinsen, Impedance basics
It is important in our field to have a common basic understanding of the fundamental concepts. We will talk about things like impedance and admittance and material properties like resistivity, conductivity and permittivity. More exotic things like memristor and memcapacitors will also be mentioned. Lastly we will go through different electrode systems and their respective sensitivity field features.

Sverre Grimnes, Modeling
Why do we need models, what are they for? Is it to have a simplified picture of the behavior of the modeled thing or process? E.g. a simple electronic circuit having the same impedance spectrum as the tissue. Is it to understand better or being able to give prognoses? I will discuss the most common models: Coulomb, Maxwell, Debye, Cole, Cole-Cole, the transducing and the Universal model

Friedrich Kremer, Charge transport and electrode polarization
Charge transport at the interfaces between electronic and ionic conductors is a topic of special technological and fundamental importance in contemporary science. The quest for quantitative understanding of the extent and nature these interfaces is necessitated by the need for more efficient and optimal power sources among others. For practical purposes, the dynamics of the charge carriers around these boundaries needs to be understood over the broadest range of frequencies and temperatures possible. Due to its ability to measure the complex dielectric function (and hence the complex conductivity) over many orders of magnitude in frequency and in a wide temperature interval, Broadband Dielectric Spectroscopy (BDS) turns out to be an ideal experimental tool in this field.

Richard Bayford, EIT basics
Electrical impedance tomography (EIT) is a often referred to as a relatively new imaging method that has evolved over the past 30 years. It has the potential to be of great value in clinical diagnosis; however, EIT is a technically difficult problem to solve in terms of developing hardware for data capture and the algorithms to reconstruct the images. This lecture looks at the development of EIT and how it has evolved. It focuses on its clinical applications, examining hardware for the collection of data and reconstruction algorithms to generate images.

Jan Gimsa, AC-electrokinetic phenomena

The lecture considers DC- and AC-electrokinetic effects which can be used for the characterization or manipulation of suspended objects, media or interfaces.

The classical DC-electrokinetic effects refer to particle and medium movements which are either induced by an external DC field (electrophoresis and electroosmosis) or by external forces. External forces may induce either sedimentation or streaming potentials. Because DC-electrokinetic effects are based on fixed surface charges they yield information on these charges and the Zeta potential.

Also in AC-electrokinetics, particle and medium movements are observed, though at frequencies from the low Hz up to the GHz ranges. The effects can be exploited in spectroscopy methods, measuring various frequency-dependent force effects. AC fields induce charges at the structural interfaces of particles and biological cells. The interaction of these charges with the inducing field generates forces which are the basis for AC-electrokinetic effects such as electrorotation, dielectrophoresis (DP), electroorientation, electrodeformation, field-trapping and travelling-wave DP. In media, local heating may alter the dielectric properties, leading to "smeared interfaces" which may be polarized by AC fields similar to structural interfaces. This may result in medium convections which can be exploited for pumping or mixing.

Andrea Robitzki, Lab-on-Chip: Bioimpedance Based High Content-Screening of Organotypic Tissues

The tutorial will address the development and use of interdigital microarrays, high dense electrode and microcavity arrays for the characterization of organotypic tissues via impedance spectroscopy. The focus of the report will be (i) the coupling of e.g. viable brain slice cultures on microelectrode arrays consisting of semi-conductive electrode materials for impedance spectroscopy, (ii) field potential recording and (iii) high resolution microscopy on one and the same sensor. For a high sensitive recording the biomimetic interface on the microelectrode is the relevant key point for an intensive cell-electrode contact. Switching between different biophysical measurement methods on the same device demands on a feasible microcontroller directing the sharing of same frequency ranges without any disturbance of the impedance or filed potential signals. Therefore, the presented technique is of great benefit for the pharmaceutical industry, especially in respect of functional lead identification or efficacy of active pharmaceutical ingredients. The cellular dielectric spectroscopy (CDS) or electric impedance spectroscopy (EIS) can be used to measure frequency dependent alterations of passive electrical properties of complex tissues. For this purpose, an alternate voltage current is applied to a biological sample e.g. a tissue slice whereby the current flows from an active working electrode through and beneath the organotypic tissue to a counter electrode. Under these non-invasive conditions the tissue itself reacts as a resistor and capacitor affecting the recorded impedance. Depending on the frequency and the dielectric properties of tissue structure and architecture, it is possible to monitor different biological or physiological processes described by an electric circuit model

Uwe Pliquett, Time Domain Spectroscopy

A frequency sweep for assessment of an impedance spectrum is time consuming and therefore not appropriate for monitoring material with fast changing passive electrical properties. A way out is the use of a broad bandwidth excitation signal and tracing its response as a function of time. Depending on the explicit application, especially on the frequency range of interest, transmission or reflection measurements arranged in single or differential configuration are possible. Measurements within the ß-dispersion region with sub-millisecond resolution are possible but better precision is achieved by averaging over several periods of a periodic excitation signal. The lecture will start with the basics of time domain spectroscopy followed by the most popular excitation signals and conclude with selected applications.

Mart Min, Signal Processing

Experimental characterization of bioimpedance needs measurements in a wide frequency range (impedance spectroscopy). For the performing of measurements, synthesis and generation of an excitation or stimulus signal (current of voltage) and extracting of information from the response signal (voltage of current) is required. One classical way is to have a sine wave excitation with controllable frequency and use a phase-sensitive detection of the sine wave response signal. However, the other waveforms are more effective in many cases where the fast spectral analysis of the impedance is of interest. In this case the waveforms like multi-sine and chirp/chirplet signals and their binary and ternary modification can be much more effective. Also the methods and algorithms for the processing of response signal as short time and fast Fourier Transforms (STFFT) and other are in use as the most suitable ones. Such and similar questions will be discussed in the signal processing tutorial.

Peer review statement

All papers published in this volume of *Journal of Physics: Conference Series* have been peer reviewed through processes administered by the proceedings Editors. Reviews were conducted by expert referees to the professional and scientific standards expected of a proceedings journal published by IOP Publishing.

Sensitivity field distributions for segmental bioelectrical impedance analysis based on real human anatomy

A A Danilov[1], V K Kramarenko[2], D V Nikolaev[3], S G Rudnev[1], V Yu Salamatova[4], A V Smirnov[3], Yu V Vassilevski[1,2]

[1]Institute of Numerical Mathematics, Russian Academy of Sciences
[2]Moscow Institute of Physics and Technology
[3]Scientific-Research Centre MEDAS
[4]Scientific-Educational Centre of the Institute of Numerical Mathematics, Russian Academy of Sciences
all - Moscow, Russian Federation

Email: a.a.danilov@gmail.com

Abstract. In this work, an adaptive unstructured tetrahedral mesh generation technology is applied for simulation of segmental bioimpedance measurements using high-resolution whole-body model of the Visible Human Project man. Sensitivity field distributions for a conventional tetrapolar, as well as eight- and ten-electrode measurement configurations are obtained. Based on the ten-electrode configuration, we suggest an algorithm for monitoring changes in the upper lung area.

1. Introduction

Segmental bioelectrical impedance analysis (segmental BIA) is commonly used for various purposes such as the assessment of body composition and abdominal adiposity, as well as for monitoring of body fluids redistribution under various physiological and pathological conditions [1-4]. Available equipment is produced by a number of manufacturers including InBody (Korea), Omron (Japan), RJL Systems (USA), Tanita (Japan), Medas (Russia) and other (see, e.g., [5,6]). For validation of existing measurement schemes, the development of new ones and for accurate data interpretation, it is important to know relative contribution of tissues and organs to the result of bioimpedance measurements of the particular body segment. Here, we aimed at the construction and visualization of sensitivity field distributions for segmental BIA using anatomically accurate 3D model of the human body. Similar approach has already been used before with partially segmented models of the human body, e.g., in impedance cardiography [7,8].

2. Methods

The segmented model of the Visible Human Project (VHP) man [9] containing 30 materials was constructed. For this, partially segmented model of the VHP man torso [10] was initially used with the subsequent improvements as described in details in [11,12] and with the addition of segmented head and extremities.

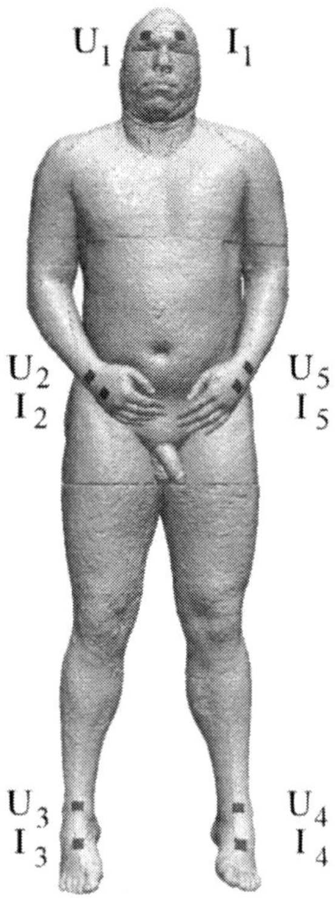

Fig. 1. A scheme of current (I) and potential (U) electrodes location on the geometrical model of the VHP man for simulation segmental BIA measurements.

Along with conventional tetrapolar wrist-to-ankle measurement configuration, two schemes of segmental BIA were considered: an eight-electrode one with the placement of current and potential electrodes 5 cm apart on the dorsal surfaces of the wrists and ankles, and also ten-electrode scheme with an additional electrode pair located on the forehead as shown in Fig.1. Accordingly, five pairs of thin bilayer square objects 23×23 mm in size simulating electrode properties were added on the forehead and distal parts of arms and legs of the segmented model. Possibilities for the assessment of body regions using various combinations of current and voltage electrode pairs as depicted in Fig. 1 are shown in Table 1. A thin nonconducting layer was added between arms and torso surfaces of the model to avoid tissue contacts. Clearly, the considered 8- and 10-electrode configurations enable measurement of at least 5 (arms, legs, torso) and 6 (arms, legs, torso, and head-neck-upper torso area) body segments, respectively.

After the addition of electrodes, an adaptive unstructured tetrahedral mesh was generated and post-processed in order to improve its quality using mesh cosmetics algorithms from the Ani3D library [13]. The resulted mesh contained 470 thousands vertices and 2.7 millions tetrahedra. In it, most anatomical features of the human body were preserved while keeping a feasible number of mesh elements. BIA measurements at the electrical current frequency 50 kHz were simulated using finite element mathematical model (for details, see [11]). The electrical conductivity parameters for labeled tissues were taken from [14] and the references therein. Sensitivity field distributions were obtained according to Geselowitz formula [15].

Table 1. Some combinations of current (I) and potential (U) electrodes used for the assessment of various body regions in segmental BIA according to the 10-electrode measurement configuration.

Body region	Combination of electrodes	
	I	U
Right arm, right leg, torso	2, 3	2, 3
Torso	5, 4	2, 3
Torso	5, 3	2, 4
Head, neck, upper torso	1, 2	1, 5
Right arm	2, 3	2, 1
Right arm	2, 3	2, 5
Left arm	5, 1	5, 4
Left arm	5, 4	5, 2
Right leg	3, 4	3, 2
Left leg	4, 5	4, 3
Head, neck, torso	1, 4	1, 3

3. Results

Current density fields were calculated from the finite-element model, and the corresponding sensitivity field distributions were obtained for various configurations of electrode sites. Figure 2 shows body regions providing 98% of measurement sensitivity for various schemes of electrodes placement: a conventional tetrapolar scheme (a) and some of those provided by the eight- (c) and ten-electrode configurations (d). (The integral of the sensitivity over the volume area represented in Figures 2a, 2c, and 2d is equal to 98% of the total integral of the sensitivity.) Also, an integral projection of sensitivity field for the tetrapolar scheme can be seen in Figure 2b: the higher values of the projection correspond to the more intense color.

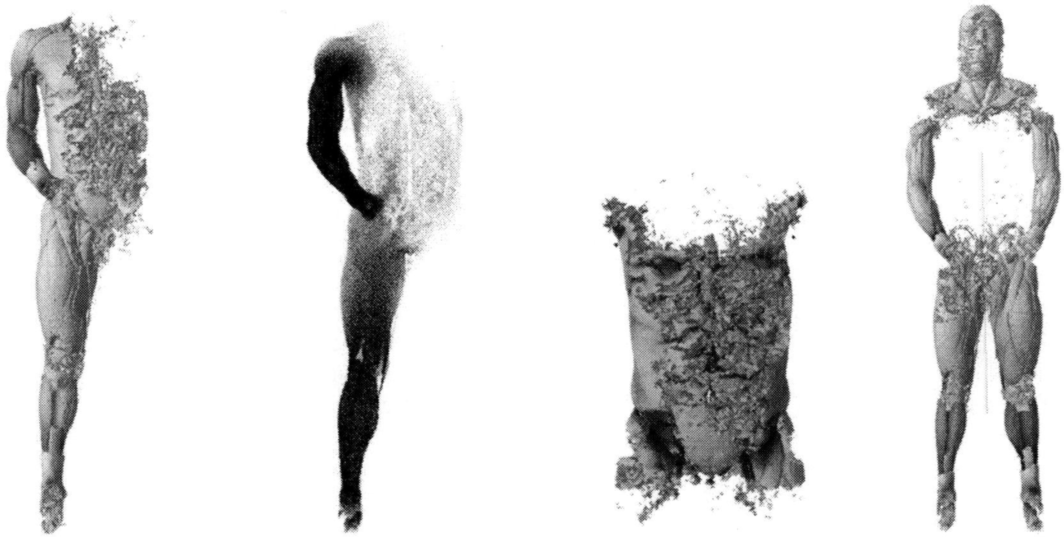

a b c d

Fig. 2. (a) The regions of high sensitivity and (b) the frontal integral projection of sensitivity density function for a conventional tetrapolar measurement scheme; (c) the regions of high sensitivity for the eight-electrode configuration, torso measurements; (d) a sum of high sensitivity regions for the ten-electrode configuration, measurements of head-neck-upper torso area and of extremities.

The regions of high sensitivity for the parallel and crosswise schemes of torso measurements according to Table 1 were essentially the same with the difference lying within 1%. The same applies to measurements according to 10-electrode configuration providing the volume difference within 1.5% compared to the above-mentioned schemes. Our data show that, given the considered electrodes configuration and body position, about 95-96% of its volume can be assessed, with the exception of distal parts of foots and hands, as well as lateral parts of shoulders (Fig. 2d). The latter feature is explained by the uncommon, from the viewpoint of the conventional BIA measurements, position of arms of the VHP man, and is expected to be eliminated with the recommended body position. The specific position of the anatomical model of the body, in particular the location of the hands, is prescribed by the position of male cadaver during VHP data acquisition. Some future work is planned to create a model with the conventional hands location.

Figure 3 demonstrates a sufficiently wide space present between the non-intersecting regions of high sensitivity of torso and head-neck-upper torso area measurements with the body contours depicted in light gray, and lungs and heart shown in dark grey as a reference. The subtraction of the impedances of torso and head-neck-upper torso region from the total impedance of head, neck and torso (we can define it for any appropriate electrode placement, e.g., according to $I_1I_4U_1U_3$ scheme, see Table 1) provides the impedance of intermediate space that involves upper lung region. Therefore, theoretically, application of the ten-electrode measurement scheme provides a method for the assessment hydration changes of this important area. Experimental testing of this assumption gave the impedance values in the range between 3 and 7 Ohms.

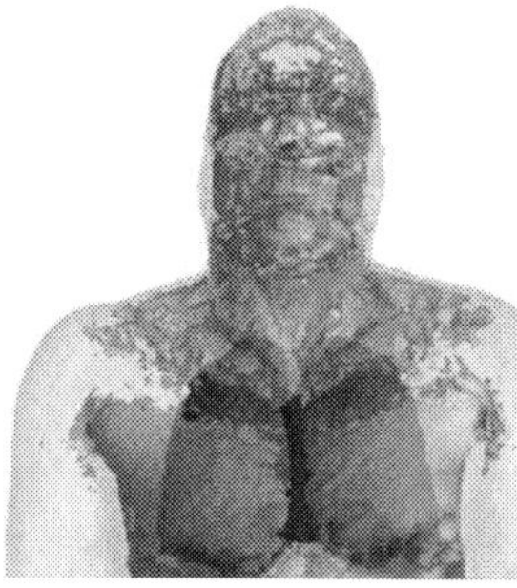

a

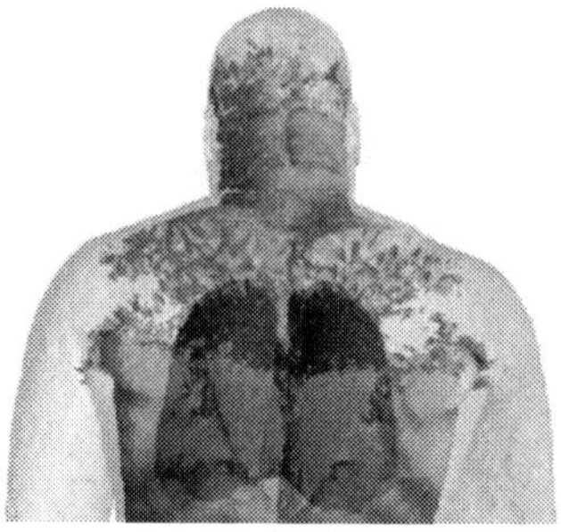

b

Fig. 3. The regions of high sensitivity of torso and head-neck-upper torso area measurements as provided by the ten-electrode configuration: a - frontal view, b - rear view.

4. Conclusion

In this study, sensitivity field distributions for the conventional tetrapolar measurement scheme as well as for the segmental BIA as represented by the eight- and ten-electrode configurations are constructed using high-resolution 3D model of the VHP man. Our data provide accurate interpretation of the results of segmental BIA. The advanced mesh generation approach along with numerical modeling procedure can serve as a valuable tool for evidence-based approach to the development and validation of novel BIA techniques and measurement schemes.

Acknowledgements
This work was partially supported by the Russian Foundation for Basic Research (grants no. 11-01-00971, 12-01-31223), and the Federal Program "Academic and pedagogical staff of innovative Russia". We thank two anonymous reviewers for useful comments.

References

[1] Organ L W, Bradham G B, Gore D T and Lozier S L 1994 *J. Appl. Physiol.* **77** 98-112
[2] Fuller N J, Fewtrell M S, Dewit O, Elia M and Wells J C 2002 *Int. J. Obes.* **26** 684-691
[3] Ward L C 2012 *Curr. Opin. Clin. Nutr. Metab. Care* **15** 424-429
[4] Bosy-Westphal A, Schautz B, Later W, Kehayias J J, Gallagher D and Müller M J 2013 *Eur. J. Clin. Nutr.* **67** S14-S21
[5] Nikolaev D, Smirnov A and Tarnakin A 2001 in: Proc. XI Int. Conf. on Electrical Bio-impedance. June 17-21, Oslo, Norway. 381-384
[6] RJL Systems 2010 http://www.rjlsystems.com/pdf-files/segmental_bia.pdf
[7] Kauppinen P K, Hyttinen J A and Malmivuo J A 1998 *Ann. Biomed. Eng.* **26** 694-702
[8] Yang F and Patterson R 2010 in: 32nd Annual Int. Conf. of the IEEE EMBS. Buenos-Aires, Argentina, August 31- Sept. 4.
[9] The Visible Human Project. http://www.nlm.nih.gov/research/visible/visible_human.html
[10] Höhne K H, Pflesser B, Pommert A, Riemer M, Schubert R, Schiemann T, Tiede U and Schumacher U 2001 *Meth. Inform. Med.* **40** 83-89
[11] Danilov A A, Nikolaev D V, Rudnev S G, Salamatova V Yu and Vassilevski Yu V 2012 *Russ. J. Numer. Anal. Math. Modelling* **27** 431-440
[12] Danilov A A, Salamatova V Yu and Vassilevsky Yu V 2012 *J. Phys.: Conf. Series* **407** 012004
[13] 3D generator of anisotropic meshes. http://sourceforge.net/projects/ani3d
[14] Gabriel C, Peyman A and Grant E 2009 *Phys. Med. Biol.* **54** 4863-4878
[15] Geselowitz D B 1971 *IEEE Trans. Biomed. Eng.* **18** 38-41

Towards an accurate bioimpedance identification

B Sanchez[1], **E Louarroudi**[2], **R Bragos**[1], **R Pintelon**[2]

[1]Electronic and Biomedical Instrumentation Group, Department of Electrical Engineering, Universitat Politecnica de Catalunya (UPC), 08034 Barcelona, Spain
[2]Department of Fundamental Electricity (ELEC), Vrije Universiteit Brussel (VUB), B-1050 Brussels,Belgium

E-mail: benjamin.sanchez@upc.edu

Abstract. This paper describes the local polynomial method (LPM) for estimating the time-invariant bioimpedance frequency response function (FRF) considering both the output-error (OE) and the errors-in-variables (EIV) identification framework and compare it with the traditional cross− and autocorrelation spectral analysis techniques. The bioimpedance FRF is measured with the multisine electrical impedance spectroscopy (EIS) technique. To show the overwhelming accuracy of the LPM approach, both the LPM and the classical cross− and autocorrelation spectral analysis technique are evaluated through the same experimental data coming from a nonsteady-state measurement of time-varying *in vivo* myocardial tissue. The estimated error sources at the measurement frequencies due to noise, σ_Z^{n}, and the stochastic nonlinear distortions, σ_Z^{NL}, have been converted to Ω and plotted over the bioimpedance spectrum for each framework. Ultimately, the impedance spectra have been fitted to a Cole impedance model using both an unweighted and a weighted complex nonlinear least square (CNLS) algorithm. A table is provided with the relative standard errors on the estimated parameters to reveal the importance of which system identification frameworks should be used.

1. Introduction

Most of the spectral analysis techniques used for estimating the bioimpedance frequency response function (FRF) from broadband Electrical Impedance Spectroscopy (EIS) measurements are still limited today. One of the main limitations of many approaches relies on the fact that no information is gathered about the measurement noise and the non-linear distortions. In fact, it is widely common to just simply calculate the impedance frequency response as the division of the voltage and current Fourier coefficients. In other cases, the identification/measuring approach considered does not exploit the periodic nature of the excitations and leakage errors are introduced on the impedance spectrum when using the discrete Fourier transform (DFT) [1]. In the last step, data are usually fitted to a model, e.g. the Cole impedance model [2], which is an empirical complex nonlinear function model in the angular frequency $\omega = 2\pi f$

$$Z(\omega, \theta) = R_\infty + \frac{R_0 - R_\infty}{1 + \left(j\frac{\omega}{\omega_c}\right)^\alpha}, \quad \text{with } \theta = [R_0, R_\infty, \omega_c, \alpha]^T. \tag{1}$$

From the optimal curve-fit parameters θ^{opt}, the Jacobian ($\mathbf{J} \in \mathbb{C}^{F \times n_\theta}$) is calculated as

$$\mathbf{J}_{ij} = \left.\frac{\partial Z(\omega_i, \theta_j)}{\partial \theta_j}\right|_{\theta_j = \theta_j^{\mathrm{opt}}}, \tag{2}$$

where n_θ and F stand, respectively, for the number of parameters and measured frequencies. An estimate of the covariance matrix of the parameters follows then easily form (2), viz.

$$\text{Cov}\left(\theta^{\text{opt}}\right) \approx \left(2\,\text{Re}\left\{\mathbf{J}^H\mathbf{W}\mathbf{J}\right\}\right)^{-1}, \tag{3}$$

where $\mathbf{W}$ is a diagonal matrix with $\mathbf{W}_{ii} = \text{Var}\left(Z\left(\omega_i\right)\right)^{-1}$ as diagonal elements, i.e. the identity matrix for the unweighted CNLS approach. The asymptotic standard error (SE) for the optimal parameter vector θ^{opt} equals then

$$\text{Std}(\theta^{\text{opt}}) = \sqrt{\text{diag}\left(\text{Cov}\left(\theta^{\text{opt}}\right)\right)}, \tag{4}$$

where diag (X) extracts the diagonal elements of X. The SE is a powerful measure of how unexplained variability in the data propagates to the variability in the solution, and is a measure for the stochastic error in the model parameters (not the fitted spectrum) when modeling impedance data.

This paper focuses on the question which time-invariant spectral analysis techniques yield an accurate identification of electrical bioimpedance? To objectively assess the quality of the identification process, we use as quality tools the standard errors (4) in the Cole model parameters. We consider two periods (2 ms) of current $i(t)$ and voltage $v(t)$ time-domain signals originating from the *in situ* measurement of *in vivo* time-varying myocardium. The excitation signal is a periodic multisine signal ($F = 26$ frequencies, 1 ms, 1 kHz $\rightarrow$ 1 MHz). The error that is present in the measurements is mainly due to noise, non-linear distortions, inherent feedback conditions and the transients in the response. We process these data including the errors, first, by using the classical spectral analysis approach based on cross- and autocorrelation and, second, with the local polynomial method (LPM). In both cases, both an output-error (OE) and Error-In-Variables (EIV) identification framework are considered.

2. The Local Polynomial Method (LPM)

Briefly stated, the general LPM solution solves an over-determined set of equations in a least square (LS) sense for each spectral component of interest. This LS problem can be calculated at the excited frequencies using the Moore-Penrose algorithm or the Singular Value Decomposition.

2.1. Output-Error Local Polynomial Method (OE-LPM)

The main idea of the LPM theory is based on considering an impedance system where the measured DFT voltage spectrum $V\left(k\right)$ equals the DFT current spectrum $I\left(k\right)$ multiplied with the bioimpedance FRF $Z\left(k\right)$, plus a transient term $T\left(k\right)$ and additive noise $N\left(k\right)$, viz.

$$V\left(k\right) = Z\left(k\right)I\left(k\right) + T\left(k\right) + N\left(k\right).$$

Then, using the smoothness property of $Z(k)$ and $T(k)$, the bioimpedance FRF $Z\left(k\right)$ and the transient term $T\left(k\right)$ can *locally* be parametrized with a pth order polynomial around the excited frequency bin k considering n neighboring spectral points around k, this is the small frequency band $[k-n, k-n+1, \cdots, k+n]$ (with $2n+1$ being the width of the LPM window). The estimated bioimpedance spectrum $\hat{Z}\left(k\right)_{\text{OE-LPM}}$ is then obtained by solving the set of equations given by $V_k = K_I(k)\,\theta(k)_{\text{OE-LPM}}$, with $V_k = \left(V(k-n)\cdots V(k)\cdots V(k+n)\right)^T$ and $\theta(k)_{\text{OE-LPM}} = \left(Z\left(k\right), z_1(k), \ldots, z_p(k), T\left(k\right), t_1(k), \ldots, t_p(k)\right)^T_{\text{OE-LPM}}$. $K_I(k)$ is a complex matrix that depends on the Fourier coefficients of the measured current spectrum $I\left(k\right)$. Note that $\theta(k)$ contains the bioimpedance FRF $Z(k)$, the polynomial bioimpedance parameters $z_i(k)$'s and the transient parameters $T(k)$, $t_i(k)$'s. The solution for this OE problem is found as the first element in $\theta(k)_{\text{OE-LPM}}$, namely $\hat{Z}\left(k\right)_{\text{OE-LPM}} = \left(\hat{\theta}\left(k\right)_{\text{OE-LPM}}\right)_1 = \left(\left(K_I(k)^H K_I(k)\right)^{-1} K_I(k)^H V_k\right)_1$. Figure 1 (top) illustrates the block diagram for computing the bioimpedance FRF with the OE-LPM. The OE-LPM results are summarized in figure 2 (C).

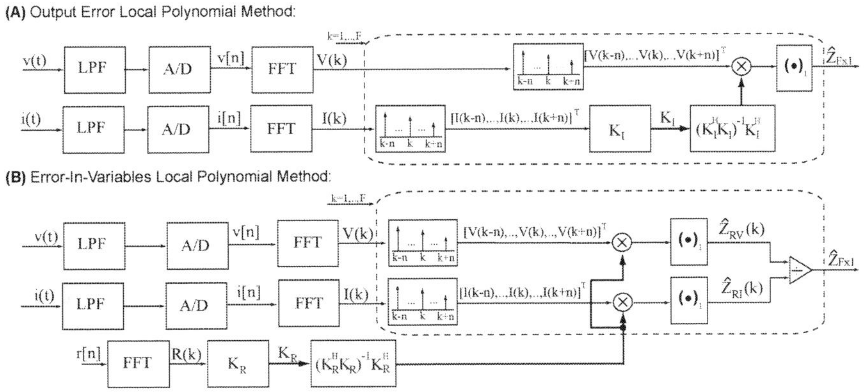

Figure 1. Schematic of an impedance measurement system implementing the OE-LPM (top) and the EIV-LPM (bottom) solutions; $R(k)$, $I(k)$ and $V(k)$ denote, respectively, the DFT of the reference $r[n]$ –e.g. (periodic) broadband signal (multisine)–, the current $i[n]$ and the voltage $v[n]$ samples for the F measured frequencies.

2.2. Errors-In-Variables Local Polynomial Method (EIV-LPM)

Contrary to the LPM-OE approach where the current signal $i(t)$ is assumed to be known and used to calculate the K_I matrix, the errors-in-variables (EIV) framework computes the LS solution w.r.t. the noisy voltage and current signals. The bioimpedance FRF is then determined as $\hat{Z}\,(k)_{\text{EIV-LPM}} = \hat{Z}_{RV}\,(k)\,/\hat{Z}_{RI}\,(k)$, where $Z_{RI}\,(k)$ is the bioimpedance spectrum from the reference spectrum $R(k)$ to the current spectrum $I(k)$, namely $\hat{Z}_{RI}\,(k) = \left(\left(K_R(k)^H K_R(k)\right)^{-1} K_R(k)^H I_k\right)_1$, and $Z_{RV}\,(k)$ from the reference spectrum $R(k)$ to the voltage spectrum $V(k)$, namely $\hat{Z}_{RV}\,(k) = \left(\left(K_R(k)^H K_R(k)\right)^{-1} K_R(k)^H V_k\right)_1$ with $X_k = [X(k-n)\cdots X(k)\cdots X(k+n)]^T$, $X = V, I$. Note that now the regression matrix $K_R(k)$ does not depend on the current nor the voltage measurements. Figure 1 (bottom) shows a schematic implementing this framework. The EIV-LPM results are shown in figure 2 (D).

3. Experimental myocardial bioimpedance identification and conclusions

The EIV results shown in figure 2 (B) originates from the estimation of the FRF from the reference multisine excitation to the voltage and current channels without taking into account the transient error in the data. The reader is referred to [3] for the technical explanation on the cross and autocorrelation OE-EIV frameworks (see figure 2 (A)-(B)) without transient suppression (see [4] for more details about the OE-LPM and the EIV-LPM estimators). To make a fair comparison, the measurement frequencies were chosen to be the same in number for the cases (A)-(D) in figure 2 ($F = 22$ fitted frequencies). As it may be observed, the largest difference between the unweighted and the weighted total noise-variance (including non-linear distortions) CNLS is in the case of the classical OE spectral analysis. The fact of using the reference (i.e. allowing to work in feedback conditions), the EIV framework significantly improves the quality in both the bioimpedance and the parameter estimates. Finally, the LPM estimator shown in figure 2 have a superior accuracy in both the bioimpedance spectrum and parameter estimates due to the (very) good transient suppression.

4. Acknowledgements

This work was supported in part by grants from the Spanish Ministry MICINN SAF2011-30067-C02-02, by the Fund for Scientific Research (FWO-Vlaanderen), by the Flemish Government

(Methusalem), and by the Belgian Government through the Interuniversity Poles of Attraction (IAP VII) Program.

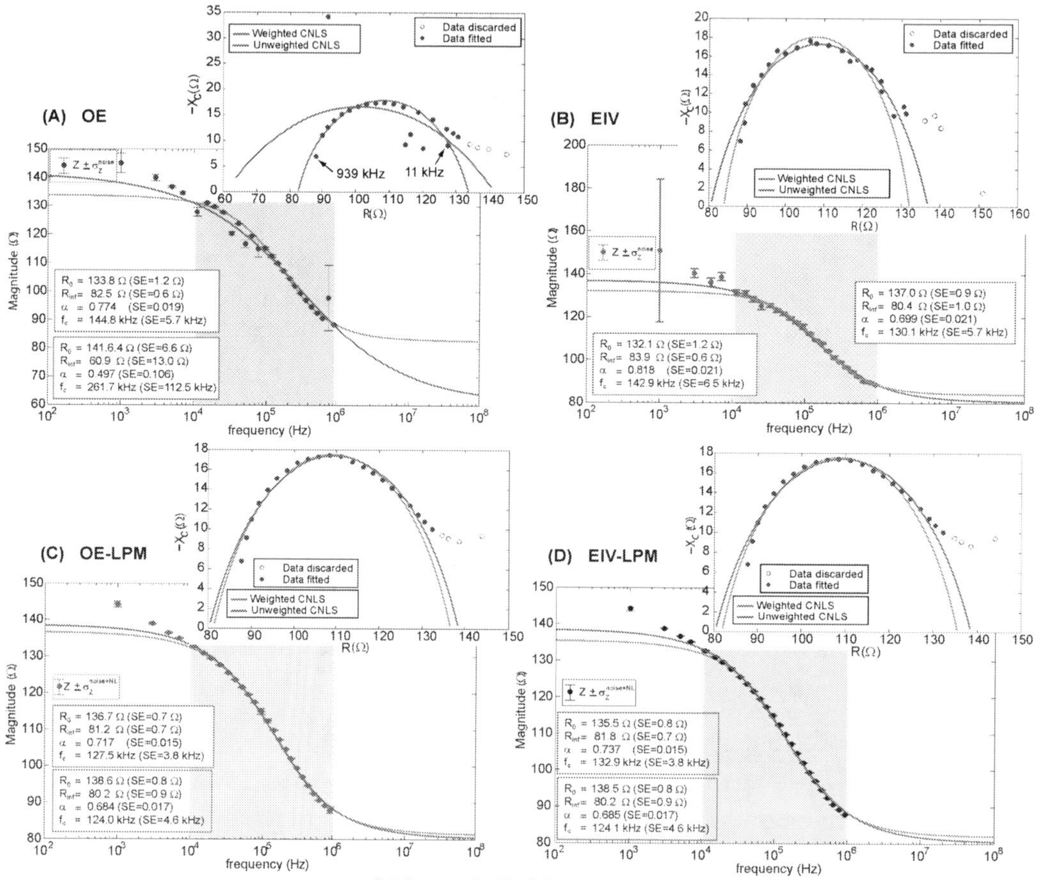

Relative error for the Cole parameters

	Unweighted CNLS				Weighted CNLS			
	OE	EIV	OE-LPM	EIV-LPM	OE	EIV	OE-LPM	EIV-LPM
R_0	4.69%	0.66%	0.58%	0.58%	0.86%	0.90%	0.56%	0.58%
R_{inf}	21.34%	1.24%	1.12%	1.12%	0.75%	0.67%	0.86%	0.85%
α	21.33%	3.00%	2.48%	2.48%	2.45%	2.57%	2.09%	2.03%
f_c	42.97%	4.38%	3.71%	3.71%	3.94%	4.57%	3.03%	2.88%

Figure 2. Accuracy of the bioimpedance and Cole parameters using either the cross- and autocorrelation spectral analysis technique or the Local Polynomial Method (LPM): (A) output-error (OE), (B) errors-in-variables (EIV), (C) OE-LPM and (D) EIV-LPM frameworks.

References

[1] Troltzsch U, Kanoun O and Trankler H R 2006 *Electrochim. Acta* **51** 1664 – 1672
[2] Cole K S 1940 *Cold Spring Harbor Symposia on Quantitative Biology* **8** 110–122
[3] Pintelon R and Schoukens J 2012 *System identification: a frequency domain approach* (Piscataway: IEEE Press (2nd edition))
[4] Sanchez B, Schoukens J, Bragos R and Vandersteen G 2011b *IEEE Trans. Biomed. Eng.* **58** 3376–3385

A model-based approach for analysis of intracellular resistance variations due to body posture on bioimpedance measurements

Sören Weyer, Mark Ulbrich, Steffen Leonhardt

Chair for Medical Information Technology,
Helmholtz Institute for Biomedical Engineering,
RWTH Aachen University, Aachen, Germany

E-mail: weyer@hia.rwth-aachen.de

Abstract. Bioimpedance spectroscopy is a known option for measuring body fluid volume. However, it is prone to a variety of influence factors which prevent a wider use. One of these influencing factors is the body posture. It could be shown that the average intracellular resistance percentage changes when the subject changes position from lying to standing. Most authors explain this phenomenon by fluid shifts. Another possible reason is the stray capacitance between the body and the ground, because if a certain fraction of the injected current follows other paths than between the potential electrodes, the result will be wrong. This paper analyses the influence of different body postures on the measured intracellular resistance and the posture depending capacity. For this purpose, FEM simulations are used. Subsequently, an electrical equivalent model with capacitances was developed. With this model, it is possible to correct the measured impedance and to neglect the influence of the stray capacitance.

1. Introduction

Due to the important role of body fluid, it is one of the major clinical challenges to assess the fluid status and its distribution accurately in human. In recent years, a method called Bioimpedance Spectroscopy (BIS) has been successfully established. It is non-invasive, rapid and it is low-cost. Clinical observation is often focused on the quantity of extracellular fluid (ECF), but also measures body composition such as intracellular fluid (ICF) are taken into account for the determination of a hydration status. BIS is prone to a variety of influence for example the body posture. The immediate influence of body posture on BIS was already recognized by Kushner et al. some years ago [1]. Most authors argue that the influence of body position on BIS measurements is caused by a redistribution of body fluids [1, 2, 3, 4]. A change from a vertical to a horizontal position produces a fluid shift from the extremities to the torso [5, 6] causing an impedance increase in the extremities and an impedance decrease in the torso [4]. As the torso contributes up to only 5 % of the whole body impedance [4], the total measured body impedance will increase [2, 4]. The change from a horizontal to a vertical position produces the opposite effect. Most fluid shifts occur in the ECF, since ICF is only slightly affected [7]. Nevertheless, it could be shown that the average ICF resistance percentage changes when the subject becomes supine [2]. Typical changes for the intracellular resistance for the extremities are small (below 3.7%). However, they are large (21.4%) for the trunk and also quite large

(6.1%) for the wrist-to-ankle measurement. Finally, because ICF is not that able to move under gravity than ECF, this variation could not only be caused by gravity. Within this work, the effect of the body posture on stray capacitances between the body and the ground, which also influences BIS measurements, is analysed.

2. FEM-Model

At bioelectrical impedance spectroscopy measurements two fundamental stray capacitances exist with the ground. First, there are stray capacitances between current electrodes, potential electrodes and connections leads. Shielding and guarding cables can largely reduce the effect of those stray capacitances, thus, their effect on the results is considerably small. The other stray capacitance C_s is formed between the human body an the ground. C_s can cause a significant error in the measurement, because C_s conducts a small part of the current emanated by the person to the ground and thus changing the measured impedance. C_s basically depends on body size and its closeness to grounded objects.

To analyse the influence of the stray capacitance - without being influenced by the fluid shift or other physiological processes - a finite element model is introduced. CST EMStudio® from Computer Simulation Technology, Darmstadt, Germany was used to perform the simulations. The excitation frequencies were between 10 kHz and 1 MHz. Since the wavelengths are higher than the human body the low frequency electroquasistatic solver has been employed. The 3D model of the human body has an accurate shape and posture. Important tissues have been realised, such as muscle, bone and various organs. An average permittivity and conductivity of the organs has been assigned to corresponding tissue with data found in literature [8]. The current is virtually injected between wrist and ankle corresponding to a whole body measurement. The model comprises over two million mesh-cells. To obtain accurate results for small geometry changes, a finer local mesh has been used for shape changing organs. With this model, different postures are chosen for simulation: tall standing, lying in bed (shown in fig. 1) and completely electrical isolated. The human finite element model position remains constant in all set-ups. Only the ground plane varies between right under the feet, with a small distance of 1 cm, and a dorsal placed ground plane with a distance of a standard bed height of approximately 45 cm.

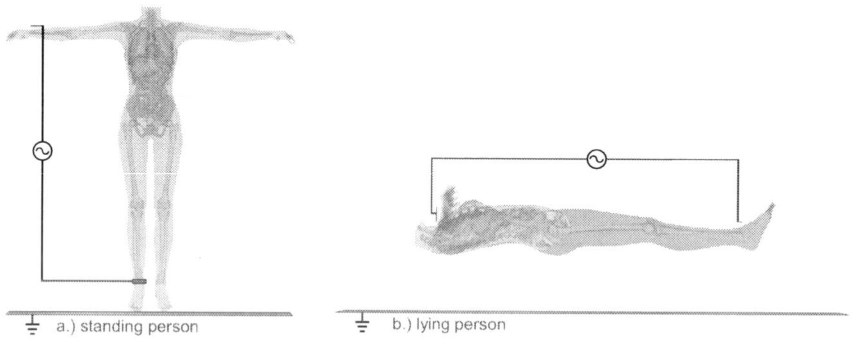

Figure 1. a.) FEM-model of a standing person, b.) FEM-model of a lying person

3. Circuit-Model

The circuit in Figure 2 models a tetrapolar bioimpedance measurement. The basis of the electrical model equivalent are the models of error in measurements proposed by Aliau-Bonet and Scharfetter [9, 10]. The circuit model is based on applying a single ended current source

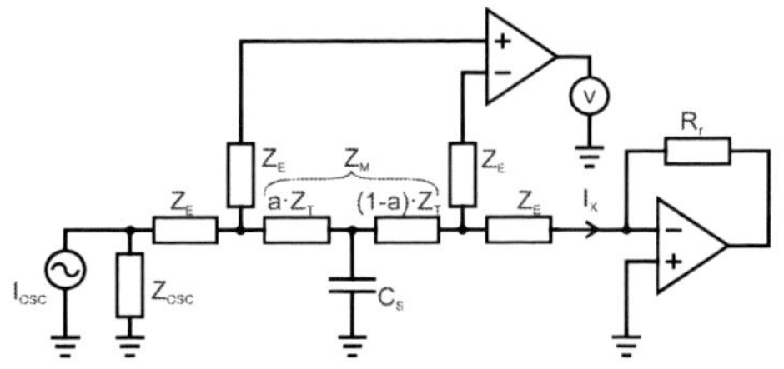

Figure 2. Simplified circuit for a BIS measurement system, modified from [9]

(I_{osc}, Z_{osc}). The electrode-skin contact transition impedances is modelled by Z_E. Due to the tetrapolar measurment setup this influence factor can be neglected. The tissue impedance is represented by the impedance Z_T, which is divided by the capacitance C_S. The parasitic capacitance C_s expresses all the parasitic capacitances that create an alternative pathway for the current to leak away from the tissue. The factor a ($0 < a < 1$) represents a coupling factor between the body and the ground [9]. The factor models the parasitic capacitances as a lumped one. Therefore, the current which flows through the tissue is smaller than the reference measurement current which produces an underestimate of the impedance of the tissue being measured (Z_m).

4. Results

With the help of the circuit model and simulated impedance values, the influence of stray capacitance can be examined. In Figure 3 the simulated complex impedance is shown. It illustrates the typical behaviour of depressed semicircular arc within the measured frequency interval. To calculate the resistance of intracellular and extracellular body water using the

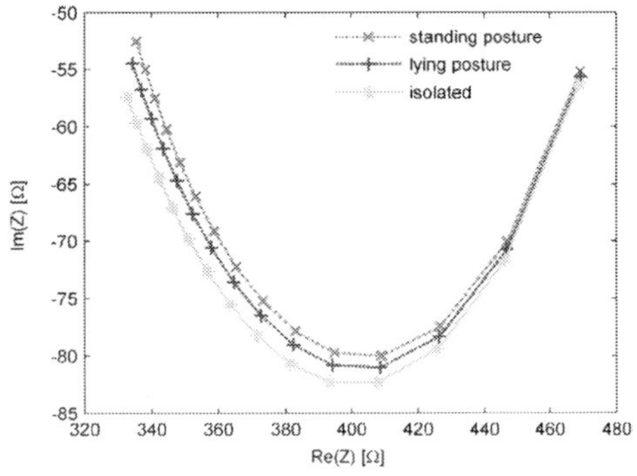

Figure 3. Simulated impedance of a standing, a lying and an electrical isolated Person

simulated whole-body impedance, the well-known Cole equation is used. For the data analysis through the Cole parameters, the Cole equation is approximated by a fitting algorithm to the simulated data, as a very common practice within BIS [11]. The extracted parameters for the resistances and the heuristic factor α from the Cole equation is for an electrical isolated person $R_e = 501.68$, $R_i = 741.82$ and $\alpha = 0.40$; in case of a standing person $R_e = 498.45$, $R_i = 813.07$ and $\alpha = 0.39$ and in case of a lying person $R_e = 499.81$, $R_i = 784.88$ and $\alpha = 0.39$ are observed. The simulated impedances are lower than in real measurements. A reason for this might be the over-estimation of muscle tissue, but this should not influence the position-depending impedance changes. The percentage changes of the simulated impedances are 0.65% for R_e, -8.75% for R_i and 1.78% for α between a standing person and an electrical isolated person and a percentage changes of a lying person and an electrical isolated are 0.37% for R_e, -5.49 for R_i and 0.98% for α. The percentage changes of the intracellular resistance are in the range of the observed R_i at real measurements by Fenech et al. [2]. With the help of the circuit model it is possible to describe the influence of the stray capacity. With the help of Kennelly's star-delta theorem, it is possible to deduce from Figure 2 the following equation:

$$Z_m = a \cdot (1 - a) \cdot Z_T^2 \cdot j\omega C_s + Z_T \tag{1}$$

With the assumption of a symmetrical coupling from each point of the body to ground (a=0.5), the coupling influence can be rewritten as $C_s' = 0.25 \cdot C_s$. That solves the equation (1) with respect to Z_T results in the following equation:

$$Z_T = \frac{2\sqrt{j\omega C_s' \cdot Z_m + 1} - 2}{j\omega C_s'} \tag{2}$$

With this formula it is possible to correct the impedance curve over the entire frequency range. Due to the close gap between the body and the ground plane, the capacitance of a standing person is larger than of a supine lying person. With eq. (2) and a value for the stray capacitance, the measured impedance can be corrected. In the given scenario C_s' is $18pF$ for the supine and $32pF$ for the standing posture. With the FEM-model and the equivalent electrical model, it is possible to analyse and correct the influence of different body postures on the measured intracellular resistance and the posture depending capacitance.

5. Conclusion

Stray capacitance form the body to ground is a possible source of error in tetrapolar bioimpedance spectroscopy, as obtained by many authors. Using a FEM model, the influence of different body postures on the measured impedance are simulated, without considering the physiological effects of gravity. By considering only the capacitive coupling from the body segment between potential electrodes to ground, a simplified circuit model is obtained that predicts frequency depending measurement errors. This paper demonstrates that the influence of the body position on the intracellular resistance is mainly explainable by a variation of the stray capacitance.

Acknowledgment

The project has been selected under the operational programme co-financed by the European Regional Development Fund (ERDF) Objective 2 "Regional Competitiveness and Employment" 2007-2013), North Rhine-Westphalia (Germany).

References

[1] Kushner R F, Gudivaka R and Schoeller D A 1996 *The American journal of clinical nutrition* **64** 423–427
[2] Fenech M and Jaffrin M Y 2004 *IEEE transactions on bio-medical engineering* **51** 166–175
[3] Scharfetter H, Monif M, László Z, Lambauer T, Hutten H and Hinghofer-Szalkay H 1997 *Kidney international* **51** 1078–1087
[4] Zhu F, Schneditz D, Wang E and Levin N W 1998 *Journal of applied physiology (Bethesda, Md. : 1985)* **85** 497–504
[5] Hinghofer-Szalkay H and Moser M 1986 *The American journal of physiology* **250** 68–75
[6] Berg H E, Tedner B and Tesch P A 1993 *Acta physiologica Scandinavica* **148** 379–385
[7] Maw G J, Mackenzie I L and Taylor N A 1995 *Acta physiologica Scandinavica* **155** 157–163
[8] Gabriel S, Lau R W and Gabriel C 1996 *Physics in medicine and biology* **41** 2271–2293
[9] Aliau-Bonet C and Pallas-Areny R 2012 *IEEE transactions on bio-medical engineering* **59** 3405–3411
[10] Scharfetter H, Hartinger P, Hinghofer-Szalkay H and Hutten H 1998 *Physiological measurement* **19** 247–261
[11] Weyer S, Röthlingshöfer L, Walter M, Leonhardt S and Bensberg R 2012 *Ninth International Conference on Wearable and Implantable Body Sensor (BSN)* pp 22–27

Theoretical and Experimental Estimations of Volumetric Inductive Phase Shift in Breast Cancer Tissue.

C A González[1,2*], L M Lozano[1], M C Uscanga[3], J G Silva[1,2] and S M Polo[1]
[1] Universidad del Ejército y Fuerza Aérea/EMGS, SEDENA, D.F. México.
[2] Instituto Politécnico Nacional/Escuela Superior de Medicina, México D.F., México.
[3] Clínica de Especialidades de la Mujer/SEDENA, México D.F., México.

* E-mail: cgonzalezd@ipn.mx

Abstract. Impedance measurements based on magnetic induction for breast cancer detection has been proposed in some studies. This study evaluates theoretical and experimentally the use of a non-invasive technique based on magnetic induction for detection of patho-physiological conditions in breast cancer tissue associated to its volumetric electrical conductivity changes through inductive phase shift measurements. An induction coils-breast 3D pixel model was designed and tested. The model involves two circular coils coaxially centered and a human breast volume centrally placed with respect to the coils. A time-harmonic numerical simulation study addressed the effects of frequency-dependent electrical properties of tumoral tissue on the volumetric inductive phase shift of the breast model measured with the circular coils as inductor and sensor elements. Experimentally; five female volunteer patients with infiltrating ductal carcinoma previously diagnosed by the radiology and oncology departments of the Specialty Clinic for Women of the Mexican Army were measured by an experimental inductive spectrometer and the use of an ergonomic inductor-sensor coil designed to estimate the volumetric inductive phase shift in human breast tissue. Theoretical and experimental inductive phase shift estimations were developed at four frequencies: 0.01, 0.1, 1 and 10 MHz. The theoretical estimations were qualitatively in agreement with the experimental findings. Important increments in volumetric inductive phase shift measurements were evident at 0.01MHz in theoretical and experimental observations. The results suggest that the tested technique has the potential to detect pathological conditions in breast tissue associated to cancer by non-invasive monitoring. Further complementary studies are warranted to confirm the observations.

1. Introduction

Breast Cancer (BC) early detection techniques have been proposed by measurements of tissue electrical properties based on the principle that the hyper-vascularization of malignancies promotes changes in Electrical Impedance (EI) of the tissue. Bioelectrical measurements through magnetic induction at a single frequency with noninvasive coils have been proposed as a valuable alternative to monitor, non-invasively, the health of organs and tissues [1], [2], [3] and [4]. Other authors have agreed to propose non-contact measurements for the development of alternative electrical imaging Magnetic Induction Tomography (MIT) [5] and [6]. Recently; MIT has been proposed to obtain three-dimensional images on the basis of a matrix-free reconstruction method [7]. Human breast tissue tumor shows an increase in electrical conductivity from 4 to 8 mS/cm compared to normal tissue [8]. Our group has proposed bioelectric measurements through multi-frequency magnetic induction as a valuable alternative to monitor, non-invasively, growths in breast tissue. We have developed a prototype inductor-sensor low intensity magnetic field at non-ionizing frequencies; the system uses the technique of Magnetic Induction Spectroscopy (MIS) to measure the volumetric electrical properties of tissue by magnetic fields at multiple frequencies. The aim of this study was to evaluate analytically and experimentally the use of MIS for detection of breast cancer tissue conditions associated to its volumetric electrical conductivity changes through inductive phase shift measurements.

2. Methods

2.1. Biophysical concept to detect non-invasively BC.

We have proposed two coupled coils of different radii in an inductor-sensor arrangement coaxially centred and spaced in such a way that a typical breast tissue volume is placed along the centre line. An alternating current Ie^{jwt} is injected to the inductor coil to generate a primary magnetic field **B** and to induce Eddy currents in the tissue as well in the sensor coil, the induced currents promotes a perturbation of the primary magnetic field (**ΔB**) as a function of the bulk conductive electrical properties of the breast and tumour volumes, and generates an "inductive phase shift" (Δθ), which is detected as a composite magnetic field **B+ΔB** in the sensor coil. The phase shift of the total magnetic field is obtained through the voltage phase difference between inductor and sensor coils [6].

2.2. Three-dimensional simulation study.

A pixel model was designed on the basis of tridimensional polygons (meshes), which use orthogonal bi-dimensional views as reference guides. We used the software SolidWorks® (SolidWorks Inc. 2008, Concord, MA USA) as 3D CAD platform design. The breast and tumor were simulated in realistic dimensions. The scale used was 1:10. The pixel model includes healthy breast and tumoral tissues as structural components (subdomains). Circular inductive and sensor coils were placed along the model in such a way that the breast-tumor volume occupied a concentric position between the coils (see figure 1). The 3D CAD model was export to multiphysics commercial software for biophysical simulation as STL file format.

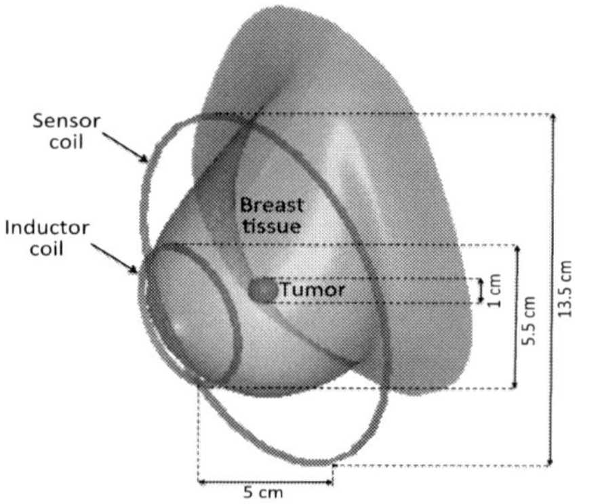

Figure 1. 3D breast-tumor pixel model.

2.2.1. Simulation criteria. Breast and tumor volume tissues were considered to be homogeneous and isotropic with electrical properties taken from [9] (Table 1). The model was surrounded by a sphere of 20 cm of radius that had electrical parameters correspond to the air. The inductor and sensor coils were taken to have electrical properties of copper (σ=5.998x10^7 S/m and ε_r= 1). A sinusoidal current in the inductive coil was considered as a magnetic field generation signal. To investigate the effect of the tumor we simulate induced current densities inside the breast-tumor model by 1 A/m^2 of alternating currents at four frequencies (0.01, 0.1, 1 and 10 MHz) and we estimate the Δθ spectrum in the sensor coil.

Table 1. Electrical parameters for breast-tumor tissues simulation.

Freq (MHz)	*Conductivity (mS/cm)*		*Relative Permitivity*	
	Breast	*Tumor*	*Breast*	*Tumor*
0.01	2.8	6.0	800	20000
0.1	2.8	6.1	100	5000
1	3.0	6.3	25	1100
10	3.2	8.2	12	300

2.2.3. Time-harmonic and quasi-static assumption. In the *time-harmonic* case we omit the displacement currents in the Ampere's equation. To simplify initial conditions; we consider a non-moving geometry ($\vec{v}$=0) and no external electric potential gradient (∇V=0) on the border of the breast pixel model. In addition, we have assumed that the border of the sphere surrounding the system defined in above section is grounded. Thus; we can obtain:

$$(j\omega\sigma-\omega^2\varepsilon)\vec{A}+\nabla\times(\mu^{-1}\nabla\times\vec{A})=\vec{J}^e \tag{1}$$

where: σ is the medium conductivity (S/m), ω is the angular frequency (rad/s), $\vec{J}^e$ is a current density generated externally (A/m^2), μ is the medium permeability (Henry/m) and ε is the medium permittivity (farad/m). In this study equation (1) was solved to obtain the magnetic vector potential ($\vec{A}$).

2.2.4. Inductive phase shift estimation. We consider the breast-tumor model and coil configuration shown in figure 1 with an alternating current Ie^{jwt} flowing through the inductor coil. The presence of a conductive sample (the breast model) produces $\Delta\theta$ in the sensor coil current as a function of the structural composition and its electrical properties of the sample (σ, ε). In this study the tumor volume represents a different structural compositions, which is detected trough changes in the phase of the induced current at the sensor coil, and it could be separated in its complex coefficients. We define a basal induced current argument in the sensor coil as $(\omega t + \theta)$ and the argument influenced by the tumor volume as $(\omega t + \theta_1)$. To estimate the inductive phase shift ($\Delta\theta$) we can use the argument differences at specific frequency and time according to the following expression:

$$\Delta\theta=(\omega t+\theta_1)-(\omega t+\theta)=\theta_1-\theta \tag{2}$$

2.2.5. Mathematical analysis. The mathematical analysis was performed with COMSOL Multiphysics (COMSOL 3.3a, Inc. Boston Ma. USA, 2005) to solve (1) for the magnetic vector potential. Boundary conditions of the surrounded sphere correspond to the air and represent electric and magnetic insulation. The induced complex current densities in the sensor coil were calculated by post-processing (subdomain integration). Those values were used to estimate spectra of $\Delta\theta$ at the evaluated frequencies homogenized with respect to induction without the breast model (in the air).

2.3. Experimental Study.
2.3.1. Experimental Inductive Spectrometer Prototype. An experimental inductive spectrometer to use MIS measurements in a typical breast tissue volume was designed and constructed (figures 2 and 3). The system consists of five modules: digital synthesizer, transceiver, phase detector, data acquisition and data processing. A description of the different modules is shown in [10]. In brief; the inductor coil generates the primary magnetic field **B** and it induces a current in the sensor coil by magnetic induction, a perturbation of the primary magnetic field ($\Delta\boldsymbol{B}$) generates an inductive phase shift ($\Delta\theta$) as a function of the bulk electrical properties of the breast and tumour volumes, which is detected as a composite magnetic field $\mathbf{B}+\Delta\mathbf{B}$ in the sensor coil. The prototype is programmed-controlled by PC.

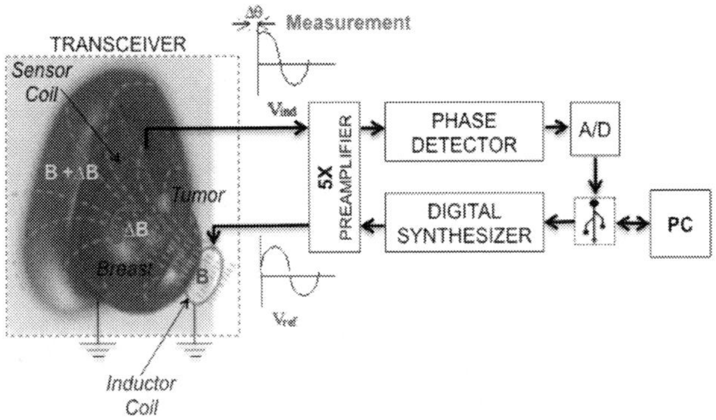

Figure 2. Block diagram of the experimental inductive spectrometer for non-invasive detection of BC and the measurement concept.

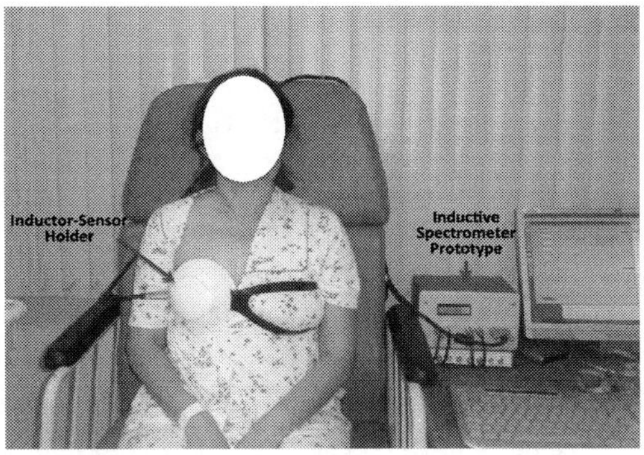

Figure 3. Inductive spectrometer prototype manufactured at our laboratory and the experimental set-up in patients.

2.3.2. Experimental Measurements. Five female volunteer patients with infiltrating ductal carcinoma previously diagnosed by the radiology and oncology departments of the Specialty Clinic for Women of the Mexican Army were measured by the experimental inductive spectrometer described above and the use of an ergonomic inductor-sensor coil holder designed to estimate the volumetric $\Delta\theta$ in human breast tissue. Spectra of $\Delta\theta$ were estimated as homogenized data with respect to the average of three-air measurements at the frequencies described in previous subsections. A picture of the experimental set-up is shown in figure 3. The study was conducted according to the principles expressed in the Declaration of Helsinki and was approved by the "Research" and "Bioethics" Committees of the Institution. The volunteer patients were gently informed and signed an informed consent.

3. Results
Figure 4 shows a graphical contrast between the simulated induced current density in the 3D breast-tumor pixel model and the mastography of a representative patient measured by the experimental inductive spectrometer, the figure is intended to illustrate the theoretical and experimental measurements conditions to produce the volumetric inductive phase shift spectra.

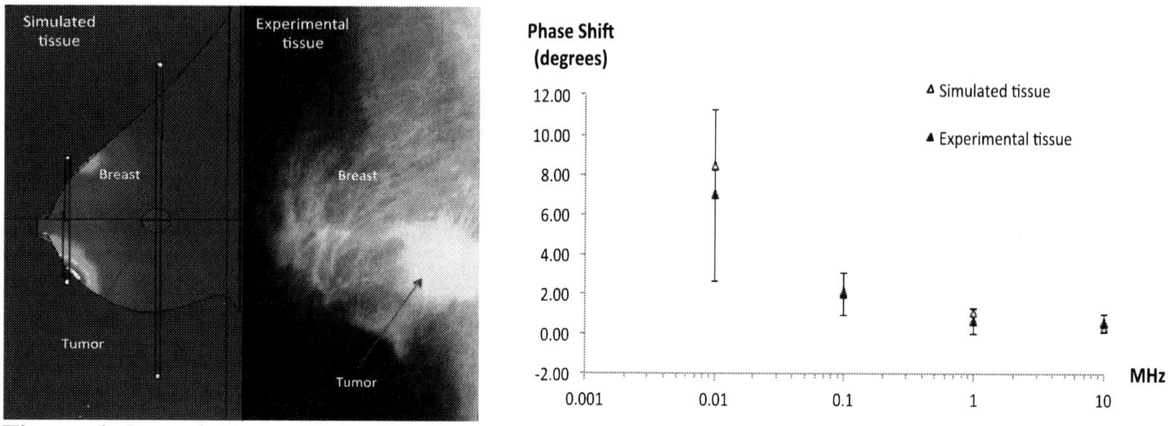

Figure 4. Lateral views of simulated induced current density in breast tissue and a representative mastography of a patient with a well-defined high-density nodule characteristic of BC measured by MIS (left). Simulated and experimental spectra of the $\Delta\theta$ estimation in BC tissues (right).

4. Discussion.
Theoretical estimations are qualitatively in agreement with experimental measurements. Interestingly; wider dispersion of the experimental $\Delta\theta$ is evident as the frequency decreases, particularly at 0.01 MHz, which correspond to an overlap alfa-beta dielectric dispersion regions for tissue, and where the component of the cell structure emerges, it indicates that in the lowest frequency evaluated the $\Delta\theta$ could be associated to hyper-vascularization and fibrous-cyst conditions characterized in BC as well as different tumor positions observed in the patients, and it could explain in part the dispersion observed with respect to a theoretical estimation for a well defined and positioned tumor.

5. Conclusion.
Theoretical estimations were qualitatively in agreement with the experimental findings. Important increments in volumetric $\Delta\theta$ measurements were evident at 0.01MHz. MIS measurements have the potential to detect pathological conditions in breast tissue associated to cancer by non-invasive monitoring. Further complementary studies are warranted to confirm the observations.

References
[1] Scharfetter H, Ninaus W, Puswald B, Petrova G I Kovachev D and Hutten H 1999 Inductively coupled wideband transceiver for bioimpedance spectroscopy (IBIS) *Ann N Y Acad Sci* **873**:322-34
[2] Griffiths H, Steward W R and Gough W 1999 Magnetic induction tomography - a measuring system for biological materials *Ann N Y Acad Sci* **873**:335-45
[3] Griffiths H 2001 Magnetic induction tomography *Meas Sci Tech* **12**:1126-31
[4] Watson S, Williams R J, Gough W A and Griffiths H 2008 A magnetic induction tomography system for samples with conductivities less than 10s m^{-1} *Meas Sci Tech* **19**:045501-11
[5] Al-Zeiback S and Saunders N H 1993 A feasability study of in vivo electromagnetic imaging *Phys Med Biol* **38**:151-60
[6] Korjenevsky A V and Cherepenin V A 1999 Progress in realization of magnetic induction tomography *Ann N Y Acad Sci* **873**:346-52
[7] Wei H Y and Soleimani M 2012 Three-dimensional magnetic induction tomography imaging using a matrix free krylov subspace inversion algorithm *Prog Electromagn Res* **122**: 29–45
[8] Burdette E C 1982 Electromagnetic and acoustic properties of tissues in pyisical Aspects of Hyperthermia *G H Nussbaum ed (American Association of Phycisist in Medicine)* **8**: 105-50
[9] Duck F A 1990 Physical properties of tissue *Academic Press Limited* (London NW1) 197-204
[10] Gonzalez C A, Silva J G, Lozano L M and Polo S M 2012 *JEMAA* **4(8)** 317-25.

Simulation studies of statistical distributions of cell membrane capacities and an ellipse model to assess the frequency behaviour of biological tissues

Willard Gerritsen[1], Robin van den Berg[1], Jan Mulder[2], Rudolf M. Verdaasdonk[1] and Jan H. Meijer[1]

[1] Department of Physics and Medical Technology, VU University Medical Center, Amsterdam, the Netherlands
[2] Physical Practica & Didactics, VU University, Amsterdam, the Netherlands

jh.meijer@vumc.nl

Abstract. The frequency behaviour of biological tissues is commonly described by a Cole model reflecting a single-cell bio-impedance model extended with an exponent α. However, for this parameter α there is no physical or biological substrate, which impedes an interpretation. The present study confirms by computer simulations of tissue models that the factual frequency behaviour can be explained by assuming a distribution of the electrical impedance properties of cells and of the capacitive coupling between cells. This behaviour is modelled mathematically by an ellipse. A mathematical procedure is presented to estimate this ellipse from experimental data by a least square method. A model parameter β is introduced, representing the ratio of the axes of the ellipse. A higher value of β means a larger variation in cell properties, which makes a patho-physiological interpretation of changes possible.

1. Introduction

A frequently used single-cell bio-impedance model consists of a unit of two resistors and one capacitor (see figure 1). The model is used to describe the frequency behaviour of biological tissues in the middle frequencies, called *beta-dispersion*. In this model R_e represents the extra-cellular resistance, R_i the intra-cellular resistance and C the capacity of the cell membrane. The complex impedance of this model is given by:

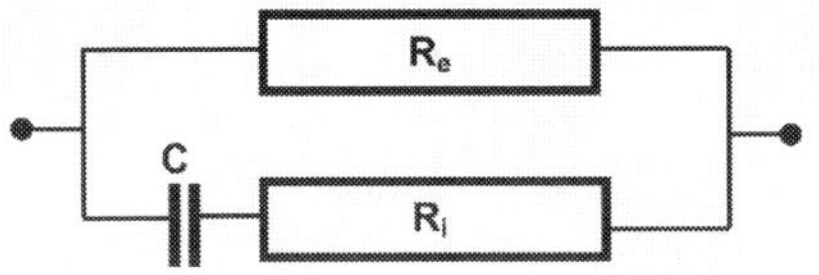

Figure 1. Single cell unit impedance model consisting of three elements.

$$Z_{tot} = \frac{R_0 + j\omega\tau R_\infty}{1 + j\omega\tau} \quad (1)$$

In which ω is the angular frequency: $\omega = 2\pi f$. The time constant $\tau = R_i C$. R_0 is the impedance at the low frequency limit, $R_0 = R_e$, and R_∞ is the impedance at the high frequency limit and is given by:

$$R_\infty = \frac{R_e R_i}{R_e + R_i} \quad (2)$$

A graphical representation of the frequency behaviour of this simple model in the complex plane results in a semi-circle.

In practice, however, the shape of the graph in the complex plane of tissue impedance differs from this simple model behaviour. The factual shape is more "flattened". A commonly used mathematical model of the factual frequency behaviour of the impedance of biological tissues is the Cole model [1]. In this model an adjusting parameter α enhances the agreement with the experimental data. This mathematical formulation of Cole model is:

$$Z_{tot} = \frac{R_0 + (j\omega\tau)^\alpha \cdot R_\infty}{1+(j\omega\tau)^\alpha} \qquad (3)$$

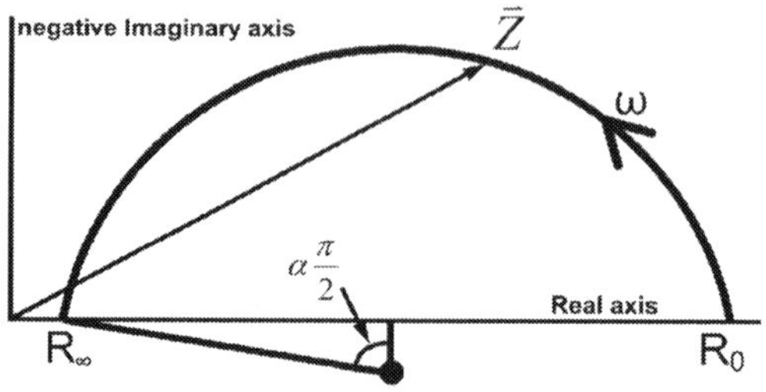

Figure 2. Frequency behaviour of the Cole model in the complex plane. (The negative imaginary axis is chosen upwards).

The parameter α takes care of the problem of "flattening" of the semi-circle by shifting the centre of the semi-circle below the real axis, as is shown in figure 2. However, this model poses a new problem, since there is no physical or biological substrate for the mathematical parameter α. This means that no interpretation is possible when α changes in the tissue under study, for instance in healthcare, or differs from similar tissues.

In 1953 Herman Schwan presented an interpretation of the observed frequency behaviour in the *beta-dispersion* region[2, 3]. A combination of multiple cell units with distributed time constants τ is understood to be responsible for this behaviour. This interpretation was investigated in the present study by computer simulations in which two or more cell units, up to thousands, were coupled in series and parallel. Further, in the present study a semi-ellipse is proposed to fit the frequency behaviour in order to bypass the problem of the interpretation of the parameter α. In the relevant range of frequencies both models are expected to differ little in numerical value. A parameter β is introduced that describes the variation in cell properties and that can be interpreted (patho-)physiologically.

2. Experimental method

Computer simulations were programmed in Wolfram's Mathematica. This allowed unlimited elements in series and/or parallel to be chosen. The values of R_e, R_i and C were all standard set on 1, unless the parameter was selected to vary. Simulations were performed using specifically chosen values for a specific variable. Also randomized values from one or more Gaussian distributions were used.

To study the effect of varying respectively R_e, R_i and C, simulations with a two cell-unit model were done. The values of cell elements were varied respectively. Subsequently the model was extended to tens, hundreds and thousands of cell units in series and into models with cell units in series and parallel.

3. Results

3.1 Computer simulation of the frequency behaviour of biological tissues

Simulations with two cell units model with two different values for the elements resulted in a "flattening" of the complex impedance graph (see figure 3). It was found that especially variation of C resulted a flattening of the curve. If the chosen C's were very different, the curve indented in the middle.

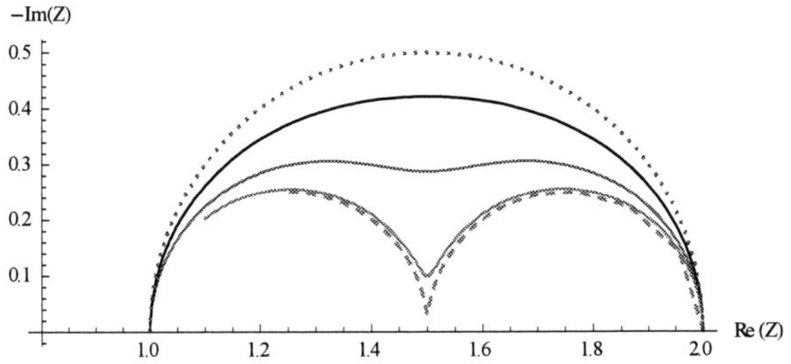

Figure 3. The computed impedance behaviour of a biological tissue model with two cell units. The plot of 5 values of the ratio of the first to the second capacitor is shown: $C_2/C_1 = 1$ (purple, dotted line), 0.5 (black), 0.1 (blue), 0.01 (brown), 0.001 (red, striped).

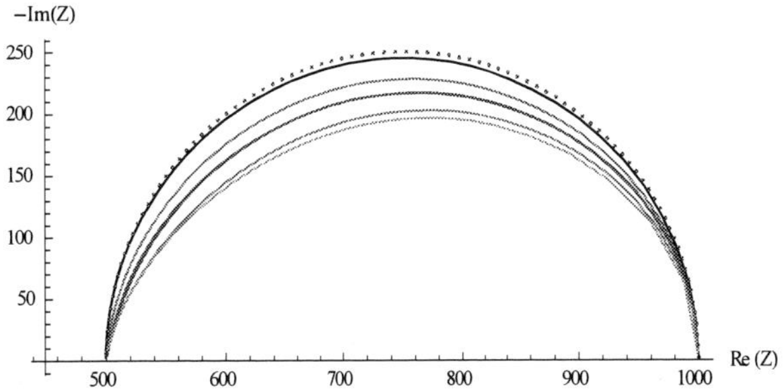

Figure 4. With increasing ranges of values of C the semi-circle flattens. The plotted values are: sd = 0 (purple, dotted line), 0.2 (black), 0.4 (brown), 0.6 (blue), 0.8 (purple), 1 (red), 2 (orange). Other parameter values $R_e = R_i = 1$ (n = 1000 elements).

Simulations with larger numbers of cell units, using series of randomized numbers with a sufficient standard deviation, resulted in a "flattened" curve as well (see figure 4). If two series of randomized numbers with a different mean were combined, flattening was pronounced. Varying R_e and/or R_i also resulted in flattening, however variation of C had the largest influence.

3.2 Method to find the best fitting ellipse for experimental data and establishing the ratio a/b.

From the simulations above it follows that the factual shape of the complex impedance curve deviates from a circle. In the present study it is assumed that the "flattened" form of the curve can be approximated by an ellipse.

The ellipse has two main axes, a real axis (long, a) and an imaginary axis (short, b). The ratio of the main axes a/b is considered to be a measure for the wideness of the distribution and is therefore physiologically interpretable. The ellipse can be fitted to measurement data by means of a least squares method, which results in a unique solution.

It is assumed that the centre of the ellipse is on the base line (real axis). Therefore the ellipse can be written as:

$$\left(\frac{x-x_0}{a}\right)^2+\left(\frac{y}{b}\right)^2=1 \tag{4}$$

This is rewritten as:

$$y^2=b^2-\left(\frac{b}{a}\right)^2 x_0^2+2\left(\frac{b}{a}\right)^2 x_0 x-\left(\frac{b}{a}\right)^2 x^2 \tag{5}$$

Three parameters are defined: $\kappa=-\left(\frac{b}{a}\right)^2, \lambda=2\left(\frac{b}{a}\right)^2 x_0, \mu=b^2-\left(\frac{b}{a}\right)^2 x_0^2$ (6)

Finding the best fitting ellipse is reduced to finding the optimal solutions for $\kappa x_i^2+\lambda x_i+\mu=y_i^2$. (7)

To solve this equations, two matrices are defined: $$A = \begin{bmatrix} x_1^2 & x_1 & 1 \\ x_2^2 & x_2 & 1 \\ \vdots & \vdots & \vdots \\ x_n^2 & x_n & 1 \end{bmatrix}, \quad B = \begin{bmatrix} y_1^2 \\ y_2^2 \\ \vdots \\ y_n^2 \end{bmatrix}, \qquad (8)$$

in which n is the number of data sets. Using any mathematical program solutions can be calculated for AX = B using the least squares method in which $$X = \begin{bmatrix} \kappa \\ \lambda \\ \mu \end{bmatrix}. \qquad (9)$$

A new parameter β can be defined: $$\beta = \frac{a}{b} = \frac{1}{\sqrt{-\kappa}} \qquad (10)$$

to describe the frequency behaviour of biological tissues.

4. Discussion

The computer simulations confirmed the interpretation that the frequency behaviour of biological tissues originate from a distribution in the electrical impedance model parameters of tissue cells. A wider distribution corresponds to a "flatter" graph in the complex plane which resembles a lower α in the Cole model, although the graph in the complex plane differs from a semi-circle. This distribution of tissue properties can be interpreted physiologically: Biological tissues usually consist of more than one type of cells. Differences in the properties of the cells, e.g. in cancer, may influence the variation in these impedance parameters. Even within one type of cells a distribution in age and therefore a distribution in stage of development and size will be present. Further, cells are not arranged within tissues with a mathematical regularity. This will result in a statistical distribution in capacitive coupling between the cells. For instance, the volume of interstitial fluid and/or inflammation may influence this coupling.

To bypass the problem of absence of a physiological substratum for the Cole parameter α and the theoretical absence of an indication to fit a semi-circle, an ellipse model is designed to fit measured impedances and a procedure is presented extract to model parameters. A parameter β is introduced, that represents the ratio of the long axis to the short axis of the ellipse. The computer simulations show that tissues with a large variation in cell properties lead to a flatter ellipse, resulting in a higher value of β. Thus a higher value of β means less uniformity. These considerations make patho-physiological interpretations of differences and variations in frequency behaviour possible. Therefore, this interpretation offers a variety of clinical applications. Future studies have to validate this ellipse model by comparing the quality of the fit with the Cole model in experimental data.

References

[1] Cole K S 1972 *Membranes, Ions and Impulses* 2nd printing (University of California Press, Berkeley and Los Angeles)

[2] H P Schwann and K Li 1953 Capacity and conductivity of body tissues at ultrahigh frequencies *National Convention Record, i.r.e., Proc. Of I.R.E.* **41** 1735

[3] Grimnes S and Martinsen Ø G 2008 Bioimpedance and bioelectricity basics (2nd edition) *Elsevier. Amsterdam, Boston, Heidelberg*

Journal of Physics: Conference Series **434** (2013) 012006 doi:10.1088/1742-6596/434/1/012006

Electrical localization of weakly electric fish using neural networks

Greg Kiar[1], Yasin Mamatjan[1], James Jun[2], Len Maler[2], Andy Adler[1]

[1]Systems and Computer Engineering, Carleton University, Ottawa, Canada

[2]Faculty of Medicine, University of Ottawa, Canada

gkiar@sce.carleton.ca

Abstract. Weakly Electric Fish (WEF) emit an Electric Organ Discharge (EOD), which travels through the surrounding water and enables WEF to locate nearby objects or to communicate between individuals. Previous tracking of WEF has been conducted using infrared (IR) cameras and subsequent image processing. The limitation of visual tracking is its relatively low frame-rate and lack of reliability when visually obstructed. Thus, there is a need for reliable monitoring of WEF location and behaviour. The objective of this study is to provide an alternative and non-invasive means of tracking WEF in real-time using neural networks (NN). This study was carried out in three stages. First stage was to recreate voltage distributions by simulating the WEF using EIDORS and finite element method (FEM) modelling. Second stage was to validate the model using phantom data acquired from an Electrical Impedance Tomography (EIT) based system, including a phantom fish and tank. In the third stage, the measurement data was acquired using a restrained WEF within a tank. We trained the NN based on the voltage distributions for different locations of the WEF. With networks trained on the acquired data, we tracked new locations of the WEF and observed the movement patterns. The results showed a strong correlation between expected and calculated values of WEF position in one dimension, yielding a high spatial resolution within 1 cm and 10 times higher temporal resolution than IR cameras. Thus, the developed approach could be used as a practical method to non-invasively monitor the WEF in real-time.

1. Introduction

Weakly electric fish (WEF) emit an electric organ discharge (EOD), which travels through the surrounding water and enables WEF to locate nearby objects or to communicate with other individuals. From studying the temporal patterns of EOD, the potential exists to track the location and movement of WEF, which could allow us to understand the sensory motor brain function of vertebrates. WEFs EOD are a time varying current dipole between the head and tail that discharges at a rate of approximately 50 Hz [1]. The head and tail region of the WEF always have opposite current flows, but the polarity can switch depending on the pulse cycle of the EOD. Much like an actual current dipole, the strength of the electric field produced is proportional to the length of the WEF. Previously, the movement of WEF has been detected using infrared (IR) cameras. IR cameras lack reliability when there is a visual obstruction, and thus the tracking of WEF is inconsistent [1]. Here, we aim to use neural networks (NN) to provide an alternative non-invasive means of tracking WEF using their EOD.

Published under licence by IOP Publishing Ltd

2. Methodology

2.1. *Overview*

We aimed to investigate if WEF can be accurately localized from recordings of their electrical outputs. The method employed was three-fold. Stage I was to create a finite element method (FEM) model of the system with similar stimulation patterns to the WEF. The model (Stage II) was then validated against a phantom electrical impedance tomography (EIT) based system. Once the FEM model had been validated, a database was compiled of corresponding voltage readings and position coordinates. The method of neural networks (NN) was tested with this data. In stage III, real WEF data was acquired and tested using the NN based tracking method as chosen from the FEM system.

2.2. *Stage I: FEM Model*

In order to create the FEM model of the phantom system, Electrical Impedance Tomography and Diffuse Optical Tomography Reconstruction Software (EIDORS) was used [2]. The FEM model tank was a scaled down version of the phantom tank to reduce computation time. The phantom tank has a height-diameter ratio of 9:7, and the FEM model has a ratio of 1:4. The reason for the reduction in this ratio in the FEM model was because the measurement protocol consisted of positions in the horizontal plane. The FEM tank contained one row of 32 electrodes. In this model, the electrodes were not paired across the tank, but all referenced to a common ground. In order to compare values with those of the phantom, opposite channels could be subtracted from one another. The WEF in the FEM model was created as a cylinder with charged electrodes on the head and tail, with an insulating section in the middle. The charge on the head was positive and the tail was negative like the opposing poles of an actual WEF [3].

2.3. *Stage II: Verification of FEM Model*

In order to test the tracking method, a database of position and voltage pairs needed to be collected. The database needed to contain a large number of test cases, and have very high precision in the position of the fish. The tank was equipped with four rows of 32 evenly spaced electrodes, and a ground pin at the centre of the base; one level and 16 electrodes were used and connected directly opposite the tank in pairs, providing eight channels for measurements. The phantom WEF was 0.9 mm in diameter and 17 cm in length, in a tank with 28 cm diameter. A measurement protocol consisting of 21 different positions of the phantom was used to obtain consistent and highly inclusive data throughout the tank. Since the WEF containing tank that the phantom system was based on was shallow, the measurement protocol consisted of positions exclusively in the horizontal plane, in line with the electrodes.

2.4. *Stage II: Neural Networks on FEM Data*

Using the FEM data, the inverse solution was accomplished using NN. Measurements were made around the FEM model tank, and the electrode voltages for each trial had pseudorandom noise added up to ±10 % of the maximum channel voltage. The position data was packaged into coordinates of the head and tail of the fish in Cartesian coordinates, with the real length of the fish as 23 cm. A cascade forward network was declared using the Matlab NN toolbox and was trained with the data acquired from the simulation in the format of 32 signed voltages as inputs, and two sets of (x, y) coordinates as outputs.

2.5. *Stage III: Experimental Data*

The measurement data was acquired using a restrained WEF (with length of 23 cm) within a tank (1.5 m in diameter) surrounded by eight graphite electrodes and filled with water (17.5 cm in depth). In order to obtain accurate WEF datasets, the WEF was restrained near the surface of the tank with a Styrofoam harness [1]. The measurement protocol consisted of displacements in the x direction and orientations ranging from 0 to 180 degrees, all centered on the x axis. Since there was no y displacement, the y coordinates only varied with WEF orientation. The measurement data was acquired based on (i) positions for full tank, (ii) positions for half of the tank. The WEF dataset contained 299 data points compared to the 1196 of the simulated data, so interpolation had to be performed to maintain a high quality result. The full tank dataset contained 299 data points, which were interpolated to be 29,900 data points. The half tank dataset, which consisted of the same 299 data points, which was interpolated to be the 15,210 positive x data points. The computer used for processing had 6 GB of memory and a quad-core processor running at 2.67 GHz.

2.6. Stage III: Neural Network on Real WEF Data

Voltage values were taken across symmetrically paired electrode channels, thus a mirroring issue presented itself across the x-axis of the tank. Since voltages are measured between two points, the voltage read across a channel when the WEF was at x = ±2 cm, for example, would be the same. In order to proceed with using the NN and obtaining reasonable results, it was necessary to only train and test points that fell on either side of x = 0. The positive x-axis was selected. The NN was trained with two datasets in order to show the constraints of the physical system on which the experiments were performed.

3. Results

Though the performance of the NN was analyzed for the FEM data, it was largely for verification purpose, so only the real WEF data analysis of the NN is shown below.

Figure 1a shows the performance of a pseudo-randomly generated NN with full tank data. As calculated from Figure 1a, the percentage of calculated head or tail positions, which fell within 1 cm of the true location, was 54.45 % and that the results had a mean squared error (MSE) of 5.53 mm. This is a very high MSE and low percentage of accuracy, neither of which is desirable.

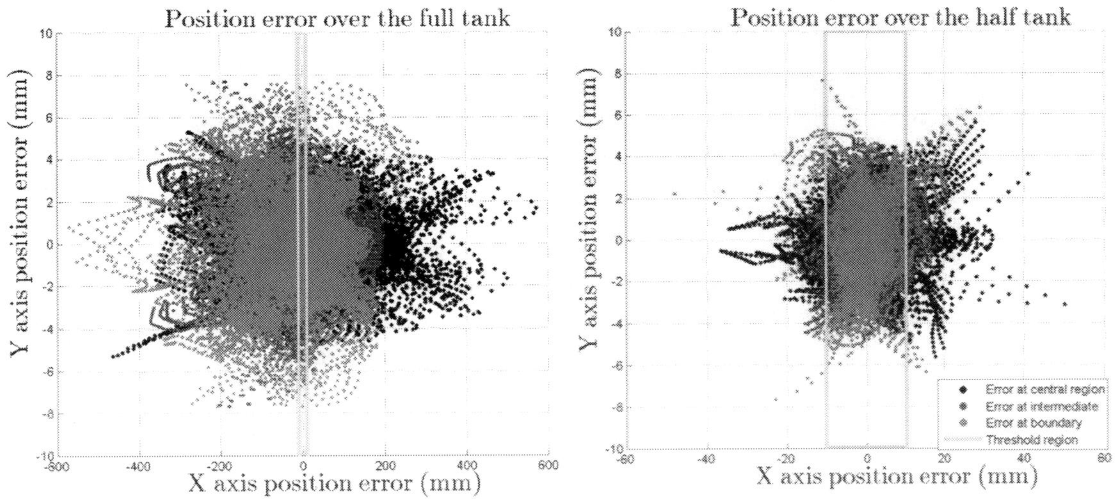

Figure 1. Difference between position of the WEF and equivalent calculated positions as generated by the Neural Network. (A) The performance of the full tank trained network. (B) The performance of the half tank trained network. Black dots stands for errors in the central region, blue dots are for errors in the intermediate region and red dots are for errors in the boundary region. The green box outlines 1 cm of error in each direction.

In the case of computing positions for half of the tank, we saw an improvement. The data set contained 299 data points, which was interpolated to be 15210 data sets. Shown in Fig. 1B is the performance of a pseudo-randomly generated NN trained with the half tank data set. We can see from this plot that the percentage of the calculated head and tail positions within 1 cm is 94.2 %, with a MSE value of 0.02 mm. The x-coordinate errors are much less in Fig. 1B.

The average computation time for the 15210 sample dataset was 4.7 msec. This computes to a 213 Hz frame rate. The speed at which the NN operates increases with respect to number of consecutive operations and peaks at approximately 213 Hz.

4. Discussion and Conclusion

A verified FEM model was required in order to test possible strategies of WEF localization. The average difference of voltages was 0.129 V between the model and the phantom system. This value had a standard deviation of 0.181. On a range of 0 - 2.5 V these values are relatively low, so it was concluded that the FEM model acted as a reasonable substitute for the phantom system and could be used for investigating three dimensional tracking in future research.

The full tank data shown in Fig. 1A had an accuracy of 54.45 % within 1 cm, and a MSE of 5.53 mm. This accuracy is very low. The largest errors in the x direction were nearly 60 cm. All regions of the tank show consistently high errors. The poor quality of these results was anticipated because of the symmetry of the electrode channels. In order to remove the symmetry issue, a non-symmetric electrode channel (not paired at 180°) could be introduced or a tracking algorithm employed.

The modification made to improve the tracking accuracy was eliminating positions on the $x < 0$ half of the tank. If the mirroring issue was resolved for the entire tank, results as seen below could be expected for the full tank. Fig. 1B shows the distribution of errors for the half tank solution. The accuracy of the half tank was 94.3 % within 1 cm and a MSE of 0.02 mm. This accuracy is higher than that of the IR cameras, which have a 1 cm resolution. It can be seen on Fig 1B that the area of the tank in which there is lowest accuracy is the centre third at 89.9 %. The middle and outer third sections have accuracies of 96.6 % and 96.1 %, respectively. The accuracy is lowest at the centre of the tank because the EODs from the WEF have travelled the furthest distance, and are therefore have more interference.

The current means of tracking WEF uses IR cameras with a frame rate of approximately 20 Hz. The frame rate of the NN for a large trial was 213 Hz. This frame rate is nearly 10x that of the previous means of tracking. Regardless of the length of the trial, determined by number of samples analyzed, the frame rate of the NN did not vary significantly. Testing 10 samples yielded a frequency of 209 Hz, and testing over 15,000 samples yielded a frequency of 213 Hz.

The use of neural networks for localization of WEF appears to be a viable alternative to IR camera tracking. NNs advantages over IR cameras are both an increase in the temporal resolution or frame rate, and increased spatial resolution. The NN can localize the WEF within 1 cm in one dimension for 94.3 % of cases with a MSE of 0.02 mm. The frame rate of the NN was approximately 213 Hz, which was a temporal resolution of 10 times that of the IR camera. State of the art IR cameras have a frame rate of 1 kHz, but the image processing is much more costly to compute than the NNs raw data analysis and has a lower spatial resolution than NN. NNs could serve as a practical real-time method of tracking WEF non-invasively.

References

[1] Jun, J. J., Longtin, A., and Maler, L. (2011) Precision measurement of electric organ discharge timing from freely moving weakly electric fish. *J. Neurophysiol* 107, pp. 1996-2007.

[2] Adler A., Lionheart W.R.B. (2006) Uses and abuses of EIDORS: An extensible software base for EIT. *Physiol. Meas.*, 27:S25-S42.

[3] C. Assad, et. al. (1999) Electric organ discharge and electric image during electrolcation. *J. Exp Biol* pp. 1185-1193.

Simulation of 'pathologic' changes in ICG waveforms resulting from superposition of the 'preejection' and ejection waves induced by left ventricular contraction

VV Ermishkin, VA Kolesnikov, EV Lukoshkova, and RS Sonina

Russian Cardiology Research Center, Moscow, Russia

v.v.erm@mail.ru, elena.lukoshkova@gmail.com

Abstract. The impedance cardiography (ICG) is widely used for beat-to-beat noninvasive evaluation of the left ventricular stroke volume and contractility. It implies the correct determination of the ejection start and end points and the amplitudes of certain peaks in the differentiated impedance cardiogram. An accurate identification of ejection onset by ICG is often problematic, especially in the cardiologic patients, due to peculiar waveforms. Using a simple theoretical model, we tested the hypothesis that two major processes are responsible for the formation of impedance systolic wave: (1) the changes in the heart geometry and surrounding vessels produced by ventricular contraction, which occur during the isovolumic phase and precede ejection, and (2) expansion of aorta and adjacent arteries during the ejection phase. The former process initiates the preejection wave W_{pE} and the latter triggers the ejection wave W_{Ej}. The model predicts a potential mechanism of generating the abnormal shapes of **dZ/dt** due to the presence of preejection waves and explains the related errors in ICG time and amplitude parameters. An appropriate decomposition method is a promising way to avoid the masking effects of these waves and a further step to correct determination of the onset of ejection and the corresponding peak amplitudes from 'pathologically shaped' ICG signals.

1. Introduction

A reliable and accurate method for detection of the ejection onset is crucial for evaluation of the pre-ejection period (PEP), an important indicator of the changes in contractility during functional tests. To this aim, many algorithms [1, 2] have been elaborated but none of them is universal. The problem of correct and unambiguous detection of the start of ejection by ICG is not completely solved, and the automatic detection of specific 'landmarks' in ICG signals (such as B point) remains rather complicated. Moreover, in the case of specific ICG waveforms met in the cardiologic patients and elderly people [3. 4], the conventional algorithms and 'guidelines' used for finding the demarcating points (and even C peaks) usually fail. So, this problem turns to be not simply technical but fundamental in character.

The characteristic feature of such 'pathological' waveforms is a remarkable stepwise preejection wave superposing on the ejection wave front. By our hypothesis, there are two major processes contributing to the formation of impedance pulse wave: (1) the changes in the heart geometry and surrounding vessels which occur during the isovolumic phase of the left ventricular contraction, and (2) expansion of aorta and adjacent arteries during the ejection phase. The former process initiates the preejection wave W_{pE} and the latter triggers the ejection wave W_{Ej}. Using a simple mathematical model, we analyzed some effects of summation of these processes on the ***dZ/dt*** waveform and associated ICG parameters.

2. Model

By our assumption, the ICG impedance waveform represents an algebraic sum of the preejection wave and the ejection wave, i.e., $\Delta Z = (W_{pE} + W_{Ej})$. An asymmetric bell-shaped function $W(A, b, c, t) = A \cdot t^b \cdot e^{-ct}$ was chosen to simulate the contribution of each of these waves (figure 1, ***A***: W_{pE} – blue curve, W_{Ej} green curve). The result of summation of the big bell W_{Ej} with the small bell W_{pE} can be seen on panel ***B***. The first and the second time derivatives of all curves are shown below (***A*** and ***B***). The formulas describing the analytical functions and their derivatives are given on the right panel (figure 1). Note, that the first derivative ***dW/dt*** of each bell is biphasic. It attains the maximum at a time t_{max}, which points to the steepest part of the bell front, crosses zero line at t_0 (corresponds to bell's maximum), then runs down to the minimum at t_{min} indicating the steepest site on the downslope, and, afterwards, goes again to zero. Such a trend of ***dW/dt*** curve is typical for any bell-shaped curve with sigmoid up- and downslopes. Noteworthy, that the formulas for calculation of the derivatives of the chosen function ***W(A, b, c, t)*** are very simple (figure 1). Another intriguing property of this function is that the time t_0 lies strictly on a half way from t_{max} to t_{min}. This allows obtaining realistic-like ***dZ/dt*** curves by setting not the abstract parameters ***A, b,*** and ***c***, but the more meaningful time and amplitude characteristics: t_{min}, t_{max}, and $(dW/dt)_{max}$. For the ejection wave these parameters have the following meaning: t_{min} (or, more exactly, the period from $t = 0$ to t_{min}) is an analog of the left ventricular ejection time, $(dW/dt)_{max}$ is an analog of the C-wave amplitude, and t_{max} indicates the position of $(dW/dt)_{max}$ point on the time scale and relates to the steepness of C-wave front. In our model the start of the big bell (W_{Ej}) was fixed at 100 ms on the time scale, while the time shift Δt between the small and big bells (starting from the point $t = 0$ set for each bell function separately) varied from -100 to 0 ms (i.e., function W_{pE} started earlier than W_{Ej} by Δt). The impact of the preejection waves on the shape of the resultant wave ΔZ and its derivatives, as well as the associated changes in the timing of ***dZ/dt*** and d^2Z/dt^2 peaks and their amplitudes were analyzed.

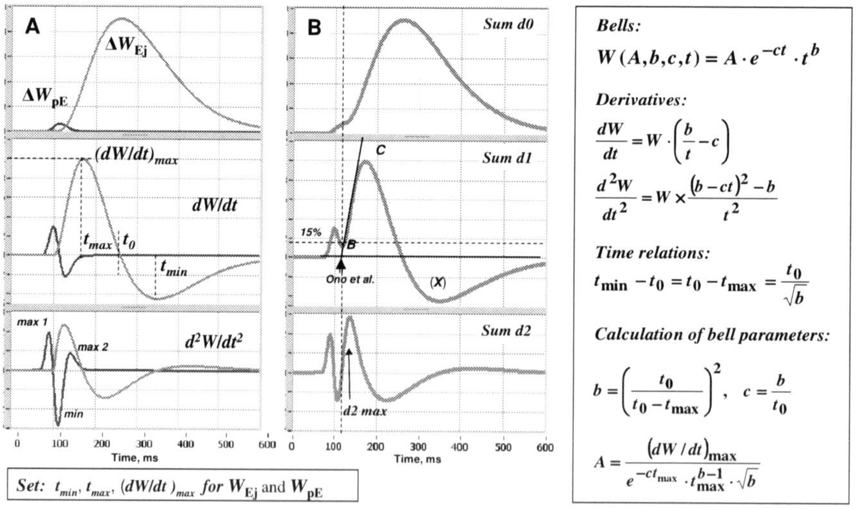

Figure 1. Simulation of ICG waves by a sum of two bell-shaped functions.

The points of interest were the C-wave maximum, the B-point, maximum peak of the second derivative (***d2 max***), and other fiducial marks used for identification of the onset of ejection (the end of PEP), for example, the cross-point of zero line with the inclined line which approximates the major part of the C-wave upstroke (by Ono et al [2]; figure 1, ***B***). We also used the similar method to find the 'true start' of each bell (not to be confused with the set point $t = 0$, a zero time to calculate each bell function).

3. Results

3.1. ICG waveform distortion induced by preejection waves

The model predicts the appearance of double-humped derivatives ***dZ/dt*** and d^2Z/dt^2 of the resultant wave. Figure 2 (a-g) shows the changes in the simulated ICG waveform representing the sum of two bell-shaped functions with the fixed ***dW/dt*** peak heights (1:5), as the small bell (preejection) approaches and superimposes onto the big bell (ejection). Although little changes are evident in the resultant **ΔZ** waveform, the shape of ***dZ/dt*** alters from ***b*** to ***f*** remarkably. The most pronounced changes are seen in the shape of d^2Z/dt^2. Thus, even small amplitude preejection waves, if they are rather steep, can produce a noticeable distortion of the resultant ***dZ/dt*** and d^2Z/dt^2 waveforms. Such phenomenon we observed earlier in some young and middle-age people, especially, during functional tests [4]. Summation of W_{pE} and W_{Ej} waves can also lead to elevated B-point in ***dZ/dt*** (figure 2, d-f).

When the amplitude of C-wave decreases or, vice-versa, the preejection wave increases and W_{pE} downslope becomes steeper or comparable to the upstroke of W_{Ej}, the resultant **ΔZ** waves will acquire the features typical of the cardiologic patients, with the first ('preejection') peak in both derivatives (***dZ/dt*** and d^2Z/dt^2) being nearly equal or even greater than the second peak corresponding to ejection (figure 2, g-h). In such situations most algorithms used for ICG processing fail to work. Some decomposition technique will be better to use in these cases.

It is important that the combination of this two-bell model with the non-linear fitting of the curves based on Levenberg-Marquardt algorithm [5] offers two-bell simulation of a large body of really recorded abnormal ICG ***dZ/dt*** shapes with a good precision ($r^2 > 0.99$).

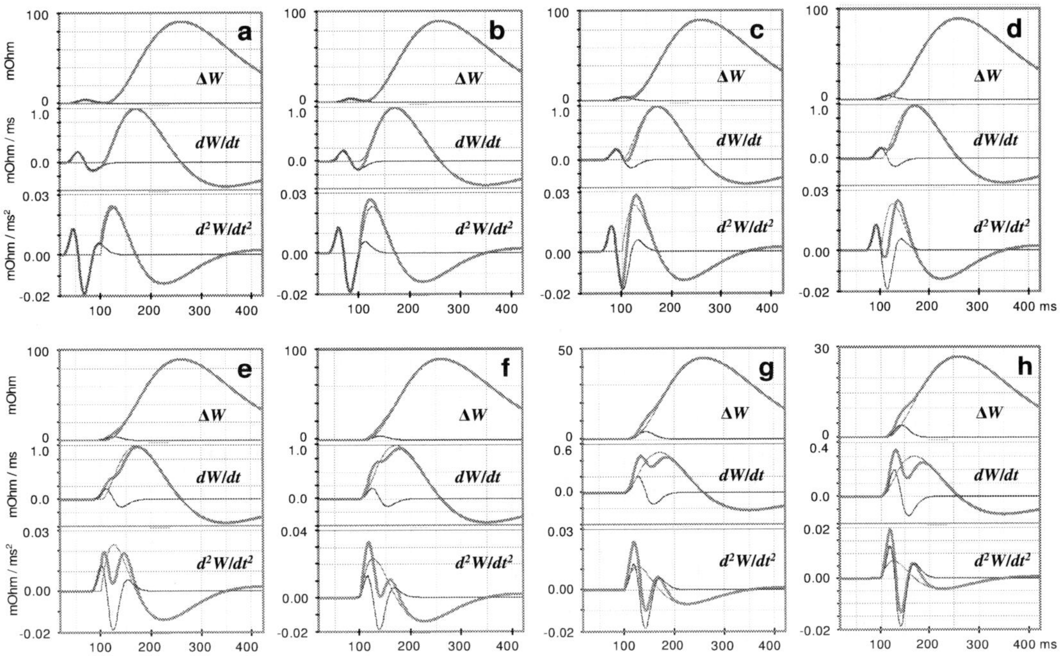

Figure 2. Transformation of ICG waves (red) induced by preejection waves (blue) started at various times before the ejection waves (green).

*3.2. Changes in timing of ejection onset and amplitudes of **dZ/dt** and d^2Z/dt^2 peaks*

The waves W_{pE} not only mask but shift the 'mark' points which are used for detection of PEP end. Figure 3, ***A*** shows the changes of the resultant ***dZ/dt*** waveforms as wave W_{pE} approaches to wave W_{Ej}. The colored lines are the traces of the corresponding 'landmarks': B-point, defined by a complex algorithm (red) and by maximum of the 3-rd derivative [1] (magenda), maximum of the d^2Z/dt^2 (***d2 max***, green), and maximum of the ***dZ/dt*** (blue) traces. Three black traces on the left indicate the start (by Ono et al [2]), peak and the minimum points of dW_{pE}/dt. The other black traces correspond to the big bell W_{Ej} which does not move. They indicate the start point and peaks of dW_{Ej}/dt and d^2W_{Ej}/dt^2. Below, on panel ***B***, a top view of the scaled-up plot ***A*** is shown. In figure 3, panel ***C*** the changes (in %) of C-wave (blue) and ***d2 max*** (green) amplitudes are plotted as functions of the time shift Δt between the starts of waves W_{pE} and W_{Ej}. The bottom plot (***C*** panel) shows the time lag between the ***d2 max*** peak (the green line in ***B***) and the corresponding peak of the ejection wave (the black vertical line in ***B***). The magenda lines on these plots reflect behavior of the absolute maximum of d^2Z/dt^2 when ***d2 max*** peak corresponding to C-wave becomes smaller than the earlier peak related to W_{pE}.

Thus, the presence of preejection waves can lead to considerable errors in the time and amplitude ICG parameter measurements resulting in erroneous assessment of stroke volume and contractility indices.

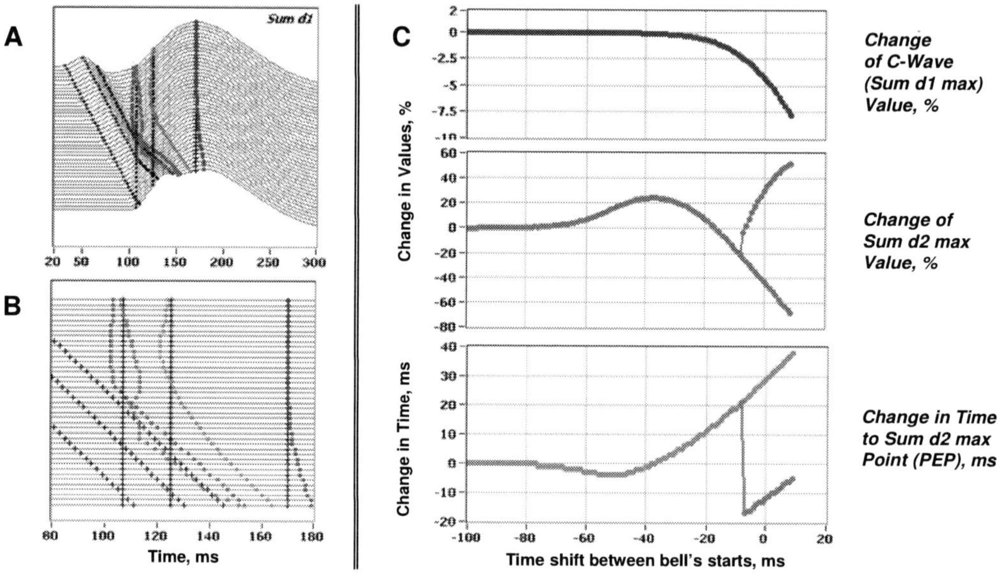

Figure 3. Changes in ICG parameters with changing time shift between the bell's starts. Step in changing time shift (time interval between horizontal lines in **A** and **B**) is 2 ms.

4. Conclusions

A simple two-bell model can adequately depict the peculiarities of actual ICG waveforms correlating to the early phase of ventricular systole which are typical of cardiologic patients and some elderly people. The model predicts a potential mechanism of generating the abnormal shapes of ***dZ/dt*** due to the presence of the preejection waves and explains the related errors in the impedance time and amplitude parameter measurements. An appropriate decomposition method is a promising way to avoid the masking effects of these waves and a further step to correct determination of the start of ejection and the corresponding peak amplitudes from 'pathologically shaped' ICG signals.

Acknowledgements

This study was supported by Russian Foundation for Basic Research, Grant No 10-04-01765-a.

References

[1] Kubiček W G, Kottke J, Ramos M U *et al* 1974 *Biomed. Eng.* **9** 410–6
Sherwood A, Allen M T, Fahrenberg J *et al* 1990 *Psychophysiology* **27** 1–23
Lozano D L, Norman G, Berntson G G *et al* 2007 *Psychophysiology* **44** 113–9
Ermishkin V V, Lukoshkova E V, Khayutin V M *et al* 2007 *IFMBE Proc.* **17** 655–8
[2] Ono T, Miyamura M, Yasuda Y *et al* 2004 *Tohoku J. Exp. Med.* **203** 17–29
[3] Mohapatra S N 1981 *Noninvasive Cardiovascular Monitoring by Electrical Impedance Technique* (London: Pitman Medical) p 286
Deshpande A K *et al* 1990 *J. Postgrad. Med.* **36** 207–12
Bour J and Kellett J 2008 *Eur. J. Intern. Med.* **19** 399–405
[4] Ermishkin V V, Kolesnilov V A, Lukoshkova E V *et al* 2012 *J. Phys.: Conf. Series* **407** 012016
[5] Press W H, Teukolsky S A, Vetterling W T, Flannery B P 1992 *Numerical recipes in C: the art of scientific computing* (Cambridge University Press) p 994

Journal of Physics: Conference Series **434** (2013) 012008 doi:10.1088/1742-6596/434/1/012008

Simulating Non-Specific Influences of Body Posture and Temperature on Thigh-Bioimpedance Spectroscopy during Continuous Monitoring Applications

A H Ismail and S Leonhardt
Philips Chair of Medical Information Technology, RWTH Aachen University, Pauwelsstrasse 20, 52074 Aachen, Germany

E-mail: ismail@hia.rwth-aachen.de

Abstract. Application of bioimpedance spectroscopy (BIS) for continuous monitoring of body fluid volumes is gaining considerable importance in personal health care. Unless laboratory conditions are applied, both whole-body or segmental BIS configurations are subject to non-specific influences (e.g. temperature and change in body position) reducing the method's accuracy and reproducibility. In this work, a two-compartment mathematical model, which describes the thigh segment, has been adapted to simulate fluid and solute kinetics during change in body position or variation in skin temperature. The model is an improved version of our previous one offering a good tradeoff between accuracy and simplicity. It represents the kinetics of fluid redistribution, sodium-, potassium-, and protein-concentrations based on simple equations to predict the time course of BIS variations. Validity of the model was verified in five subjects (following a sequence of 7 min supine, 20 min standing, and 40 min supine). The output of the model may reduce possible influences on BIS by up to 80 %.

1. Introduction

Recent advances in personal health care demonstrated the feasibility of applying bioimpedance spectroscopy (BIS), as an alternative for gold standard methods, to determine human body composition (e.g. fat / fluid content). Beyond non-invasivity and simplicity of the method, our obvious advantage would be the option for real-time continuous monitoring if integrated into daily clothes via textile electrodes. Such a scenario would be highly favourable from a dialysis patient point-of-care, as it would offer a bridge between inter- and intra-dialectic health monitoring.

Despite aforementioned advantages, non-specific factors (as changes in skin temperature, body posture, and / or consumption of food and beverage) can limit the method's accuracy and reproducibility. Accordingly, several authors have suggested application of standardize "laboratory" conditions (include performing the measurements in supine position after a period of recumbence up to 10 min and, when possible, always at the same time of the day [1]) and appropriate physiological models [2, 3] to achieve adequate detection limit for clinical measurement of hydration status. Unfortunately, these conditions cannot always be fulfilled, and alternative tools are required to maintain the reliability of the BIS method.

In this sense, this article presents a two-pool physiological model, representing the thigh segment, to simulate potential changes in BIS-measured data associated with changes in body posture and / or skin temperature. The model is an improved version of our previous one [4] offering a good compromise between accuracy and simplicity. It includes additional solutes (Na^+, K^+, Cl^-) with their excreted osmotic pressure, enhanced description of lymphatic system, modelling of protein transport across capillary membrane, and influence of variations in skin temperature on physiological parameters

2. Material and methods

Influence of body posture on body fluids: Changing body's orientation in space leads to different hydrostatic blood pressures due to gravity altering the hydrostatic and osmotic pressures balance between body compartments. As a result, fluid redistribution takes place. Maw *et. al.* [5] have showed that this fluid redistribution is limited to extracellular space and no significant change in intracellular fluid is expected. Based on all aforementioned factors, in this work the influence of body posture on BIS was simulated only extracellular (between plasma and interstitial spaces) using a two-pool model as described below.

Two-pool model: Fluid redistribution between interstitial (IntF) and plasma (Pl) compartments under the influence of changes in body posture are described considering the fluid's filtration through capillaries (Q_{Cap}) and the action of lymphatic pump (Q_{Lym}), as following

$$\frac{dV_{Pl}(t)}{dt} = -\frac{dV_{IntF}(t)}{dt} = -Q_{Cap}(t) + Q_{Lym}(t) \tag{1}$$

Following the Starling equations [6, 7], filtration rate across capillaries is calculated based on hydrostatic and osmotic pressure gradients in both compartments [8], given

$$Q_{Cap}(t) = K_f\left[\left(P_{Cap}(t) - P_{IntF}(t)\right) - \left(\Pi_{Pro,Pl}(t) - \Pi_{Pro,IntF}(t)\right) - \sum_s \left(\Pi_{s,Pl}(t) - \Pi_{s,IntF}(t)\right)\right] \tag{2}$$

where P represents the hydrostatic pressure in the capillaries (P_{Cap}) and in the interstitial compartment (P_{IntF}), and K_f is the fluid filtration coefficient. The osmotic pressure in each compartment includes that excreted by protein (π_{Pro}) and by small ions (π_s) respectively, where the subscript s accounts for the ions of sodium (Na^+), potassium (K^+), and chloride (Cl^-). Protein osmotic pressure was determined as a function of protein concentration (C_{pro}) as follows [9];

$$\Pi = 2.1 \cdot C_{Pro} + 0.16 \cdot C_{Pro}^2 + 0.009 \cdot C_{Pro}^3 \tag{3}$$

The hydrostatic pressure in IntF is determined by the actual interstitial fluid volume (V_{IntF}) and the tissue's compliance. In order to calculate the change in (P_{IntF}), the curve presented in [10] was expressed in terms of relative amount of IntF's volume to pressure, and a mathematical approximation for P_{IntF} were obtained using fitting curve algorithm

$$\begin{aligned} P_{IntF}(t) &= \frac{-1}{0.3 \cdot \left[V_{IntF,rel}(t) - 0.61\right]} + 5.17 \\ V_{IntF,rel}(t) &= \frac{V_{IntF}(t)}{V_{IntF}(0)} \end{aligned} \tag{4}$$

Protein transport rate across the capillary membrane was modelled according to [8, 11] considering protein reflection coefficient (η) and protein permeability-surface area product (PS)

$$Q_{Pro} = Q_{Lym}(1-\eta)\left[\frac{\left(C_{Pro,Pl} - C_{Pro,IntF}\right) e^{-\left(Q_{Lym}\frac{(1-\eta)}{PS}\right)}}{1 - e^{-\left(Q_{Lym}\frac{(1-\eta)}{PS}\right)}}\right] \tag{5}$$

The lymphatic system was described as a nonlinear relationship between lymphatic flow rate (Q_{Lym}) and the interstitial pressure (P_{IntF}) considering lymphatic flow sensitivity (LS), according to [8, 12]

$$Q_{L:ym} = \begin{cases} Q_{Lym,norm} + LS(P_{IntF} - P_{IntF,norm}) & |\forall \quad P_{IntF} \geq P_{IntF,norm} \\ Q_{Lym,norm} \cdot \left(\dfrac{P_{IntF} - P_{IntF,ex}}{P_{IntF,norm} - P_{IntF,ex}} \right) & |\forall \quad P_{IntF,norm} \geq P_{IntF} \geq P_{IntF,ex} \\ 0 & |\forall \quad P_{IntF} \leq P_{IntF,ex} \end{cases} \tag{6}$$

where $P_{\text{IntF,ex}}$ is the interstitial pressure corresponding to the excluded interstitial volume, and $Q_{\text{Lym,norm}}$ and $P_{\text{IntF,norm}}$ are the normal steady state value of Q_{Lym} and P_{IntF}, respectively.

Influence of temperature on BIS: Body response to change in skin temperature involves different mechanisms as vasodilatation and changes in cardiac output. Levick and Michel [13] reported different capillary pressures (P_{Cap}) in relation to change in skin temperature (T_s). Based on their values, a mathematical approximation linking both parameters was drawn in this work as follows

$$P_{Cap} = 2.4 \cdot K_T \cdot T_s - 47.1 \qquad |\forall \quad 20 \leq K_T \cdot T_s \leq 40 \tag{7}$$

where K_T is a temperature coefficient defined as the ratio between interstitial (T_{IntF}) and skin (T_s) temperatures. K_T is applied for the thigh segment at which thickness of fat layer cannot be ignored and may cause that T_{IntF} and T_s is not equal. Additionally, conductivity of electrolytes depend on temperature as shown by [14], which can be expressed in case of IntF using reference conductivity ($\sigma_{\text{IntF},0}$) and temperature ($T_{s,0}$) values, respectively.

$$\sigma_{IntF} = \sigma_{IntF,0} \cdot \left[1 + 0.02\% \cdot K_T \cdot (T_s - T_{s,0})\right] \tag{8}$$

Study design: Thigh segmental BIS was performed on 5 healthy male test subjects (2 for control) aged between 22 and 26, using a commercial bioimpedance device (Xitron Hydra 4200, Xitron Technologies Inc., San Diego, CA, USA) and tetra-polar standard hydrogel-aluminum BIS electrodes (Fresenius Medical Care, Bad Homburg, Germany). A sequence of 7 min supine, followed by 20 min standing, and then 40 min supine position (scenario 1) was applied to simulate influence of body posture on BIS. To study effect of temperature on BIS, ambient temperature was heated-up starting from 25 ± 1°C to 40 ± 1°C and then cooled down to 10 ± 1°C, each within 50 min phase (scenario 2), and variations in skin temperature were detected using an infrared camera (Vario CAM hr head, Infratec, Dresden, Germany). Measurements were done every 5 min (scenario 1) or 17 min (scenario 2), and each one was an average of five runs made every 3 sec. Cole-Cole model [15] was fitted to the measured data to extract extracellular resistance (Re) values.A period of at least 4 hours without eating or drinking and 48 hours without heavy physical exercises were prerequisites for the tests.

3. Results and discussion

The thigh segment's extracellular resistance (Re) increases during the supine position and very quickly decreases during the vertical position with measured ΔRe (% Re) of -4.55 ± 0.31 (standing 20 min) and 5.12 ± 0.22 (supine 40 min) for all subjects. The simulations are in agreement with the measured data. The mean values of difference between measured and simulated data, in % of initial measured Re, for all subjects are 0.67 ± 0.21 % (standing 20 min) and 0.89 ± 0.16 % (supine 40 min), respectively. During ambient temperature's change, measured ΔRe (% Re) was -5.71 ± 2.81 (heat-up) and 12.92 ±5.39 (cool-down), while simulated values were relatively higher with -8.74 ± 5.78 (heat-up) and 15.47 ± 8.14 (cool-down). An example from two selected test subjects is given in figure 1. The presented results suggested the capability of the model to reproduce influences of selected external factor on thigh-BIS data. Nevertheless, the applicability of the model could be extended to other body segments if appropriate adaptation was applied. Sensitivity analysis (data not shown) suggested that slope of the mathematical approximation (equation 7) is the most important element affecting the accuracy of the temperature model. As this approximation is based on published results of only two subjects (the authors themselves [13]), this could explain the relatively higher obtained simulation error. Finally, if the presented model were to be applied for determining effective changes

Journal of Physics: Conference Series **434** (2013) 012008 doi:10.1088/1742-6596/434/1/012008

in thigh's fluid during specific amount of time, while removing non-specific influences and given actual body posture profile, error in BIS could be reduced by up to 79.45 %.

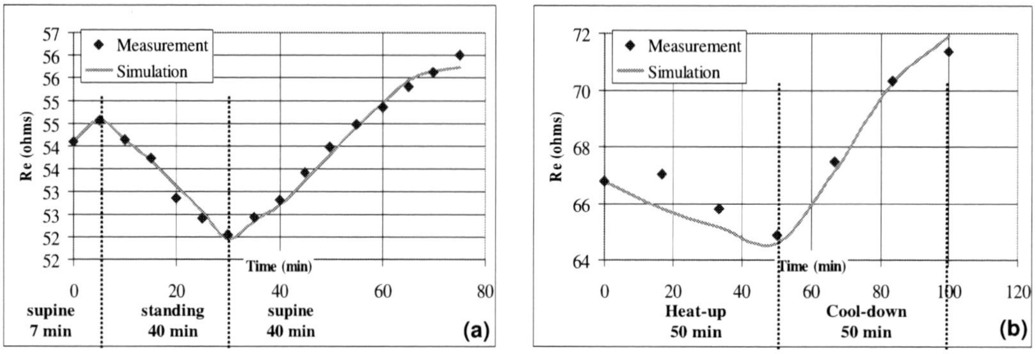

Figure 1. Measured and simulated thigh-Re values under the influence of changes in body posture (a) and ambient temperature (b).

4. Conclusion

The application of a physiological model for reducing artefact in BIS data is highly attractive, as it may enable measuring scenarios where BIS accuracy is questionable. Based on our results, the model reduces the influences of changes in body posture and temperature on BIS by up to 80 %. Despite that, further improvements to the presented method are still necessary, especially if the human thermoregulatory response (passive and active system), validity of the model for all body segments, and the individual difference in body composition are considered.

References

[1] Fresenius Medical Care 2006 Body composition monitor (BCM) operating instructions.

[2] Krämer M, Rode C, and Wizemann V 2006 *Kidney Int.* **69** 1609.

[3] Chamney P W, Krämer M, Rode C, and Wizemann V 2002 *Kidney Int.* **61** 2250.

[4] Medrano G, Eitner F, Walter M, and Leonhardt S 2010 *Med. Biol. Eng. Comput.* **48** 531.

[5] Maw G, Mackenzie I, and Taylor N 1995 *Acta Physiol Scand* **155** 157.

[6] Guyton A C 1986 *Textbook of Medical Physiology 7th ed.* (Philadelphia: W. B. Saunders Co.) pp 382 – 392.

[7] Levick J R 1995 *Changing perspectives on microvascular fluid exchange. In: Cardiovascular Regulation* ed D Jordan, and J Marshall. (London: the Physiological Society, Portland Press) pp 127–132.

[8] Fernandez Canete J and Del Saz Huang P 2010 *Comput. Biol. Med.* **40** 740.

[9] Hamilton W 1965 *Handbook of physiology* ed P Dow (Washington DC: Waverly Press) pp 961–1034.

[10] Guyton A, Granger H, and Taylor A 1971 *Physiol. Rev.* **51** 527.

[11] Bresler E H and Groome L J 1981 *Am. J. Physiol.* **241** 468

[12] Chapple C, Bowen B D, Reed R K, Xie S L, and Bert J L 1993 *Comput. Methods Programs Biomed.* **41** 33.

[13] Levick J R and Michel C C 1978 *J. Appl. Physiol.* **274** 97.

[14] Grimnes S and Martinsen O 2008 *Bioimpedance and bioelectricity basics.* (New York: Academic Press).

[15] Cole K and Cole R 1941 *J. Chem. Phys.* **9** 341.

Journal of Physics: Conference Series **434** (2013) 012009 doi:10.1088/1742-6596/434/1/012009

Effect of a spherical object in 4 electrode Focused Impedance Method (FIM): measurement and simulation

R Abir[1], F J Pettersen[2,3], O G Martinsen[2,3] and K S Rabbani[1]

[1]Department of Biomedical Physics & Technology, University of Dhaka, Dhaka, Bangladesh

[2] Department of Clinical and Biomedical Engineering, Oslo University Hospital HF, Oslo, Norway

[3]Department of Physics, University of Oslo, Norway

E-mail: raihan.abir@bmpt.du.ac.bd,rabbani@univdhaka.edu

Abstract: Focused Impedance Method (FIM) gives enhanced localized sensitivity at the centre of a zone defined by a simple system of electrodes, of which a 4-electrode version with electrodes at the corners of a square region has been studied in detail in the present work. The present work studies the effect of a large sphere whose diameter almost equals the dimensions of the central focused zone, or, the Focused Impedance. The sphere is placed at different positions with respect to the centre of the system at the electrode plane. The study has been made using a phantom in which the electrodes are fixed on a side wall while an insulating ball is hung at various positions inside the saline and moved with respect to the electrodes in their vicinity. The same was then simulated by providing appropriate parameters in COMSOL multiphysics, a software package utilizing Finite Element Method, by providing appropriately matching parameters. The measured impedance decreases to the background value as the ball is moved away from the centre in the electrode plane or along the depth, which is expected. The sensitivity also decreases with an increase in electrode spacing. The behaviours agreed very well with that obtained using COMSOL multiphysics package..

1. Introduction

Focused Impedance Method (FIM) is a recent technique, innovated by one of the authors at University of Dhaka, which can localize a zone of interest in a volume conductor and shows quite a promise in providing practical results with a minimum of instrumentation complexity [1]. Three version of FIM were developed, using 8, 6 and 4 electrodes respectively.FIM is attractive because of its ability to localize a small region using a few electrodes and simple instrumentation [2]. Utilising the 3D sensitivity of FIM it is possible to detect objects at shallow depths, typically of the order of the separation of the electrodes [3]. FIM has potential applications in many physiological studies and in the detection and diagnosis of diseases and disorders of the human body [4]. Particularly it may be useful in studying stomach emptying, stomach acid secretion, localized lung ventilation, bladder emptying, etc., where temporal changes occur that may be distinguished through a difference measurement. Again, through changes in the frequency of measurement it may be possible to identify organs or localized tissues that have distinctive frequency spectrum from the surrounding tissues. Thus it may be possible to characterise breast tumours and to monitor tumour ablation at shallow depths using multiple frequency measurements.

Among the three electrode configurations, 4 electrode FIM (abbreviated as 4-FIM) is simple in electrode configuration and useful in many applications. The concept was initially developed with the help of equipotential lines as shown in Figure 1. Here 4 electrodes, 1,2,3,4 are fixed in a square

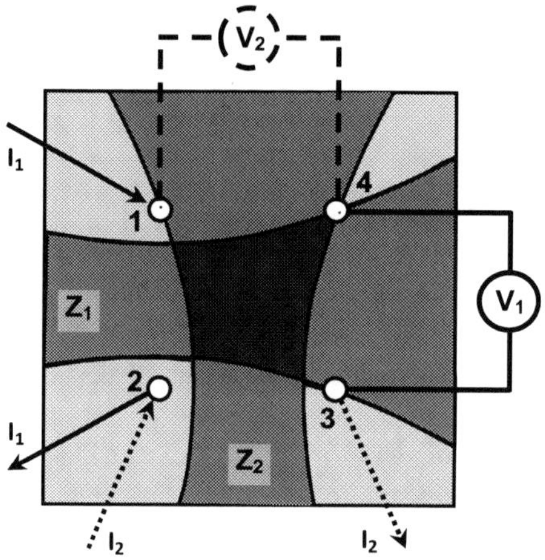

Fig.1. Basic concept of 4-electrode FIM

formation. Firstly, current is driven through electrodes 1 & 2 while potential is measured across electrodes 3 & 4. This gives a transfer impedance Z_1 which is essentially due to the impedance of the horizontal shaded region between the equipotential lines passing through electrodes 3 & 4. Then current is driven through electrodes 2 & 3 while potential is measured across electrodes 4 & 1. This gives a transfer impedance Z_2 which is essentially due to the impedance of the vertical shaded region. The Focused Impedance FZ is defined as the average of the two impedances, $FZ = (Z_1 + Z_2) / 2$.

Previously Geselowitz's lead field method and Finite Element method (FEM) were used to assess the point sensitivity in 4-FIMsystem [5, 6] which matched the above prediction well except an increased sensitivity at the four electrodes, which however decreased with depth (into 3D), eventually matching the predicted shape at about one third the electrode spacing.

The effect of five important variables namely electrode spacing, object size, object resistivity, background resistivity and depth of the object were studied in detail to have an understanding of the system [6]. The effect of an object at various positions along the horizontal and verticalaxis was also studied. Results of these studies verified the focusing effect of FIM and gave useful insight into the impedance change with depth.

The present work studied the effect of a large sphere whose diameter almost equals the dimensions of the central focused zone, or, the Focused Impedance. The study has been made using a saline phantom with electrodes fixed to a wall. An insulating ball is hung at various positions inside the saline and moved with respect to the electrodes in their vicinity. The same was then simulated by providing appropriate parameters in COMSOL multiphysics, a software package utilizing Finite Element Method, by providing appropriately matching parameters.

2. Methods and measurements

FIM measurements were carried out at 50kHZ using a saline filled 3D phantom, made of transparent acrylic, of dimension 30cm x 30cm x 30cm (related to actual depth of water tank). Electrodes having a diameter of 1cm and a depth of 2mm into the saline were fixed on one side while a spherical insulating object under measurement was hung from the top using a thin thread, guided by calibrated scales. The spacing of the current drive and pick up electrodes was 7cm initially (sides of the square), changed later as indicated. Electrical conductivity of the saline was measured using a conductivity cell of known geometry and was found to be 0.052 S/m. A tennis ball wrapped using black insulating tape was used as the spherical object. It had a diameter of about 6.5 cm, slightly less than the dimension of the square focused zone defined by the electrode spacing. As mentioned before, to get the focused impedance, an alternating current of constant amplitude was driven through two adjacent pairs of electrodes and the potential was measured across the other two. Then the configuration was rotated by 90^0 and again a similar measurement was taken. Because of the constant current, the potentials are proportional to the respective transfer impedances. The average of the two transfer impedances gave the focused Impedance, FZ.Measurements were taken using a Maltron Bodyscan 920 tetra polar impedance measurement equipment.

To perform simulation using COMSOL, the dimensions of the phantom and the object mentioned above were inserted. The conductivity of saline was taken from the measured value at 0.052 S/m. The conductivity of the insulating ball was taken at 10^{-14} S/m.

3. Results

The results of the studies are given through the plots in Figures 2 to 4. The measured impedance values are given by the dark lines while the COMSOL simulated values are given by the grey lines. For Figure 2, the object was moved along the horizontal axis parallel to the electrode plane with the centre of the ball at 3.5cm inside the phantom (considered as the depth of the ball, with its surface at 0.25cm depth inside the wall). The electrode separation was fixed at 7cm. For Figure 3, the ball was moved into the 3D having its centre always at the centre of the focused zone, theelectrode separation again being 7cm. For Figure 4, the ball was kept at the centre at a depth of 3.5 cm, but the electrode separation was changed.

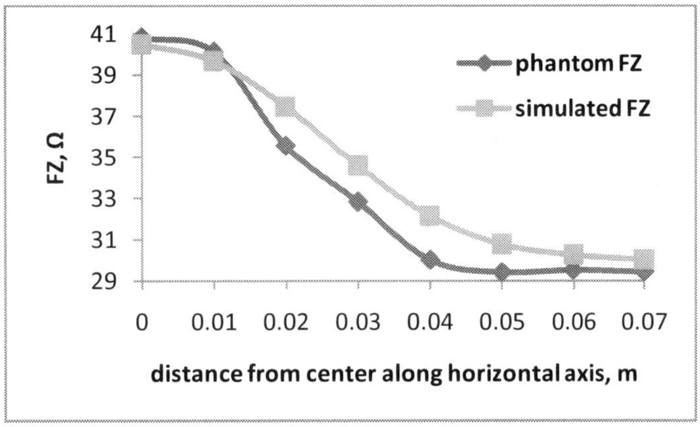

Fig.2. Measured and simulated focused impedance FZ when the ball was moved along the horizontal direction, keeping the depth same, at 3.5 cm.

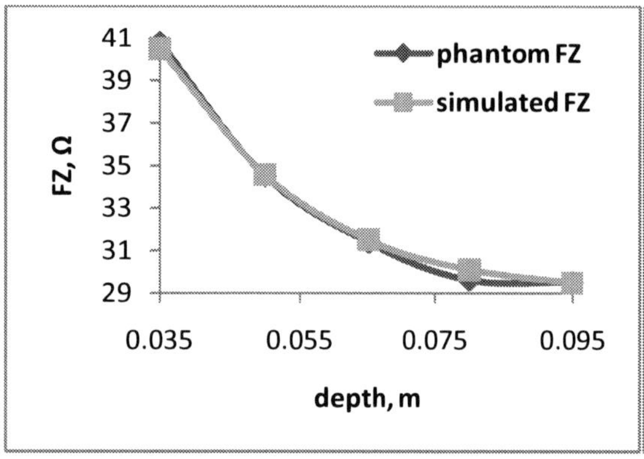

Fig.3. Measured and simulated focused impedance FZ when the ball was moved along the depth with the ball always at the centre of the focused zone.

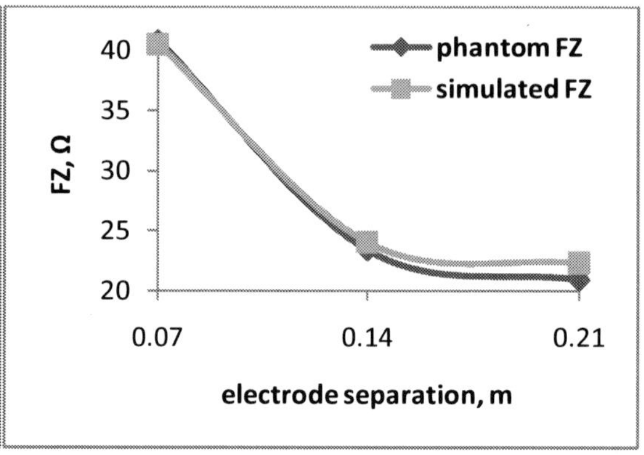

Fig.4. Measured and simulated focused impedance FZ when the ball was kept fixed at the centre of the focused zone while the electrode spacing was increased.

4. Discussions

In all the three plots the results of the actual phantom measurements agree well with that obtained using COMSOL simulation. However, there is some discrepancy in Fig.2 which may be due to experimental errors, or other specific reasons yet to be found. The measurements have been repeated several times and therefore, the values can be relied upon. It may be mentioned that not considering the 2mm height of the electrodes, COMSOL gave significantly different results. Therefore one needs to be very careful with the dimensions of the relevant items while working with COMSOL.

Acknowledgement

International Science Programme (ISP) of Uppsala University for funding support.

References

[1] Rabbani K S and Karal M A S 2008 A new four-electrode Focused Impedance Measurement (FIM) system for physiological study*Ann. Biomed. Engg. (Springer)***36** 1072-1077

[2] Rabbani K S, Sarker M, Akond M H R and Akter T 1999 Focused Impedance measurement (FIM) - A new technique with improved zone localization *Ann. New York Acad. Sci.***873**408-420

[3] Iquebal H M and Rabbani K S2010 3D sensitivity of 6-electrode Focused Impedance Method (FIM)*J. Phys.: Conf. Ser.(IOP Publishing)***224** No.1. http://iopscience.iop.org/1742-6596/224/1/012156/pdf

[4] Rabbani K S and Kadir M A 2011 Possible applications of Focused Impedance Method (FIM) in biomedical and other areas of study*Bangladesh Journal of Medical Physics***4** 67-74

[5] Islam N, Rabbani K S and Wilson A 2010 The sensitivity of focused electrical impedance measurements*Physiol. Meas.***31** S97-S109 http://iopscience.iop.org/0967-3334/31/8/S08/pdf

[6] Abir A R, Pettersen F J and Rabbani K S 2012 Sensitivity study for a 4-electrode Focused Impedance Method (FIM) using Finite Element Method analysis(submitted) *Physiol. Meas.*

Applicability of an effective conductivity approach in modeling thoracic impedance changes

M Lewandowska[1], B Truyen[2], C Boca[2], J Wtorek[1]

[1]Gdansk University of Technology, Poland; [2]Vrije Universiteit Brussel, Belgium

E-mail: maglew@biomed.eti.pg.gda.pl

Abstract. This paper describes numerical simulations of the influence of conductivity changes inside a volume conductor on impedance changes measured on its surface. A simple model based on the finite element method has been developed to estimate an applicability of the effective conductivity theory in human chest modeling. The model consisted of a cylinder with two concentric spheres inside. Simulations were performed for two cases: first the geometry was changing and material properties were constant within each subdomain, next, the geometry was constant but conductivity values were changing for each phase of cardiac cycle. In the considered range of geometry changes dependence between impedance changes calculated for two models was linear. Performed simulations showed that effective conductivity approach can be utilized when studying dynamic processes involved by volume changes of internal organs.

1. Background

Impedance measurement of heart activity has a great value when trying to evaluate its mechanical function. However, since the mechanism of bioimpedance signal formation is very complex the problem of signal sources identification and separation is still present. The application of an anatomically detailed finite element chest model allows for spatial sensitivity analysis. Having the spatial sensitivity distribution one can assess the impact of components from spatially separable sources. When trying to analyze impedance changes during cardiac cycle the dynamic model has to be generated. The first method to obtain a dynamic model is to generate several static geometric models with finite element mesh, each of them corresponding to the different cardiac cycle phase. Because of the complexity of the human chest it is time consuming and complicated method. The purpose of the article is to examine whether changes in the geometry of individual organs may be replaced by effective conductivity changes inside organs volume.

2. Methods

2.1. Effective conductivity theory

The basic model for a dielectric mixture consists of isotropic dielectric spheres with permittivity ε_i embedded in an isotropic dielectric environment with permittivity ε_e. If permittivity values of that two components and the fraction f of the total volume that is occupied by the inclusion phase are known, the average permittivity can be calculated using Maxwell Garnett formula [1].

Since the relation between permittivity and conductivity states that [2] $\varepsilon = \sigma/(j\omega\varepsilon_0)$, the Maxwell Garnett formula can be converted as follows:

$$\sigma_{eff} = \sigma_e + 3f\sigma_e \frac{\sigma_i - \sigma_e}{\sigma_i + 2\sigma_e - f(\sigma_i - \sigma_e)} \tag{1}$$

where: σ_{eff} – effective conductivity, σ_e – conductivity of environment, σ_i – conductivity of inclusion.

The equation (1) describes the case of homogenous inclusion. Considered model, however, contains inhomogeneous structure – a layered sphere (figure 1).

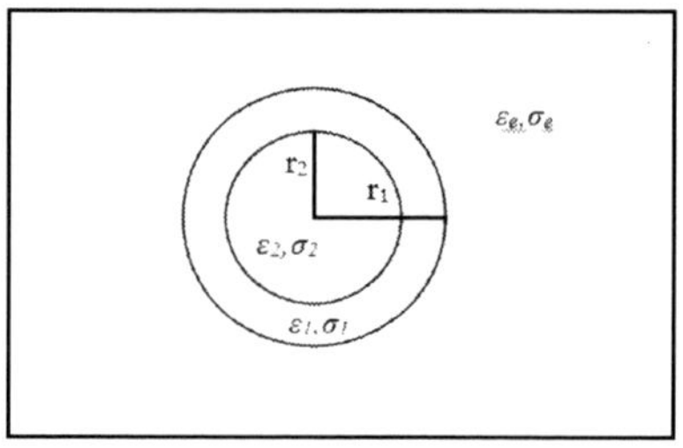

Figure 1. Homogenous medium with shell-sphere inclusion.

To obtain effective conductivity of that structure, the effective conductivity σ_{eff}' of shell-sphere combination is calculated from equation (1) assuming that $f=(r_2/r_1)^3$, then the result is used to calculate effective conductivity σ_{eff} of the inclusion-medium system:

$$\sigma_{eff} = \sigma_e + 3f\sigma_e \frac{\sigma_{eff}' - \sigma_e}{\sigma_{eff}' + 2\sigma_e - f(\sigma_{eff}' - \sigma_e)} \tag{2}$$

where f is the fraction of the total volume that is occupied by the inclusion phase with radius r_1.

Although the theory in most cases is applied to mixtures and heterogeneous materials with many small inclusions, it is reasonable to apply it also in the considered model:

a) the mixing formula satisfies the limiting processes for vanishing inclusion phase

$$f \to 0 \;\Rightarrow\; \sigma_{eff} \to \sigma_e \tag{3}$$

and vanishing host medium

$$f \to 1 \;\Rightarrow\; \sigma_{eff} \to \sigma_i \tag{4}$$

therefore there are no restrictions on the fraction f of the total volume occupied by the inclusion,

b) the theory was constructed by finding the solution of the Laplace equation, considered model also applies quasi-static approach,

c) the ratio between the size of inhomogeneities and the wavelength of the electromagnetic field of 20 kHz frequency for considered human body tissues is less than $1.15 \cdot 10^{-3}$ [3], so it is much smaller than unity that satisfies the homogenization condition [1].

2.2. The model

The simplified finite element model of a human chest was built (figure 2). It consisted of a cylinder of 0.4 m height and 0.12 m radius with two concentric spheres inside (left ventricle). Simulations were performed for two different sets of model's parameters. First, the geometry was changing and material properties were constant within each subdomain. Next, the geometry was constant but conductivity values were changing for each phase of cardiac cycle.

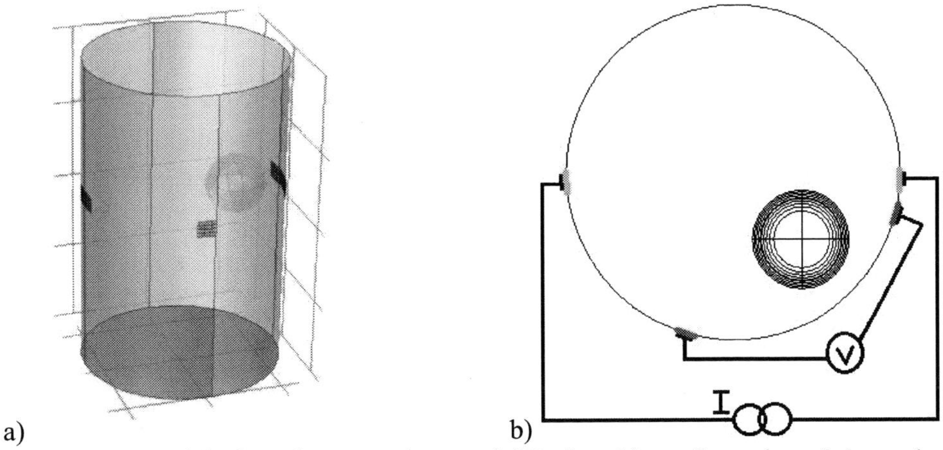

Figure 2. The model of a volume conductor: a) 3D view, b) configuration of electrodes.

2.2.1. Changing geometry

Analyses were performed for five different phases of the cardiac cycle. The first phase corresponds to systole and the last corresponds to diastole. The volume V_b of the inner sphere of radius r_2 changes from $4.46{\cdot}10^{-5}$ m^3 to $1.13{\cdot}10^{-4}$ m^3 that corresponds to changes of a blood volume inside a ventricle during heart cycle [4]. The radius r_1 of the second sphere changes so as to ensure constant volume of the difference between outer and inner sphere (table 1). That "shell" corresponds to left ventricle wall with constant volume of heart muscle V_m and variable thickness that changes from 10^{-2} m in systole to $6.62{\cdot}10^{-3}$ m in diastole.

Table 1. Effective conductivity values for different phases of cardiac cycle.

phase of the heart cycle	**V_b [m^3]**	**r_2 [m]**	**r_1 [m]**	**σ_{eff}' [S/m]** $\left(f=\frac{r_2^3}{r_1^3}\right)$	**σ_{eff} [S/m]** $\left(f=\frac{V_{b\,diastole}}{V_{cylinder}}=0.0114=const\right)$
systole	$4.46\cdot10^{-5}$	0.022	0.0320	0.2907	0.2050
.	$5.79\cdot10^{-5}$	0.024	0.0330	0.3054	0.2270
.	$7.36\cdot10^{-5}$	0.026	0.0341	0.3203	0.2560
.	$9.20\cdot10^{-5}$	0.028	0.0353	0.3350	0.2951
diastole	$1.13\cdot10^{-4}$	0.030	0.0366	0.3492	0.3492

The model consists of three subdomains with different material properties. Conductivity of the V_b region corresponds to blood conductivity $\sigma_2 = 0.5$ S/m. The second subdomain, "shell" volume V_m, has heart muscle conductivity $\sigma_1 = 0.22$ S/m [3]. $\sigma_e = 0.1$ S/m was assigned to the remaining volume of the cylinder.

2.2.2. Changing effective conductivity

To evaluate whether, for the purpose of thorax impedance modeling, changes of the blood volume inside a ventricle can be replaced with changes of the effective conductivity inside the considered region, Maxwell Garnett formula was applied. First, the effective conductivity σ_{eff}' of the blood-muscle structure was calculated (equation 1), then the result was used to calculate effective conductivity σ_{eff} of the ventricle-thorax system (equation 2). Obtained results are presented in table 1.

2.3. Simulation

Changes of impedance value between two electrodes on the cylinder surface (four-electrode technique, configuration shown in figure 2b) were calculated for each "cardiac cycle phase," first, for changing volumes and constant conductivities within each subdomain, next, for constant spheres volume but variable effective conductivity values inside the modeled left ventricle.

3. Results

In the considered range of geometry changes dependence between impedance changes calculated for two models – with changing geometry (ΔZ_{geom}) and that with changing effective conductivity (ΔZ_{eff}) – was linear, $\Delta Z_{eff} = 0.98\ \Delta Z_{geom}$. The relation between impedance changes and blood volume changes was linear for two considered models (figure 3). This proves that changes of geometry parameters can be replaced with effective conductivity changes while maintaining constant volume of considered organs. Relationship between ΔZ_{eff} and ΔZ_{geom} was analyzed for different values of blood $\sigma_2 \in (0.43;0.73)$ and lungs $\sigma_e \in (0.04;0.22)$ conductivities (figure 4) [5]. Obtained correlation coefficient between ΔZ_{eff} and ΔZ_{geom} ranged from 0.993 to 0.993 that indicated strong correlation between that two results.

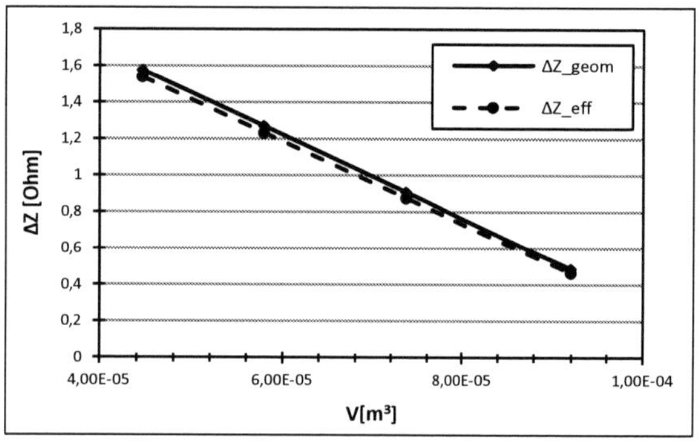

Figure 3. Impedance changes calculated for considered models, four phases of cardiac cycle ($\sigma_e = 0.1$ S/m, $\sigma_1 = 0.22$ S/m, $\sigma_2 = 0.5$ S/m).

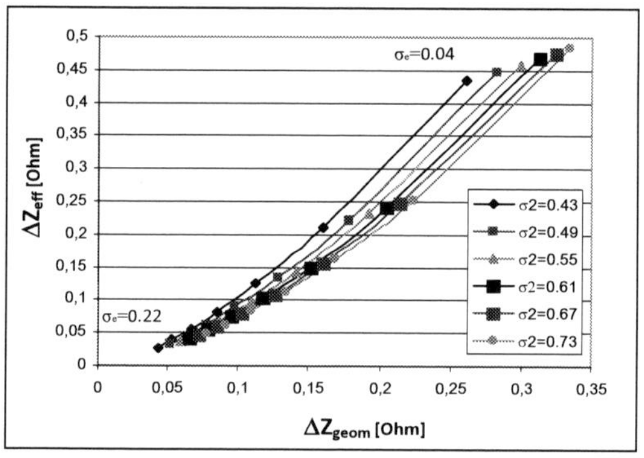

Figure 4. Relation between impedance changes connected with effective conductivity changes and geometry changes for 42 different combinations of tissue conductivity values.

4. Conclusions

Performed simulations indicate that effective conductivity approach can be utilized in cases of relatively big models when studying dynamic processes involved by volume changes of internal organs.

References

[1] Sihvola A H 1999 *Electromagnetic Mixing Formulas and Applications* (London: Institution of Engineering and Technology)
[2] Asami K 2006 Dielectric dispersion in biological cells of complex geometry simulated by the three-dimensional finite difference method *J. Phys. D: Appl. Phys.* **39** 492
[3] Calculation of the dielectric properties of body tissues, IFAC-CNR, Florence (Italy), 1997-2007 http://niremf.ifac.cnr.it/tissprop/htmlclie/htmlclie.htm
[4] Trail T A, Gibson G and Brown D J 1978 Study of left ventricular wall thickness and dimension changes using echocardiography *Br. Heart J.* **40(2)** 162-9
[5] Geddes L A and Baker L E 1967 The specific resistance of biological material - a compendium of data for the biomedical engineer and physiologist *Med. Biol. Eng.* **5(3)** 271-93

A model of end-expiratory lung impedance dependency on total extracellular body water

J Suchomel[1] and V Sobota[1]

[1] Czech Technical University in Prague, Faculty of Biomedical Engineering, Sitna Sq. 3105, 272 01 Kladno, Czech Republic

E-mail: jan.suchomel@fbmi.cvut.cz

Abstract. Electrical impedance tomography (EIT) is an attractive method for clinical monitoring of patients during mechanical ventilation. This study evaluates lung impedance measurements using Dräger PulmoVista 500 EIT system on an animal model. Mechanically ventilated model was created. Vital signs were monitored as well as mechanical ventilation parameters. Extracellular fluid balance and blood volume were handled as follows: 30-40 % of total blood volume were removed and returned back, 0.5 to 1 litre of Ringer's solution was injected afterwards. The quantity of injected fluids was recorded for each animal. During this process thoracic electrical impedance measurement was performed. Recorded data from PulmoVista 500 EIT system were analysed using the official Dräger EIT Data Analysis Tool. The dependency of end-expiratory lung impedance on the change of fluid balance was observed. The relation between end-expiratory (minimum impedance value) frames and changes of fluid balance is shown. Preliminary results strongly support the expectation that electrical impedance of thorax can be affected by total extracellular fluid change.

1. Introduction

Electrical Impedance Tomography (EIT) can produce tomographic images of tissue resistivity. This method is based on the visualization and quantification of tissue impedance determined by injecting small electrical currents and measuring the resulting voltages on the surface of the torso [1]. Lungs have high electrical resistivity relative to other tissues because of the air content [2]. In inspiration the ratio of air to condensed matter in lungs increases and the structures as alveolar walls become thinner. Electrical current applied to lung tissue flows through the condensed matter hence resistivity increases with inspired air volume [3].

However, the EIT has many problems which arise from the three dimensional spread of current, the need of many electrodes and the fact that the body size and the shape affect measured potentials as well as resistivity distribution [4]. Regarding this, the total impedance of the thorax could be affected by fluid balance of the body.

This preliminary study describes the possibility of utilization of PulmoVista 500 EIT system (Dräger Medical, Lübeck, Germany) for monitoring of lung impedance dependency on the fluid balance of mechanically ventilated animal. Although the system is designed for monitoring of lung ventilation and operates only with relative impedance changes (displayed as arbitrary units – AU) it has a potential to be a suitable tool for the purposes of this study.

2. Methods and material

The study protocol was approved by the Institutional Animal Care and Use Committee of the First Faculty of Medicine, Charles University in Prague. The study was performed at an accredited animal laboratory of the Department of Physiology, First Faculty of Medicine, Charles University in Prague, in accordance with Act No. 246/1992 Coll., on the protection of animals against cruelty.

Five large female pigs (*Sus scrofa domestica*) four to five months old with an average body weight of 52 kg (Tab. 1) were used in the study.

2.1. Premedication, anesthesia, preparation

Animals were premedicated by azaperone (2 mg/kg IM), followed by anesthesia with atropine sulphate (0.02 mg/kg IM) and ketamine hydrochloride (20 mg/kg IM). When placed on the operating table, initial boluses of propofol (2 mg/kg IV) and morphine (0.1 mg/kg IV) were administered. Dräger PulmoVista 500 electrode belt (size S) with 16 electrodes and one reference electrode was placed on the animal. ECG gel has been used several times to establish a good connection between the animal and the electrode belt. The signal quality has been checked by the EIT system. Animals were orotracheally intubated by a cuffed endotracheal tube (I.D. 7.5 mm) and connected to the conventional ventilator Hamilton G5 (Hamilton Medical AG, Bonaduz, Switzerland) in INTELLiVENT-ASV mode. An ear vein was cannulated and a continuous infusion of propofol (8 to 10 mg/kg/h IV) combined with morphine (0.1 to 0.2 mg/kg/h IV) and heparin (40 to 50 U/kg/h IV) was initiated to maintain anesthesia. The drugs dose was adjusted to keep the animals in the stage of surgical anesthesia throughout the procedure. The depth of anesthesia was checked according to the reflex responses and to the brain monitoring bispectral index (BIS).

2.2. Normoventilation

Conventional ventilator Hamilton G5 was used in the (S)CMV volume controlled ventilation mode with respiratory rate 18 bpm, FiO_2 21%. The initial tidal volume was set to 8.5 mL/kg of the actual body weight and was titrated to reach normocapnia ($PaCO_2 = 40 \pm 1$ mmHg). Reaching the normocapnia the EIT system has been calibrated to get a baseline for measurement (Fig. 1 and 2).

Table 1. Parameters of experiment realizations.

No.	Weight of animal (kg)	Calculated volume of blood (mL)	Blood loss (mL)	Blood transfusion (mL)	Ringer's solution (mL)
1	54	3564	1260	1000	2250
2	50	3300	1155	1000	1500
3	54	3564	1260	1000	1500
4	50	3300	1155	1000	1000
5	54	3564	1050	1000	2000

2.3. Study protocol

Total blood volume was calculated as 6.6 % of total body weight. Regarding to the study protocol the first 10 % of total blood volume was removed from vein. Another 5 % steps were performed up to 30-40 % of total blood volume after a short stable phase. All blood losses were recorded by the EIT system. The returning of blood was performed in two steps by 500 ml per 30 minutes to vein. After blood returning a fluid challenge was made by 1000 ml of Ringer's solution in 30 minutes to vein. The whole returning phase and the fluid challenge were recorded by the EIT system.

2.4. Evaluation methods

Recorded data from PulmoVista 500 were collected and the waveforms of relative impedance were reconstructed. Based on the protocol, the time intervals of fluid change were determined and the dependency of ΔEELI on the fluid change was calculated using linear regression (Fig. 4). The EIDORS software was used for reconstruction of selected frames to demonstrate the effect of the fluid change on baseline values in long term records.

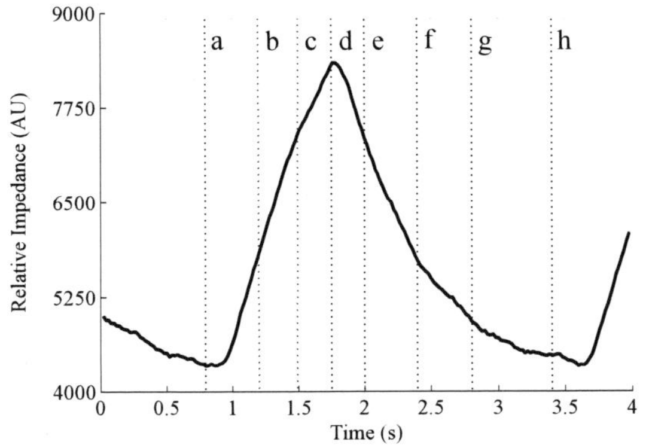

Figure 1. Relative impedance waveform of one single breath as recorded by PulmoVista 500. Points *a-h* correspond with respective EIDORS reconstructions in Fig. 2. Frame with the smallest value of relative impedance is used as the reference one (baseline) and is displayed as black. In this case is frame *h* the baseline frame. ΔEELI curve is generated by connecting the minimum impedance values (in this case *a* and *h*).

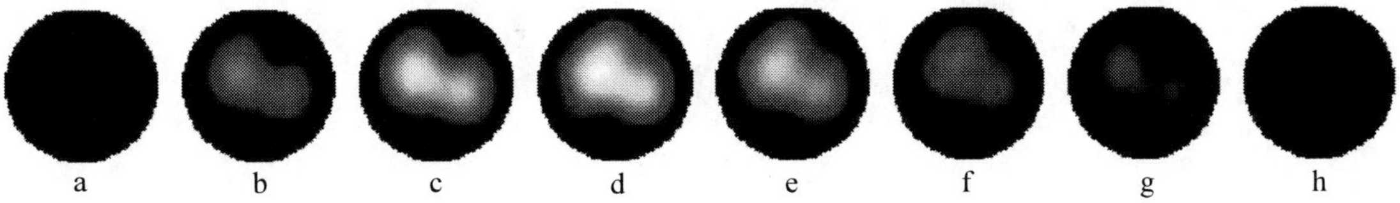

Figure 2. EIDORS reconstructions of points of relative impedance waveform from Fig. 1.

Table 2. Mean values and standard deviations of ΔEELI changes.

Fluid	Mean (AU/L)	SD (AU/L)	SD (%)
Ringer's solution	-4403	566	12,9
Blood	-4797	1241	26,0

3. Results

First preliminary results show that PulmoVista 500 is a suitable tool for monitoring of the dependency of lung impedance on the fluid balance. The change of ΔEELI caused by blood transfusion was recorded five times (volume 500 mL, dosing rate 1000 mL/h), the change caused by application of Ringer's solution was recorded seven times (volumes in range 500 to 1000 mL, dosing rate 1000 or 2000 mL/h) in five realizations of the experiment. Mean decrease of ΔEELI due to blood transfusion was −4797 AU/L (SD 26.0 %), due to application of Ringer's solution −4403 AU/L (SD 12.9 %, Tab. 2).

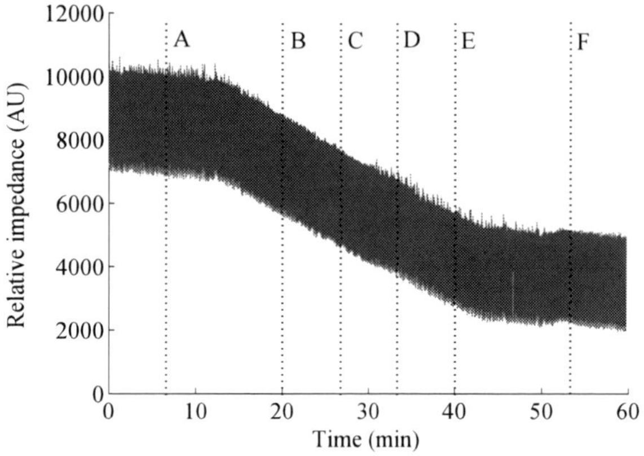

Figure 3. A record of relative impedance waveform (blue) during an application of Ringer's solution. ΔEELI (red) is counted as connections of minimum impedance values (see Fig. 1).

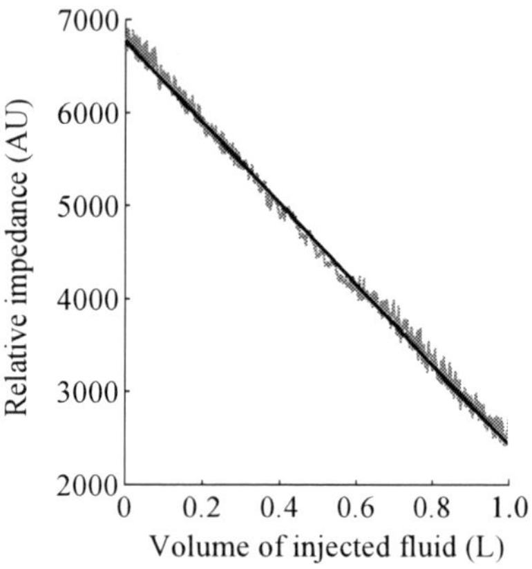

Figure 4. ΔEELI dependency on the change of injected fluid (Ringer's solution, dosing rate 2000 mL/h, volume 1000 mL).

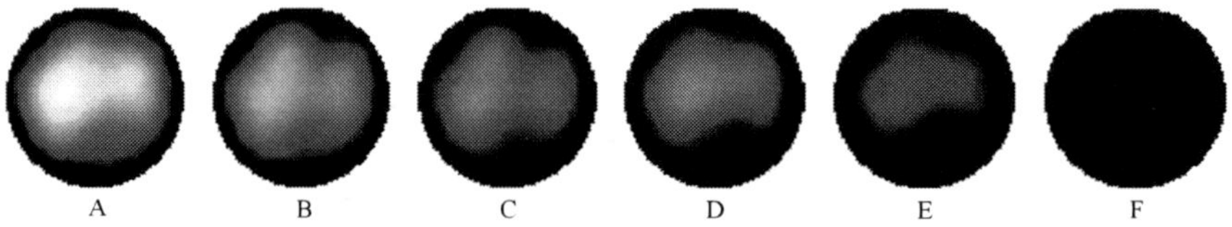

Figure 5. EIDORS reconstructions of points of relative impedance waveform from Fig. 3.

4. Discussion

The results show an obvious dependency of ΔEELI on change of fluid balance. As is shown in Fig. 3, according to the application of Ringer's solution or blood transfusion the mean of relative impedance waveform as recorded by PulmoVista 500 decreases and so does the respective ΔEELI curve. Although the results indicate the system is not capable to distinguish whether the change was caused by blood transfusion or by application of Ringer's solution there is a potential for using the system to diagnose pulmonary diseases characterized by oedema or contusion.

References

[1] Leonhardt S and Lachmann B 2012 *Intensive Care Medicine* **38**(12) 1917–1929 ISSN 0342-4642
[2] Duck F 1990 *Physical properties of tissue: a comprehensive reference book* (Academic Press) ISBN 9780122228001
[3] Witsoe D and Kinnen E 1967 *Medical and biological engineering* **5**(3) 239–248 ISSN 0025-696X
[4] Holder D 2010 *Electrical Impedance Tomography: Methods, History and Applications* Series In Medical Physics And Biomedical Engineering (Taylor & Francis) ISBN 9780750309523

Anatomically Accurate Infant Head Models for EEG Source Localization

Jasmine Song[1], Sergei Turovets[1,2], Pavel Govyadinov[1,2], Chelsea Mattson[1], Phan Luu[1,3], Kirk Smith[4], Fred Prior[4], Linda Larson-Prior[4] and Don M. Tucker[1,3]

[1]Electrical Geodesics, Inc. 1600 Millrace Dr. Suite 200, Eugene OR 97403, USA
[2]Neuroinformatics Center, University of Oregon, Eugene OR 97403, USA
[3]Department of Psychology, University of Oregon, Eugene OR 97403, USA
[4]Electrical and Optical Imaging, Neuroimaging Laboratory, Washington University, St. Louis, MO, USA

E-mail: jsong@egi.com

Abstract. Children differ from adults in head size, skull morphology, and tissue conductivity. We conducted a simulation to examine the error of source localization when a rescaled adult head model and different skull conductivities are used for EEG source localization in children. We have proven by simulation that source localization accuracy is the best with an infant specific head model including the age specific skull structure and conductivity.

1. Introduction

Dense array electroencephalography (dEEG) is an important noninvasive imaging modality for elucidating normal cognitive development in children and brain monitoring in neonates intensive care units. At present, the use and validation of dEEG and Electromagnetic Source Localization (ESL) in infants and children is hindered by the lack of pediatric head models. To generate an ESL solution, two independently specified problems must be solved: (1) the forward problem which specifies how currents move from their site of generation in cortex to the scalp, and (2) the inverse problem which is highly underspecified, and works from the recorded scalp potentials to the source of those signals. While many studies have shown that properly constrained realistic head models based on the high resolution MRI generate more accurate ESL solutions, the computability and necessity of additional variables such as tissue inhomogeneity and anisotropy are less well understood. While this is true of the current head models for normal adult populations, it is even more critical for pediatric populations where both the shape and density of the skull and the volume and composition of the brain are changing rapidly during first few years of life. Therefore, accurate ESL requires an accurate lead field matrix (LFM), specifying the forward lead fields, from each cortical source to each head surface electrode. Children differ from adults in head geometry, skull thickness, and tissue conductivity. We conducted a simulation to examine the error when an adult skull model and different skull conductivities are used in pediatric ESL. Specifically, we analyze three pediatric head models based on 1) a 7 months old infant MRI coregistered with the same subject CT, 2) the same MRI with warped adult CT atlas skull and adjusted thickness, and 3) an adult head rescaled to the 7 month old infant size.

Published under licence by IOP Publishing Ltd

The effects of the geometry (like presence or absence of fontanelles) and the skull conductivity specifications were examined. All three models were analyzed for six conductivity values ranging from the lowest value reported in the literature [1] 0.004 S/m through the average adult value 0.018 S/m to the average scalp conductivity (effectively no skull) and 18 (3×6) LFMs have been generated by our in-house finite difference forward solver for ESL. We have chosen a set of representative dipoles near fontanelles on the infant cortical surface in the model as the "synthetic ground truth" EEG data assuming 0.1 S/m to be the true infant skull conductivity value, and localized the sources with the minimum norm (MN) and sLORETA [2] distributed source localization methods using the rest of LFMs for 3 geometries with 6 skull conductivity specifications.

2. Methods

2.1. Forward problem

MRI image segmentation and registration were conducted with the EGI MRI/CT software package, BrainK[3] that is capable of cortical surface extraction and skull data warping to subject specific head shapes obtained from MRI scans. Sensor positions and head shapes are provided by the Geodesic Photogrammetry System (GPS) [4]. For the ground truth and reference, a true infant model(True) was created using subject specific MRI/CT data acquired at Washington University [5] and sensor positions for a 7 month-old baby (figure 1). After the segmentation and cortex tessellation, the oriented dipoles grid was constructed and LFMs were generated with our in-house finite difference solver [6]. We calculated 18 different LFMs, comparing 3 geometries and 6 skull conductivities. For comparison an adult skull was warped (Warped) to match subject specific MRI and an adult model (Adult) was simply rescaled to match the pediatric sensor positions, retaining the adult tissue morphology. In both cases the fontanelle structure was lost (figure 2). The ground truth EEG was calculated assuming a uniform skull conductivity of 0.1 S/m and the true infant model for several dipole locations: near the fontanelles, the eyeball and deep in the brain. The skull conductivities were varied according to the set of values 0.004 S/m, 0.018 S/m, 0.1 S/m, 0.2 S/m, 0.35 S/m and 0.45 S/m.

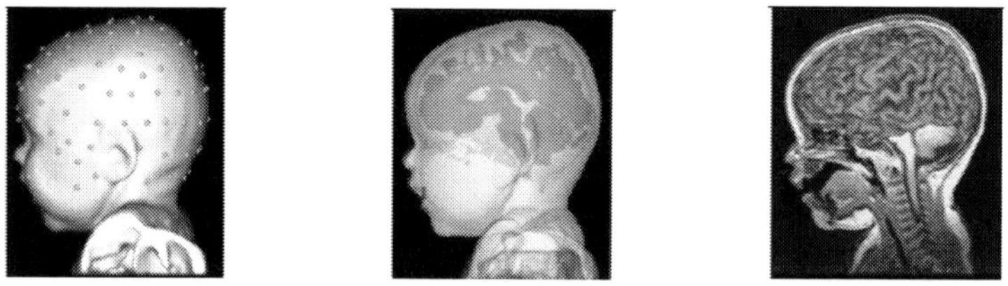

(a) Sensors on Scalp (b) Segmentation (c) Cortical surface

Figure 1. Sensors on pediatric head were registered by geodesic photogrammetry. A relative thresholding method segmented the scalp, skull, CSF, and gray and white matter. The cortical surface was extracted.

(a) True (b) Warped (c) Adult

Figure 2. Skulls of three geometries. The true model contains frontal and occipital fontanelles. Both warped and adult head models don't have fontanelles.

2.2. Inverse problem

We use the MN and sLORETA methods of ESL. The distributed source localization can be stated as following: for measured Φ and known K, find J given $\Phi = KJ$, where Φ is the electric potential at the electrodes of size N_e, J is the (unknown) amplitude of each current dipole of size N_v, and K is the lead field matrix linking the current sources to the electric potential of size $N_e \times N_v$. The MN solution is $\hat{J} = \arg\min_J\{\|\Phi - KJ\|^2 + \alpha\|J\|^2\}$, where α is a non-negative regularization parameter and $\|\cdot\|^2$ represents the square of the l_2-norm. The MN solution $\hat{J}$ is $\hat{J} = K^T[KK^T + \alpha I_{N_e}]^{-1}\Phi$. The sLORETA solution $\hat{J}^*$ is $\hat{J}^*_l = [C_{\hat{J}}]_{ll}^{-1/2}\ \hat{J}_l$, where $l = 1, \ldots, N_v$ and $C_{\hat{J}} = K^T[KK^T + \alpha I_{N_e}]^{-1}K$.

After the inverse problems are solved, the source solutions are validated by estimating the localization error distance (LED). The LED is the Euclidean distance between the locations of true dipole and the dipole with maximum intensity in the source distribution. To separate the effect of conductivity variation from differences in the geometry on source estimates, the ground truth EEG was calculated in each model assuming a uniform skull conductivity of 0.1 S/m. The sources were localized using all 6 LFMs with different conductivities. To see the effect of the head geometry, we have focused on a dipole near the rear fontanelles in the subject specific model with the skull conductivity of 0.1 S/m. The forward projection of the dipole was considered as the true EEG. We performed the source localization of this EEG data with the 3 LFMs created for 3 different geometries with skull conductivity 0.1 S/m.

3. Results

Table 1 summarizes the LEDs to assess the effect of skull conductivity. The LED is minimal, for both MN and sLORETA solutions, when the conductivity values used in the head model matches with those used to generate the synthetic EEG data. Use of the head models with the correct skull conductivity values results in the minimal LEDs. When skull conductivity values are small (skull is more resistive), the errors are larger.

The effect of head geometry on source localization can be seen in figure 3. A dipole was placed near the frontal fontanelle in the true model with skull conductivity of 0.1 S/m. Scalp potentials of the forward projections of the chosen dipole shows that the potential distribution is very focal. Since the dipole is located near the head surface, it can be easily localized. LEDs for true and warped head models are 0 mm but 5.39 mm for adult head model in both the MN and sLORETA methods. Placing the dipoles in other brain regions we showed that accurate head geometries are significantly more important for the deep sources than the shallow ones.

Table 1. The mean of LED is in mm. LFMs with 0.1 S/m obtain the minimum LED in three geometries by both MN and sLORETA.

Method	Head Geometry	Skull Conductivity (S/m)					
		0.004	0.018	0.1*	0.2	0.35	0.45
MN	True	17.3	15.5	15.1	15.1	15.2	15.2
	Warped	18.4	16.2	15.2	15.3	15.4	15.5
	Adult	19.8	18.3	13.9	14.9	16.3	16.6
sLORETA	True	3.7	0.3	0	0	0	0.0
	Warped	7.6	1.1	0	0	0.0	0.1
	Adult	10.6	4.6	0	0.1	0.5	1.0

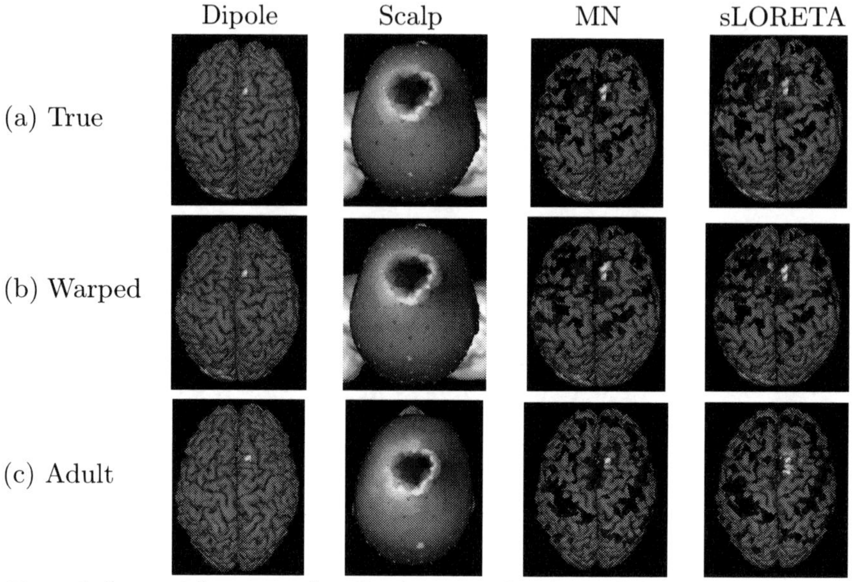

Figure 3. From left to right: a probe dipole near the frontal fontanelle in the true model is placed at the same location in both comparison cases; scalp potential of the forward projections of the probe dipole; localized sources using MN and sLORETA.

4. Conclusions

Accurate ESL with infants and young children requires electrical lead fields (head models) constructed to match both the unique geometries and skull conductivities of the appropriate age child instead of using rescaled adult head models. The actual conductivity values for infant skull are not well known presently, but unlikely to be in the adult range of low values. Therefore further studies are needed and methods like bounded EIT [6] can be used for noninvasive regional tissue conductivity estimates in infants.

Acknowledgments

This study was supported by National Institute of Health grant number R43NS067726-02.

References

[1] Roche-Labarbe N, Aarabi A, Kongolo G, Gondry-Jouet C, Dumpelmann M, Grebe R and Wallois F 2008 *Human Brain Mapping* **29** 167–176

[2] Pascual-Marqui R D 2002 *Methods and Findings in Experimental and Clinical Pharmacology* **24** 5–12

[3] Li K, Malony A and Tucker D M 2006 *First International Conference on Computer Vision Theory and Applications (VISAPP), Portugal, February* pp 354–364

[4] Russell G S, Eriksen K J, Poolman P, Luu P and Tucker D M 2005 *Clinical Neurophysiology* **116** 1130–40

[5] Larson-Prior L, Prior F, Smith K, Mattson C, Song J, Nolan T S, Luu P, Tucker D and Turovets S 2013 *IEEE Explore, EMBC, June 2013*

[6] Salman A, Turovets S, Malony A and Volkov V 2008 *International Workshop on OpenMP (IWOMP 2005/2006) LNCS 4315* pp 119–130

A Low-power and Compact-sized Wearable Bio-impedance Monitor with Wireless Connectivity

Seulki Lee[1], Salvatore Polito[1], Carlos Agell[1], Srinjoy Mitra[2], Refet Firat Yazicioglu[2], Jarno Riistama[3], Jörg Habetha[3], and Julien Penders[1]

[1]Holst Centre / imec, High Tech Campus 31, 5656 AE, Eindhoven, The Netherlands
[2]imec, Kapeldreef 75, B-3001, Leuven, Belgium
[3]Philips Research, Eindhoven, The Netherlands

Seulki.Lee@imec-nl.nl

Abstract. In this paper, we present a new bio-impedance monitor for wearable and continuous monitoring applications. The system consumes less than 14.4mW when measuring impedance, and 0.9mW when idling. Its compact size (4.8cm × 3cm × 2cm) makes it suitable for portable and wearable use. The proposed system has an accuracy of 0.5Ω and resolution of 0.2Ω on both resistance (R) and reactance (X) measurements, for impedance ranging between (j0.7)Ω to (54+j5)Ω with $2.9<\phi<5.7$. We also report the results of the system validation using passive loads as human tissue model, and show our wireless and miniaturized bio-impedance monitoring system has comparable performances with a reference system.

1. Introduction

Bio-impedance has been considered as one of the promising parameters for several healthcare applications [1], such as body composition, cardiac output, respiration, and body fluid monitoring applications. Several bio-impedance monitors have been developed and used in preliminary tests [2]-[4]. Most of them focus on small size and wireless connectivity for convenience. However, none of them also considers low power consumption for portable usage. As a result, there is no wearable system currently available for continuous bio-impedance monitoring in daily life. In this paper, a new bio-impedance monitor is introduced which enables low power, small size, and wireless connectivity at once. Its initial validation with passive loads shows that it can successfully replace the existing bulkier bio-impedance monitors.

2. Low-power Wireless Bio-impedance Monitor

2.1. System Architecture

The proposed bio-impedance monitor consists of a sensor and a controller module. The sensor module is based on a low-power analog-front end ASIC with square-wave current generator designed by imec. Using square-wave for both stimulating current and the demodulator clock allows to reduce ASIC power dissipation below 25μW. A microcontroller (TI MSP430F1611) is used for multi-frequency control, A-to-D conversion, and interface to the controller module via SPI. The controller module provides embedded processing, wireless connectivity, and data storage functionality. It includes a microcontroller (TI MSP430F1611), 2.4GHz Radio (Nordic nRF24L01), and micro SD card so that it can stream or store data according to the application. Figure 1 shows the overall architecture of the proposed bio-impedance monitor and its photograph. Two modules are stacked and connected to each other via a 10-pin connector. One battery (3.7V, 150mAh) is physically placed between them and electrically connected to the power management unit of the controller module so that operating power is distributed by it. The load is biased at half supply voltage, i.e. 0.6V in this case, during whole measurements in order to hang the input common mode voltage to a value within the dynamic range of the analog-front end.

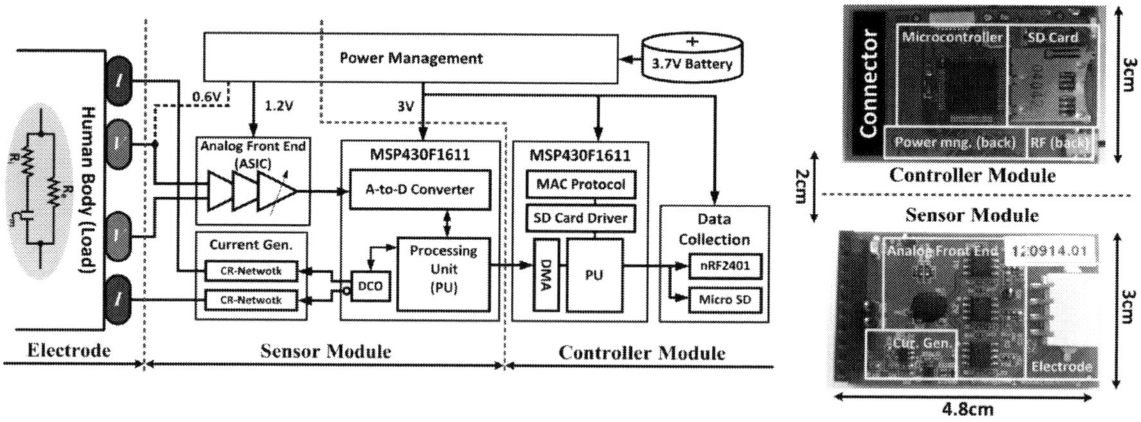

Figure 1 System architecture and its photograph

2.2. System Implementation Results

As shown in Figure 1, the overall size of the system is 4.8cm (W) × 3cm (L) × 2cm (H), which is, to our knowledge, the most compact bio-impedance monitor ever reported [2]-[4]. The system dissipates 7.5mW for sensing and 6.9mW for controlling, streaming and storing so that it can be continuously used for over 30 hours on a 150mAh battery with 3.7V. In order to improve measurement accuracy (less than 0.5Ω error) and resolution (less than 0.5Ω step) which are important specifications of a bio-impedance monitor, the sensor module of the proposed bio-impedance monitor includes on-board calibration performed before every measurement. It consists of 3 phases measured with shorted, small ($R_{CL} + jX_{CL}$), and large ($R_{CH} + jX_{CH}$) loads, while the resistance (R) and reactance (X) are calculated from extra/intra-cellular resistance (R_e and R_i) and membrane capacitance (C_m) [5].

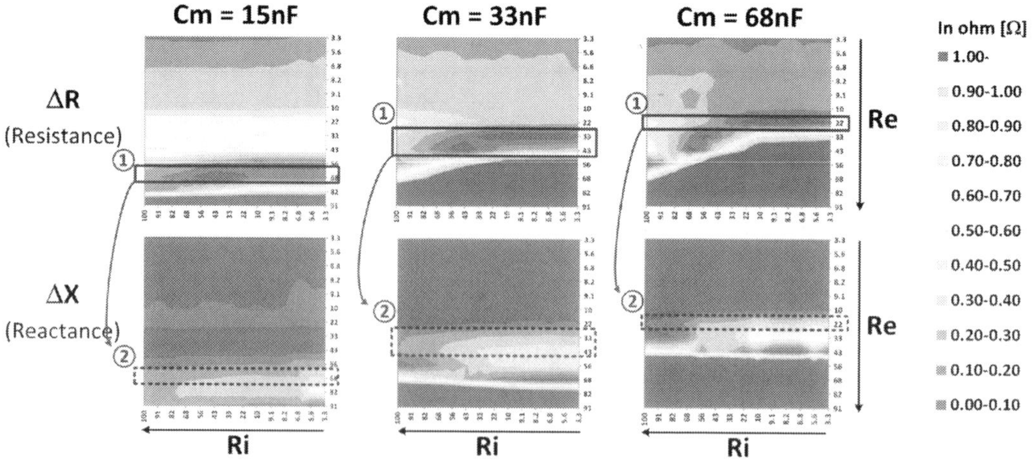

Figure 2 Bio-impedance measurement error (ΔR and ΔX) map according to load values

Figure 2 shows the R and X error (ΔR and ΔX) from measurements according to load values. ΔR and ΔX are calculated based on the difference between known R and X of loads and measured values.

Using these graphs, the range for R where ΔR is smaller than 0.5Ω is found (blue box numbered as ① in Figure 2). Then, ΔX in the same range is investigated whether it is smaller than 0.5Ω or not (blue dotted box numbered as ② in Figure 2), showing that ΔX is slightly larger than ΔR but the values are still below 0.5Ω. It is also derived from measurements that the target load range of the system with less than 0.5Ω error is directly related to R_{CL}, R_{CH}, X_{CL} and X_{CH} values as presented in Table 1. By setting their values to 22, 32, 0.7, and 3.2, respectively, the proposed monitor achieves 0.5Ω accuracy and 0.2Ω resolution at 10kHz for impedance ranging between (j0.7)Ω to (54+j5)Ω with $2.9<\phi<5.7$.

Table 1 Relations between measurable bio-impedance range and calibration values

Parameter	General condition	Bio-impedance range in this work (at 10kHz)
R [Ω]	$R \le R_{CL} + R_{CH}$	$R \le 54$
X [Ω]	$X_{CL} \le X \le \left(1 + \frac{R_{CL}}{R_{CH}}\right) \cdot X_{CH}$	$0.7 \le X \le 5$
$\frac{R}{X}$	$\frac{R_{CH}}{X_{CH}} \le \frac{R}{X} \le \frac{1}{2}\left(\frac{R_{CL}}{X_{CL}} + \frac{R_{CH}}{X_{CH}}\right)$	$10 \le \frac{R}{X} \le 20$

3. Validation with Passive Loads

3.1. Experimental Set-up

The system performance is verified using passive loads as a model for human tissue consisting of R_e, R_i, and C_m [5]. The Philips Bio-impedance Monitor [6] is used as a reference system, and the readings from the two systems are compared. Since single tone measurement can be seen as the subset of multi-frequency measurement, its validation is performed with 27 different frequencies between 10kHz and 211kHz for potential usage to the wider bio-impedance monitoring applications. After impedance measurement and calibration, the results are fitted to a Cole-Cole plot using the Least Squares Method over the target frequencies. Extracted parameters R_e, R_i, and C_m are compared with those from the reference system.

3.2. Validation Results

In this validation test, 49 different combinations of load values in the range of 22 to 43Ω for R_e and R_i and 22 to 49nF for C_m were chosen. Although some of them are somewhat out of range with respect to Table 1, their results are still considered as valid input to the analysis since multi-frequency measurement is less susceptible to the individual ΔR and ΔX from each single frequency measurement thanks to its reduced weight divided by the total number of frequencies. All pairs of (R, X) from whole frequency measurements are plotted together in one R-X plane, followed by fitting to the Cole-Cole plot to extract the R_e, R_i, and C_m value from it. Figure 3 shows two examples of multi-frequency measurement results. Each dot and line in the plot indicates theoretically calculated R and X from known values of R_e, R_i, and C_m (*Theoretical* in the legend, dot in blue color), measured and calibrated R and X using the proposed bio-impedance monitor (*Measured*, dot in red color), and fitted result to the Cole-Cole plot (*C-C Fitted*, line in red color). Even though few measured points are still included for analysis which are out of line and have larger error from the theoretical values than the average as found in Figure 3 (a), it does not significantly affect the fitting to Cole-Cole plot and extraction of R_e, R_i, and C_m. These erratic points are occasionally observed in high frequency measurements above 120kHz, where an initial phase of the summation vector of R and X becomes equal to or slightly larger than 90°. Since they are varied according to the load impedance R and X, it is not easy to exclude them from the every automatic analysis. Therefore, when the proposed bio-impedance monitor is adopted for single frequency measurement applications, the upper limit of available frequency should be set below 120kHz, which is low enough for the bio-impedance range in Table 1.

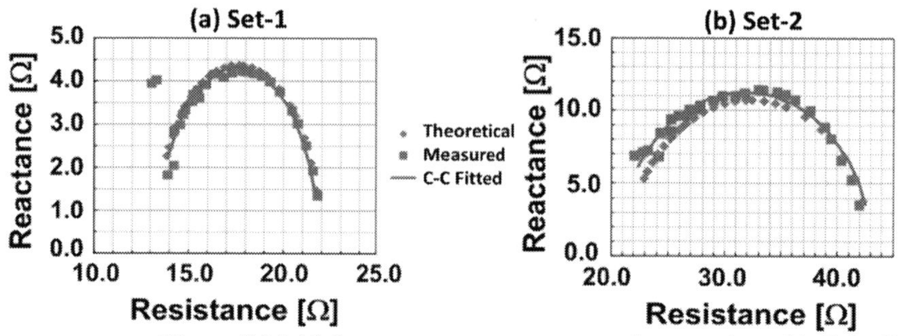

(c) Fitted R_e, R_i, C_m

	Set-1		Set-2	
	Theo.	Fitted	Theo.	Fitted
R_e	22	21.9	43	43.1
R_i	33	33.8	43	39.3
C_m	49nF	49nF	33nF	31nF

Figure 3 Multi-frequency measurement result examples with Cole-Cole plot fitting

When compared with the reference system, the average error computed from the 49 different load values is 0.18%, 7.2%, and 7.2% for R_e, R_i, and C_m, respectively. An accurate estimation of R_e is a quite promising result for many bio-impedance applications more related to R_e or Z, which is equal to $\sqrt{R^2 + X^2}$ and mostly determined by R_e, than R_i and C_m, such as shown in [7]-[8]. The relatively large errors in R_i and C_m is mostly caused by the lower maximum frequency used in the proposed monitor than the reference system, and could be improved by adopting a higher bandwidth analog front end or a stimulating current with reduced high frequency harmonics, e.g. sine current, while this would increase power consumption of the system.

4. Conclusions and Further Works

We reported a new bio-impedance monitor with compact form factor and wireless connectivity, validated against a reference system. Thanks to its low power consumption, convenience, and accuracy, the proposed system can be a good candidate to replace bulkier existing bio-impedance monitoring devices for wearable and continuous bio-impedance monitoring applications. The remaining work includes human body test for several applications, and investigation of the correlation factors between bio-impedance and biophysical phenomena.

References

[1] S. Grimnes and Ø. G. Martinsen, *Bioimpedance & Bioelectricity Basics*, 2nd Ed., London, Academic Press, 2000.

[2] Y. Yang, et al., "Design and preliminary evaluation of a portable device for the measurement of bioimpedance spectroscopy," *Physiol. Meas.* 27, pp. 1293-1310, 2006.

[3] G. Medrano, et al., "Bioimpedance Spectroscopy with textile Electrodes for a continuous Monitoring Application," *4th Int. Workshop on Wearable and Implantable Body Sensor Networks*, pp. 23-28, 2007.

[4] T. Vuorela, "Technologies for Wearable and Portable Physiological Measurement Devices," *Doctoral Thesis*, Tampere University of Technology, 2011.

[5] E. T. McAdams and J. Jossinet, "Tissue impedance: a historical overview," *Physiol. Meas.* 16, A1-A13, 1995.

[6] Philips Bio-impedance Monitor, http://www.research.philips.com

[7] L. Beckmann, et al., "Monitoring of body fluid in patients with chronic heart failure using Bioimpedance Spectroscopy," *World Congress on Medical Physics and Biomedical Engineering*, IFMBE Proceedings Volume 25/7, pp 532-535, 2009.

[8] V.-P. Seppä, et al., "Assessment of Pulmonary Flow Using Impedance Pneumography," *IEEE Transaction on Biomedical Engineernig*, Vol. 57, No. 9, pp. 2277-2285, Sep., 2010.

Journal of Physics: Conference Series **434** (2013) 012014 doi:10.1088/1742-6596/434/1/012014

Minimal implementation of an AFE4300-based spectrometer for electrical impedance spectroscopy measurements

B Sanchez[1], A Praveen[2], E Bartolome[2], K Soundarapandian[2], R Bragos[1]

[1]Electronic and Biomedical Instrumentation Group, Department of Electrical Engineering, Universitat Politecnica de Catalunya (UPC), 08034 Barcelona, Spain
[2]HealthTech Product Line, Texas Instruments, Dallas, 75243 USA.

E-mail: benjamin.sanchez@upc.edu

Abstract. The AFE4300 is a new low-cost on-chip impedance spectrometer developed by Texas Instruments able to handle multiple four electrode interface measurements. In this work, we present a brief description and characterization of this device and, besides its interesting features as a body-composition impedancemeter system; we evaluate its potential to develop minimal implementations for other biomedical applications. As the case study presented in this paper, its use to monitor ventilatory time-varying bioimpedance.

1. Introduction

Over the past several years, electrical bioimpedance has gain popularity for its potential use in a wide range of applications [1]. Impedance-based devices have been developed for many applications [2], e.g. body composition analysis [3], electrical impedance myography [4] or impedance cardiography [5] to name but a few. Although in practice most studies are conducted with high-performance devices, in some cases, minimum size instrumentation is required, as for example in wearable bioimpedance applications. For that, Texas Instruments has released the AFE4300, a new system on-chip fully integrated electrical impedance spectrometer initially designed for body composition determination in commercial weight scales.

The AFE4300 is a 4-electrode measurement device (see the principle of operation in figure 1) that enables multiple impedance measurement configurations through the internal multiplexers. To perform single or multi-frequency impedance spectroscopy measurements, the user can select either the full wave rectifier (FWR) or quadrature and phase (I/Q) demodulators. The peak-to-peak value of the injected sinusoidal current can be limited with an external resistor. The excitation signal is internally generated using an integrated Direct Digital Synthesizer (DDS) and the frequency can be modified by registers. The voltage drop is sensed through the high and low voltage channels. The clock for the device is generated from an external reference clock.

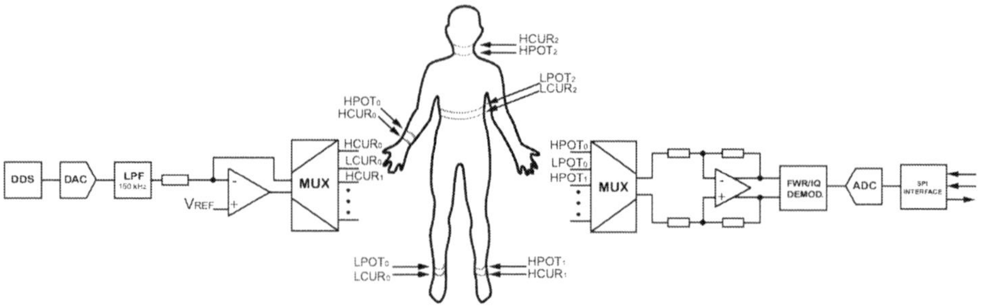

Figure 1. AFE4300 functional block diagram for tetrapolar single- or multi-frequency body impedance measurements.

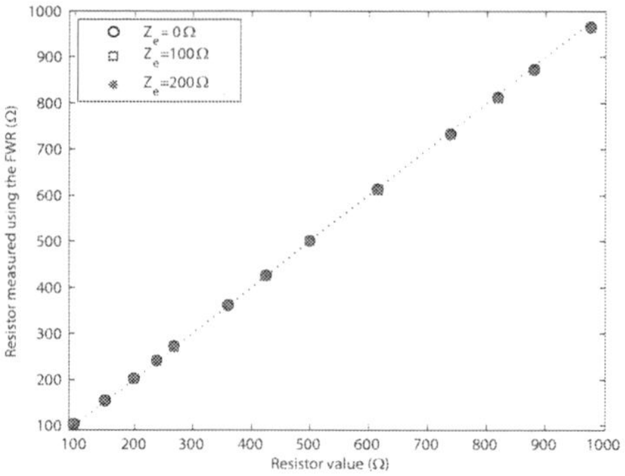

Figure 2. Response of the FWR demodulator as a function of load resistance (measured at 8 kHz) and electrode impedances $Z_e = [0, 100, 200]$ (Ω).

2. Experimental results

2.1. Characterization

With the purpose of evaluating the performance of the AFE4300 to measure impedance, we proceed to perform a series of experimental measures using the evaluation board provided by the manufacturer. In order to determine the linearity range of the device, we measure using the FWR demodulator a set of pure resistors from 100 Ω to 1 kΩ (1%) without considering electrode impedance resistors ($Z_e = 0\Omega$). This is done using a decade resistance box manufactured by Danbridge type CDR4/BCDE. We calibrate the FWR-voltage output with a two-point method performing a two-reference resistors measurements, 701.5 Ω and 196.8 Ω (1%), which takes into account the gain $[\Omega/V]$ and offset $[V]$ factors. As shown in figure 2, the AFE4300 works in a linear region for the range of resistors measured. As regards the I/Q data, data are calibrated with the same two-reference resistors.

As for the robustness in front of electrode impedance, two electrode resistor values ($Z_e = 100\Omega$ and $Z_e = 200\Omega$) are also tested. The dummy electrode resistors are connected to the AFE4300 four-electrode impedance measurements terminals. The values of electrode impedance mismatch (50% mismatch of an electrode impedance) on the voltage sense channels are also tested (data not shown). As it may be observed in figure 2, the AFE4300 is able to handle with electrode impedances around hundreds of ohms behaving linear with the unknown impedance measured.

2.2. Impedance measurements in phantoms

Figure 3 (A) shows the ability of the AFE4300 to perform impedance spectroscopy measurements measuring a dummy circuit compared to a commercial HP4192 impedance analyzer. The

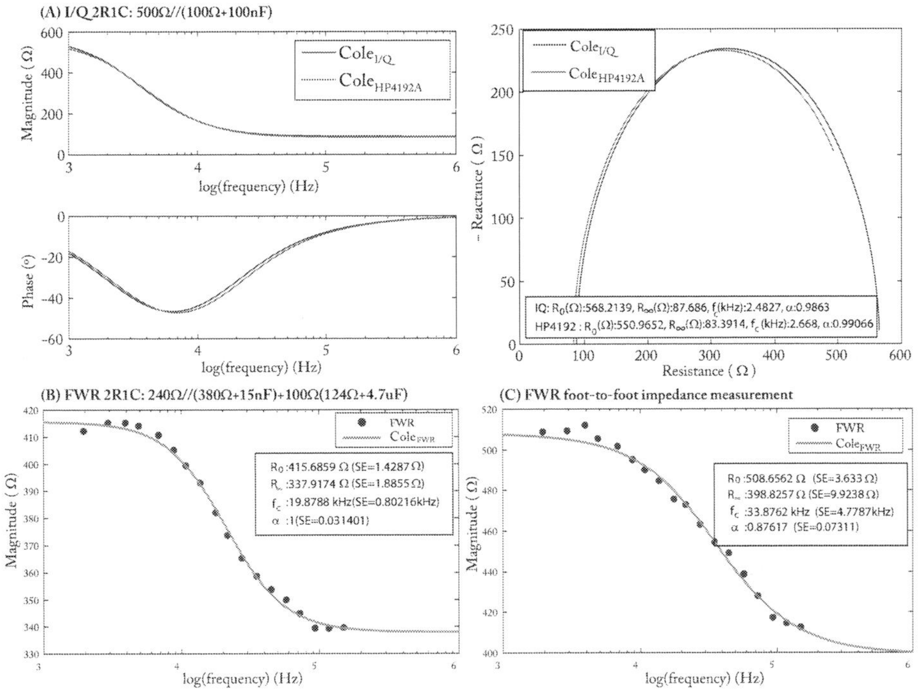

Figure 3. Accuracy in multi-frequency impedance measurements using the I/Q (A) and FWR demodulator in phantoms (B) and foot-to-foot body impedance measurements (C)..

Table 1. Predicted Total Body Water (TBW), Fat-Free Mass (FFM) and Fat Mass (FM). Refer text for details.

	Subject 1		Subject 2	
	SFB7	AFE4300	SFB7	AFE4300
TBW (L)	45.48	48.1	42.24	43.0
FFM (kg)	62.13	65.7	66.87	58.7
FM (kg)	26.67	23.1	28.59	27.7

impedance measurement was based on a 4 electrode topology using I/Q demodulator and data were processed with MATLAB^{TM} [6].

2.3. Body impedance measurements for body composition assessment

In table 1 are shown the reduced parameters for body composition assessment via the multifrequency Hanai-Cole approach for two male caucasian subjects, 29 and 50 years old, with body mass indices (BMI) of 23.84 kg/m^2 and 28.83 kg/m^2 respectively. As for the measurements performed with the SFB7 (Impedimed, San Diego, CA, USA), the measurement configuration were the standard distal body impedance analysis (BIA) configuration to foot-to-foot with the patient in supine decubitus position (3 kHz $\rightarrow$ 500 kHz). The four electrodes were placed on the right and left foot in the third metatarso-phalangeal and in the articulation, 6 cm apart. Disposable pregeled Ag/AgCl electrodes were used (3M Red Dot). As for the AFE4300 measurements, the device was connected to the electrodes of a Tanita TBF-611 commercial weight scale and the subjects were in upright position. The AFE4300 current level applied was

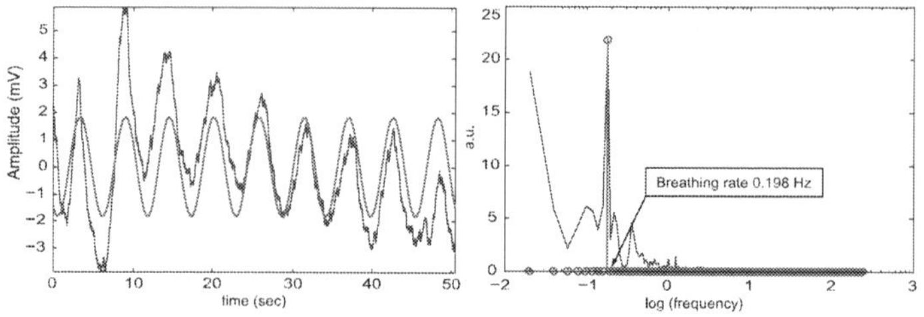

Figure 4. Time course evolution and frequency dependencies of the voltage signal: measured (blue) and its periodic reconstruction (red); DC-values are not shown in the DFT-plot for convenience.

of 100 uA. The standard errors (SE) of the Cole model parameters were estimated from the fitting in figure 3 (B) – (C). The relative errors in the estimation of R_0 and R_∞ are 0.34% – 0.55% and 0.71% – 2.49% respectively. The frequency sweep electrical impedance spectroscopy measurement performed is measured at 18 frequencies within the range 2 kHz → 100 kHz (see an example in figure 3 (C)).

2.4. Respiratory activity monitoring

To monitor time-varying bioimpedance-based respiratory activity, we measure the hand-to-hand impedance using Shieldex P180 textile electrodes manufactured by Statex placed on a handlebar. A current injection (100 uA, 32 kHz) was applied through the high potential - current electrodes and the voltage was measured with the low potential - current electrodes. The textile electrodes are fabric made of 78% Nylon, 22% elastomer and plated with 99.9% conductive silver with an average surface resistivity 5 Ω per square with two directional stretchability (wrap-weft). The respiratory characteristics of the sampled (f_s=475 sps) voltage signal is shown in figure 4.

The periodic reconstruction of the voltage signal measured shown in figure 4 (left, blue) is obtained estimating the mean periodicity of the signal in the frequency domain. The discrete Fourier transform (DFT) is calculated an plotted in figure 4 (right, blue). Further, only the harmonics corresponding to an integer number of the mean periodicity were retained, and the leftovers of the DFT spectrum were set to zero (right, red). The reconstructed periodic time varying (PTV) domain voltage signal was then obtained using the inverse of the reconstructed DFT (iDFT) spectrum (left, red).

3. Conclusions

Ultimately, the results show the ability of the AFE4300 to acquire the stationary and nonstationary behavior of bioimpedance. Unlike the AD5933 from Analog Devices, the AFE4300 does not need an external front-end to measure at 4-electrodes [7]. The reduced size and complexity of the electronics of the device widens the possibilities for the measurement of electrical impedance with minimal implementations.

References

[1] Schwan H P 1999 *Ann. New York Acad. Sci.* **873** 1–12
[2] Martinsen O G and Grimnes S 2008 *Bioimpedance and Bioelectricity Basics, Second Edition* (Academic Press)
[3] Jaffrin M Y and Morel H 2008 *Med. Eng. & Phys.* **30** 1257–69
[4] Rutkove S B 2009 *Muscle nerve* **40** 936–46
[5] Kubicek W G, Karnegis J N, Patterson R P, Witsoe D A and Mattson R H 1966 *Aerospace medicine* **37** 1208–12
[6] MATLAB 2012 *version 8.0.0 (R2012b)* (Natick, Massachusetts: The MathWorks Inc.)
[7] Seoane F, Ferreira J, Sanchéz J J and Bragós R 2008 *Physiol. Meas.* **29** S267–78

Journal of Physics: Conference Series **434** (2013) 012015 doi:10.1088/1742-6596/434/1/012015

Bioimpedance-Based Wearable Measurement Instrumentation for Studying the Autonomic Nerve System Response to Stressful Working Conditions

J Ferreira[1,3], L Álvarez[2], R Buendía[1,2,3], D Ayllón[2], C. Llerena[2], R Gil-Pita[2] and F Seoane[1,3]

[1] School of Engineering, University of Borås, SE-501 90 Borås, Sweden
[2] Department of Signal Theory and Communications, University of Alcala, ES-28871, Madrid, Spain
[3] School of Technology and Health, Royal Institute of Technology, SE-14152, Stockholm, Sweden

E-mail: javier.ferreira@hb.se

Abstract. The assessment of mental stress on workers under hard and stressful conditions is critical to identify which workers are not ready to undertake a mission that might put in risk their own life and the life of others. The ATREC project aims to enable Real Time Assessment of Mental Stress of the Spanish Armed Forces during military activities. Integrating sensors with garments and using wearable measurement devices, the following physiological measurements were recorded: heart and respiration rate, skin galvanic response as well as peripheral temperature. The measuring garments are the following: a sensorized glove, an upper-arm strap and a repositionable textrode chest strap system with 6 textrodes. The implemented textile-enabled instrumentation contains: one skin galvanometer, two temperature sensors, for skin and environmental, and an Impedance Cardiographer/Pneumographer containing a 1 channel ECG amplifier to record cardiogenic biopotentials. The implemented wearable systems operated accordingly to the specifications and are ready to be used for the mental stress experiments that will be executed in the coming phases of the project in healthy volunteers.

1. Introduction

In the last decade, extensive research efforts in the field and advances in textile materials and instrumentation technology have supported the development and implementation of measuring garments for wearable measurement systems. Nowadays, there are commercially available simple wearable Heart Rate (HR) monitors for fitness e.g. polar heart rate strap, in addition also body worn multiparametric measurement systems can be found e.g. EQ02 life monitor manufactured by Equivital. The availability of such wearable measurement systems fosters the development of personalized health monitoring applications not only for chronic patients and home-care but also for personal working in dangerous conditions.

The ATREC project aims to perform real time analysis of mental stress of the Spanish armed forces during military activities using sensorized garments and wearable measurement systems. The first phase of the project aims to identify non-invasive indicators suitable for assessing on the autonomic

Published under licence by IOP Publishing Ltd

nervous system and for that customized wearable instrumentation with sensorized garments have been made. The project is funded by the Spanish Ministry of Defense through the "*COINCIDENTE*" program (DGAM DNR 1003211003500) under the supervision of the Metrology and Human Factors Department, of "*La Marañosa*" Institute of Technology.

2. Methods and Material

In order to record the physiological signals that would allow studying the response of the autonomic nervous system to stressful situations in a non-invasive manner i.e. cardiac biopotentials, Galvanic Skin Response (GSR), skin temperature and respiration rate [1-3], the following textile-enabled instrumentation and sensor-based garments have been implemented.

2.1. Measuring Devices

Two devices have been built: one to measure GSR and temperature and a second device to record cardiogenic biopotentials and impedance plethysmography measurements. Both are battery operated with 900 mAh Lithium-Ion batteries, and are provided with an internal 4GB SD memory card to allow long-term recordings and off-line analysis.

2.1.1. GSR and Body Temp. This device enables skin impedance monitoring as well as ambient and body temperature monitoring. The reduced dimensions 50x35x15 mm enable its integration in a wearable measurement system like a glove or a measuring strap.

2.1.2. ECG amplifier and EBI plethysmographer. This device enables the recording of ECG and thoracic electrical bioimpedance (TEB) at a single frequency. The aim is to obtain the heart rate from the ECG signal and the respiration rate from TEB signal. Depending on the placement of the electrodes, the TEB measurement can provide also information about cardiac function, *e.g. heart rate*. Such reduced dimensions 50x35x20 mm make it suitable for its integration in a garment.

2.2. Sensor-based Garments

The following garments have been confectioned with textile electrodes made of conductive Shieldex® Fabric P130+B and integrating DS1825 temperature sensors manufactured by Dallas Semiconductor.

2.2.1. Glove for GSR and skin Temperature. Two textile electrodes have been integrated in the proximal phalanx of the index and the middle fingers of the glove for measuring the GSR and a temperature sensor has been placed in the tip of the ring finger of the glove to sense the peripheral skin temperature. Figure 1.a shows a glove with the GSR and Body Temp measurement device attached to the wrist with a strap.

2.2.2. Upper-Arm Strap for GSR and superficial Temperature. An upper arm strap with two textile electrodes integrated to sense the galvanic skin response in the skin surface between the electrodes has been confectioned. A temperature sensor is also integrated in the inner lining of the strap in order to

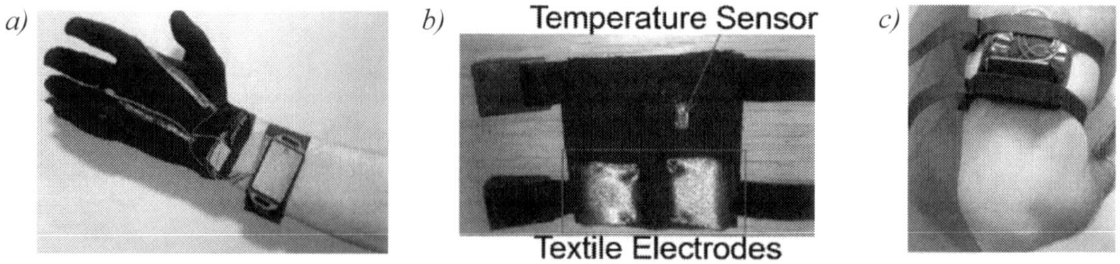

Figure 1. Garments for GSR and temperature measurements a) glove and b) and c) upper arm strap

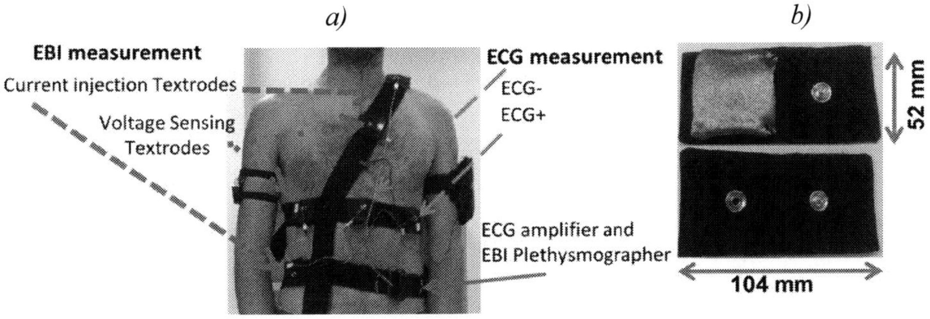

Figure 2. Repositionable chest straps system: a) ECG & EBI measurement systems, b) textrode

measure the skin temperature. Figure 1.b shows the sensorized upper arm strap, also with the GSR and body temp measurement device attached to the arm in Figure 1.c.

2.2.3. Repositionable Chest Straps System for Cardiac and Respiration Recordings. A chest strap garment with repositionable textile electrodes to record 1-lead ECG from 2 textrodes and tetrapolar TEB measurements with other 4 textrodes has been developed. The possibility of placing the electrodes in any place along the horizontal and vertical straps enables the performance of different kinds of TEB measurements. Depending on actual placement of the textrodes around the surface of the thorax and abdomen, the TEB measurement will have more or less cardiac and respiratory components

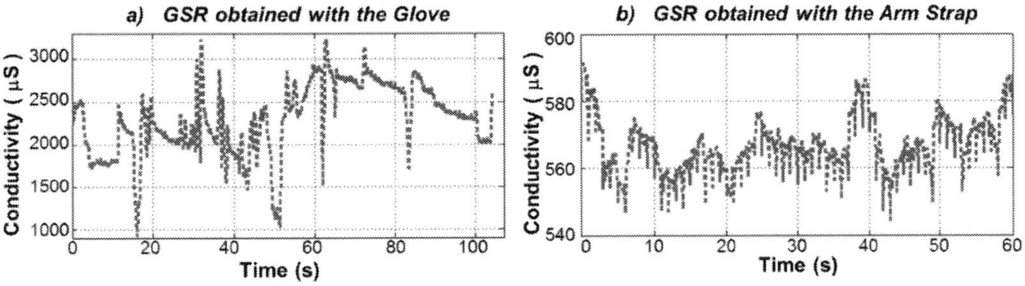

Figure 3. Galvanic skin response measurements: a) Glove, b) Upper arm strap

allowing us to perform a multi-parametric signal recording if required. The mobility of the electrodes is achieved by implementing a textile clamp using press-studs, see Figure 2.b.

3. Experimental Results

This section presents the measurements recorded with the devices and the sensorized garments.

3.1. GSR and Temperature

Skin temperature and GSR were recorded in several subjects. While skin temperatures produced very stable measurements around 34 °C with both garments, GSR measurements exhibited much more variability. Especially in the case of the GSR values recorded with the glove that showed remarkable jumps, most probably related with movement artifacts from the hand and fingers, see Figure 3.a. The GSR values obtained with the arm strap present much lower variability see Figure 3.b.

3.2. Thoracic Measurements

3.2.1. ECG recordings. Surface cardiac biopotentials were recorded in several subjects with the measurement system of Figure 2.a. The ECG recording plotted in Figure 4.a shows 60 seconds of ECG recording. As shown in the 4 seconds zoomed segment the recording is very fair, allowing for a straight detection of the R complex for heart rate assessment. The first 25-30 seconds show slow

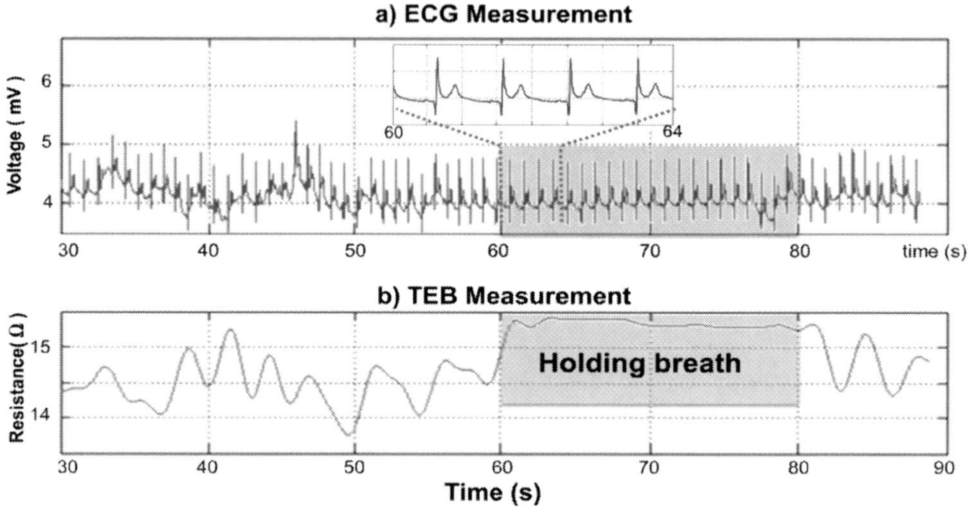

Figure 4. In a) ECG recording, in b) Transthoracic Electrical Bioimpedance recording

activity related with respiratory movement. In the plot it is possible to see that during the seconds 60 to 80 while the subject kept the breath, the ECG recording was completely stable.

3.2.2. Thoracic EBI recordings. TEB recordings were obtained with the measurement system presented in Figure 2.a. in several subjects. A 60 seconds segment of the TEB recording obtained simultaneously with the ECG recording plotted in Figure 4.a is shown in Figure 4.b. The respiration activity is easily observable in the plot where the resistance increases and decreases accordingly to the respiration cycle. Note that the changes stop while the subject was keeping the breath for 20 seconds, between seconds 60 and 80.

4. Discussion and Conclusions

The preliminary measurements show that the GSR measurements obtained with the arm strap are more stable than the recordings obtained with the glove. This was expected since the robustness of the arm strap against movement is expected much better that the gloves with the textrodes on the fingers. In general, the devices implemented and the sensorized garments are functioning as expected.

The proper functioning of the implemented garments and devices allow us to conclude that the systems are ready for the next stage of this experimental phase: data recording in volunteers performing specifically design activities to cause stress. If the ATREC project is successful, a novel application based on EBI technology and textrode garments might be implemented further like in [4].

References

[1] Bernardi L, Wdowczyk-Szulc J, Valenti C, Castoldi S, Passino C, Spadacini G and Sleight P 2000 Effects of controlled breathing, mental activity and mental stress with or without verbalization on heart rate variability *J. Am. Coll. Cardiol.* **35** 1462–9.

[2] De Santos Sierra A, Sánchez Ávila C, Casanova JG and del Pozo GB 2011 A stress-detection system based on physiological signals and fuzzy logic *IEEE Transactions on Industrial Electronics*, **58** 4857–4865.

[3] Zhai J, Barreto AB, Chin C and Li C 2005 Realization of stress detection using psychophysiological signals for improvement of human-computer interactions In *Proceedings IEEE SoutheastCon*, 415–420.

[4] Marquez JC, Seoane F and Lindecrantz K 2013 Textrode functional straps for bioimpedance measurements-experimental results for body composition analysis *Eur. J. Clin. Nutr.* (S1):S22-S7.

Journal of Physics: Conference Series **434** (2013) 012016 doi:10.1088/1742-6596/434/1/012016

A wireless, compact, and scalable bioimpedance measurement system for energy-efficient multichannel body sensor solutions

J Ramos, J L Ausín, A M Lorido, F Redondo and J F Duque-Carrillo

Department of Electrical, Electronics and Control Engineering, University of Extremadura, 06006 Badajoz, Spain

E-mail: jlausin@unex.es

Abstract. In this paper, we present the design, realization and evaluation of a multichannel measurement system based on a cost-effective high-performance integrated circuit for electrical bioimpedance (EBI) measurements in the frequency range from 1 kHz to 1 MHz, and a low-cost commercially available radio frequency transceiver device, which provides reliable wireless communication✦. The resulting on-chip spectrometer provides high measuring EBI capabilities and constitutes the basic node to built EBI wireless sensor networks (EBI-WSNs). The proposed EBI-WSN behaves as a high-performance wireless multichannel EBI spectrometer where the number of nodes, i.e., number of channels, is completely scalable to satisfy specific requirements of body sensor networks. One of its main advantages is its versatility, since each EBI node is independently configurable and capable of working simultaneously. A prototype of the EBI node leads to a very small printed circuit board of approximately 8 cm^2 including chip-antenna, which can operate several years on one 3-V coin cell battery. A specifically tailored graphical user interface (GUI) for EBI-WSN has been also designed and implemented in order to configure the operation of EBI nodes and the network topology. EBI analysis parameters, e.g., single-frequency or spectroscopy, time interval, analysis by EBI events, frequency and amplitude ranges of the excitation current, etc., are defined by the GUI.

1. Introduction

The shifting of eHealth from desktop platforms to wireless and mobile configurations enables a distributed and pervasive care model for health and wellness management through the use of miniaturized information and communication technologies [1]. A pervasive health system will ultimately need information about individuals and their surrondings, which will be delivered by embedded sensors. In this sense, wireless sensor network (WSN), which has settled as a leading technology for various applications, exhibits one of its potential uses in the form of body sensor networks (BSNs) for remote monitoring of patients under their natural physiological state [2].

Despite the high degree of development achieved by present solutions that ensure rigorous and reliable monitoring of certain vital signs, e.g., ECG, heart rate, etc., there is still lack of solutions designed to meet the demand of sensors which can be adapted to the particular characteristics of each individual for accurate and noninvasive monitoring of physiological parameters, emotional states as well as for reducing the impact of medical therapies in chronic patients. Bioelectrical impedance analysis has qualities very attractive for personalized monitoring of the physical and mental state of patients and citizens, besides being a noninvasive preventive diagnosis technique [3].

✦ Patent pending

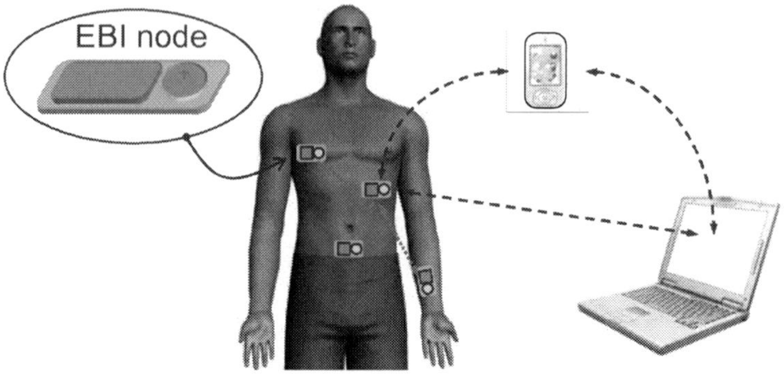

Figure 1. Proposed wireless multichannel electrical bioimpedance spectrometer.

2. Wearable bioimpedance-based body sensor network

The term electrical bioimpedance (EBI), or simply bioimpedance, is used to describe the response of a biological material to the flow of an applied alternating electrical current with given amplitude and frequency [4]. Although EBI technology can be applied in many fields, the detection of physiological events and anatomical structures is one of its major working areas. In certain cases, the information provided by a single EBI measurement may not be enough to resolve complex problems or reliably produce physiological parameters. An alternative approach where multiple EBI signals are recorded in various locations might provide more reliable information. Most of the actual multielectrode methods for measuring bioimpedances aim to the design of fixed-frequency electrical impedance tomography (EIT) systems. Unfortunately, EIT solutions are unsuitable or excessively complex for other purposes such as the time-domain analysis of EBI magnitude [5] and the knowledge of spectral information for tissue characterization [6]. In addition, the difficulty associated with constructing high-fidelity multi-channel, multi-frequency data acquisition instruments has limited widespread development of multiple spectroscopic electrical impedance measurements in different locations. Conventional multichannel systems utilize multiple single-lead EBI measurement systems connected to the same organism, whose excitation signals may overlap, rendering it difficult to extract specific and reliable information.

In this work, we present a wireless multichannel spectroscopy measurement system based on a high-performance EBI sensor. This solution facilitates the measurement of separated bioimpedances in a human body with maximum accuracy. As shown in Figure 1, wireless EBI nodes are placed over the patient's areas of interest in order to carry out independent bioimpedance measurements and transfer the records to the BSN coordinator. The coordinator is, in turn, wirelessly connected to the patient's end device, where data are processed. The collected records are available to be transferred to a PC for further analysis. To follow the skin surface, the node can be fixed over a patch or over a strap.

3. Low-power wireless EBI sensor architecture

A simplified block diagram of the proposed wireless EBI node is outlined in Figure 2a. The EBI sensor comprises four electrodes, which allow tetra-polar impedance measurements in order to eliminate the skin impedance contribution. The EBI sensor, which was designed in 0.35-μm CMOS technology (Figure 2b), injects an ac excitation current into the biological material under test (MUT) which has an unknown impedance Z_x, detects the voltage drop across the MUT (V^+ and V^-), and processes these two signals so as to provide two dc voltages proportional to the magnitude, V_m, and the phase, V_p, respectively, of the MUT bioelectrical impedance by using a gain/phase detector. Thanks to the EBI sensor architecture, the extraction of the two dc voltages does not depend on the amplitude of the excitation current, which is extremely important at high frequencies where the output impedance of the current source is usually lowered. This reduces the complexity of the circuitry to perform further processing. A microcontroller unit (MCU) supervises the operation of the whole EBI node, controls the amplitude and the frequency of the excitation current, and converts the ensuing dc voltages (V_m, V_p) to a digital code by using the inbuilt analog-to-digital converter. Besides, the MCU

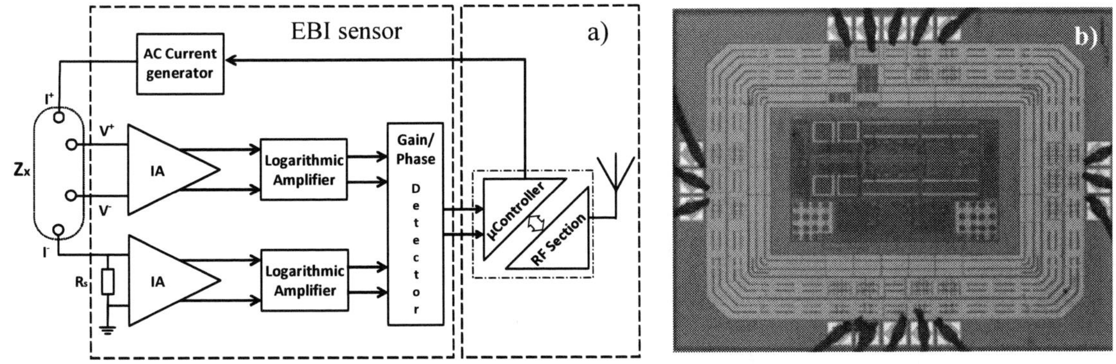

Figure 2. (a) Block diagram of the proposed measurement unit. (b) Die photograph of the EBI sensor.

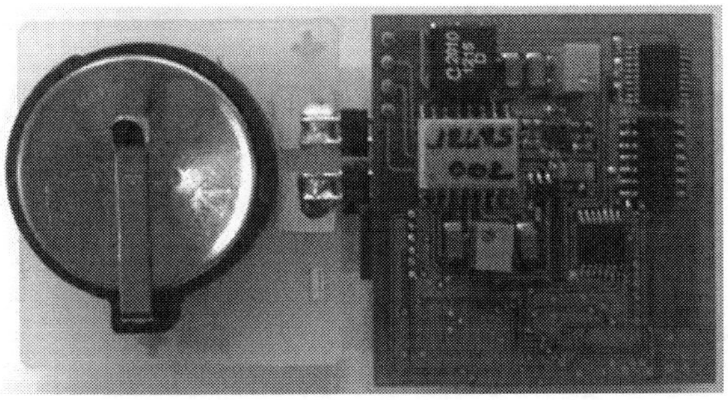

Figure 3. Photograph of the realized wireless EBI node.

Table 1. Experimental performance of the EBI sensor

Measuring system	Tetrapolar
AC current excitation	• 5 μA to 1 mA • 1 kHz to 2 MHz
Magnitude measurement range	1 Ω to 3.5 kΩ
Phase measurement range	0° to 90°
Analog front-end power consumption	1.6 mW

Table 2. Programmable parameters of the proposed EBI-BSN node

Parameter	Options
EBI analysis	• Single-frequency (SF-BIA) • Bioimpedance spectroscopy (BIS)
Frequency	• Single value (SF-BIA) • Sweep and interval (BIS)
Num. of analyses/ sweep	• Single (SF-BIA, BIS) • Analysis time interval (SF-BIA, BIS)
Amplitude excitation current	• Single value (SF-BIA. BIS) • Automatic tuning (SF-BIA, BIS)

supports calibration tasks that aim at avoiding systematic measurement errors. The obtained digital words are transmitted by a radiofrequency section following a short-range wireless protocol (ZigBee).

4. Experimental results

Figure 3 is a photograph of the realized wireless EBI node. The heart of the proposed wireless section is the commercial solution CC2430 System-on-Chip from TI/Chipcon. The CC2430 is optimized for long-term battery operation and includes the CC2420 transceiver and an efficient 8051-based microcontroller, which implements the whole digital processing stage. The experimental performance of the developed EBI node is summarized in Table 1. Every EBI node can be independently adjusted (Table 2) so as to optimally measure the electrical properties of the underlying tissue in a given body location. A suitable graphical user interface (GUI), EBI-WSN, was developed in Matlab (7.10.0, The MathWorks Inc., Natick, MA, 2010) to control the operation of the EBI nodes and to show how the BSN works. Along with the sites for EBI measurements, the topology and the complexity of the BSN

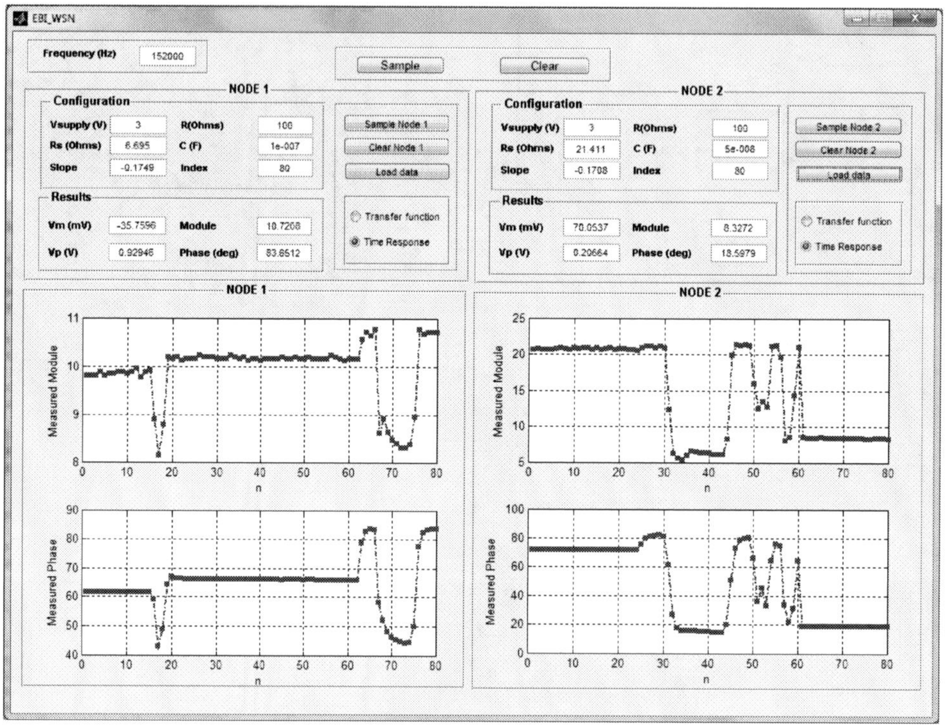

Figure 4. Front panel of the designed Matlab-based GUI.

that result more appropriate for the study of a particular physiological event can be also defined. The effectiveness of the developed GUI environment for time-domain measurement of two independent bioimpedances, i.e., two channels, is shown in Figure 4. The analysis and graphical representation of the measured results transmitted by EBI nodes is also included in the software.

5. Conclusions

The reliability of electrical bioimpedance (EBI) technology for diagnosis may be improved by utilising multiple simultaneous independent measurements. To aid in this, this work presents the design and implementation of a novel energy-efficient wireless multichannel EBI spectrometer, which has the potential to improve the accuracy of actual applications, as well as to develop new ones.

6. Acknowledgement

This work has been supported by the Spanish R&D Plan under Grant TEC2010-19019.

References

[1] Varshney U 2009, Pervasive healthcare computing: EMR/EHR, wireless and health monitoring, Springer.
[2] Xijun C, Meng MQ-H and Hongliang R 2005, Design of sensor node platform for wireless biomedical sensor networks in *Proc. IEEE Eng. Med. Biol. Annual Conference*, 4662–65.
[3] Jossinet J 2005, Bioimpedance and p-Health. In: Nugent CD *et al.* (eds) Personalised health management systems, I.O.S. Press.
[4] Grimnes S and Martinsen ØG 2008, Bioimpedance and bioelectricity basics, Academic Press.
[5] Gracia J, Seppä V-P, Viik J and Hyttinen J 2012, Multilead measurement system for the time-domain analysis of bioimpedance magnitude. *IEEE Trans. Biomed. Eng.* **59** 2273–80.
[6] Hartov A *et al.* 2000, A multichannel continuously selectable multifrequency electrical impedance spectroscopy measurement system. *IEEE Trans Biomed. Eng.* **47** 49–58.

CMOS based sensor for dielectric spectroscopy of biological cell suspension

S Guha, K Schmalz, C Meliani and Ch Wenger
IHP, Im Technologie Park, Frankfurt (Oder), Germany

E-mail: guha@ihp-microelectronics.com

Abstract. In this work we investigate the use of microwave frequency range to measure the concentration of cells in a biological cell suspension. A theoretical model is discussed and the advantage of high frequency, which is to avoid dispersion mechanisms due to the cell parameters at lower frequencies (for example membrane capacitance), has been described. Interdigitated capacitor (IDC) has been proposed as the sensor for analysing the concentration of a cell species in the suspension. The read-out circuit is a VCO using the IDC and a pair of inductors as resonator. The capacitance of the IDC which is the function of the permittivity of the biological cell suspension determines the resonant frequency of the LC tank oscillator. Thus the concentration of cells in a solution, affecting its permittivity, is read out as the frequency of the oscillator.

1.Introduction

Dielectric spectroscopy has been effectively used for decades to characterize biomaterials, giving information about membrane capacitance of cells, intra-cellular organells, etc [1]. However characterization of concentration of a particular cell species in a suspension, based on dielectric measurements is yet to be explored in a broader aspect. In this paper we explain the proposed theory of dielectric measurement to characterize the concentration of cells in a suspension and also describe a CMOS sensor which is used for the above purpose.

A new era of biosensors has been ushered in with the establishment of solely electrical measurement based label free sensing techniques. Typical biosensors, especially based on optical principle are affinity based and require selective target analyte combination for functioning [2]. These sensors are faced with the challenge of a complex read out mechanism usually based on fluorescence detection [3]. The advantage of simple read out circuits coupled with the ease of fabrication of CMOS or BiCMOS circuits have therefore made electrical biosensors very lucrative.

In this work we propose a solely electrical measurement based biosensor to detect the concentration of cells in a biological cell suspension operating at a frequency around 12 GHz. Jaric et al performed capacitive measurements of cells at 2 GHz [4]. However it has been reported that dispersion mechanisms of biomaterials especially DNA can exist up to 5 GHz to 7 GHz [5]. Hence we selected a frequency of operation above this frequency and slightly below the characteristic frequency of water (17 GHz). Interdigitated capacitor (IDC) has been used as the sensor element. The IDC structure is coupled with inductors, to form an LC oscillator whose resonant frequency is a function of the permittivity of the solution underneath the sensor. The control circuit is a pair of cross coupled transistors, similar to a cross coupled VCO, with an output buffer stage. Thus a dielectric change in the solution due to change in concentration of cells in it is detected as the frequency change of the oscillator.

2. Dielectric Dispersion Modelling

Biological cell suspensions show three noted dielectric dispersions over the frequency range of 1 Hz to 100 GHz, namely α dispersion, β dispersion and γ dispersion respectively [5]. The α and β dispersions are principally low frequency phenomenon up to few hundred megahertz, arising mainly because of interfacial polarisation and Maxwell Wagner polarization mechanism respectively. However in this work mainly γ dispersion of the cell suspension is taken into account as the operating frequency of the sensor is centred around 12 GHz, close to the characteristic frequency of water at 17 GHz. The γ dispersion of a cell is caused by the polarization of the water molecules within it. The effective dielectric constant of the cell suspension is given by the Debye equation,

$$\epsilon_{sol} = \epsilon_a + \frac{\Delta\epsilon}{1+(\frac{f}{f_c})^2} \quad (1)$$

where ϵ_{sol} is the effective dielectric permittivity of the solution, ϵ_a is permittivity of the suspending medium at frequency f, also given by a similar Debye equation, $\Delta\epsilon$ is the dielectric dispersion due to the presence of the cells, f_c is the characteristic frequency of the dispersion process and f is the measuring frequency. The static (f=0) permittivity change of the suspending medium is given as,

$$\epsilon_s = \epsilon_a + \Delta\epsilon \quad (2)$$

$\Delta\epsilon$ is called the dielectric dispersion and is given by cellular parameters as,

$$\Delta\epsilon = \frac{9PrC_m}{\epsilon_0} \quad (3)$$

where r is the radius of the cell and C_m is the membrane capacitance as shown in fig.1. P is the volume fraction of the cells, which is dependent on the concentration of cells N. In this work we have considered a standard yeast cell of diameter 10 μm and Cm=1μF/cm^2.

$$P = \frac{4\pi r^3 N}{3V_{Channel}} \quad (4)$$

$V_{channel}$ is the volume of the fluidic channel containing the cell suspension.

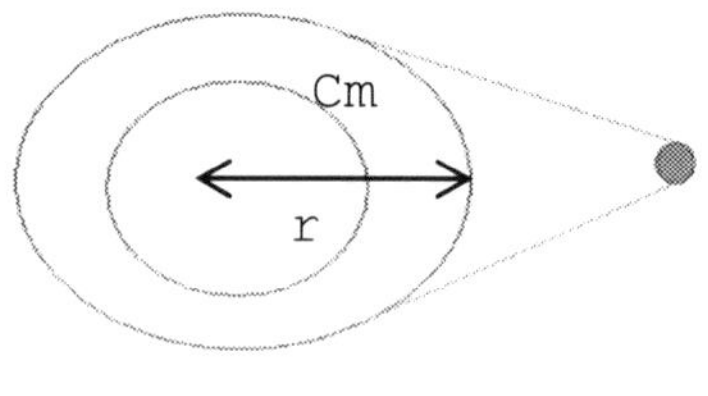

Figure 1. Single shell model of yeast cell showing cell radius and membrane capacitance

It can be seen from equation (4) that the volume fraction for same number of cells N can be increased by reducing the volume of the channel. Debye equation also determines the permittivity of water and is given as

$$\epsilon_w = \epsilon_\infty + \frac{\epsilon_s - \epsilon_\infty}{1+(\frac{f}{f_c})^2} \quad (5)$$

where ϵ_s is the static permittivity of the water (suspending medium) and ϵ_∞ is the high frequency permittivity. Therefore using equation (2),(3) and (4), the permittivity of the solution can be written as

$$\in_{[cell-sol]}=\in_{\infty}+\frac{\in_a+\frac{3\pi r^4 C_m N}{\in_0 V_{channel}}-\in_{\infty}}{1+(\frac{f}{f_c})^2} \tag{6}$$

The permittivity of the cell suspension with respect to the concentration of cells is shown in fig.2. At the operating frequency, 12 GHz, the slope $\frac{\partial \in}{\partial N}$ is approximately 2, that is, for single cell the permittivity changes by 2.

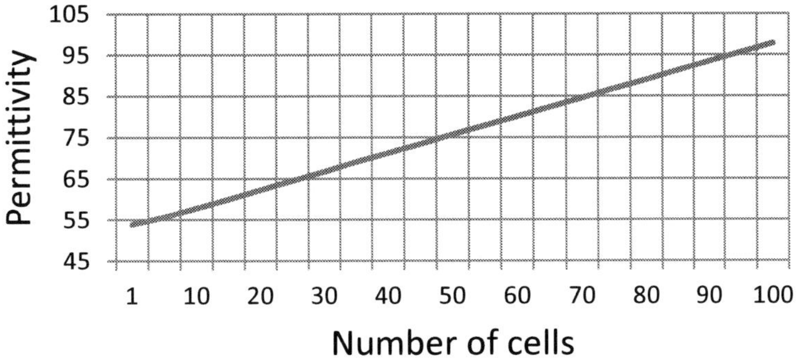

Figure 2 Permittivity with respect to number of cells in the biological suspension.

3. Sensor Circuit Design

As mentioned above interdigitated capacitor is used for sensing the permittivity of biological cell suspension. The resonant frequency of the IDC coupled with two inductors L is dependent on the capacitance of the IDC with the standard relation,

$$f=\frac{1}{2\pi(LC)^{0.5}} \tag{7}$$

The ratio of the capacitance of the IDC with air as the dielectric material to the capacitance with cell suspension gives the information about the permittivity of the cell suspension,

$$\frac{C_{IDT[Air]}}{C_{IDT[sol]}}=\frac{F(\in_{air})G(geom)}{F(\in_{sol})G(geom)} \tag{8}$$

where *G(geom)* is the function of the geometry of the IDC obtained from conformal mapping techniques [6] and $F(\in)$ is the function of permittivity. From equation (7) and equation (8), the ratio of the resonant frequencies can be obtained as,

$$\frac{f_{air}}{f_{sol}}=(\frac{F(\in_{sol})}{F(\in_{air})})^{0.5} \tag{9}$$

The IDC is fabricated from the top metal 2 of 250nm process of IHP. The core VCO circuit is shown in fig. 3 along with the IDC. The transistors M1, M2 and M3 form the current source for the oscillator and the transistors M4 and M5 are used as the cross coupled pair for negative resistance to sustain the oscillations.

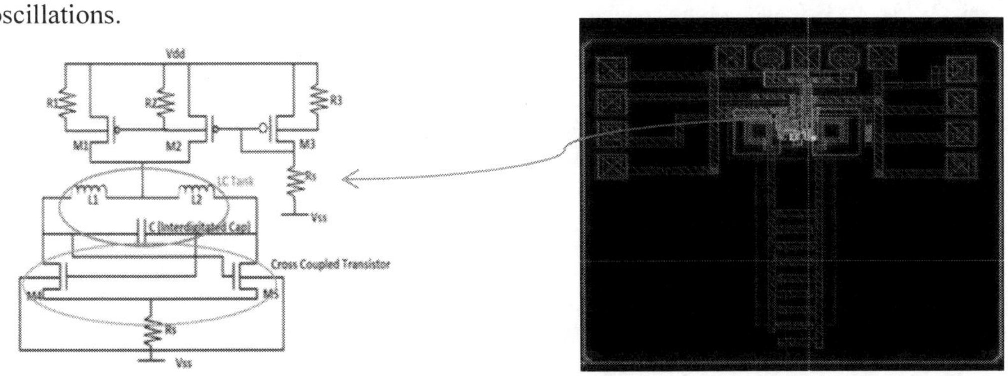

Figure 3 (a) The core VCO control circuit for the oscillator. **(b)** Circuit layout with the IDC sensor

The sensors are often passivized using a thin layer of SiO_2 or Si_3N_4 of about 300 nm thickness. This influences the sensitivity of the measurement and had to be carefully taken into consideration while designing the VCO and the IDC. Fig.4 shows the FEM simulated variation of capacitance of IDC due to change of permittivity of a biological cell suspension flowing through a capillary placed on top with different passivation thickness. No major effect due to the passivation is noted from the slope of the curves up to 1μm thickness of passivation which makes this concept a good candidate for several practical applications.

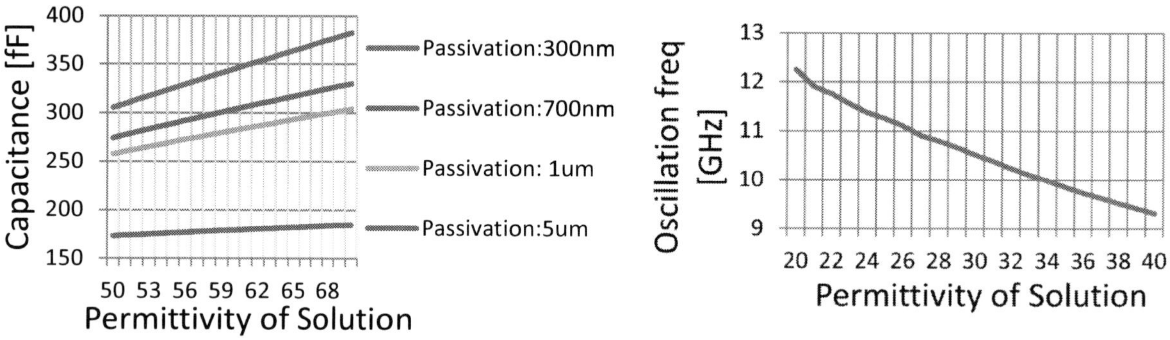

Figure 4 (a) Variation of capacitance of IDE with respect to permittivity of biological suspension **(b)** Variation of resonant frequency with permittivity of solution

Fig 4 (b) shows the variation of resonant frequency with the permittivity of the solution. Approximately 130 MHz, of frequency shift is noted for a unit change of permittivity for ideal case.

4. Conclusion

The theory of dielectric permittivity measurement to determine the concentration of a cell species in a biological cell suspension is explored and established in this work. The frequency range around 17 GHz has been identified to be relevant to perform this measurement as it avoids the dispersions mechanisms at lower frequencies (α dispersion, β dispersion). At this frequency, concentration of the cells is the dominant parameter for change in permittivity of the cell suspension. An easy and effective read out circuit based on a cross coupled VCO topology has been proposed and designed. A sensitivity of 130 MHz per unit permittivity has been obtained in simulation. Furthermore, the influence of the passivation on top of the chip has been investigated; it has been shown that no major effect is noted for a thickness up to 1μm. The Typical passivation thickness for the used process is 300nm. Hence the almost constant sensitivity up to a passivation thickness of 1μm is a highly relevant aspect for the practical applications of such a sensor and makes it a realistic solution for use in most biological environments.

References

[1] Harris C and Dell D 1983 "Radio frequency dielectric properties of yeast cells measured with rapid automated frequency domain dielectric spectrometer" *J. Bio-electrochem. and Bioenergetics* **11** 14-28

[2] Daniels J and Pourmand N 2006 "Label Free Impedance Biosensors: Opportunities and Challenges" *J.Electroanal.* **19** 1239-57

[3] Nikolic-Jaric M, Romanuik S, Ferrier G A, Bridges G E, Butler M, Sunley K, Thomson D J and Freeman M R 2009 "Microwave frequency sensor for detection of biological cells in microfluidic channels" *J. Biomicrofluidics* **3**

[4] Tainaka K, Sakaguchi R, Hayashi H, Nakano S, Liew F and .Morii T 2010 "Design Strategies of FluorescentBiosensors Based on Biological Macromolecular Receptors" *J. Sensors* **10** 1355-76

[5] Asami K 2002 "Characterization of biological cells by dielectric spectroscopy" *J. Non-Crystalline solids* **305** 268-77

[6] Ramajo L, Ramajo D, Reboredo M, Santiago D and Casto M 2008 " Evaluation and modelling of Integral capacitor produced by Interdigitated Comb electrode" *Material Research* **11** 471-76

Journal of Physics: Conference Series **434** (2013) 012018 doi:10.1088/1742-6596/434/1/012018

Removing respiratory artefacts from transthoracic bioimpedance spectroscopy measurements

I Cuba-Gyllensten[1,2,5], F Abtahi[2], AG Bonomi[1], K Lindecrantz[1,3], F Seoane[1,4] and O Amft[5]

[1] School of Technology and Health, Royal Institute of Technology, 141 52 Huddinge, Sweden.
[2] Philips Research Europe, High Tech Campus 34, 5656AE, Eindhoven, the Netherlands.
[3] School of Engineering, University of Borås, 501 90, Borås, Sweden.
[4] Department of Clinical Science, Intervention and Technology, Karolinska Institute, 141 86 Stockholm, Sweden.
[5] ACTLab, Signal Processing Systems, TU Eindhoven, 5600MB Eindhoven, the Netherlands.

E-mail: illapha.cuba.gyllensten@philips.com

Abstract. Transthoracic impedance spectroscopy (TIS) measurements from wearable textile electrodes provide a tool to remotely and non-invasively monitor patient health. However, breathing and cardiac processes inevitably affect TIS measurements, since they are sensitive to changes in geometry and air or fluid volumes in the thorax. This study aimed at investigating the effect of respiration on Cole parameters extracted from TIS measurements and developing a method to suppress artifacts. TIS data were collected from 10 participants at 16 frequencies (range: 10 kHz - 1 MHz) using a textile electrode system (Philips Technologie Gmbh). Simultaneously, breathing volumes and frequency were logged using an electronic spirometer augmented with data from a breathing belt. The effect of respiration on TIS measurements was studied at paced (10 and 16 bpm) deep and shallow breathing. These measurements were repeated for each subject in three different postures (lying down, reclining and sitting). Cole parameter estimation was improved by assessing the tidal expiration point thus removing breathing artifacts. This leads to lower intra-subject variability between sessions and a need for less measurements points to accurately assess the spectra. Future work should explore algorithmic artifacts compensation models using breathing and posture or patient contextual information to improve ambulatory transthoracic impedance measurements.

1. Introduction

On-body measurements of impedance spectroscopy provide a tool for non-invasive monitoring of important health markers such as hydration and oedema, which are key elements to enable novel and effective management strategies for chronic conditions [1]. To achieve this in a telehealth home-monitoring setting the impedance spectroscopy measurements must be robust in presence of noise and/or slight deviations from the assumed measurement protocol.

To this end we have carried out a data collection study to quantify and characterize some disturbances that are thought to adversely affect measurements of transthoracic impedance spectroscopy (TIS). One such contributor for TIS measurements is the subject's respiration [2] during the measurement period. Respiration changes the structure of the alveoli and the thorax, which poses a challenge for TIS estimations based on a frequency sweep techniques were the

Journal of Physics: Conference Series **434** (2013) 012018 doi:10.1088/1742-6596/434/1/012018

measurement time is slower than the physiological change [3]. This could lead to errors in the estimated TIS parameters and interpreted patient fluid status.

2. Methods

Ten healthy participants, 5 male and 5 female, were included in the analysis. The volunteers were all screened so that no one had any breathing abnormality, cardiac dysfunction or any other objection to take part in the study. Prior to the trial, each subject was properly informed and provided written informed consent. The whole procedure was reviewed and approved by the internal ethical review board and covered by the regional MEC.

2.1. Study population

An effort was done to recruit participants in a large age-range. Half of the recruited subjects were female and half of the subjects were older than 30 (four older than 50). The characteristic of the subjects were (Age: 38 ± 15 years, BMI: 24.3 ± 4.7 kg/m^2).

2.2. Devices, calibration and synchronization

Tetrapolar measurements were made using the BIM (Philips Technologie Gmbh, Aachen, Germany) which applies a sinusoidal current at 16 different frequencies sequentially (10 kHz-1MHz) at a rate of 5Hz. Wetted textile electrodes were then placed on either side of the ribcage at the approximate level of the tenth rib and held in place with an adjustable vest. The measurements obtained should relate to the impedance of the thorax.

To measure the respiration participants were fitted with a sized mask, which connected to a heated pneumotach (Hans Rudolph Inc., Shawnee, KS USA) with data-logging provided by a DAQ (NI USB 6353, Texas Instruments, Dallas, TX USA) at a sample rate of 1kHz. The system used factory set calibration for the Voltage to Flow conversion. Tidal volume was then estimated by integrating the flow measurements over time. To account for the slight drift in the pneumotach the output was calibrated using a 3 liter calibration syringe at the start and end of recording and at fiduciary end-expiratory points.

To account for any clock drift between the devices the time stamps were linearly adjusted to synchronize detected ECG peaks. Signals were then downsampled to the sampling frequency of the BIM.

2.3. Study protocol

Participants enrolled in this study took part in a larger study protocol which also included sessions to estimate artifact arising from movements and increased cardiac activity. The presented protocol is the selected part that was used for this analysis according to table 1.

Table 1. Part of the study protocol used for the presented analysis

Posture	Breathing instructions
Lying down, session 1	Paced shallow breathing at 10 and 16 bpm (60s each).
Lying down, session 2	Paced deep breathing at 10 and 16 bpm (60s each).
Reclining, session 1	Paced shallow breathing at 10 and 16 bpm (60s each).
Reclining, session 2	Paced deep breathing at 10 and 16 bpm (60s each).
Sitting, session 1	Paced shallow breathing at 10 and 16 bpm (60s each).
Sitting, session 2	Paced deep breathing at 10 and 16 bpm (60s each).

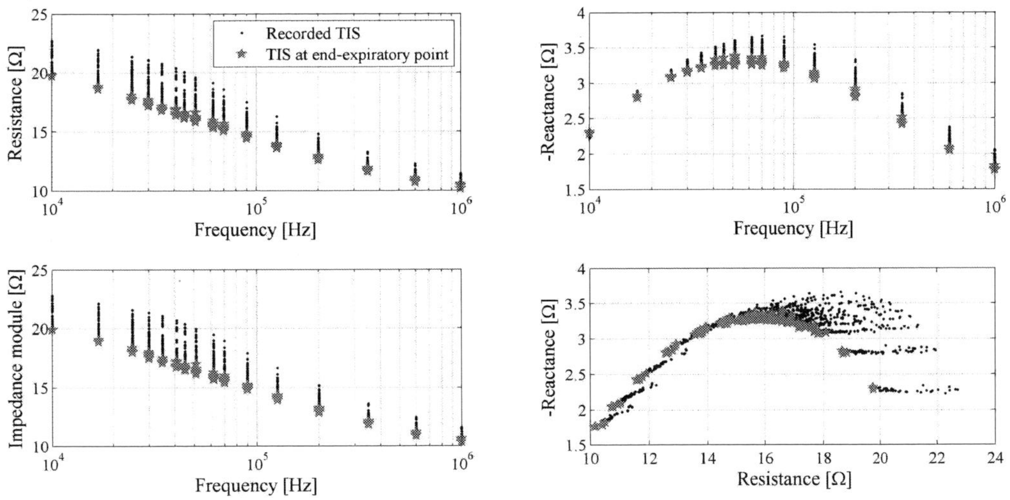

Figure 1. TIS data from participant 4 while lying down. The black dots show the data for the full recording (shallow and deep breathing at both 10 and 16 bpm). The red stars indicate the recorded data at the end-expiratory points extracted from the spirometer.

2.4. Impedance Spectroscopy Modelling

The Cole function (equation 1) was fitted to each of the TIS measurements by non-linear least squares. Minimization was done on the squared sum of the absolute difference between the recorded complex impedance and the model value for the frequency [4][5]. Two models were estimated using this minimization scheme: one by using all the input data for one session and a second using only the data at exhalation (these points were estimated from the recorded volume measurements).

$$Z = R_{\text{inf}} - \frac{R_{\text{inf}} - R_0}{1 + (j\frac{f}{f_c})^{\alpha}} \tag{1}$$

3. Results

Figure 1 shows the transthoracic bioimpedance during the measurement of an example subject separated to show all data and data at the end-expiratory points. It is clear that transthoracic impedance can be highly affected by respiration. Indeed R_0 seem to change with almost 4 Ohms (from 20 to 24) between maximum inhalation and exhalation for this specific subject.

The population-mean of the fitted Cole parameters for the two methods and intra-subject variability between the different breathing sessions is presented in table 2. These findings show that the high variability in the recorded impedance at different stages of the respiratory cycle creates difficulty in finding a representative Cole fitting even though these fittings had a substantially larger amount of data points (544 points per session) when compared to the fittings which only used the data at the end-expiratory points (median 72 points per session). On average the difference between the Cole parameters for the two sessions were four times larger when respiration was not taken into account.

Table 2. Intra-subject variability measured by absolute difference between two breathing sessions of the estimated Cole parameters using either: all data or only data at exhalation for the session. Population mean and standard deviation are shown within parenthesis.

Cole parameter	Estimation at exhalation	Estimation using all data
R_0	0.33 ± 0.26 Ω (mean 26.44 ± 4.64 Ω)	1.24 ± 1.65 Ω (mean 27.11 ± 4.36 Ω)
R_{inf}	0.30 ± 0.25 Ω (mean 12.10 ± 2.66 Ω)	1.05 ± 1.50 Ω (mean 12.02 ± 3.21 Ω)
f_c	1.90 ± 1.36 kHz (mean 53.81 ± 12.30 kHz)	8.26 ± 17.25 kHz (mean 58.83 ± 20.82 kHz)
α	0.0250 ± 0.0226 (mean 0.607 ± 0.049)	0.0779 ± 0.0762 (mean 0.587 ± 0.085)

4. Discussion and Conclusion

Reducing artifacts from on-body measurements provides more accurate non-invasive assessments of important health markers. This analysis shows that one such artifact that can be reduced is respiration. TIS using frequency sweep techniques usually average over the recorded data span [3] to ensure frequency coverage when the physiological changes are faster than the frequency changes. However, this analysis suggests that averaging is not a suitable technique, especially if respiration patterns are prone to changes.

Furthermore, it is known that the impact of respiration on impedance varies at different frequencies [2], which can be seen in the example data. An averaging technique might therefore produce a spectrum which does not correspond to any respiratory position and might provide sub-optimal values for a Cole fitting procedure.

Cole model fitting could be improved by taking into account the respiratory cycle leading to lower intra-subject variance of TIS parameters. Future work should look into effective algorithms to accomplish this, also in presence of other sources of errors such as movements. The population studied in this analysis, although varied, was healthy and unlikely to be targeted for monitoring. There is therefore also need to control such algorithms against the specific population, which is intended for monitoring.

Acknowledgments

This work was supported by the EU Marie Curie Network iCareNet under grant number 264738.

References

[1] Reiter H, Muehlsteff J and Sipila A 2011 Medical application and clinical validation for reliable and trustworthy physiological monitoring using functional textiles: Experience from the HeartCycle and MyHeart project *Conf. Proc. IEEE Eng. Med. Biol.* 3270-3

[2] Brown BH, Barber DC, Morice AH and Leathard AD 1994 Cardiac and respiratory related electrical impedance changes in the human thorax *IEEE Trans. Biomed. Eng.* **41** 729-34

[3] Sanchez B, Schoukens J, Bragos R and Vandersteen G 2011 Novel Estimation of the Electrical Bioimpedance Using the Local Polynomial Method. Application to In Vivo Real-Time Myocardium Tissue Impedance Characterization During the Cardiac Cycle *IEEE Trans. Biomed. Eng.* **58** 3376-85

[4] Kun DS and Peura RA 1999 Selection of measurement frequencies for optimal extraction of tissue impedance model parameters *Med. Biol. Eng. Comput.* **37** 699-703

[5] Buendia R, Gil-Pita R and Seoane F 2011 Cole Parameter Estimation from the Modulus of the Electrical Bioimpeadance for Assessment of Body Composition. A Full Spectroscopy Approach *J. Electr. Bioimp.* **2** 72-8

The effect of the random distribution of electronic components in the output characteristics of the Howland current source

P Bertemes-Filho and A Felipe

State University of Santa Catarina, Dept. Electrical Engineering,
Campus Universitario Prof. Avelino Marcante, Joinville, 89219710, Brazil

E-mail: bertemes@joinville.udesc.br

Abstract. When a Howland source is designed, the components are chosen so that the designed source has the desired characteristics. However, the operational amplifier limitations and resistor tolerances causes undesired behaviours. This work proposes to take in account the influence of the random distribution of the commercial resistors in the Howland circuit over the frequency range of 10 Hz to 10 MHz. The probability density function due to small changes over the resistors was calculated by using an analytical model. Results show that both output current and impedance are very sensitive to the resistor tolerances. It is shown that the output impedance is very dependent on the open-loop gain of the Opamp rather than the resistor tolerances, especially at higher frequencies. This might improve the implementations of real current source used in electrical bioimpedance.

1. Introduction

Because of its simplicity, stability and other advantages, the voltage controlled current source (VCCS) has been widely used in many applications, such as single-electrode capacitive sensors [1], electrical impedance tomography (EIT) systems both for industrial and medical applications [2] and bioimpedance analysis (BIA) for tissue characterization [3]. Most VCCS circuits in BIA use the Howland current source (HCS) [3]. The first HSC circuit was proposed by Howland in 1962 [4] for converting a voltage into a current. However, it suffers from output voltage compliance [5]. Therefore, modified version circuit have been proposed and widely used [6].

The most important requirement in EIS systems is to assure that the injecting current, also called source current, has constant amplitude over a wide frequency range, which should be obtained by a high output impedance circuit [7]. However, stray capacitances [7] and non-idealities of the operational amplifiers used for the design [1] reduce the current amplitude and introduce phase shift errors at higher frequencies. However, some of these requirements are essentially in conflict with each other. For example, the frequency of the current source is practically limited to 100 kHz when the output impedance is required to be sufficiently larger, say larger than 1 MΩ [8]. In most publications, the analysis of the circuit is based on either simplified ideal op-amp circuits or using simulation tools, e.g. PSpice or Multisim from NI [8]. In many cases, however, the formulas they used are not suitable because the calculation errors are too large with a high-frequency current source [8]. Furthermore, the formulas do not take into account the mismatch between the resistors used to design such a current source. The HCS circuit is sensitive upon this mismatching [6].

The objective of this work is to investigate the probability density function and the sensibility of a current source over the frequency range of 10 Hz to 10 MHz due to resistor tolerances.

2. Materials and Method

2.1. Modeling of the modified Howland circuit

Figure 1 shows the modified Howland current source (HCS) circuit used for modeling the output characteristics taking into account the effects of the resistors tolerances. If all resistors are perfectly matched and the operational amplifier (Opamp) has got a large gain, then the output current is given by –Vin/r (assuming R_2=R_3=R_5=R, R_4=r and R_1=R+r). This approximation yields a good result if the Opamp gain is sufficiently high and the load impedance is small. In order to get high output impedance it is necessary to trim the resistors and choose FET inputs amplifiers. On the other hand, the non-ideal characteristics of the Opamps (i.e., input impedance Z_{in}, output resistance R_o and open-loop gain A) reduce frequency bandwidth of both output current and impedance.

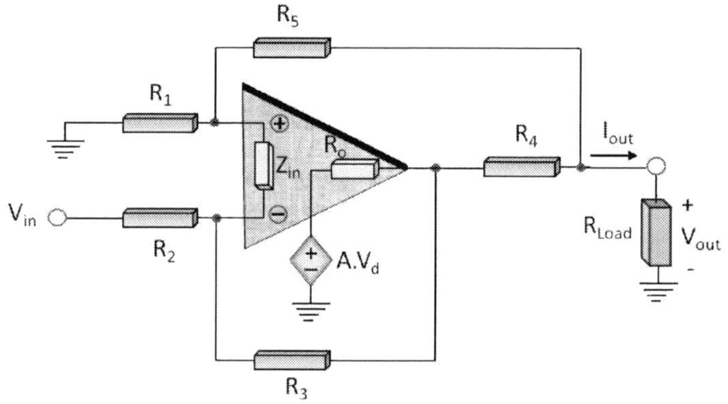

Figure 1. Schematic diagram of the modified Howland current source with grounded load [modified from 9].

The circuit can be characterized by the nodal equations (1), where V_1 is the voltage at non-inverting input of the Opamp, V_2 is the inverting input of the Opamp and V_3 is the voltage at the output of the Opamp. After determining the voltages, the output current can be calculated by equation (2). The transconductance gain is calculated by I_{out}/V_{in} when V_{out} is grounded. On the other hand, the modulus of the output impedance Z_{out} is calculated by V_{out}/I_{out} when V_{in} is grounded.

2.2. Probability Distribution Function

The probability density was computed by using the R programming language, a random were taken randomly by producing different Howland sources. The characteristics of each source were calculated at 50 discrete frequencies per decade in the frequency range from 10Hz to 10MHz. The density function of the transconductance gain was estimated by the "r" function density and then the output impedance distribution is calculated and plotted as transparent lines for each source, as well as the maximum impedances are also calculated and plotted using a red line.

The MHCS circuit, as shown in figure 1, was designed by using R=100 kΩ, r=1 kΩ, V_{in} is a sine wave of 1 Vp and two Opamps, such as OPA657 and uA741 (both manufactured by Texas Instruments), where the technical characteristics can be found in the manufacturer website. It was considered random variations of ±1% and ±10% over the resistor values.

$$\begin{cases} \dfrac{V_1 - V_{out}}{R5} + \dfrac{V_1}{R1} + \dfrac{V_1 - V_2}{Z_{in}} = 0 \\ \dfrac{V_2 - V_{in}}{R2} + \dfrac{V_2 - V_1}{Z_{in}} + \dfrac{V_2 - V_3}{R3} = 0 \\ \dfrac{V_3 - V_2}{R3} + \dfrac{V_3 - V_{out}}{R4} + \dfrac{V_3 - (V_1 - V_2)A}{R_o} = 0 \end{cases} \tag{1}$$

$$I_{out} = \frac{V_3 - V_{out}}{R4} + \frac{V_2 - V_{out}}{R5} \tag{2}$$

The output impedance Z_{out} can be calculated from equations (1) and (2), which is fully described in [9]. If the open-loop gain A is assumed to be very high, then Z_{out} can be given by (3).

$$Z_{out}\,|_{A=\infty} = \frac{R_2R_4R_5 + R_1R_2R_4}{R_2R_5 + R_2R_4 - R_1R_3} \tag{3}$$

3. Results

Figure 2 shows the transconductance gain probability density of the current source designed by using the Opamp OPA657. It is shown in terms of a gray scale, darker points are more likely to be the resulting transconductance gain. It can also be observed that the transconductance gain is flat from 10 Hz to approximately 1 MHz and this does not depend on the resistor tolerance.
Figure 3 shows the results of the output impedance modulus for both OPA657 and uA741 Opamps in the frequency range from 10 Hz to 10 MHz. The red curves show the maximum output impedance Z_{out}. It can be observed that the MHCS circuit designed by using the OPA657 has got higher Z_{out} at over all frequencies. Z_{out} is approximately 1 kΩ at 1 MHz by using the Opamp uA741 whereas 100 kΩ for the OPA657, however any resistor combination may lead the circuit to have high output impedance in a wider band.

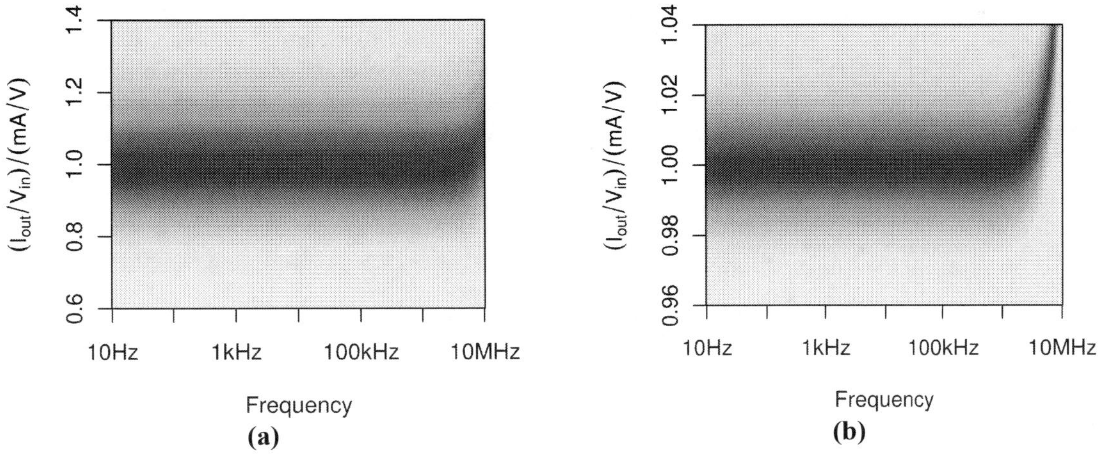

Figure 2. Probability distribution of the transconductance gain upon frequency, using a load of 1 kΩ. **(a)** δ of 10%. **(b)** δ of 1%.

4. Discussions

It was shown in figure 2 that the frequency dependency is more relevant than the ones caused by the resistor tolerances. The transconductance gain can be approximately calculated by the ratio $R_3/(R_2R_4)$, thus if the tolerances of the resistors are δ_3, δ_2 and δ_4, respectively, the tolerance of the transconductance gain will have a tolerance of approximately $\delta_1+\delta_2+\delta_4$. It implies that the use of a wideband operational amplifier will produce a very stable transconductance gain. On the other hand, the output impedance Zout will be significantly changed by the tolerance of the resistors, as showed in figure 3. Any combination of resistors with a tolerance of ±1% may produce either significant reduction in the output impedance or large variation in the transconductance gain, which might explain why the load current variations may also be large. If δ is the nominal tolerance of each resistors, then the minimum output impedance from (3) can be calculated by the ratio between "rR" and "2δ(R+r)". This means that the minimum Zout is inversely proportional to the resistor tolerances.

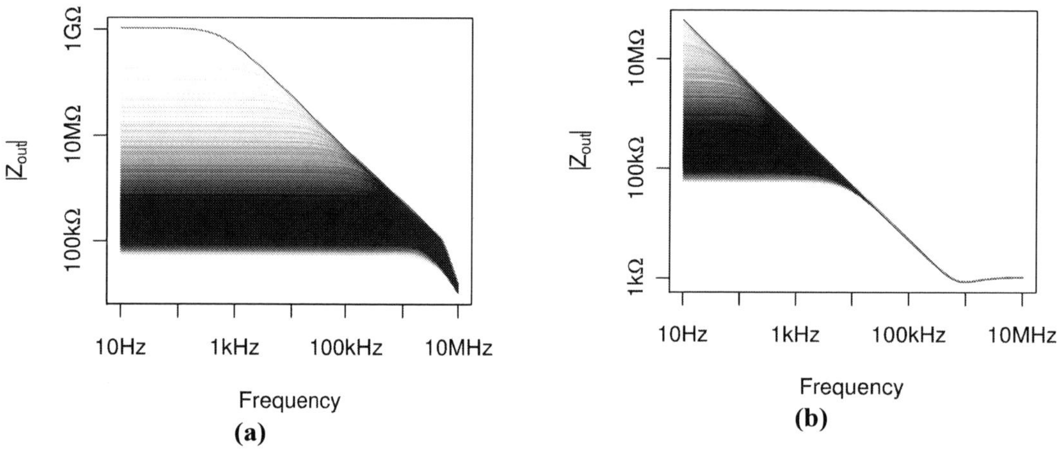

Figure 3. Probability distribution of the output impedance modulus upon frequency, using resistor tolerance of 1%. **(a)** OPA655. **(b)** uA741.

5. Conclusion

It was shown that both I_{out} and Z_{out} of the MHCS circuit is significantly sensitivity to the resistor tolerances used to design it. In order to get maximum Z_{out}, the resistors cannot be exactly matched as the maximum Z_{out} is limited by the open-loop gain, then it can be concluded that the difficulty of getting high Z_{out} at higher frequencies is due to the reduction of the open-loop gain of the Opamp. On the other hand, at lower frequencies, high Z_{out} may be achieved by matching the resistors in the circuit.

Acknowledgment

We thank the State University of Santa Catarina (UDESC) for the financial support.

References

[1] Chen D X, Deng X, Yang W Q 2010 *Rev. Sci. Instrum* **81** 1-3
[2] Frounchi J, Dehkhoda F, Zarifi M Z 2009 *European Journal of Scientific Research,* **27** 56-65
[3] Bertemes-Filho P 2002 *Tissue Characterisation using an Impedance Spectroscopy Probe* (Sheffield: The University of Sheffield Press) p. 96
[4] Sheingold D H 1964 *Lightning Empiricist* **12** 1-8
[5] Bertemes Filho P, Lima R G, Amato M B P et al. 2006 *XX Brazilian Congress Biomed. Eng.* (São Paulo) vol 1 p 1167-70
[6] Seoane F, Macías R, Bragós R, Lindecrantz K 2011 *Meas. Science and Technology* **22** 1-11
[7] Seoane F, Macías R, Bragos R, Lindecrantz K 2011 *Meas. Sci. Technol.* **22** 11p
[8] Bertemes-Filho P, Brown B H, Wilson A J 2000 *Physiological Measurement* **21** 1-6
[9] Bertemes-Filho P, Negri L H, Felipe A, Vincence V C 2012 *Journal Phys.: Conf. Series* **407** 1-8

Measurements of Electrode Skin Impedances using Carbon Rubber Electrodes - First Results

Steffen Kaufmann, Gunther Ardelt and Martin Ryschka

Centre of Excellence for Technology and Engineering in Medicine (TANDEM), Lübeck University of Applied Sciences, Mönkhofer Weg 293, 23562 Lübeck, Germany

kaufmann@fh-luebeck.de, gunther.ardelt@fh-luebeck.de, ryschka@fh-luebeck.de

Abstract. Non-invasive bioimpedance measurement as a tool in biomedical engineering and life sciences allows conclusions about condition and composition of living tissue. For interfacing the electronic conduction of the instrumentation and the ionic conduction of the tissue, electrodes are needed. A crucial point is the uncertainty arising from the unknown, time-varying and current density depend Electrode Skin Impedance (ESI). This work presents ESI measurements using carbon rubber electrodes on different human test subjects. The measurements for this work are carried out by employing a high accuracy Bioimpedance Measurement System (BMS) developed by the authors group, which is based on a Field Programmable Gate Array (FPGA) System on Chip (SoC). The system is able to measure magnitude and phase of complex impedances using a two- or four-electrode setup, with excitation currents from 60 µA to 5 mA in a frequency range from about 10 kHz to 300 kHz. Achieved overall measurement uncertainties are below 1 %.

1. Introduction

Bioimpedance is the electrical, frequency depending complex impedance of living tissue. Bioimpedance is usually measured by injecting a predefined auxiliary current into the test subject and by measuring the corresponding voltage drop. From bioimpedance measurements conclusions can be drawn about the physiological composition, condition and functionality of the investigated tissue [1, 2]. Non-invasive bioimpedance measurements via surface skin electrodes are affected by the Electrode Skin Impedance (ESI), which is the combination of the impedance of the electrode interface and the impedance of the outer epidermal layer of the skin. The layered structure of skin can electrically be modeled as a lossy capacitor [1, 2]. The outer epidermis (lat. stratum corneum) consists of a stack of nearly dry and dead epithelial cells with a typical overall thickness of 10-20 µm, which significantly contributes to the overall ESI. The total impedance of the current path across two electrodes and the tissue in between is considerably determined by the ESI. Therefore the value of the ESI limits the excitation current by a given compliance range of a current source.

For biomedical applications additionally to classical silver chloride electrodes, carbon rubber electrodes can be used. First introduced in the nineteen sixties, carbon rubber electrodes can be classified as electrochemical inert, washable and reusable. Moreover, rubber electrodes can be applied directly to the human skin, whereby they adapt well to different shapes, due to their elasticity. Therefore they are predestined for long term measurements [2, 3]. So far carbon rubber electrodes are commonly used in low frequency applications below 1 kHz like EEG, ECG or muscle stimulation [3]. However for bioimpedance measurements frequency ranges above 10 kHz are typical. For this frequency range ESI measurements are yet rarely reported.

This work presents ESI measurements using carbon rubber electrodes on different human test subjects with excitation frequencies of about 10 kHz to 300 kHz. The measurements are carried out by a newly developed high accuracy Bioimpedance Measurement System (BMS). The BMS is able to provide adjustable excitation currents from 60 μA to 5 mA in the desired frequency range and measure impedance magnitudes with an overall uncertainty of less than one percent and phases with an overall uncertainty of less than one degree [4].

2. Materials and Methods

The used BMS consists of an embedded FPGA System on Chip (SoC) and a host PC. The BMS drives a predefined alternating current through the test subject and measures the resulting voltage drop to calculate magnitude and phase of the investigated bioimpedance. To allow a high precision impedance measurement the actual current through the subject is measured via a low side shunt-resistor. Figure 1 shows the principle block diagram of the used BMS.

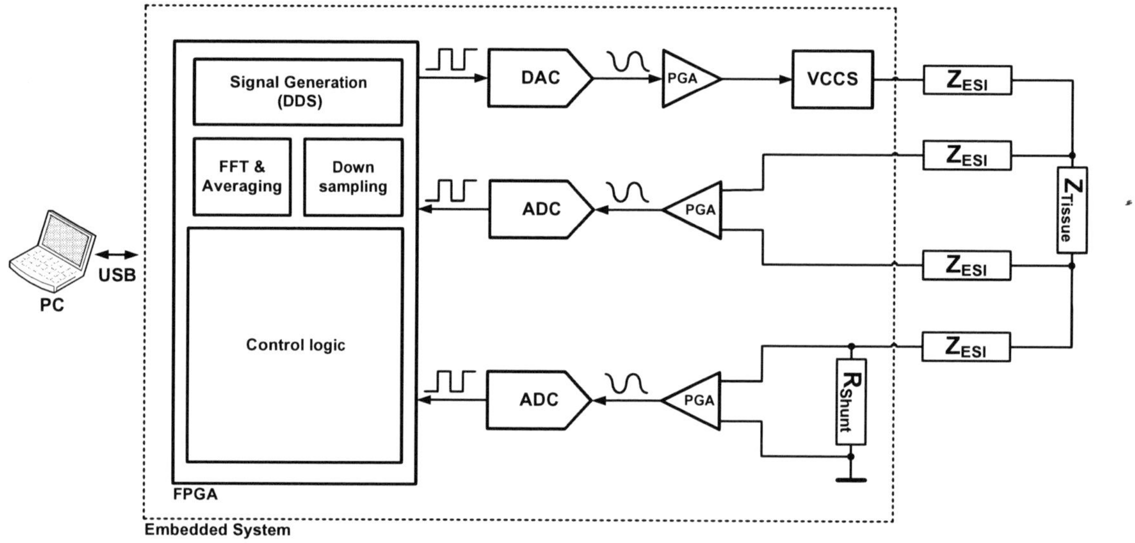

Figure 1. Principle block diagram of the used Bioimpedance Measurement System (BMS) [4].

A harmonic excitation signal is generated inside the FPGA (LFXP2-8E, Lattice Semiconductor) via a Direct Digital Synthesis (DDS) technique. This signal is fed into a 16 bit Digital to Analog Converter (DAC, LTC1668, Linear Technology), which drives after amplification a Voltage Controlled Current Source (VCCS) [4] to generate the excitation current. The current is adjustable by the DDS and a Programmable Gain Amplifier (PGA, AD8250 Analog Devices) in a range of 60 μA to 5 mA. The frequency range extends from about 10 kHz to 300 kHz. The acquisition of the voltage drop across Z_{Tissue}, as well as the shunt voltage drop is realized by a dual Analog to Digital Converter (ADC, LTC2296, Linear Technology) in combination with two additional PGAs for maintaining an optimal Signal to Noise Ratio (SNR) at different loads. Inside the FPGA the measured signals are demodulated, frequency decomposed, filtered and subsequently transmitted via a high speed USB link to the host PC. On the host PC the measurement data are further conditioned and analyzed with MathWorks MATLAB.

The BMS has been verified with respect to systematic and statistic measurement uncertainties, via known calibration impedances [4]. With reference impedances between 20 Ω and 1 kΩ the achieved Signal to Noise Ratio (SNR) is about 80 dB, with a Spurious Free Dynamic Range (SFDR) of about 60 dB. The overall measurement uncertainty depending on load and frequency is about 0.05 % to 1 % for the magnitude and 0.01 ° to 1 ° for the phase.

For the measurements commercial available carbon filled silicone rubber electrodes with a size of 38 mm x 45 mm are used (Carbon Electrode 573 and Reflex 690, Uni-Patch Incorporated). Assuming a serial alignment of the tissue impedance with two ESI, the ESI is given by $Z_{ESI} = 0.5 \cdot (Z_{Total} - Z_{Tissue})$ (see figure 1). Z_{Total} and Z_{Tissue} can be obtained by successive two and four-electrode measurements.

All performed measurements are carried out on the left ventral forearm with four linear arranged electrodes with a distance of 2 cm each, following a defined protocol for electrode application. Each series of measurements is taken at 18 different frequencies between 12 kHz and 293 kHz with a sinusoidal excitation current of 500 μA. The ambient conditions of the measurements are room temperature (about 20 °C) and relative humidity of about 35 %.

In the first test series three different compositions of the ESI (1. dry skin, 2. for one minute pre-moistened skin with 0.9 % sodium chloride solution and 3. dry skin with pre-gelled electrodes) are measured over a time period of 3.5 hours. Based on the results of the first test series in the second test series the ESI of the pre-moistened skin was repeated on fifteen different subjects. The subjects feature mixed gender and ages between 26 and 68 years. For the second test series only a single ESI measurement, eight minutes after electrode application is realized.

3. Results and Discussion

Figure 2 shows the results of the first test series for a single male subject. Shown is the measured ESI (magnitude |Z| and phase φ) over time and frequency measured on the left ventral forearm.

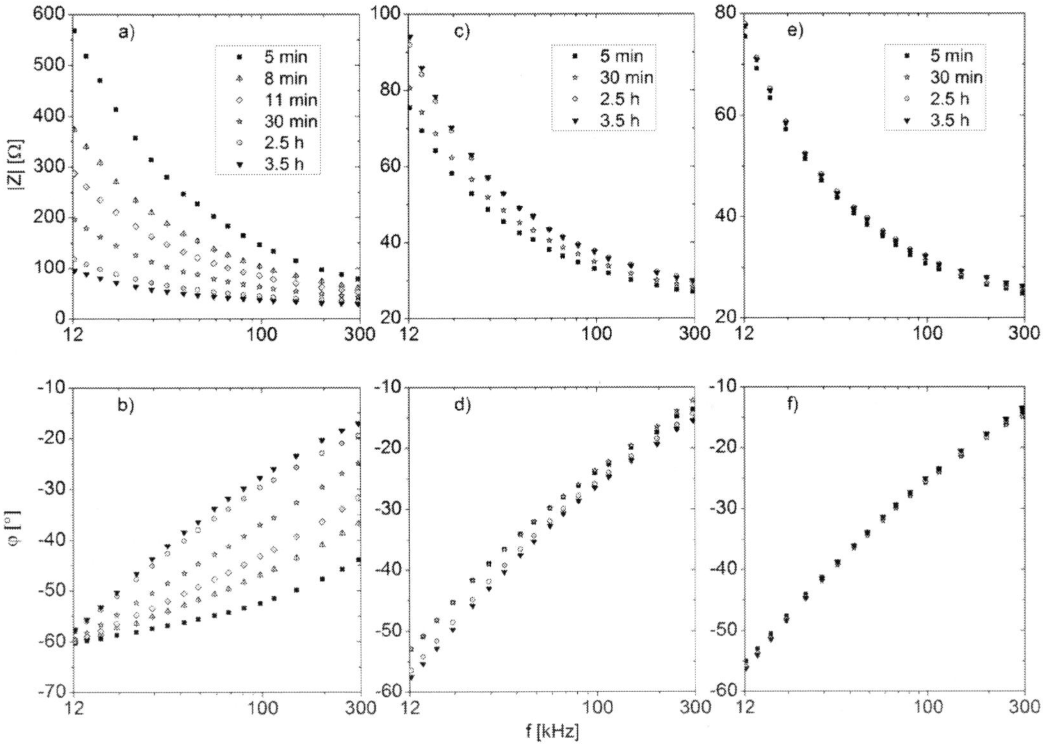

Figure 2. Electrode Skin Impedance (ESI) magnitude |Z| and phase φ over time and frequency, measured on the left ventral forearm- a, b) dry skin and dry electrodes (Carbon Electrode 573), c, d) with dry electrodes and sodium chloride solution pre-moistened skin (Carbon Electrode 573), e, f) dry skin and pre-gelled electrodes (Reflex 690).

In subfigure a) the ESI magnitude decreases, in dry condition, within 3.5 hours from about 570 Ω to about 100 Ω at 12 kHz, and from about 80 Ω to 30 Ω at 293 kHz respectively. Subfigure b) shows the corresponding phase measurements. At smaller frequencies the phase variance over time is low but increases over frequency. At 293 kHz the phase changes from -44 ° after 5 min to -17 ° after 3.5 hours. This change of magnitude and phase of the ESI over time of initial dry skin is most likely caused by moisture accumulation under the rubber electrodes. This assumption is supported by the ESI measurement results shown in subfigure c) and d), where the skin was initially pre-moistened. It is apparently that here in difference to the initial dry condition, magnitude and phase of the ESI increase over time. This contrasting behavior could be explained by a developing equilibrium of the moisture content under the electrodes. Pre-gelled electrodes seem to shift and sustain this moisture equilibrium, causing smaller ESI values, as subfigure e) and f) imply. This effect can be explained by the moisture buffer feature of gels.

In Figure 3 the ESI of fifteen different subjects over frequency, eight minutes after skin moistening with a 0.9 % sodium chloride solution, are given.

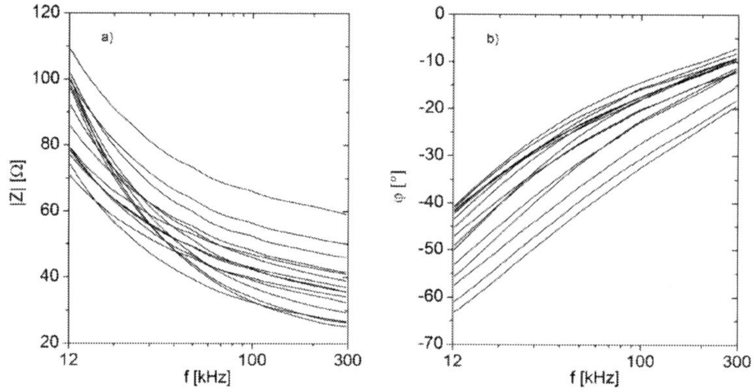

Figure 3. Electrode Skin Impedance (ESI) of fifteen different subjects over frequency with a 0.9 % sodium chloride solution moistened skin eight minutes after electrode (Carbon Electrode 573) application - a, b) Impedance magnitude |Z| and phase φ.

The results show magnitudes of 70 Ω to 110 Ω at 12 kHz and 25 Ω to 59 Ω at 293 kHz respectively. Similar variation of -63 ° to - 41 ° at 12 kHz and -19 ° to - 7 ° at 293 kHz can be found for the phases. For tissue impedance measurements this ESI values cannot be neglected, but can be managed by appropriate current sources and high input-impedance amplifiers. For improved spatial resolution of impedance measurements (e. g. in Electrical Impedance Tomography) smaller electrodes are preferable. This will obviously increase the ESI magnitude and the corresponding excitation current density. However, in a range from 7 μA/cm² to 300 μA/cm² (corresponding to excitation currents from 125 μA to 5 mA with Carbon Electrode 573), the current density shows no effect on the measured ESI magnitude.

Acknowledgment
This work is financed by the program for the Future-Economy out of the European Regional Development Fund (ERDF).The authors would also like to thank Analog Devices, Lattice Semiconductor and Linear Technology for their support in terms of free samples during the development process.

References
[1] Grimnes S and Martinsen O G 2008, *Bioelectricity and Bioimpedance Basics* (Academic Press)
[2] Webster J G (ed.) 2009, *Medical Instrumentation - Application and Design 4th edition* (Wiley)
[3] Nelson H E Jr, Smith M B, Bowman B R and Waters R L 1980 *Electrode effectiveness during transcutaneous motor stimulation* (*Arch Phys Med Rehabil* 61:73–77)
[4] Kaufmann S, Ardelt G and Ryschka M 2012 *A high accuracy Bioimpedance Measurement System - System Design and first Measurements* (Proceedings of the 5th International Workshop on Impedance Spectroscopy, Chemnitz, Germany)

Journal of Physics: Conference Series **434** (2013) 012021 doi:10.1088/1742-6596/434/1/012021

A flexible electrode array for muscle impedance measurements in the mouse hind limb: A tool to speed research in neuromuscular disease

J Li and S B Rutkove

Department of Neurology, Beth Israel Deaconess Medical Center, Harvard Medical School, 330 Brookline Avenue, TCC-810, Boston, MA, 02215, USA

E-mail: srutkove@bidmc.harvard.edu

Abstract. Electrical impedance myography (EIM) is a bioelectrical impedance technique focused on the assessment of neuromuscular diseases using tetrapolar surface arrays. Recently, we have shown that reproducible and sensitive EIM measurements can be made on the gastrocnemius muscle of the mouse hind limb and that these are sensitive to disease alterations. A dedicated array would help speed data acquisition and provide additional sensitivity to disease-induced alterations. A flexible electrode array was developed with electrode sizes of 1mm X 1mm by Parlex, Inc. Tetrapolar electrode sets were arranged both parallel to (longitudinal) and orthogonally to (transverse) the major muscle fiber direction of the gastrocnemius muscle. Measurements were made with a dedicated EIM system. A total of 11 healthy animals and 7 animals with spinal muscular atrophy (a form of motor neuron disease) were evaluated after the fur was completely removed with a depilatory agent from the hind limb. Standard electrophysiologic testing (compound motor action potential amplitude and motor unit number estimation) was also performed. The flexible electrode array demonstrated high repeatability in both the longitudinal and transverse directions in the healthy and diseased animals (with intraclass correlation coefficients of 0.94 and 0.89, respectively, for phase angle measured transversely). In addition, differences between healthy and diseased animals were identifiable. For example, the 50 kHz transverse phase angle was higher in the healthy as compared to the SMA animals ($16.8° \pm 0.5$ vs. $14.3° \pm 0.7$, respectively) at 21 weeks of age ($p = 0.01$). Differences in anisotropy were also identifiable. Correlations to several standard neurophysiologic parameters also appeared promising. This novel flexible tetrapolar electrode array can be used on the mouse hind limb and provides multidirectional data that can be used to assess muscle health. This technique has the potential of finding widespread use in the evaluation of drug therapies in neuromuscular animal disease models.

1. Introduction

Generalized neuromuscular diseases encompass a wide variety of disorders ranging from amyotrophic lateral sclerosis (ALS) to muscular dystrophy to generalized polyneuropathy. A theme common too many of these disorders is that there is no readily available therapy and that much basic research continues to rely on animal models. In fact, for many of these disorders one or more animal models have been developed. For example, in the case of ALS, while one genetic model has been used for years [1], other models become available as new genetic causes for the disease are identified [2, 3].

In addition to seeking mechanisms underlying the pathogenesis of these diseases, these models are also used for preclinical drug testing [4]. Such drug testing is used to determine preliminarily the safety of the drug, the doses needed, and the degree of therapeutic benefit offered. Current approaches for evaluating these diseases include functional measures, such as animal grip strength or walking speed as well as electrophysiologic measures [5, 6].

We have been exploring the development of new methods for assessing muscle function in these animals based on localized measurements of muscle impedance via electrical impedance myography (EIM) [7]. As part of that effort we initially began by studying larger rodents (rats) using a modified approach of that which we had been performing on humans [8]. However, for several reasons, the rat model is limited in value and thus we began performing measurements on mouse disease models, focusing the gastrocnemius-soleus muscles. Recent work showed that a fixed stainless steel 4-electrode array provided repeatable data and was also able to differentiate normal from ALS-affected muscle [9]. However, in an effort to improve our impedance measurements, we developed a multielectrode array capable of measuring in multiple directions relative to the major muscle fiber direction. Here we test that array in a group of healthy animals as well as a group with spinal muscular atrophy (SMA), a genetic pediatric neuromuscular disease that causes generalized weakness.

2. Methods

2.1. Mice
All studies were approved by the Beth Israel Deaconess Medical Center Institutional Animal Care and Use Committee. Breeding colonies of SMA (strain: B6.129-Smn1tm5(Smn1/SMN2)Mrph/J) mice were established from animals obtained from Jackson Laboratories (Bar Harbor, Maine). SMA mice were genotyped by tail snip and were weaned at 28 days. A total of 7 SMA mice and 11 littermate controls (control) were studied at 21 weeks of age.

2.2. Mouse setup
Animals were placed under 1% isoflurane anesthesia delivered through a nose cone with a heating pad underneath the limb to maintain consistent temperature. A depilatory agent (Nair face cream, Princeton, NJ) was applied to the skin to remove all fur; the skin was cleaned with 0.9% saline solution. The leg and animal were positioned as previously described [9].

2.3. Electrical impedance measurement
A tetrapolar electrode array was developed in conjunction with Parlex, Inc, (Methuen, MA). Figure 1a shows the electrode array in comparison to a penny. It is small enough to fit over the gastrocnemius muscle. In addition, as shown in Figure 1b, a schematic of that array, both longitudinal (current flowing along the muscle fiber direction) and transverse (the current flow perpendicular to the muscle fiber direction) configurations can be measured with the combinations of using different current and voltage electrodes. In this study, the set of current electrodes 1 and 2 and voltage electrodes 1 and 2 was utilized in the transverse direction; while the longitudinal direction was made by using current electrodes 3 and 4 and voltage electrodes 3 and 4. The circles are electrical vias connecting the contact side (top) with the electrode side (bottom). EIM measurements were performed with the EIM Vet1102 1kHz-10 MHz system designed by Convergence Medical Devices, Inc, (Medford, MA).

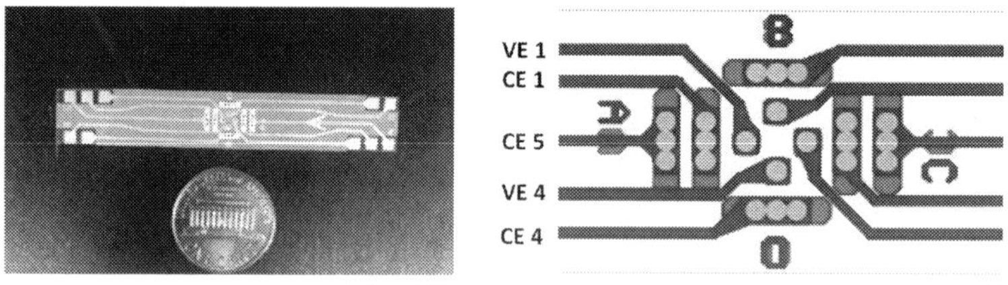

(a) (b)

Figure 1. Views of the electrode array. (a) The entire view of the electrode in comparison to a penny; (b) A close-up view of the tetrapolar electrode array. VE: voltage electrode; CE: current electrode.

2.4. Standard electrophysiology
Compound motor action potential (CMAP) and motor unit number estimation (MUNE) were performed using a TECA Synergy T2 EMG Monitor System (Viasys, Madison, WI) on the left hind limb stimulating the sciatic nerve at the sciatic notch and recording via disposable ring electrodes around the entire distal leg, as previously described [10].

2.5. Reproducibility measurements
For assessment of technique repeatability, after an initial set of measurements was made, the electrode array was lifted up from the animal, and then replaced to the leg surface; the measurements repeated.

2.6. Data analysis
Reproducibility was assessed by calculating intra-class correlation coefficients (ICC). All data were summarized as mean ± standard error. Anisotropy difference is defined as the difference between transverse and longitudinal directions. Unpaired t-tests were used to compare the means between groups; Spearman correlation was used to assess the relationship between EIM phase and CMAP amplitude. For all analyses, significance was accepted at $p < 0.05$, two-tailed.

3. Results

3.1. Reproducibility
Excellent reproducibility was obtained with this technique in control and SMA mice at 50 kHz on both longitudinal and transverse configurations. Test-retest plots for one of the major EIM parameters, phase, are shown in Figure 2 a-d. High values of intra-class correlation coefficients (ICC) from the 2 groups of animals were achieved in both directions; for example, in transverse direction, the ICC values for control and SMA mice were 0.94 and 0.89 respectively.

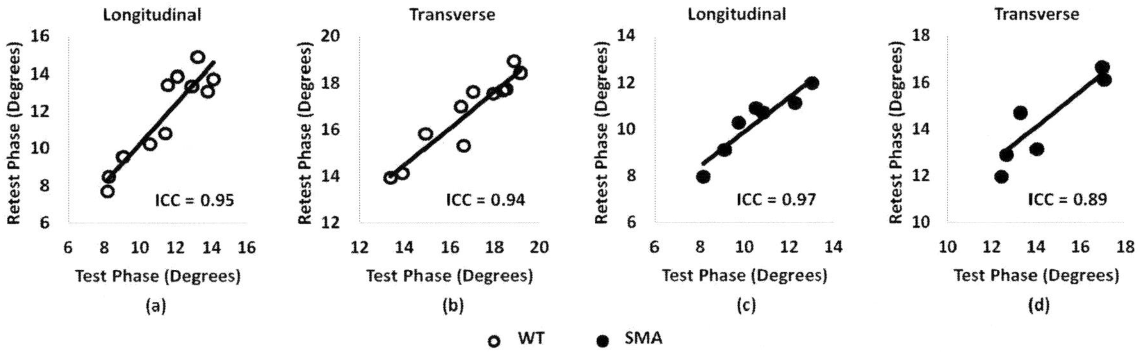

Figure 2. The test-retest plots of phase at 50 kHz for the control and SMA mice in both longitudinal and transverse configurations. (a) 11 control mice: longitudinal, (b) 11 control mice: transverse, (c) 7 SMA mice: longitudinal, (d) 7 SMA mice: transverse. It is noted that 2 points overlap in the plot (d).

3.2. EIM differences between control and SMA animals at 50 kHz
As shown in Figure 3 a-f, smaller values for EIM parameters and CMAP amplitude were found in the SMA animals as compared to the control animals. However, no difference for MUNE or girth was found between the 2 groups of animals. For example, transverse phase was 14.3° ± 0.7 for the SMA animals and 16.8° ± 0.5 for the control animals ($p = 0.01$). In addition, the anisotropy difference for phase was 5.2° ± 0.4 in the control animals and 3.9° ± 0.3 ($p = 0.02$) in the SMA animals. CMAP amplitude showed significantly differences between the 2 groups animals (44.5 ± 6.3 mV for the SMA mice and 62.4 ± 4.3 mV for the control mice, $P = 0.04$). However, MUNE for the control mice (61.3 ± 24.9 motor units) was not different from the SMA animals (26.0 ± 3.3 motor units) with p as 0.19. Neither was girth (6.9 ± 0.1 mm for control mice vs. 6.8 ± 0.1 mm for SMA mice, $p = 0.39$).

3.3. Correlation between transverse phase and CMAP amplitude
Spearman correlation coefficient was 0.68 with p as 0.09 for the 7 SMA animals.

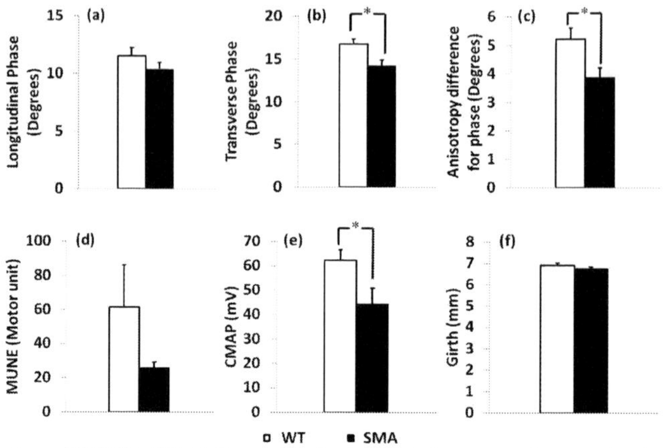

Figure 3. EIM parameters, MUNE, CMAP amplitude, and girth comparing control and SMA animals. (a) longitudinal phase, (b) transverse phase, (c) anisotropy difference for phase, (d) MUNE, (e) CMAP amplitude, (f) girth. * indicates $p < 0.05$.

4. Discussion

These results show that a flexible 4-electode array can provide EIM data that is reproducible both in the longitudinal and transverse directions. These reproducibility values are similar to those observed for rigid electrodes, as previously described [9]. In addition, the data can differentiate normal animals from those with SMA. Finally, the data appear to correlate with the CMAP amplitude, a standard measure of nerve and muscle function. Taken together, these results suggest that such a flexible electrode can differentiate normal muscle from that affected by SMA.

The main goal of this work was not to demonstrate the value of the technique for drug evaluation, since in order for that to be true, we would need to include a therapeutic intervention, which was not performed in this instance. However, doing so would be a logical next step, comparing the EIM data obtained to standard function, electrophysiologic and possibly pathological data.

One major advantage over other forms of electrophysiologic testing is that EIM is likely to be sensitive to conditions were standard electrophysiologic measurements would be expected to be relatively insensitive such as in myopathy or disuse atrophy [11, 12].

In summary, these data support the potential value of a flexible electrode array in animal testing, providing high-quality, repeatable data that can be used by scientists investigating disease mechanisms and by pharmaceutical companies seeking data as to potential drug efficacy.

Acknowledgments
This work was funded by the SMA foundation.

5. References

[1] Azzouz M, Leclerc N, Gurney M, Warter JM, Poindron P and Borg J 1997 *Muscle Nerve* **20** 45-51.
[2] Nalbandian A, Llewellyn KJ, Badadani M, Yin HZ, Nguyen C, Katheria V, Watts G, Mukherjee J, Vesa J, Caiozzo V, Mozaffar T, Weiss JH and Kimonis VE 2012 *Muscle & nerve*
[3] Wu LS, Cheng WC and Shen CK 2012 *The Journal of biological chemistry* **287** 27335-44.
[4] Scott S, Kranz JE, Cole J, Lincecum JM, Thompson K, et al 2008 *Amyotrophic Lateral Sclerosis* **9** 4-15.
[5] Weydt P, Hong SY, Kliot M and Moller T 2003 *Neuroreport* **14** 1051-4.
[6] Shefner JM, Cudkowicz ME and Brown RH, Jr. 2002 *Muscle Nerve* **25** 39-42.
[7] Rutkove SB 2009 *Muscle Nerve* **40** 936-946.
[8] Ahad M and Rutkove SB 2009 *Clinical Neurophys* **120** 1534-8.
[9] Li J, Staats WL, Spieker A, Sung M and Rutkove SB 2012 *PloS one* **7** e45004.
[10] Shefner JM, Cudkowicz M and Brown RH, Jr. 2006 *Muscle Nerve* **34** 603-7.
[11] Tarulli A, Esper G, Lee K, Aaron R, Shiffman C and Rutkove S 2005 *Neurology* **65** 451-452.
[12] Tarulli AW, Duggal N, Esper GJ, Garmirian LP, Fogerson PM, Lin CH and Rutkove SB 2009 *Arch Phys Med Rehabil* **90** 1806-10

Journal of Physics: Conference Series **434** (2013) 012022 doi:10.1088/1742-6596/434/1/012022

Extensive ionic partitioning in interfaces that membranous and biomimetic surfaces form with electrolytes: Antitheses of the gold-electrolyte interface

Terry Chilcott, Chuan Guo and Hans Coster

School of Chemical and Biomolecular Engineering, The University of Sydney, Sydney 2006, Australia

terry.chilcott@sydney.edu.au

Abstract. Maxwell-Wagner modeling of electrical impedance measurements of tetradecane-electrolyte systems yielded three interfacial layers between the tetradecane layer and the bulk electrolytes of concentration ranging from 1-300 mM KCl whereas the gold-electrolyte system yielded only one layer. The conductivity and thickness for the surface layer were orders of magnitude different from that expected for the Gouy-Chapman layer and did not reflect dependencies of the Debye length on concentration. Conductivity values for the three layers were less than those of the bulk electrolyte but exhibited a dependency on concentration similar to that expected for the bulk. Thickness values for the layers indicate an interface extending ~10^6 Å into the bulk electrolyte, which contrasts with the gold-electrolyte interface that extended only 20-30 Å into the bulk. Maxwell-Wagner characterizations of both interfaces were consistent with spatial distributions of ionic partitioning arising from the Born energy as determined by the dielectric properties of the substrates and electrolyte. The distributions for the membranous and silicon interfaces were similar but the antitheses of that for the gold interface.

1. Introduction

Silicon and gold have proven suitable substrates for supporting molecularly thin bio-sensitive surfaces in biosensors that on one side make contact with aquatic samples but on the other side provide connectivity to electronic devices that transduce the binding of targeted molecules in the samples. Such constructs introduce dielectric asymmetry not featured in biological systems in which a thin membrane separates two extensive aquatic phases (dielectric constant of ε_{water} = 79) and perform the very binding events that the constructs are attempting to mimic. Essentially, the substrate in a biosensor replaces one of the aquatic phases, and its dielectric properties ($\varepsilon_{gold} = \infty$ or $\varepsilon_{silicon}$ = 11.5) will influence differently the ionic concentrations in the single aquatic phase and the interfacial region where recognition of targeted molecules begins.

The magnitude and extent of such dielectric influences are characterized herein in a theoretical study and impedance measurements of electrolytes in contact with a tetradecane monolayer attached to silicon. These characterizations are compared with those in the literature of electrolytes in contact with gold substrates and bilayer membranes.

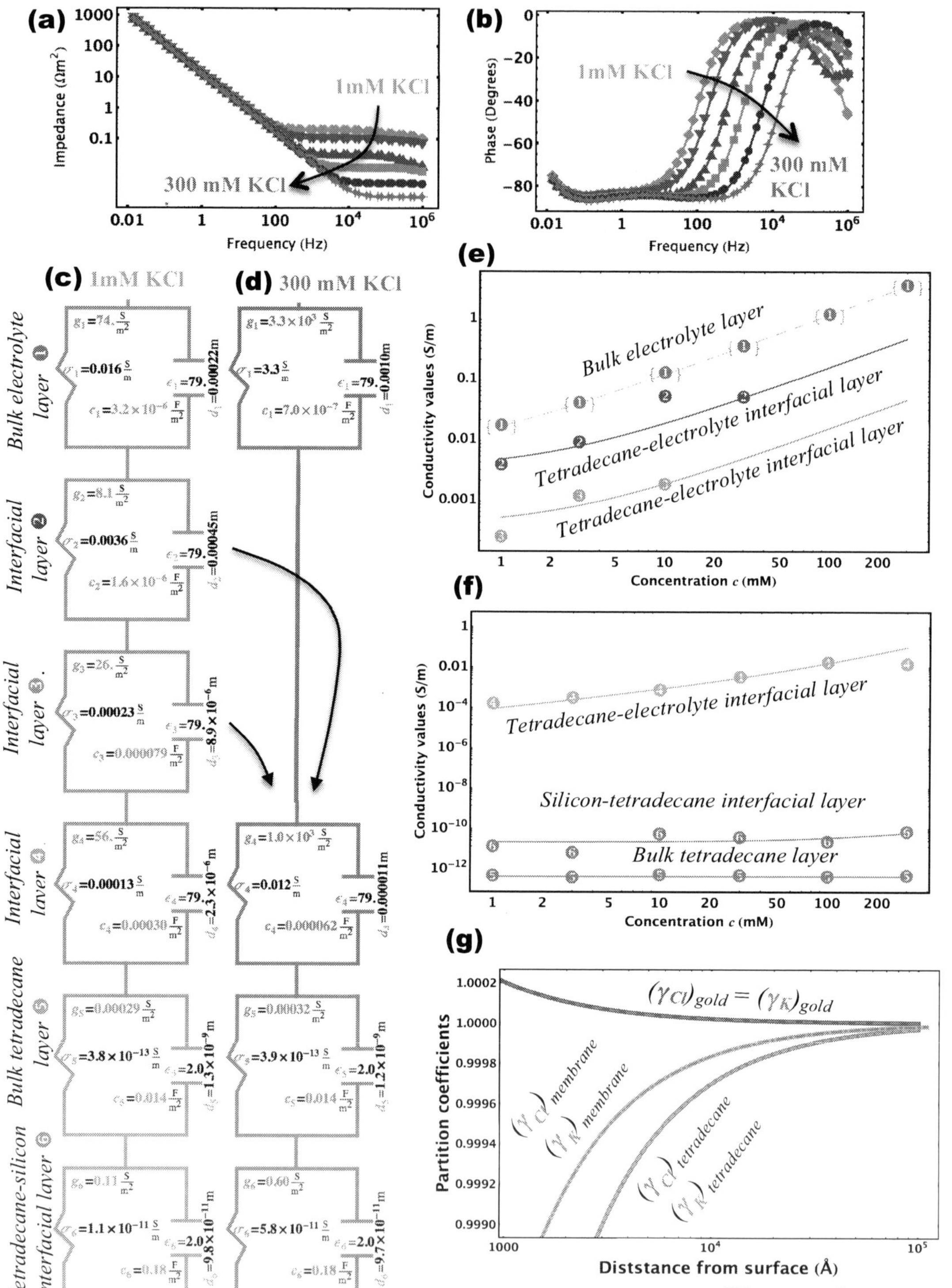
(a)
Impedance (Ωm²)
1mM KCl
300 mM KCl
Frequency (Hz)
(b)
Phase (Degrees)
1mM KCl
300 mM KCl
Frequency (Hz)
(c) 1mM KCl
(d) 300 mM KCl
Bulk electrolyte layer ❶
Interfacial layer ❷
Interfacial layer ❸
Interfacial layer ❹
Bulk tetradecane layer ❺
Tetradecane-silicon interfacial layer ❻
(e)
Conductivity values (S/m)
Bulk electrolyte layer
Tetradecane-electrolyte interfacial layer
Tetradecane-electrolyte interfacial layer
Concentration c (mM)
(f)
Conductivity values (S/m)
Tetradecane-electrolyte interfacial layer
Silicon-tetradecane interfacial layer
Bulk tetradecane layer
Concentration c (mM)
(g)
Partition coefficients
(γCl)gold = (γK)gold
(γCl)membrane
(γK)membrane
(γCl)tetradecane
(γK)tetradecane
Diststance from surface (Å)

Figure 1

2. Theory

Expressions for the spatial distribution of the Born energy [1] for an ionic species i of molecular radius r_i in a system comprised of a layer of thickness d and dielectric constant ε_2 interposed between a medium of dielectric constant ε_1 on one side ($-\infty<x<0$) and a medium of dielectric constant ε_3 on the other, i.e. ($d<x<\infty$) are given by;

$$\mu_i^o(-\infty < x < r_i) = \frac{q_i^2}{16\pi\varepsilon_o\varepsilon_1 d}\left\{\frac{2d}{r_i} - \frac{1}{(x/d)} + \frac{\vartheta_{3,1}}{(1-x/d)}\right\} \quad (1)$$

$$\mu_i^o(r_i < x < d - r_i) = \frac{q_i^2}{16\pi\varepsilon_o\varepsilon_2 d\vartheta_{2,1}}\left\{\vartheta_{2,1}\frac{2d}{r_i} - \sum_{n=0}^{\infty}\frac{\vartheta_{2,1}^n\vartheta_{2,3}^n}{(n-x/d)} - \frac{1}{(x/d)} - \vartheta_{2,1}^2\sum_{n=0}^{\infty}\frac{\vartheta_{2,1}^n\vartheta_{2,3}^n}{(n+x/d)}\right\} \quad (2)$$

$$\mu_i^o(d + r_i < x < \infty) = \frac{q_i^2}{16\pi\varepsilon_o\varepsilon_1 d}\left\{\frac{2d}{r_i} - \frac{\vartheta_{3,1}}{(x/d)} + \frac{\vartheta_{2,3}}{(x/d-1)}\right\} \quad (3)$$

$$\text{where } \vartheta_{1,3} \equiv -\vartheta_{3,1} \equiv \frac{\epsilon_3-\epsilon_1}{\epsilon_3+\epsilon_1}; \quad \vartheta_{2,3} \equiv -\vartheta_{3,2} \equiv \frac{\epsilon_3-\epsilon_2}{\epsilon_3+\epsilon_2}; \quad \vartheta_{1,2} \equiv -\vartheta_{2,1} \equiv \frac{\epsilon_2-\epsilon_1}{\epsilon_2+\epsilon_1} \quad (4)$$

and ε_o is dielectric permittivity in free space. These revert to expressions derived by Neumcke and Laüger [2] for a membrane separating two electrolytes when $\varepsilon_1 = \varepsilon_3 = 79$.

In the intermediate regions where the ion is partly in one medium and the other, the distribution can be approximated by the function;

$$\mu_i^o(x) = \frac{p_0+p_1x+p_2x^2}{1+q_1x+q_2x^2} \quad (5)$$

where equating boundary energy and gradient values yields the polynomial coefficients.

The partition coefficient for ionic species i is defined so;

$$\gamma_i \equiv e^{-\{\mu_i^o(x)-\mu_i^o(\infty)\}/kT} \approx \frac{C_i(x)}{C_i(\infty)} \quad (6)$$

where the approximation applies in the absence of an electrical potential distribution when the partition coefficients determine the distribution of ionic concentrations, i.e. $C_i(x)$ with respect to $C_i(\infty)$.

Spatial distributions of the partitioning coefficients in the electrolytes ($\varepsilon_3 = 79$) are shown in Figure 1(g) for a bilayer ($d = 50$ Å and $\varepsilon_2 = 3$) separating two electrolytes, a tetradecane monolayer ($d = 15$ Å and $\varepsilon_2 = 2$) separating silicon ($\varepsilon_1 = 79$) and the electrolyte as well as a Stern layer ($d = 8.08$ Å and $\varepsilon_2 = 6$) separating gold ($\varepsilon_1 = 100\text{x}\varepsilon_3$) and the electrolyte.

3. Materials and Methods

A tetradecane monolayer was covalently attached to a ~15 mm square segment of n-type silicon (111) of resistivity ~0.01Ωcm using the technique described in detail previously [3]. Briefly, the segment was thoroughly cleaned and etched in deoxygenated ammonium fluoride producing a Si-H surface that was immediately reacted with 95% 1-tetradecene in a sealed vessel of nitrogen gas under ultraviolet illumination. The oxygen-purged atmosphere favored the reaction of H-C=H termini of tetradecene with the Si-H surface which yielded a layer of tetradecane rather than silicon dioxide. Gallium-indium eutectic was applied to the underside of the segment that formed a low-ohmic contact with the working electrode of an Inphaze 3-terminal characterization chamber shielded in a Faraday cage [4].

The chamber featured a spring and gearing mechanism for lowering a soft gasket of rectilinear cross-section onto the monolayer surface where it sealed a precisely defined 'active' area (1.81 10^{-5} m^2) without substantially distorting the surface at the gasket perimeter. KCl electrolytes of concentration 1, 3, 10, 30, 100 and 300mM were perfused in turn into the inner space of the chamber wetting the 'active' surface as well as the counter and reference electrodes. An Inphaze impedance spectrometer generated a spectrum of precise sinusoidal currents ranging in frequency from 10^{-2} to 10^6 Hz that were injected in turn through the electrolytes and monolayer via the counter and working electrodes. A recessed reference electrode and the working electrode connected to the inputs of an instrumentation amplifier (input impedance $10^{13}\Omega$) sensed the voltage responses to the current stimuli. This amplifier and another simultaneously monitored the responses and stimuli that were buffered for transmission along low-impedance coaxial cables to the 24-bit ADCs of the spectrometer. The digitized data were transferred to a computer via a USB interface where the magnitude and phase of the impedances were calculated to respective resolutions of 0.002% and 0.001 degrees.

4. Results and Discussion

Impedance measurements of the tetradecane-electrolyte system are shown in Figures 1(a) and 1(b) for electrolyte concentrations ranging from 1 to 300mM. However, the measurements at high frequencies did not yield directly the known dielectric, geometrical and conductive properties of the bulk electrolytes such as are readily discernable from spectra of gold-electrolyte systems [5]. In this instance, the known properties for the electrolytes were introduced in the modeling as layer ❶ in the Maxwell-Wagner models shown in Figures 1(c) and 1(d), the total impedance of which is given by;

$$z_{MW} \equiv \sum_{k=1}^{N_{NW}} \frac{d_k}{\sigma_k + j\omega\varepsilon_k} \qquad (7)$$

where d_k denotes the thickness, σ_k the conductivity and ε_k the dielectric permittivity of the kth of N_{MW} parallel layers. In contrast, the convergences of spectra at low frequencies in Figures 1(a) and 1(b) characterized the known concentration independent properties of the tetradecane monolayer and interfacial layers it forms with silicon, characterized in the models by layers ❺ and ❻, respectively.

The properties of layer ❹ and those for layers ❷ and ❸ when required to fit the spectra, were then adjusted to obtain the least-squares-error fits of the models to the spectra also shown in Figures 1(a) and 1(b). Justification of the fits was based on χ^2-statistics [6], which minimizes;

$$Reduced - \chi^2 \equiv \frac{\chi^2}{N - 2N_{MW}} \qquad \text{where} \qquad \chi^2 \equiv \sum_{n=1}^{N} \left(\frac{Mean_n - f(\omega_n)}{Error_n} \right)^2 \qquad (8)$$

in which $f(w)$ denotes either the conductance or capacitance component of the model impedance, and $Mean_n$, $Error_n$ and ω_n denote the mean, standard deviation and frequency, respectively, of the nth of N measurements. Note that minimization of Equation (8) is countered by the number of Maxwell-Wagner layers, i.e. N_{MW}, and provided justification for the inclusion or exclusion of layers ❷ and ❸.

Figure 1(f) shows that layer ❹ features in models for all concentrations and the dependency of the conductivity of this layer on concentration is similar to that dependency of the electrolyte (layer ❶ in Figure 1(e)) and the expected dependency of the Gouy-Chapman interfacial layer. However, the thickness values for layer ❹ were ~10^5 Å (e.g. see Figures 1(c) and 1(d)) which is orders of magnitude larger than Debye thicknesses for a Gouy-Chapman layer, and did not exhibit the expected dependency on concentration of the Debye thickness. Figure 1(e) shows that the conductivity values for layers ❸ and ❷ were orders of magnitude closer to those values for the electrolyte (layer ❶) than those estimates for layer ❹ and exhibited similar dependencies on concentration to the electrolyte and layer ❹. The estimated thicknesses for layers ❸ and ❷ indicate that the tetradecane-electrolyte interface extended more than ~10^6 Å into the electrolyte for concentrations less than 30 mM.

Layers ❹, ❸ and ❷ characterize an interfacial region in which the conductivity (which is proportional to concentration) increases from values greater than that for the tetradecane monolayer (layers ❺) to values less than that of the bulk electrolyte (layer ❶). This interface extends many orders of magnitude further into the bulk electrolyte than the 20-30 Å for a gold-electrolyte interface with a concentration distribution consistent in form with that predicted by partitioning (Equations (1-6)) for this system shown Figure 1(g) and contrasting with that predicted for the gold interface reflecting constant bulk electrolyte properties. Figure 1(g) further illustrates the compatibility of silicon and membranous interfaces for recognizing, sequestering and binding biologically important molecules.

References

[1] Born MZ 1920 *Physik.* **1** 45
[2] Neumcke B and Laüger P 1969 *Biophys. J.* **9** 1160
[3] Shamsi F, Coster HGL, Chilcott TC and Jollife K 2011 *Material Chem. and Phys.* **126** 955
[4] Chilcott TC, Coster HGL and Ellis JL (2010) *American Laboratory* **42,5** 24
[5] Chilcott TC, Wong ELS, Coster HGL Coster AFC and James M 2009 *Electrochimica Acta* 54 3766
[6] Kenney JF and Keeping ES 1951 Mathematics and statistics 2nd edition "Princeton, NJ: Van Nostrand"

Electrode-Skin contact impedance: In vivo measurements on an ovine model

D T Nguyen[1], R Kosobrodov[1], M A Barry[2], W Chik[2], C Jin[1], T I Oh[3], A Thiagalingam[2] and A McEwan[1]

[1] School of Electrical and Information Engineering, The University of Sydney, New South Wales, Australia
[2] Department of Cardiology, Westmead Hospital, New South Wales, Australia
[3] Impedance Imaging Research Center, Kyung Hee University, South Korea

E-mail: d.nguyen@sydney.edu.au

Abstract.
The problem of electrical impedance between the skin and the electrode is an on-going challenge in bio-electronics. This is particularly true in the case of Electrical Impedance Tomography (EIT), which uses a large number of skin-contact electrodes and is very sensitive to noise. In the present article, contact impedance is measured and compared for a range of electrodes placed on the thorax of an ovine model. The study has been approved by the Westmead Hospital Animal Ethics Committee. The electrode models that were employed in the research are Ag/AgCl electrodes (E1), commonly used for ECG and EIT measurements in both humans and animal models, stainless steel crocodile clips (E2), typically used on animal models, and novel multi-point dry electrodes in two modifications: bronze plated (E3) and nickel plated (E4). Further, since the contact impedance is mostly attributed to the acellular outer layer of the skin, in our experiment, we attempted to study the effect of this layer by comparing the results when the skin is intact and when electrodes are introduced underneath the skin through small cuts. This boundary effect was assessed by comparison of measurements obtained during E2 skin surface contact, and sub-cutaneous contact (E5). Twelve gauge intradermal needles were also tested as an electrode (E6). The full impedance spectrum, from 500 Hz to 300 kHz, was recorded, analysed and compared. As expected, the contact impedance in the more invasive cases, i.e the electrodes under the skin, is significantly lower than in the non-invasive cases. At the frequency of 50 kHz which is commonly used in lung EIT acquisition, electrodes E3, E4 and E6 demonstrated contact impedance of less than 200 Ω, compared to more than 400 Ω measured for electrodes E1, E2 and E5. In conclusion, the novel multipoint electrodes proved to be best suited for EIT purposes, because they are non-invasive and have lower contact impedance than Ag/AgCl and crocodile clips, in both invasive and non-invasive cases. This further prompted us to design a flexible electrode belt using the novel multi-point electrodes for lung EIT on animal models.

1. Introduction

There is an increasing need in biomedical engineering to consider new electrode technologies. Movement artefacts, baseline variations and contact impedance prevent the recording of the electrocardiogram (ECG) during resuscitation [1], constrain implanted devices for functional electrical stimulation of muscles, the cochlea and the retina [2], and have restricted the use of the electroencephalogram (EEG) and impedance classification of stroke [3; 4]. Dry biosignal recordings are currently limited by changes in contact impedance, noise, motion artefacts and baseline voltage variations at the electrode to skin interface [5]. These are the limiting factors in recording ECG and fECG with a handheld device [6; 7]. For conventional bio-potential measurements, electrodes are attached to the patient's skin after preparation; this includes cleaning, shaving, mechanical abrasion to remove the acellular outer layer of the skin, and moistening. A layer of electrically conductive gel is applied between the skin and the electrodes to reduce contact impedance. Two double layers form between the electrode metal conductor, the gel and the cell membranes of active skin. These present significant resistance and capacitance which results in variable contact impedance between electrodes. High input impedance instrumentation amplifiers are successful at overcoming this contact impedance in controlled environments but skin preparation takes time and requires a compliant subject [8]. In EIT measurements the variations in contact impedance impact on both the current source and voltmeter [9].

The purpose of this paper was to compare electrodes suitable for EIT experiments in an ovine model. Animal experiments are convenient for testing the ability of new devices, particularly in the invasive situations in which we are interested such as with the use of internal electrodes or contrast agents. However the electrodes are generally more difficult to use on animals due to the increased hair and skin oils.

2. Method

This research has the approval from the Westmead Hospital Animal Ethics committee. The animal was anaesthetised and ventilated appropriately. For this experiment, the skin on the thorax of the animal was cleanly shaven. The skin was then abraded with alcohol rub in an attempt to remove the dead epidermis layers and oils. The skin was wiped with alcohol rub after each experiment to ensure the integrity of subsequent measurements.

Electrode models utilised in the experiment included:

- E1: Ag/AgCl electrodes (3M), applied directly onto the skin with gel,
- E2: Crocodile clips, pinched the skin between the clip,
- E3: Bronze multi-point electrodes, applied directly onto the skin,
- E4: E3 plated with a nickel coating
- E5: Crocodile clips, introduced through a pocket made by a cut on the skin of the animal.
- E6: 12 gauge needles, slipped under the skin horizontally.

E1 is the most frequently used electrodes in both humans and animal models for a wide range of applications from ECG to EIT. E2 is commonly used on animal models on which the skin does not adhere well to Ag/AgCl electrodes, for example sheep skin. E3 and E4 are novel multi-point dry electrodes, which are yet to be tested on an ovine model. Further, since the contact impedance is mostly attributed to the acellular outer layer of the skin, in our experiment, we attempted to study the effect of this layer by comparing the results when the skin is intact and when electrodes are introduced underneath the skin through small cuts (E5). Twelve gauge intradermal needles were also tested as an electrode (E6).

The contact impedance of each case was measured using an HIOKI IM3570 impedance analyser. The frequency range was set between 100 Hz and 300 kHz for all of the measurements. A two-terminal impedance measurement was made between electrode pairs. A fixed current of 1mA was injected through the two terminals while voltage was measured across the same terminals. The measured voltage is the average of 16 measurements at each frequency to mitigate the effect of electrode movements and other noise sources.

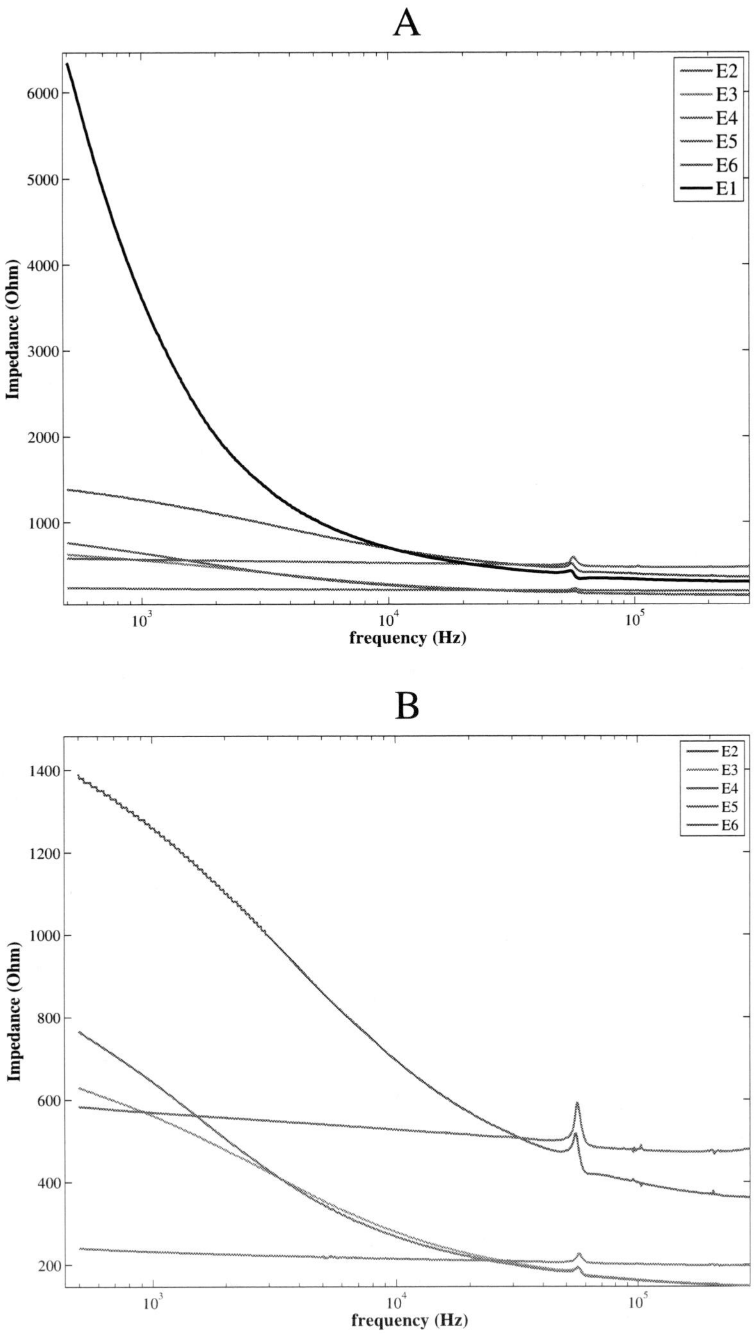

Figure 1. Electrode-Skin contact impedance spectra of various electrode models, measured on an ovine model; **A** Impedance spectra of all electrodes models E1–6; **B** Impedance spectra of electrodes models E2–6. E1 has considerably higher impedance in the lower frequency range than the other electrodes.

3. Results

As Figure 1 depicts, the contact impedance in the more invasive cases, i.e the electrodes under the skin, are significantly lower than in the non-invasive cases. At higher frequencies, that is, above 10 kHz, this difference is not as significant as at the lower frequencies. At the frequency of 50kHz which is commonly used in lung EIT acquisition, electrodes E3, E4 and E6 demonstrated contact impedance of less than 200 Ω, compared to more than 400 Ω measured for electrodes E1, E2 and E5. The multi-point electrodes performed particularly well in the typical range of frequencies for EIT, i.e. 10 kHz to 100 kHz.

4. Discussion and Conclusion

The impedance spectra shown in Figure 1 demonstrated that the Ag/AgCl electrodes clearly performed the worst, with contact impedance at 1 kHz almost double that of all other electrode models. The differences rapidly decreased up to 10kHz. This is in agreement with a wide body of previous work [10; 11] which showed that the epidermis layer mainly affects the lower frequency signals. The needle electrodes performed well despite their much smaller size and hence higher current density, while the crocodile clips in pockets, albeit highly invasive, retained a high contact impedance with a relatively flat spectrum.

In all of the impedance spectra, as seen in Figure 1, there was a spike occurred around 50 kHz, which can be attributed to instrumental interference. The experiments were conducted in a clinical settings with many other instrumentations running in parallel, such as fluoroscopy, ventilator, etc. Since the crocodile clips used in E2 and E5 had larger areas of exposed metal, the interference were more prevalent in these two electrodes types than any other.

In conclusion, the novel multipoint electrodes proved to be best suited for EIT purposes, because they are non-invasive and have lower contact impedance than Ag/AgCl and crocodile clips, in both invasive and non-invasive cases. This further prompted us to design a flexible electrode belt using the novel multi-point electrodes for lung EIT on animal models. This device helps to mitigate a number of technical problems which arise while performing EIT in vivo on an animal model, including large values of electrode contact impedance, its high variability, position errors and undetected disconnection of electrodes during the experiment.

References

[1] Li Y, Bisera J, Geheb F, Tang W and Weil M H 2008 *Critical Care Medicine* **36** 198–203

[2] Wallace G G, Moulton S E and Clark G M 2009 *Science* **324** 185–6

[3] McEwan A, Romsauerova A, Yerworth R, Horesh L, Bayford R and Holder D S 2006 *Physiol Meas.* **27** 199–210

[4] Finnigan S P, Rose S E and Chalk J B 2006 *Clinical Neurophysiology* **117 (10)** 2338–9

[5] Hamilton P S, Curley M and Aimi R 2000 *Biomed. Instrum, Technol.* **34(3)** 197–202

[6] Oostendorp T F, van Oosterom A and Jongsma H W 1989 *Med Biol Eng Comput* **27(3)**

[7] Degen T and Jäckel H 2008 *IEEE Trans Biomed Eng.* **55(6)** 1711–5

[8] Searle A and Kirkup L 2000 *Physiol Meas.* **21(2)** 271–83

[9] McEwan A, Cusick G and Holder D S 2007 *Physiol Meas.* **28** S197–215

[10] Cardu R, Leong P H W, Jin C T and McEwan A L 2012 *Physiol Meas.* **33(5)**

[11] Yamamoto T and Yamamoto Y 1986 *Med. Biol. Eng. Comput* **24** 71–7

Textile electrode characterization: dependencies in the skin-clothing-electrode interface

R Macías[1], M Fernández[1] and R Bragós[1]

[1]Department of Electronic Engineering, Universitat Politécnica de Catalunya, Campus Nord, Edifici C-4, 08034, Barcelona, Spain

E-mail: raul.macias@upc.edu

Abstract. Given the advances in the technology known as smart textiles, the use of textile electrodes is more and more common. However this kind of electrodes presents some differences regarding the standard ones as the Ag-AgCl electrodes. Therefore to characterize them as best as possible is required. In order to make the characterization reproducible and repetitive, a skin dummy made of agar-agar and a standardized measurement set-up is used in this article. Thus, some dependencies in the skin-electrode interface are described. These dependencies are related to the surface of the textile electrode, the conductive material and the applied pressure. Furthermore, the dependencies on clothing in the skin-textile electrode interface are also analyzed. Thus, based on some parameters such as textile material, width and number of layers, the behavior of the interface made up by the skin, the textile electrode and clothing is depicted.

1. Introduction

As described in [1], using smart textiles for several medical and healthcare applications is becoming more common. One of these applications are textile electrodes. This kind of electrodes can be used to monitoring any physiological parameter in different places such as a hospital, an airline seat or even inside a car. The main reasons could be that the textile electrodes not only are comfortable for the user but also capable of monitoring during long-term because seem not be irritation problems unlike the standard electrodes with hydrogel [2].

However, textile electrodes also show some drawbacks. As there is no hydrogel, the problems related to the lack of contact increase. Furthermore, several aspects of the theory of electrodes and the skin-electrode interface should be revised deeply. Therefore, based on previous studies accomplished by several authors such as [3], [4] y [5], the aim of this paper is to contribute a little bit in the field of the textile electrode characterization.

2. Methods

Based on the measurement set-up described in [3] and shown in figure 1(a), a set of tests are done to check the dependencies on the skin-textile electrode interface impedance, $Z_{skin-textile}$. These dependencies are related to the conductive material of the textile electrode, its contact surface and also the pressure applied on it and, finally, the interaction between the textile electrode and clothing. In order to characterize the textile electrodes, a skin dummy with a conductivity, σ, of $1 \cdot 10^{-3}$ μS$\cdot$cm^{-1}, is made. This conductivity matches the wet skin conductivity at a frequency of

100 kHz, [6]. The mentioned skin dummy is made of a 5% solution of agar in a saline medium, i.e. distilled water with sodium chloride (NaCl). On this agar phantom, two electrodes are placed in a 10-cm distance. Whereas one of these electrodes is used as reference electrode, the other electrode is the textile electrode under study. It is worth mentioning that the reference electrode is made of platinum and its surface is 3 cm x 5 cm. Later, 2-wire measurements are done using the HP4192A device in ratio mode (A(dB) - B(dB)) and a front-end board. The measured impedance could be evaluated as a function of not only the skin-dummy impedance but also the impedances related to the skin-textile electrode interface and the skin-reference electrode interface, i.e. $Z_{meas} = Z_{skin-textile} + Z_{skin} + Z_{skin-ref}$. Therefore, assuming that Z_{skin} and $Z_{skin-ref}$ are non-variant over the tests, the differences between measurements are just related to the skin-textile electrode interface impedance, $Z_{skin-textile}$. However, this method works only if the stability of the skin-dummy impedance can be assured. As shown figure 1(b), the in-frequency and in-time stability of the agar phantom can only be assured at the first six hours because of liquid loss. Thus, there are some options. On one hand, it is possible to make other skin dummy of any material which retain better the water, e.g. graphite instead of sodium chloride as suggested in [7]. On the other hand, it is also possible to use a different skin dummy made of agar for each test. In this case, a new agar phantom is used for each test. Finally, there is an additional option when measuring electrodes with an interposed fabric layer, which consist in wrapping the agar with a thin polyethylene layer.

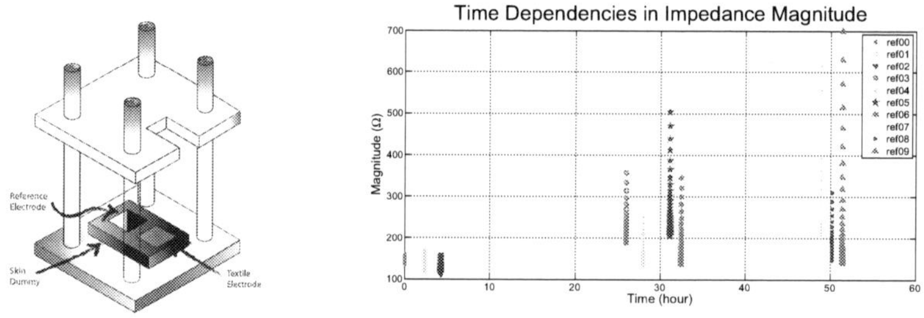

(a) (b)

Figure 1: (a) Set-up of measurements. The skin-dummy made of agar has a volume of 280 cm^3 (14 cm x 10 cm x 2 cm) and the distance between electrodes is 10 cm. (b) In-time stability of the agar phantom.

3. Results & Discussion

3.1. Dependencies on conductive material

To check the dependencies regarding the conductive material which the textile electrode is made of, the set-up is used to measure a set of textile electrodes. These electrodes have the same area, 3.5 cm x 3.5 cm, but different manufacturing, i.e. they are made of different conductive materials or made using different techniques. The mentioned electrodes are made of materials manufactured by A-Jin Electron Company. In this case, the naming of the selected products is W-290-BG, W-290-Silver, M-200-PCNR y WD-270-PCN(ATU). These products are made using different plating methods: Au, Ag, Cu+Ni+Carbon and Ni+Cu+Polyurethane(PU), respectively. An in-depth description of the materials can be looked up in the company website.

As shown figure 2(a), above a frequency of 10 kHz, all values of the measured impedance seem to match regardless of the textile electrode used. On the other hand, for lower frequencies, this fact does not happen. Thus, it appears that at frequencies above 10 kHz, the skin-electrode interface is not dependent on conductive material. For this reason, other factors as the electrode surface and the pressure applied are checked.

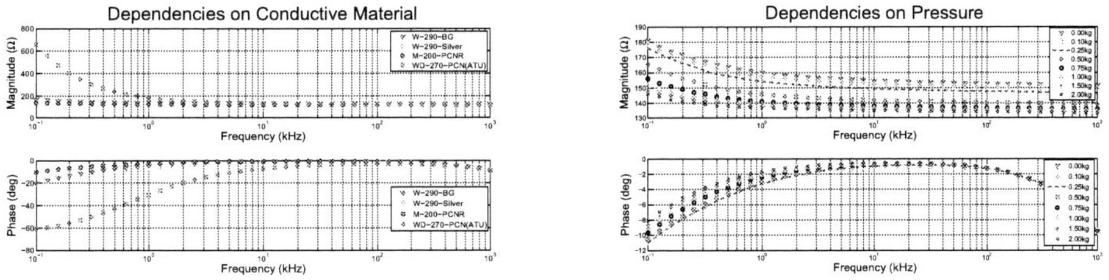

(a) Dependencies on Conductive Material **(b)** Dependencies on Pressure

Figure 2: Magnitude (Upper) and Phase (Bottom) of the measured impedances according to the conductive material,(a), and the pressure, (b).

3.2. Dependencies on pressure

Regarding to above results, the dependencies on pressure are tested. Applying a set of weights over the same textile electrode, the pressure is changed. It is worth mentioning that the electrode is based on the A-Jin Electron product named W-290-GG and its surface is 3.5 cm x 3.5 cm. As expected, the more pressure is applied, the lower the measured impedance, see figure 2(b). The main reason is because of the air gap between the textile electrode and the skin dummy is lower. Thus, whereas the capacitive component of the skin-electrode interface, C_E, increases, the skin-electrode impedance is reduced as can be checked in the equation described in [5], $Z_{electrode} = R_{sc} + R_s/(1 + j\omega R_s C_s) + R_{EL} + R_E/(1 + j\omega R_E C_E)$. Note that for weights above 0.75 kg, the variations between the measured impedances are lower because the air gap has been almost completely minimized.

3.3. Dependencies on electrode surface

To test the dependencies on clothing, four electrodes with the same conductive material but different surfaces are used. The tested surfaces are A, 2A, 3A and 4A, where A is an area of 3.5 cm x 3.5 cm. It is worth mentioning that the textile electrode is made of a material which belongs to Statex Company and is named MedTex P-130.

As shown figure 3(a), as larger the surface of the electrode, lower the measured impedance. This fits theory and the aforementioned equation because as larger the surface, more capacitance, C_E, and as mentioned previously, the skin-electrode impedance, $Z_{electrode}$, becomes lower.

3.4. Dependencies on clothing

From the electrode with a surface 2A, 3.5 cm x 7 cm, and made of the product called MedTex P-130, several measurements are done placing, previously, several clothing between the agar phantom and the textile electrode. The reason is to simulate the situation where textile electrodes are placed in a seat, e.g. in automotive applications. The several clothing tested are a thin piece of 100% cotton, a thick piece of 100% cotton and also a very thick piece of 100% acrylic. Furthermore, placing two layers is also tested, i.e. over the thick piece of cotton, the thin piece of cotton as much as the very thick piece of acrylic are tested.

Of course the behavior becomes purely capacitive, with a measured capacitance which depends inversely on the fabric thickness. The values of the measured capacitance are $2.29 \cdot 10^{-10}$

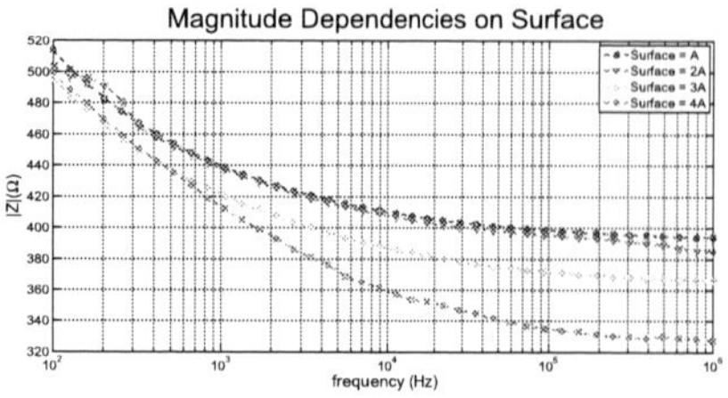

(a) Dependencies on Surface

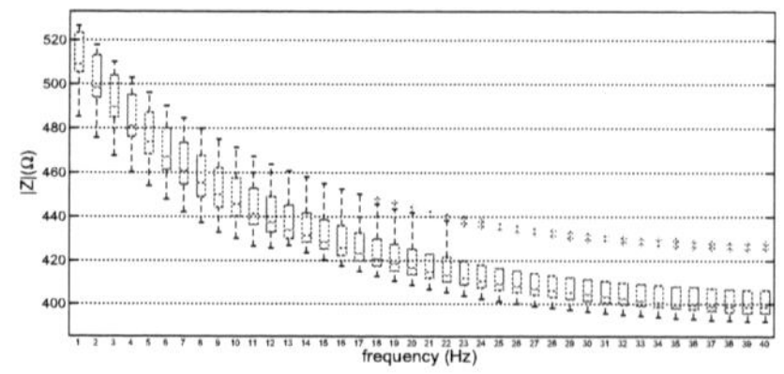

(b) Variations in Magnitude

Figure 3: (a) Dependencies on Electrode Surface. (b) Dependence of the measurement error on electrode mismatch.

F, $9.93 \cdot 10^{-11}$ F and $6.27 \cdot 10^{-11}$ F for the thin cotton piece, the thick cotton and the double layer of cotton pieces, respectively. The result is not only a huge electrode impedance but a high mismatch between electrodes. The dependence of the measurement error on the ratio between the electrode mismatch and the measurement system input impedance is known. Although in most applications this effect is attenuated by the fact that at medium-high frequencies the mismatch is low, in this case remains high at frequencies where the system input impedance has decreased due to the system input capacitance as shown figure 3(b). The solutions are then on the electronics side (driven guards, distributed front-ends, CMFB). All these issues cause that the capacitive component of the skin-electrode interface, C_E, becomes non relevant.

4. Conclusion

By this paper, a further step is expected to have done in the textile electrode characterization. Although a better phantom with low liquid loss is required for a long-term monitoring, characterizing properly textile electrodes with a phantom made of agar seems to be a good choice. The first tests show that at frequencies above 10 kHz, the electrode surface and the applied pressure are parameters more decisive than the conductive material. Furthermore, in case of there is no direct contact between the skin and the textile electrode, injecting at higher frequencies, above 100 kHz, is recommended to minimize the high value of the skin-clothing-electrode interface impedance. Nevertheless, the effects of electrode mismatch can become worse at these frequencies.

References

[1] van Langenhove L 2007 *Smart Textiles for Medicine and Healthcare: Materials, Systems and Applications* (The Textile Institute) ed L van Langenhove (Cambridge: Woodhead Publishing Limited / CRC Press LLC) p 329

[2] Wheelwright C D 1962 *NASA Technical Note D-1082: Physiological Sensors for Use in Project Mercury* [Internet] (Washington D. C.) p 37 [cited 2013 Jan 21] Available from: http://hdl.handle.net/2027/uiug.30112106591503

[3] Beckmann L, Neuhaus C, Medrano G, Jungbecker N, Walter M, Gries T and Leonhardt S 2010 Characterization of textile electrodes and conductors using standardized measurement setups *Physiol. Meas.* **31** 233–47

[4] Rattfält L, Lindén M, Hult P, Berglin L and Ask P 2007 Electrical characteristics of conductive yarns and textile electrodes for medical applications *Med. Bio. Eng. Comp.* **45** 1251–7

[5] Mühlsteff J and Such O 2004 Dry electrodes for monitoring of vital signs in functional textiles *Proc. of the 26th Annual Int. Conf. of the IEEE EMBS* (San Francisco) pp 2212–5

[6] Gabriel C, Peyman A and Grant H E 2009 Electrical conductivity of tissue at frequencies below 1 MHz *Phys. Med. Biol.* **54** 4863–78

[7] Kao T, Saulnier G J, Isaacson D, Szabo T L and Newell J C 2008 A Versatile high-permittivity phantom for EIT *IEEE Transactions on Biomedical Engineering* **55** 2601–7

Impedance as guidance for electrode placement in intraoperative monitoring of nerve fibers

L Díaz Rodríguez[1], M Varga[2], J K Wolter[2], U Pliquett[1]

[1] Institute for Bioprocessing and Analytical Measurement Techniques, Heilbad Heiligenstadt, Germany
[2] Department of Electrical Engineering and Information Technology, Technical University Dresden, Germany

E-mail: laura.diaz@iba-heiligenstadt.de

Abstract. Electrodes for intraoperative monitoring should be reliable and should not disturb the surgeon. Since any wiring could complicate the handling of other medical instruments, new developments of autonomic, wireless electrodes are preferred. This new generation of electrodes will not only stimulate the nerve, but will monitor the action potentials as well. A failure to read nerve potentials may be indicative not just of damaged nerves, but may also result from bad electrode-nerve contact. To resolve this, we have developed electrodes which are equipped with impedance measurement features, facilitating simple connectivity checks. Due to energy constraints, the system works in the time domain using a rectangular voltage wave excitation. The voltage at the electrode is sampled and transmitted via RF link to the host computer. The frequency range covered by the excitation and sampling is between 100 Hz and 50 kHz, which is sufficient not only for detecting contact failure, but also for detecting thick layers of connective tissue between the electrode and the nerve fibre. In either instance, the surgeon can be warned of a bad electrode placement.

1. Introduction

Intraoperative monitoring of nervous tissue or neuromonitoring is a well-established technique used in many neurosurgical interventions which might compromise the proper function of neural structures. Usually, the electrophysiological responses of nerves, namely action potentials, are aroused by a stimulation current, which is directly applied onto the surface of a nerve. Subsequently, the electrical response of the nervous tissue is recorded and interpreted. This kind of examination often ensures integrity of neural tissue, hence reducing the risk of post-operatory deficits.

Neuromonitoring has also been used as a help to localize special structures in the brain, as guidance for electrode placement in Deep Brain Stimulation and for making lesions in specific structures in the treatment of movement disorders and pain therapy [1]. In this regard, neuromonitoring has become an important tool, which assists the neurosurgeon by detecting and alerting from probable nerve damage, thus providing the possibility of performing the necessary corrective measures immediately.

In order to ensure the success of neuromonitoring, the contact between nerve and electrode should also be guaranteed. The loss of action potentials or the inability to stimulate a nerve might not always be caused by nerve injury but also can be caused by a bad electrode-nerve-contact or by the misplacement of the electrode. Both failures can be discarded by using impedance spectroscopy as an additional security measure.

In general, different kinds of tissues possess different electrical properties. The differences are determined by the composition of the tissue, i.e. water content, proteins or lipid content. Specifically blood and brain are good conductors while skin, fat and bones are relatively poor conductors [2]. Although nerve tissue is a good conductor, the isolating property of the myelin sheaths increases the total impedance of myelinated nerves [3]. Keeping in mind these particularities and using impedance spectroscopy different kind of tissues might be distinguished from each other.

In this work we present a novel wireless organic electrode equipped with an impedance measurement system that might be useful for further assistance of electrophysiologists and neurosurgeons in the task of placing or implanting electrodes.

2. Materials and Methods

2.1. Carbon Electrodes
The electrodes employed in our measurements were fabricated and provided by the Institute for electronic packaging at the Technical University of Dresden. One of the pursued goals was to develop non-metal electrodes, which would improve biocompatibility and would enhance electrode-nerve contact. The first electrode-prototype was fabricated with a conductive carbon paste (SD 2843 HAL, black, matt ordered from Lackwerke Peters GmbH + Co KG), which was screen printed on a paper substrate [4]. The total electrode resistance was required to be as low as possible to ensure charge transfer to the nervous tissue and therefore to ensure successful stimulation. Using the conductive carbon paste we measured an electrode resistance between 360 Ω - 450 Ω before and after gas sterilization [4].

2.2. Wireless transmission of power and data
Power and data were delivered using an inductive link between coupled coils driven at 13.56 MHz. The design and optimization of the high-efficiency power amplifier and the power link are described in previous work of the authors [5], where energy transmission is achieved at a distance between coils of 15 cm. For this application, the relative large distance of transmission was required to interfere as less as possible with the maneuver-space of the surgeon during the intervention.

Data was transmitted to the electrode using the amplitude shift keying (ASK) modulation technique. Subsequently the received data was decoded in commands to setup the output signal. Once the excitation signal is enabled, the voltage drop over the tissue is sampled and digitalized to be sent back to the host computer, where further calculations can be realized.

2.3. Time domain impedance measurement
Since the designed electrodes are meant to be used for stimulation of nerve tissue, most of the energy received is going to be required for it. Due to this energy constraint the time domain approach was adopted as the lowest energy consumption alternative. Nevertheless, time domain impedance measurement offers the advantage of a fast measurement making use of a single square signal as excitation wave which is a broad-bandwidth signal. Taking advantage of this fact and using the fast Fourier transformation (FFT) it is possible to obtain amplitude and phase information over several decades of frequency in a short time.

2.3.1. Impedance measurement circuit. A two electrode configuration was used in this work. The square wave excitation signal is generated by internal circuitry and is then applied to the tissue by means of a differential output amplifier as shown in Figure 1 (b). The simplicity of the impedance measurement circuit was a deciding factor for future work in the integration into an ASIC.

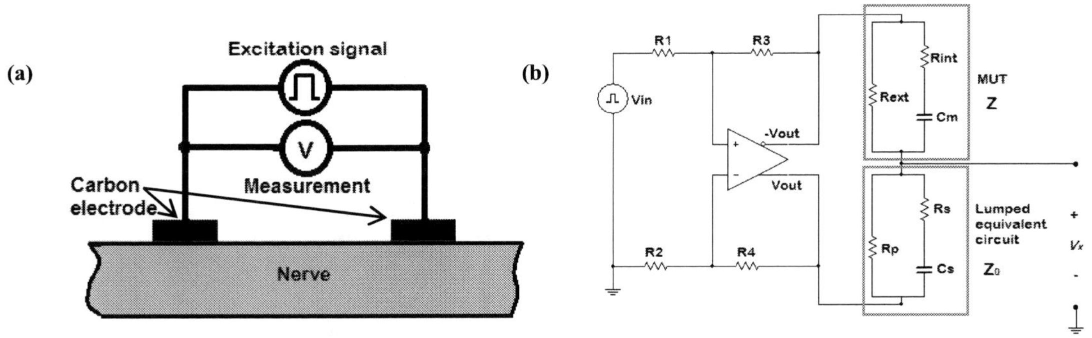

Figure 1 (a) - Two electrode impedance measurement; (b) - Circuit used for time domain spectroscopy.

In the circuit shown in Figure 1(b) the tissue impedance Z is connected in series with the known impedance Z_0, which mimics the plain electrode and is composed by resistors and capacitors. Using Ohm's law it can be shown that the voltage V_x equals:

$$V_x = V_{out}\frac{Z_0-Z}{Z_0+Z} \tag{1}$$

From Equation 1 it can be observed that, when $Z = Z_0$, voltage $V_x = 0$, when $Z \to \infty$, $V_x = -V_{out}$, in the same way, when $Z_0 \to \infty$, $V_x = V_{out}$. Rearranging Equation 1 we can calculate the tissue impedance Z:

$$Z = Z_0\frac{V_{out}-V_x}{V_{out}+V_x} \tag{2}$$

2.4. Material under test: Nervous and connective tissue

To validate the impedance measurement system, fresh nervous tissue was obtained from the spinal cord of a swine. The external membranes surrounding the spinal cord were peeled off to expose the white matter. Afterwards, the spinal cord was cut in several small pieces and these were conserved between measurements in Ringer's solution. White matter is rich in myelinated nerve fibers; therefore it is a suitable tissue for the ex-vivo experiments. The connective tissue was obtained from the hypodermis of the swine. The hypodermis is rich in fibroblasts and elastin, feature that mimics the connective tissue around the nerves.

3. Ex-vivo experiments

In order to ensure tissue vitality during the whole experiment, the impedance measurement was realized within 2 hours after excision of spinal cord and hypodermis. The carbon electrodes were coated with a thin sheath of silicon, keeping a non-isolated area for measurement of 2 x 2 mm^2 (Figure 2(a)). The purpose of the silicon isolation is to imitate implantation conditions, in which the electrode surface is only in contact with nerves of around 1-3 mm total diameter. Figure 2(b) shows a segment of spinal cord on top of the carbon electrode. This arrangement ensures that only a small part of the white matter containing the myelinated nerve fibers gets in contact with the electrode. Since the contact between electrode and tissue is built only by smoothly positioning the electrode on top or below the nerve, neither pressure nor fixation was applied to the spinal cord or to the hypodermis.

(a)

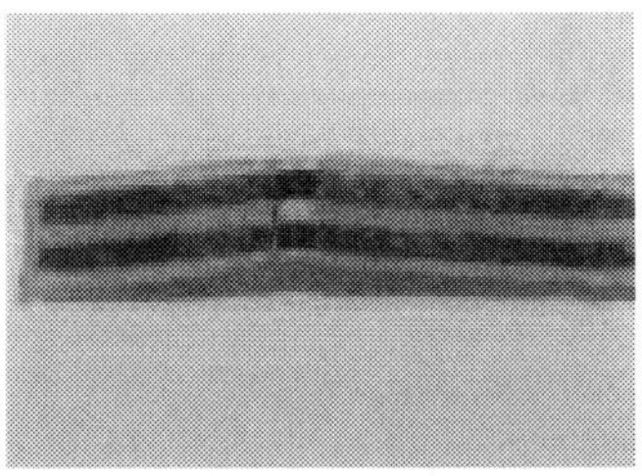

(b)

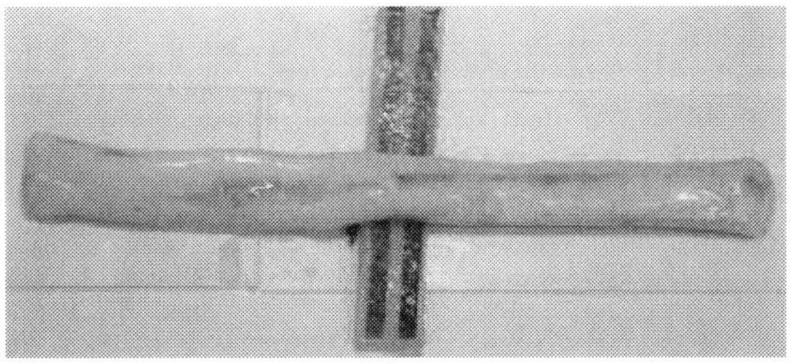

Figure 2 (a) - Carbon electrode coated with silicon, (b) - Swine spinal cord on top of the same silicon-coated carbon electrode.

4. Results and Discussion

All the measurements were done with a 1 kHz square wave. Figure 3 shows the results obtained from eight different measurements made with nerve and connective tissue. These results show the preliminary characterization of the carbon electrode. From Figure 3 (a), it can be observed that the distinction of both tissues is achieved at rather higher frequencies than expected. It can be seen that the clearest difference between tissues impedances is found in the range between 50 kHz and 1 MHz and the error bars show that different tissue can be clearly distinguished.

Figure 3 (b) shows the impedance phase of the measurements. In comparison with the standard error obtained from the impedance absolute value, the phase standard error between measurements is relatively high. An explanation of this phenomenon can be seen in Figure 3(c). Nerve tissue and connective tissue conserve a dispersion trajectory, but the capacitive contribution to the impedance is shifted in frequency. Consequently, the evaluation of multiple independent parameters yields a higher significance of the tissue discrimination.

In conclusion, although more experiments are needed, the developed impedance measurement system is a potential tool, which can be used for monitoring the electrode-nerve contact in neuromonitoring techniques.

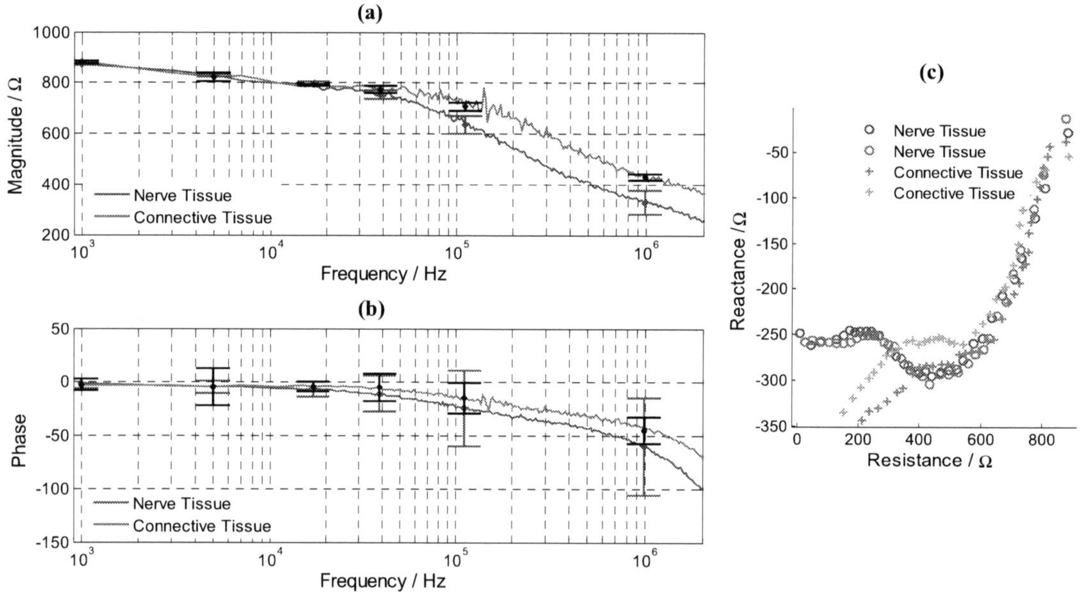

Figure 3 (a) - Impedance measurement magnitude from eight different measurements; (b) - Impedance phase of connective and nerve tissue; (c) - Nyquist plot of the frequency response of both tissues (only four measurements are shown).

References

[1] Moller AR 2011, Basis of intraoperative neurophysiological Monitoring; in: *Intraoperative Neurophysiological Monitoring*, 3rd ed. (New York: Springer Science), pp 9-22.

[2] Miklavcic D, Pavselj N and Hart F X 2006 Electric properties of Tissues; in: *Wiley Encyclopedia of Biomedical Engineering*, (John Wiley & Sons, Inc.) pp 2-12.

[3] Grimnes S and Martinsen OG 2008, Passive tissue electrical properties in: *Bioimpedace and Bioelectricity Basics*, 2nd ed. (London: Elsevier) pp 102-37.

[4] Varga M and Wolter J K 2013 Novel self-folding electrode for neural stimulation and recording *IEEE XXXIII Int. Conf. on Electronics and Nanotechnology* Submitted.

[5] Díaz Rodríguez L and Pliquett U 2012 Design and implementation of an inductive power link for intracutaneous medical implants *Dresdner Medizintechnik Symposium* 222-5.

New reusable elastomer electrodes for assessing body composition

M-V Moreno[1], L. Chaset[2], PA Bittner[2], C. Barthod[3], M. Passard[3]

1 Research Department, BioparHom, Savoie Technolac, BP 238, 73374 Le Bourget-du-Lac Cedex, France

2 University of Technology of Compiegne, 60201 Compiegne, Cedex, France

3 Laboratory SYMME, University of Savoie, Annecy, France

E-mail: marie-valerie.moreno@bioparhom.com

Abstract. The development of telemedicine requires finding solutions of reusable electrodes for use in patients' homes. The objective of this study is to evaluate the relevance of reusable elastomer electrodes for measuring body composition. We measured a population of healthy Caucasian (n = 17). A measurement was made with a reference device, the Xitron®, associated with AgCl Gel electrodes (Gel) and another measurement with a multifrequency impedancemeter Z-Metrix® associated with reusable elastomer electrodes (Elast). We obtained a low variability with an average error of repeatability of 0.39% for Re and 0.32% for Rinf. There is a non significantly difference (P T-test >0.1) about 200 ml between extracellular water Ve measured with Gel and Elast in supine and in standing position. For total body water Vt, we note a non significantly difference (P T-test > 0.1) about 100 ml and 2.2 l respectively in supine and standing position. The results give low dispersion, with R^2 superior to 0.90, with a 1.5 % maximal error between Gel and Elast on Ve in standing position. It looks possible, taking a few precautions, using elastomer electrodes for assessing body composition.

1. Introduction

The development of telemedicine requires finding solutions of reusable electrodes for use in patients' homes and abandoning gradually, for these applications, the usual gel electrodes. Many authors have worked on this issue for the development of textile electrodes [1] or metal electrodes [2]. The problem comes particularly from the interface between electrodes with skin, more than the intrinsic properties of conductivity of the electrodes [3]. The objective of this study is to evaluate the relevance of conductive elastomer electrodes for assessing body composition.

2. Material and methods

The first step of the approach was to select materials with mechanical and electrical properties suitable for the production of flexible electrodes. Then, we optimized the geometry and the relative positioning of the electrodes used for injection and reception to be able to obtain relevant measurement for the operating frequencies. Particular attention has been paid to the contact resistance between the electrode and the skin. According to various studies, we chose a strip of commercial soft conductive elastomer for the electrodes, which can be simply sizes according to the expected use. The electrode thus obtained has an excellent electrical conductivity, including for lower tensions. To implement measurement on the electrode, we only need to plug a 2 mm male rod in the conducting channel provided for this purpose. The problem was to fix the electrodes at different points of the human body for impedances measurement, intending that people can be lying down or moving as in the case of measurement on athletes. For that purpose, we designed a box with a location for the electrode and a path for the Velcro that will keep the box and the electrodes together (Figure 1). In the box are collected all connectors (injection and reception) to be linked to the impedancemeter. In the future, the embedded electronics for wireless transmissions will be implemented in the same box. Four different types of boxes were made thanks using rapid prototyping material. Moreover, these electrodes have the advantage of being inexpensive, easy to clean and flexible, that permits to adapt perfectly the measuring system to all skin types and morphologies.

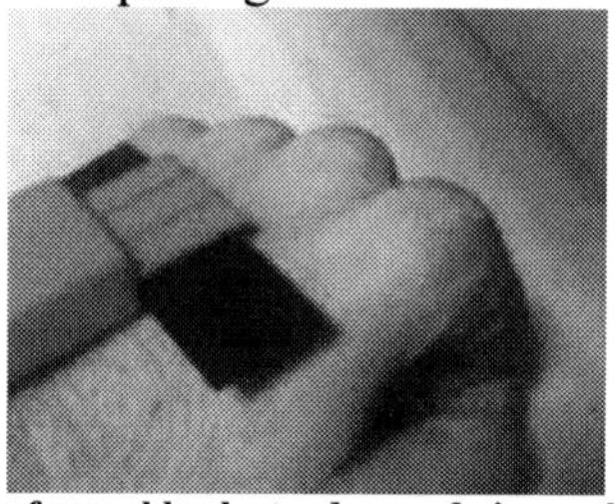

Figure 1 pictures of reusable electrodes made in conductive elastomer

A campaign of measurement has been conducted on a population of healthy Caucasian, n = 17, mean age 38.5 ± 19.0 years (from 11 to 62 years old) and average Body Mass Index 25.6 ± 3.9 kg.m-^{2}. Two devices are implemented: a reference device, the Xitron ® Hydra 4200 (Xitron Tech, San Diego) and its AgCl Gel electrodes (Gel) on the one hand and the multifrequency impedancemeter (Z-Metrix ®, Bioparhom, France) [4, 5] associated with the reusable elastomer electrodes (Elast) described before, on the other hand. Measurements were made in supine position, with the reference device – corresponding results are denoted with "x, n" subscript, and in standing and supine position with the Z-Metrix – corresponding results are denoted with "z" subscript. Before each measurement, the skin has been cleaned with water. Extracellular resistance Re and infinite Rinf, defined according to the Cole model [6] and corrected as Jaffrin et al. method [7], were calculated, as well as water volumes associated Ve [8] and Vt [9]. This was done for measurements realized for the whole body and for body segments (arm, trunk and leg).

3. Results and Discussion

First, we tested the impact of the rest time of the electrodes on the electrical characteristics without cleaning the skin with water or alcohol.

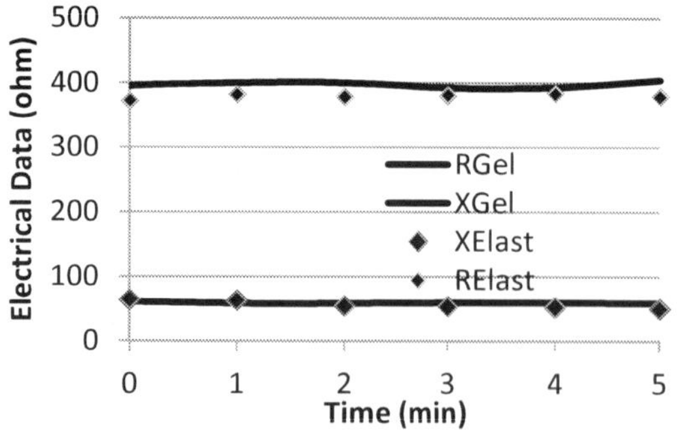

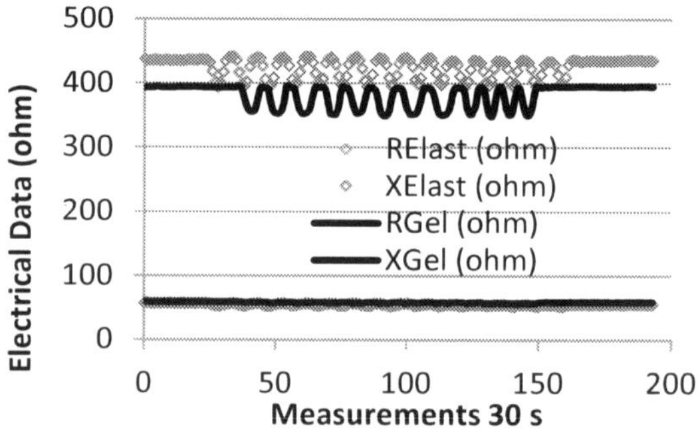

Figure 2 impact of the rest time of the electrodes on the electrical data Figure 3 impact of the body artifacts on the electrical data

Figure 2 shows an example of a monitoring during 5 min of resistance R and reactance X at 50 kHz. Gel have a low variability (1.2% and 2.0% of R on X) while Elast varies about 1.0% on R and 10.4% on X, from 64.9 to 50.4 ohm, then reaching a constant. Presumably the elastomer increases heating and moisture at the interface which reduces the reactance to the same value as that of Gel. These reusable electrodes are more reliable if you wait for about 5 minutes before doing a measurement.

Moreover, we wanted to test the impact of the body movement artifacts on the electrical characteristics. Figure 3 shows an example of a monitoring of a walk during 30 seconds after waiting 5 minutes of rest of the electrodes. Walking is an average simulation of various types of body movement artifacts (movement of arm, trunk and legs). The results were similar between Gel (an average variability of 5 ohms on X and 50 ohms on R) and Elast (2 ohms on X and 50 ohms on R). Presumably the decrease of the resistance comes from the movement of fluids in the lifting arm.

Whether Gel or Elast, the noted arm's movements generate significant variations on resistance. Beyond the interface, in telemedicine measurements, it is necessary to detect a time of rest to perform reliable measurements of bioimpedance. After the electrical qualification of Elast, we are interested in the results on body composition.

We obtain a low variability with an average error of repeatability of 0.39% for Re and 0.32% for Rinf. We obtain similar results for Re, Rinf, Ve and Vt. There is no significantly difference between Gel and Elast. The results of the paired Student T-Test for Rinf (for example), measured for the right part of the body, are reported in Table 1.

Segment	Rinf	Student T-Test
Total Body	Rinfx/Rinfz supine	0,877 Ns
	Rinfx/Rinfz standing	0,924 Ns
Leg	Rinfx/Rinfz supine	0,852 Ns
	Rinfx/Rinfz standing	0,814 Ns
Trunc	Rinfx/Rinfz supine	0,484 Ns
	Rinfx/Rinfz standing	0,585 Ns
Arm	Rinfx/Rinfz supine	0,886 Ns
	Rinfx/Rinfz standing	0,905 Ns

Table 1 Results of Paired Student T-test between Rinfx obtained with Gel and Rinfz obtained with Elas

We can observe that there is no significantly difference (P T-Test >0.1) between the extracellular resistance and the infinite resistance measured with gel electrodes and those obtained with elastomer electrodes, in supine and standing position, for total body of corporal segment.

	Gel		Elast Supine		Elast Standing	
	Veb	Vtn	Vez	Vtz	Vez	Vtz
Mean ± Sd	16.4 ± 2.9	40.2 ± 8.3	16.6 ± 3.0	40.1 ± 8.3	16.6 ± 3.2	37.8 ± 5.9
Student T-Test	Ref.	Ref.	0.603 Ns	0.635 Ns	0.362 Ns	0.877 Ns

Table 2 Results of mean, Sd and Paired Student T-test for Ve and Vt with Gel and with reusable elastomer electrodes

Analysing data in table 2, it appears that there is no significantly difference (P T-test >0.1) about 200 mL between extracellular water Ve measured with Gel and with Elast electrodes in supine and in standing position. For total body water Vt, we note a non significantly difference (P T-test > 0.1) about 100 mL and 2.2 L respectively in supine and standing position. We can suppose that the accumulation of fluids in standing position in legs increase the difference between the two devices Xitron (used as reference in supine position only) and Z-Metrix (used in supine and standing position). We can suppose that it's not related with the type of electrodes

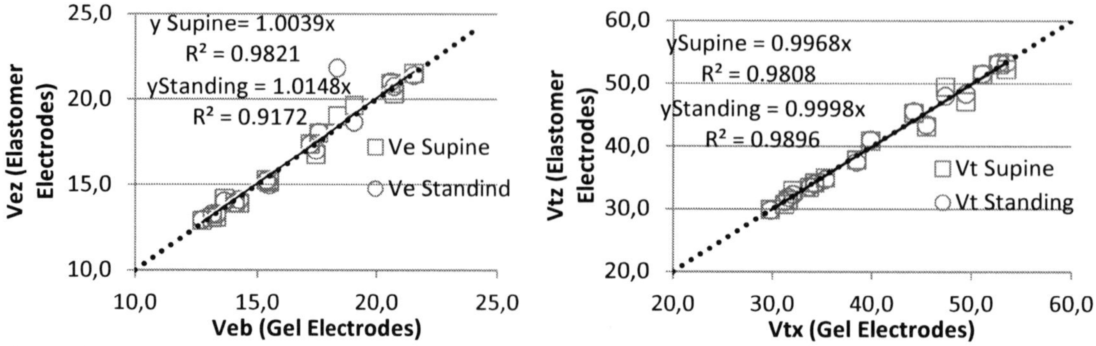

Figure 4 and 5 Graphics of linear correlation (dotted line is identity) for extracellular and total bodywater obtained with Gel and Elast electrodes

Figures 4 and 5 plots what we describe in the tables. The results give low R ² dispersion, superior to 0.90, with a 1.5 % maximal error between Gel and Elastomer on Ve in standing position.

4. Conclusion

Its looks possible, taking a few precautions (time of rest of the electrodes, detection of the movement of the body), using reusable elastomer conductive electrodes for assessing body composition. It is necessary finalizing a more ergonomic box and completely flexible. In addition, the interface must also be validated on subjects with skin with specific characteristics (elderly, athletes, children ...).

References

[1] Marquez JC 2009 Textile electrodes in electrical bioimpedance measurements - a comparison with conventional Ag/AgCl electrodes *Int Conf Proc IEEE Eng Med Biol Soc*, 1: p. 4816-9.

[2] Ragheb T, Geddes LA 1990 Electrical properties of metallic electrodes, *MBEC*, **28**, 182-186.

[3] Jossinet J, McAdams ET 1991 The skin-electrode interface impedance, *Innov. Tech. boil. Med.* , **12**, N° spe 1.

[4] Moreno MV, Ribbe E, Rebeyrol J, Vannicatte A, Krief L Evaluation of a multifrequency impedancemeter by biphotonic Densitometry, measuring independent tissue indices, in supine and standing position. Comparison with skin folds, *(In review)*).

[5] Moreno MV, Ribbe E, Rebeyrol J, Vannicatte A, Krief L Evaluation of a new impedancemeter to independently access extracellular, intracellular and total body water volumes. Application to the measurement of hydration**.** (*In review).*

[6] Grimnes S, Martinsen OG 2005 Cole electrical Impedance Model – A critique and an alternative, *IEEE Transactions on biomedical engineering*, **52**, n°1.

[7] Jaffrin MY, Moreno MV 2008 Measurements of total body water with a foot-to-foot impedancemeter. *Medical Engineering & Physics*, **30**, 483–489.

[8] De Lorenzo A, Andreoli A, Matthie JR, Withers P 1997 Predicting body cell mass with bio-impedance by using theoretical methods: a technological review. *J Appl Physiol*, **82**, 1542–58.

[9] Jaffrin MY, Fenech M, Moreno MV, Kieffer R 2006 Total body water measurements by a modification of the bioimpedance spectroscopy method. *MBEC* **44**, 873-882.

Journal of Physics: Conference Series **434** (2013) 012027 doi:10.1088/1742-6596/434/1/012027

Studying the Performance of Conductive Polymer Films as Textile Electrodes for Electrical Bioimpedance Measurements

F.J. Cunico[1,2], J.C. Marquez[1,3], H. Hilke[1], M. Skrifvars[1] and F. Seoane[1,3]

[1] School of Engineering, University of Borås, 501 90, Borås, Sweden
[2] Department of Electrical Engineering, Santa Catarina State University, 89223-100, Joinville, Brazil
[3] School of Technology and Health, Royal Institute of Technology, 141 52 Huddinge, Sweden

E-mail: fernando.seoane@hb.se

Abstract. With the goal of finding novel biocompatible materials suitable to replace silver in the manufacturing of textile electrodes for medical applications of electrical bioimpedance spectroscopy, three different polymeric materials have been investigated. Films have been prepared from different polymeric materials and custom bracelets have been confectioned with them. Tetrapolar total right side electrical bioimpedance spectroscopy (EBIS) measurements have been performed with polymer and with standard gel electrodes. The performance of the polymer films was compared against the performance of the gel electrodes. The results indicated that only the polypropylene 1380 could produce EBIS measurements but remarkably tainted with high frequency artefacts. The influence of the electrode mismatch, stray capacitances and large electrode polarization impedance are unclear and they need to be clarified with further studies. If sensorized garments could be made with such biocompatible polymeric materials the burden of considering textrodes class III devices could be avoided.

1. Introduction

The advance on conductive textile materials and the progress on textile-enabled measurement instrumentation integrating textiles and electronics producing sensorized garments are fostering the development of electrical bioimpedance (EBI) technology for personalized healthcare monitoring [1].

Textile electrodes or textrodes are rising as an application of conductive fabrics but so far mainly silver-based fabrics are the raw material to make textile electrodes. Due to economical issues like the high and highly variable price of silver, environmental issues regarding the disposal of such silver-textile material and medical device certification issues like class III classification under rule 13 [2] other alternatives are being investigated.

Certain polymer compounds present conductive properties that might allow the material to function as electrode in some cases. Some of these polymer compounds allow to be manufactured as flexible films. Such films are very textile-friendly and can be integrated in textile manufacturing processes easily.

In this work the measurement performance of three different polymeric materials as bioimpedance electrodes has been evaluated. Wrist and ankle straps were custom made with 3 different polymer films: Two films made from polypropylene and one film was made from a thermoplastic elastomer. The textrode performance was compared with tetrapolar wrist-to-ankle measurements performed with commercially available gel-based electrodes used as reference.

2. Materials and Methods

The work presented in this paper has been executed during the realization of a final degree work. In such final degree work several measurements have been performed with different skin preparation conditions and with two different bioimpedance spectrometers. This paper focuses on a limited set of experiments to assess on the feasibility of using these film materials as bioimpedance electrodes.

2.1. Electrical Bioimpedance Spectroscopy Measurements

The EBI spectroscopy (EBIS) tetrapolar measurements were obtained with the spectrometer SFB7 manufactured by ImpediMed Ltd in the frequency range of 3.096 to 999 kHz. A set of 20 measurements was performed for each electrode type with the electrodes placed on the right wrist, and right ankle with the volunteer lying in a decubitus supinus position.

2.2. Gel Reference Electrode

Electrolyte gel-based Ag/AgCl, Red Dot repositionable electrodes manufactured by 3M were used to perform the wrist-to-ankle EBIS tetrapolar use as reference for the comparison.

2.3. Polymeric Raw Materials

For this work three different kinds of polymers containing carbon black and with different electrical characteristics were selected: polypropylene PP 1380 and PP 1386 as well as thermoplastic elastomer TPE 1503. The polymers were in the granulated state and with specific process of melting they were transformed into films. The polymer materials were supplied by Premix ThermoPlastics Oy, Finland.

The PP1380 was selected after a selection process that involved preliminary EBIS measurements with several textrode materials. The PP1386 and the TPU1503 were selected because of their mechanical properties, which could contribute to facilitate the textrode manufacturing process.

Table 1. Volume and surface resistivity for the polymeric materials.

Polymer Material	Volume resistivity	Surface Resistivity
PP1380	<=5 Ωcm	<=100 Ω
PP1386	<=1000 Ωcm	<=10000 Ω
TPU1503	<=300Ωcm	<=10000 Ω

2.4. Film-Textrode Bracelets

Straps were confectioned as bracelets with a customized length for the wrist and the ankle, a pair for each site. The bracelets have been confectioned with a triple layered structure using the polymer films on the inner side, woven fabric in the outer side and foam in between, as shown in Figure 1.

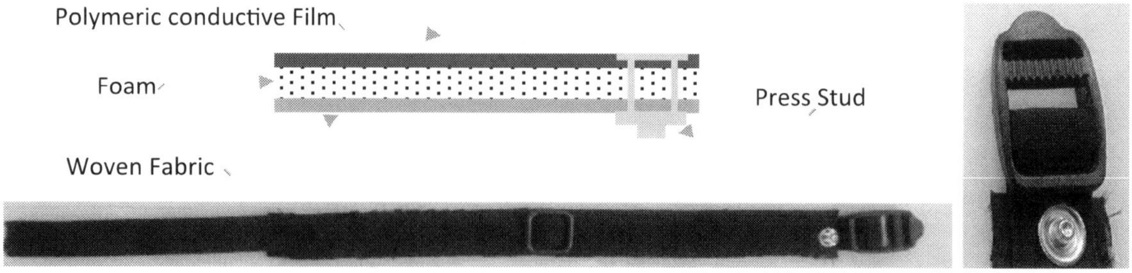

Figure1. Complete Confectioned Strap with the layered structure, the buckle and the press-stud.

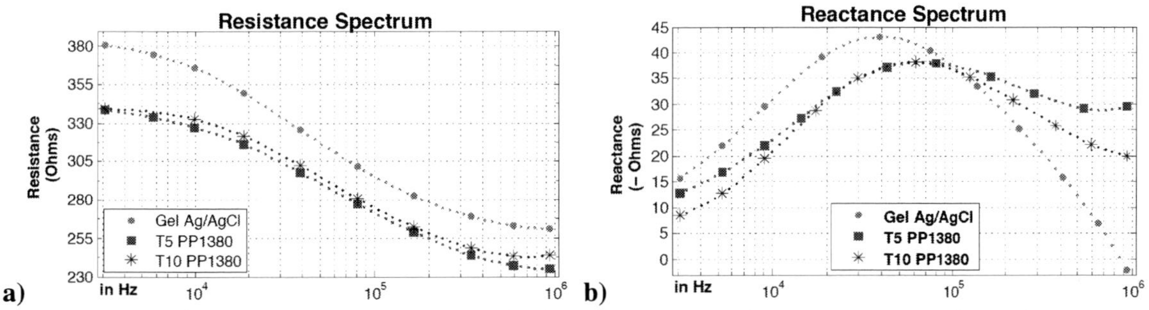

Figure 2. Impedance spectrum obtained with the reference gel electrodes and the PP1380 textrode-bracelets. Resistance spectrum in a), and reactance spectrum in b).

The straps are 2.5 cm width with an adjustable length using buckle system to fasten the bracelets around the limbs. A metallic press-stud is used to connect with the alligator clip attached to the measuring leads of the SFB7 spectrometer. The buckle and the press-stud are also shown in Figure 1.

2.5. EBIS comparison Measurements

After the textrode-bracelets were placed, wrist-to-ankle EBIS tetrapolar measurements were performed at two different times T5 and T10, after 5 minutes and 10 minutes correspondently. Both EBIS measurements were compared with the obtained with gel electrodes.

3. Experimental Results

For the textrode-bracelet made of PP 1386 and TPE 1503 no EBI spectrum was produced, the obtained impedance measurements obtained presented an oscillating resistance spectrum with negative values consistent with equipment saturation.

For the polymeric film PP 1380, the reactance spectrum obtained after 5 minutes and 10 minutes is shown in Figure 2b). In addition of the obvious difference in magnitude observed across the whole spectral spectrum, there is also a difference observed in the frequency behavior. Such difference can be observed easily in the reactance spectrum plotted in Figure 2a). In such plot, it is possible to see the different value for the characteristic frequency between the PP1380 measurements and the gel electrodes.

Both in the reactance spectrum plotted in Figure 2b) and the susceptance spectrum plotted in Figure 3, it is possible to observe an increase on the magnitude for both the T5 and the T10 measurement. Looking at the susceptance spectrum the increase at high frequencies suggests a second dispersion almost overlapping with the one at medium frequencies for T5. The observed capacitive behavior decreases for the T10 measurements. From both plots it is also possible to observe that the characteristic frequency for the reactance and the susceptance decreases with the attenuation of the deviation observed at high frequencies, *i.e. the value for the characteristic frequencies for T10 measurements are lower than for T5 measurements*.

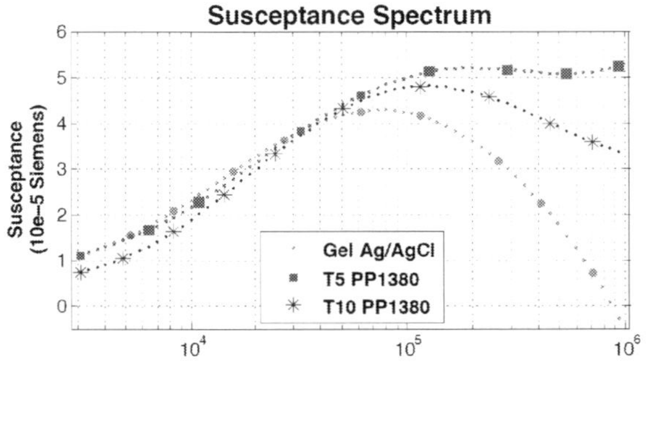

Figure 3. Susceptance impedance spectrum obtained with the reference gel electrodes and the PP1380 textrode-bracelets.

4. Discussion

The resutls presented here correspond with a single test but the resutls agree with several other test measurement performed with a complete different spectrometer and reported in the report of the final degree thesis.

The obtained results showed that only the textrode bracelets made with PP 1380 films produced EBI spectral measurements. Apparently when measurements were taken with any of the other two polymer electrode materials, the SFB7 spectrometer saturated suggesting that the skin-electrode polarization impedance, Z_{ep}, of the electrodes were larger than the load dynamic range of the meter. That the PP1380 exhibit a smaller Z_{ep} than the PP1386 and TPU1503 could be expected since the surface and volume resistivities of the PP1380 is significantly lower than the other two polymeric materials.

When using the PP 1380 films as textile electrode the obtained impedance spectra presents a certain deviation of the reactance and the susceptance at high frequencies, that is consistent with the presence of parasitic capacitances, large values of electrode polarization impedance or the combination of both. This deviation was observed already during the material selection process and it is consistent with the results presented in [3-5], which deal with poor skin-electrode interfaces causing large Z_{ep} and the presence of parasitic capacitances.

The difference observed in the value of the resistance at all frequencies is consistent with the use of electrodes around the limb and it is consistent with EBIS measurements obtained with textile electrode straps previously reported in [6].

5. Conclusions

Despite that the effects observed on the performed EBIS measurement have been observed in more than one test and in more than a single subject during the material selection process aforementioned (*unpublished*) and also EBIS measurement have shown similar deviations in works studying the use textile electrodes [3,6], and measurement artifacts caused by stray capacitances [5] and large Z_{ep} values [4], the fact that the experimental results only show measurements from a single subject and that very important issues like electrode-skin polarization impedance have not been reported remains.

Therefore the only conclusions that can be drawn from these preliminary results is that PP 1380 films have the potential to be used for the confection of sensorized garments to perform EBIS measurements and that more detailed and deep studies must be performed to confirm the suggested potential. Such studies should be aimed to fully characterize the textrode measurement performance, and its main measurement characteristics and dependencies, prioritizing the study of the skin-electrode interface.

Considering textile electrodes containing silver as class III medical devices under rule 13 bring up significant challenge for medical device manufacturer that might hinder the rising of textile-based EBI applications, thus alternative materials to silver for manufacturing textile electrodes must be found.

References

[1] Reiter H, Muehlsteff J and Sipila A 2011 *In IEEE Proc. EMBC2011* pp 3270-3273.
[2] Manual on borderline and classification in the community regulatory framework for medical devices, version 1.13 (10-2012).
[3] Marquez JC, Seoane F, Valimaki E and Lindecrantz K 2009 *In Proc. EMBC2009*. pp.4816-9,
[4] Buendí R, Bogónez-Franco P, Nescolarde L and Seoane F 2012 *Med Eng Phys* **34**(7) pp 1024-8
[5] Buendía R, Gil-Pita and Seoane F 2010 *Meas. Sci. Technol.* **21** 115802
[6] Marquez JC, Seoane F, and Lindecrantz K. 2013 *Eur J Clin Nutr.* **67** (S1):S22-S7.

Journal of Physics: Conference Series **434** (2013) 012028 doi:10.1088/1742-6596/434/1/012028

Cross sensitivity of different electrode types for impedance plethysmography under motion conditions

Sebastian Guttke, Matthias Laukner and Patrick Weber

HTWK Leipzig, Wächterstrasse 13, 04107 Leipzig, Germany

Email: sebastian.guttke@eit.htwk-leipzig.de

Abstract. The non-invasive measurement of biological signals is prone to failure when the proband is in motion. In clinical practice disposable wet adhesive Ag/AgCl electrodes are used for ECG measurements to minimize failure caused by motion. For sportive action a chest strap with conductive rubber electrodes is widespread for continuous monitoring of the heart rate. Complaints about poor comfort and friction at the skin caused by the chest strap leaded to new developments with textile electrodes. The dynamic properties of electrodes and the coupling to the skin are important to optimize signal quality. We compare new textile with standard ECG electrodes and tested their suitability to the heart-rate measurement with impedance plethysmography under motion conditions.

1. Introduction and modeling

In order to estimate the human state of health or fitness in sports and medicine one measures a few biological signals. Under motion conditions a non-invasive measurement suffers from big artefacts that can be 5 to 50 times greater in magnitude than a biosignal without motion (Fig. 1). The ECG and the bioimpedance measurement are well suited to determine the pulse rate.

The pulse wave changes the cross section of the arteries, which can be measured as a change in the transfer impedance ΔZ (compare 2008, Grimnes [1] pg. 351). The rest pulse signal in Fig. 1 is hardly recognizable against the artefacts. We compare alternative new textile electrodes from TITV Greiz e.V. to standard ECG-electrodes Vitrode C and Siemens 7269509 under motion conditions. The transfer impedance measurements are carried out at the forearm.

The use of the four-electrode configuration (tetrapolar electrode setup) provides in comparison to the two-electrode configuration great advantages, since only the transfer impedance of the tissue between the potential reading electrodes is determined. In the forearm, the magnitude of tissue transfer impedance at frequencies between f = 50...150 kHz is about $\mathrm{abs}(\mathbf{Z})_{\mathrm{Tiss}} \approx 30\ \Omega$ at a 5 cm distance and the pulse-dependent change is $\Delta\mathrm{abs}(\mathbf{Z})_{\mathrm{Tiss}} \approx 3...30\ \mathrm{m}\Omega$. This requires a high resolution impedance meter. The measured values were confirmed with impedance plethysmograph RheoScreen compact. According to our own analysis, the spectrum of the impedance pulse signal is only relevant up to a resolution of f = 20 Hz. Therefore, in the continuous pulse measurement, a sampling time of five milliseconds is sufficient and one millisecond is more than required.

An electrical model of the transition between the electrode and subcutaneous tissue [3] or [4] can be seen in Fig. 2. A liquid junction potential should be added if different electrolytes or one electrolyte with domains of different ion concentrations is present. As a simplification, the series resistances and series capacitors can be combined. The active sources are connected in series and can be summarized.

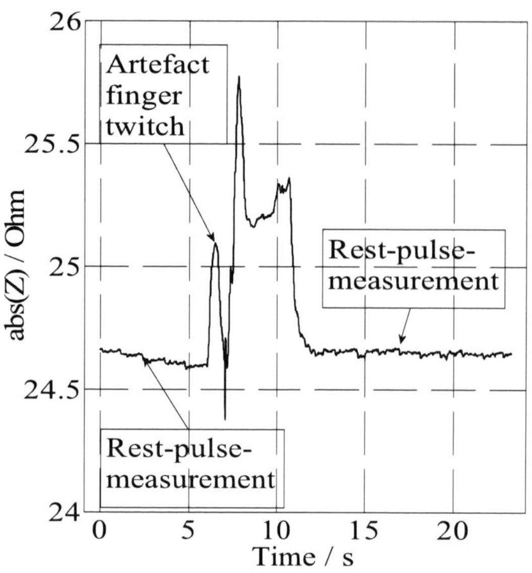

Figure 1. Transfer Impedance pulse function measured with Rheoscreen at the fore arm with motion artefacts.

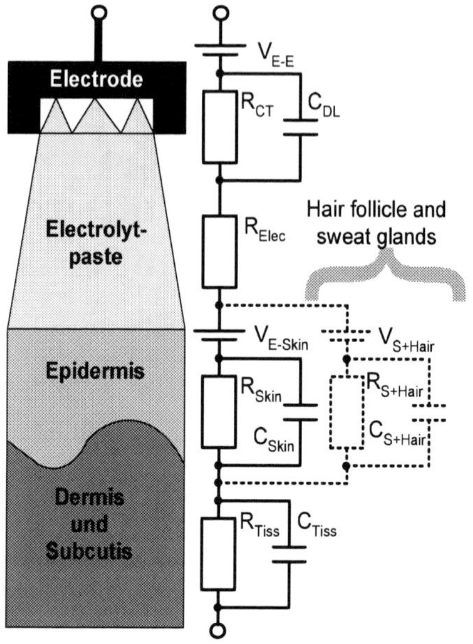

Figure 2. Skin-electrode interface and its electrical equivalent circuit, from [3] or [4].

The components of the model are V_{E-E} the electrode polarization-potential, R_{CT} the charge transfer-resistance, C_{DL} the double layer-capacity, R_{Elec} the electrolyte-resistance, C_{Skin}, R_{Skin}, V_{E-Skin} the skin equivalent circuit elements, C_{S-Hair}, R_{S-Hair}, V_{S-Hair} for sweat and hair effects, R_{Tiss} the tissue-resistance (which is dominant for $\mathbf{Z}_{Tiss}$) and C_{Tiss} the tissue-capacity. The electrolyte paste and the electrode create a galvanic cell, which induces a polarization potential (Nernst potential [1]).

$$V_{E-E} = V_1^0 + (RT/nF) \cdot \ln(a_1/a_2) \tag{1}$$

V_1^0 is the standard electrode potential with respect to the hydrogen reference electrode, R the molar gas constant, T the absolute temperature, n the number of valence electrons in the metal, F the Faraday constant, a_1 and a_2 are the activities of the ions (1 for oxidation and 2 for reduction). Under ideal conditions between two equal electrodes in the same electrolyte V_{E-E} would be zero. In practice the electrode surface always reacts to some extent with the surrounding electrolyte (e.g. sweat) and various potentials will show [2].

2. Measurement setup and calculation of the transfer impedance

For measurement of polarization potentials we used a Metrahit 30M. To suppress drifts the input impedance was reduced to Zi = 1 MΩ. The electrolyte was a saturated KCl solution at T = 23 °C. The Ag/AgCl reference electrode was a SE11 from Meinsberg, which contains saturated KCl solution.

An IVIUMStat from Ivium Technologies was used to measure all impedance spectra in two- and for-electrode configuration at the forearm without motion. An alternating current of I_{eff} = 1 mA from f = 1 kHz…500 kHz was applied. The distance between the potential reading electrodes was 5 cm.

For continuous pulse measurement under motion conditions we used a setup shown in Fig. 3 with a fixed frequency of f = 100 kHz. The factor F = 0.97 mA/V is the transfer function of the current source. The control voltage V_{ST} is also used to calculate the transfer impedance **Z** at the forearm.

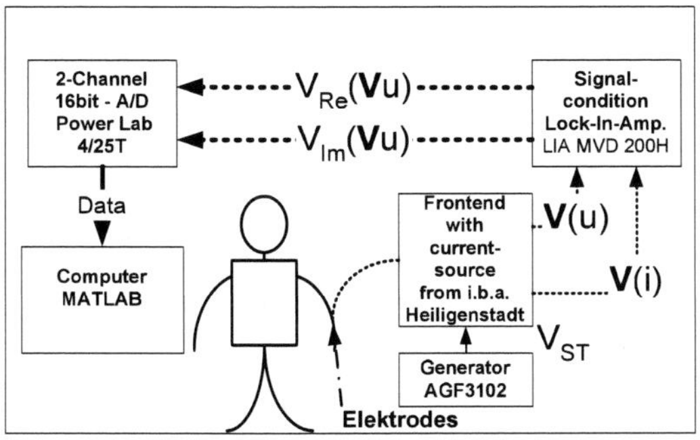

Figure 3. Setup for continuous pulse measurement.

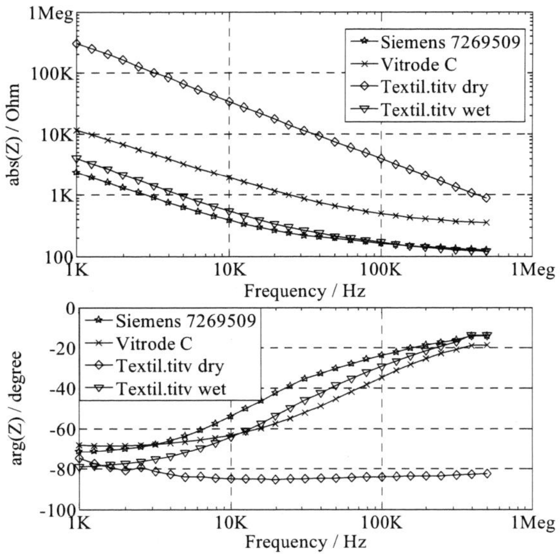

Figure 4. Impedance spectra without movement in two-electrode configuration.

$$\mathbf{Z}=abs(\mathbf{Z})\cdot\exp(i\varphi_Z)=\frac{\sqrt{\mathrm{Re}\{\mathbf{V}_U\}^2+\mathrm{Im}\{\mathbf{V}_U\}^2}}{V_{ST}\cdot F}\cdot\arctan\left(\mathrm{Im}\{\mathbf{V}_U\}/\mathrm{Re}\{\mathbf{V}_U\}\right) \tag{2}$$

The electrodes were fixed on the forearm and connected to the frontend with a 0.5m cable (AWG32) to avoid artefacts caused by the cable. Three measurements with motion (finger twitch, fist-making, jogging with the arm) were performed.

3. Results

3.1. Polarization potentials, resistance of conductive fabric and dry yarn

The voltages of Vitrode C and the textile electrode in Table 1 are in the same range. The Siemens electrode generates a smaller voltage difference. In equation (1) no pressure dependence of the polarization potential is described. In 2008, Grimnes [1] pg. 22 is published, that the pressure influences the viscosity and conductivity of the electrolyte. For skin contact, the coupling conditions are complex and the potential difference between two equal electrodes varies with different contact pressures.

Table 1. Polarization potentials measured against a Meinsberg SE11 .

Electrode	Mechanical parameters	Nernst potential
Round Ag/AgCl ecg electrode Siemens 7269509	D = 3.0 cm; A = 7.0 cm²	5 mV
Nihon Kohden Vitrode C, adhesive wet Ag/AgCl	D ≈ 1,0 cm, A = 0,8 cm²	16…20 mV
Textile electrode, TITV Greiz e.V.	D ≈ 3.0 cm; A ≈ 7 cm²	16 mV

The resistance (tetrapolar measurement with Metrahit 30M) of one meter dry yarn was R = (16.8 ± 0.1) Ω and of two meter R = (33.7 ± 0.2) Ω (titv: conductive yarn 235 dtex, R' < 20 $\Omega\cdot m^{-1}$). In addition we measured the resistance of dry conductive fabric over 3cm electrode diameter. Along the yarn we measured R = 6 mΩ and across the value was R = 30 mΩ.

3.2. Measurement of impedance spectra without movement in two- and four-electrode configuration
With the four-electrode configuration and equation (2) we determined the absolute value of transfer impedance abs(**Z**) = 30…50 Ω (Fig. 5) . In two-electrode configuration the electrode impedances at f = 500 kHz are not yet zero (Fig. 4). There is abs(**Z**) ≈ 100 Ω. The textile electrode and the Siemens electrode have comparable impedance curves. The Vitrode C has a smaller contact surface area and higher impedance values. The dry textile electrode works only as a capacitive electrode. The impedance at the frequency f = 1 kHz is high (above 200 kΩ).

3.3. Four electrode configuration under motion conditions
It is derived positive that all types of electrodes except the dry textile electrodes behave similarly in all failure modes. That speaks in particular for the use of damp textile electrodes as an alternative to metal electrodes. The disturbances are on average all at least the factor 10 to 100 greater than the changes in the measured pulse signal. If the rigid metal electrodes have a large area e.g. A = 7 cm², the coupling to the tissue is not well defined and reproducible because of tendons and uneven surface of the skin. The density of injected current could be increased at spots of higher contact pressure. In this respect the flexible textile electrodes can be advantageous. They adapt themselves to the surface of the skin and allow a more homogeneous injected current density.

We conjecture that the measured impedance leaps are mainly caused by a change of electric flow field in the forearm. A light push or pull on each of the electrodes causes a change of **Z**. Movement at the current electrodes causes impedance jumps which also indicates a change in the flow field in the tissue by modified coupling of the electrodes.

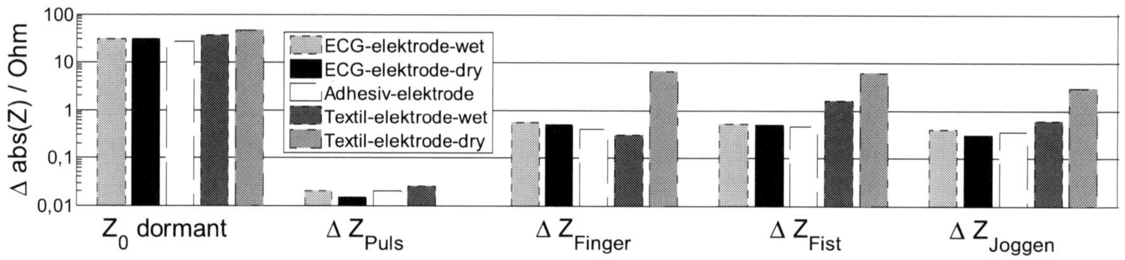

Figure 5. Comparision of transfer impedances and motion artefacts of different electrodes.

Wet textile electrodes are a good alternative to classic ECG-electrodes. Their advantage being that the pressure to the body is homogenously distributed. It could also be advantageous when small textile electrodes are selectively coupled directly above an artery. The impedance values would have of course, higher values. Instead of using electrode gel or electrolyte, we prefer an electrode that captures the body generated moisture.

References:
[1] Grimnes S and Martinsen O 2008 *Bioimpedance and Bioelectricity Basics* (Academic Press)
[2] Rattfalt L , Chedid M and Hult P 2007 *Electrical Properties of Textile Electrodes* (Proc. of the 29th Annual, International Conf. of the IEEE EMBS Internationale, Lyon, France 2007)
[3] Baba A and Burke M J 2008 *Measurement of the electrical properties of ungelled ECG electrodes* (INTERNATIONAL JOURNAL OF BIOLOGY AND BIOMEDICAL ENGINEERING, Issue 3, Volume 2) pp.89-97
[4] Meyer-Waarden K 1985 *Bioelektrische Signale und ihre Ableitverfahren.* (Stuttgart/ New York: Schattauer)
[5] Gnewuch K 2012 *Textile Elektrodensysteme für sensorische und aktuatorische Applikationen.* Textilforschungsinstitut Thüringen-Vogtland e.V., Greiz

Improved extraction of information in bioimpedance measurements

Mart Min, Toivo Paavle

Ehitajate tee 5, Tallinn, 19086, Estonia

toivo@elin.ttu.ee

Abstract. A wideband bioimpedance measurement method is proposed, which can enhance the interpretation of measurement results due to the improved resolution of monitoring. On the other hand, the corresponding measurement system uses a binary chirp waveform for excitation signal, which simplifies the signal processing hardware and does not require sophisticated software. It is shown that the binary chirp excitation has some essential advantages compared to its counterpart – the maximum length sequence (MLS) excitation.

1. Introduction

Often, the theoretical study of passive electrical properties of a biological object (bioimpedance of the living tissue or cells) is substituted by the analysis of its electrical equivalent (EBI). The EBI represents a definite set of resistive and capacitive components, changes of which (dynamics) can be described by a frequency dependent vector $\mathbf{Z}(j\omega)=|\mathbf{Z}(\omega)|\exp(j\Phi_z(\omega))$, where $\omega=2\pi f$.

This vector characterizes the EBI and approximates the state and properties of the respective object under study. Mostly, the dynamic features of this vector appear within a frequency range from some kHz to several MHz (β-region) depending on changes in the structural properties of the biological object [1]. To obtain information about the object in a wide frequency range, it is appropriate to study the impedance $\mathbf{Z}$ in frequency domain applying the Fourier Transform $\mathcal{F}(\cdot)$ to both – the output response voltage $V_z(t)$, and the excitation (stimulus) current $I_{exc}(t)$:

$$\mathbf{Z}(j\omega)=\mathcal{F}(V_z(t))\,/\,\mathcal{F}(I_{exc}(t)) \qquad (1)$$

To maximize the information from the study, the bandwidth of the excitation should coincide with the frequency range changes of the object as much as possible. Different broadband signals, scalable in frequency (multisines, chirps, maximum length sequences (MLS)) have been in use for excitation [2]. In this work, we emphasize the advantages of chirp waveforms. The chirp excitation signal can be scaled easily as in frequency, as well in time domain. Besides, the chirps have almost flat or tunable amplitude spectrum over the whole generated range, so increasing the accuracy of measurement.

However, the generation of a perfect sine-wave chirp is complicated. To reduce the complexity of required software and hardware, a binary excitation should be preferred. A good choice is the binary chirp or pseudo-chirp [3], which can be a successful rival to the MLS excitation.

To improve the quality of monitoring in the frequency domain, we suggest a 2-channel measurement method and the corresponding system. It incorporates measurement and reference channels together with spectral analysis of both, the response signal from measurement channel and the reference signal from the other channel. Finally, analysis of the phase spectrum instead of the amplitude one is proposed [4, 5].

2. Binary chirp vs. MLS

Chirp-type signals can be classified by the rule of changing their instantaneous frequency. The basic class covers so-called power-law chirps, the frequency of which changes by a power function.

A sine-wave power-law chirp of the power of n (the simplest linear chirp has n=1) and with amplitude A can be described mathematically as

$$V_{\mathrm{ch}}(t) = A\sin\left(2\pi\left(f_0 t + \beta t^{n+1}/(n+1)\right)\right), \tag{2}$$

in which $\beta = B_{\mathrm{exc}}/T_{\mathrm{ch}}$ characterizes the rate of frequency variation, $B_{\mathrm{exc}}=f_{\mathrm{fin}}-f_0$ is the excitation bandwidth between the final f_{fin} and initial f_0 frequencies, and T_{ch} is the duration of chirp pulse.

A binary chirp with two values $\pm A$, expressed mathematically as $V_{\mathrm{bin}}(t)=A\cdot\mathrm{sign}\{V_{\mathrm{ch}}(t)\}$, is illustrated in Figure 1a. For comparison, the Figure 1b depicts a fragment of a similar 11-stage MLS signal. In Figure 1c, the power spectral density (PSD) of a binary chirp and the corresponding 11-stage MLS with equal energies and the same duration $T_{\mathrm{ch}} = T_{\mathrm{MLS}} = (2^{11}-1)/f_{\mathrm{clk}}$=20.47 ms are compared at the clock frequency $f_{\mathrm{clk}} = f_{\mathrm{fin}}$=100 kHz. For the contrast, the PSD of a sine-wave chirp is shown, too.

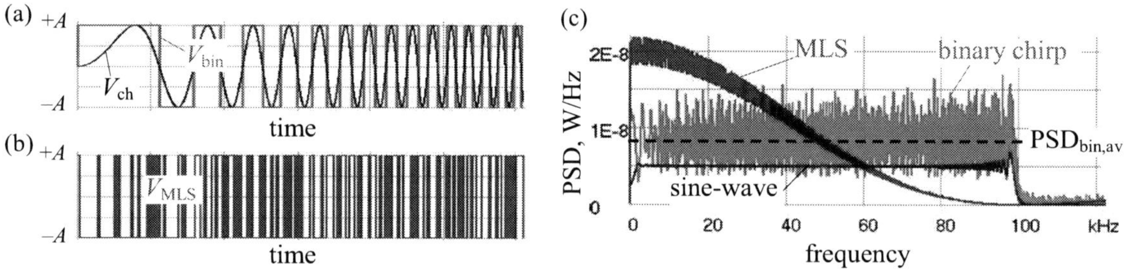

Figure 1. (a) Waveforms of sine-wave (V_{ch}) and binary (V_{bin}) chirps; (b) waveform of the 11-stage MLS signal; (c) power spectra of sine-wave and binary chirps, and the MLS with equal durations.

The envelope of the MLS spectrum is the *sinc*-function, which reduces to zero at $f=kf_{\mathrm{clk}}$ (k=1, 2, 3, etc.). It means that the main bandwidth $B_{\mathrm{exc}}=f_{\mathrm{clk}}$ of MLS is not entirely usable in measurements. On the contrary, the PSD of chirps covers the whole generated bandwidth B_{exc}. Thereby, the average $\mathrm{PSD}_{\mathrm{bin,av}}$ (see Figure 1b) of the binary chirp exceeds the PSD of the respective sinusoidal chirp $(4/\pi)^2$ times and is equal to the PSD of MLS at the frequency of $f_{\mathrm{fin}}/2$, approximately.

The drawback of binary chirp excitation is the intensively fluctuating amplitude spectrum. Fortunately, it is recouped by several benefits:

- The generation of binary chirps is simple, only a bit more complicated as the MLS. It can be done by following the current phase $\theta(t)=\int\omega(t)\mathrm{d}t$ and counting the samples. If f_0=0 and n=1, then the current phase in (2) expresses as $\theta(t)=\pi\beta t^2$. The polarity must alternate at every $\theta(t)=k\pi$ (k=0, 1, 2, …), i.e, at every integer value of βt^2. Consequently, the polarity must be changed at every i_k-th sample, where $i_k=\mathrm{round}\{f_{\mathrm{samp}}(k/\beta)^{1/2}\}$ and f_{samp} is the sampling frequency.
- Alike the MLS, the crest factor of binary chirps is 1, which means the utmost power or energy is available. It accompanies with sufficiently high energy efficiency δ_E (the percentage of generated energy within the excitation bandwidth B_{exc}), which is theoretically $\delta_E \approx 0.853$ [6].

3. Structure of the measurement system

The fluctuating spectrum complicates interpreting of the measurement results especially in the case, where the amplitude spectrum is the only source of information. Therefore, it is reasonable to catch the spectral information from the phase spectrum, which is more insensitive to the fluctuations within the bandwidth B_{exc}. Moreover, the phase spectrum does not depend on the signal levels. However, to get the phase spectrum we need a reference to evaluate the phase shift.

Furthermore, often the actual value of the impedance **Z** is not of great importance. We are more interested in observing **Z** variations caused by the changes and events inside the object. For that reason, we consider the two-channel measurement system, wherein one of the channels includes predefined reference impedance $\mathbf{Z}_{ref}$, which is approximately equal to the **Z** and is composed using *a priori* information about it. Such a structure behaves as the matched filtering, showing zero output of the phase spectrum if $\mathbf{Z}=\mathbf{Z}_{ref}$, and enabling to detect tiny deviations in the phase spectrum. Thereby, in the reference channel we can use an equivalent analog RC-circuit as well as a set of numerical values (a digital model) obtained from a previously stored response V_z.

This task can be accomplished by a simple structure with two FFT-blocks for transforming the response and reference signals V_z and V_{ref} separately (Figure 2a). The complex numbers from the both of blocks are divided for obtaining the spectra of the relative amplitudes $|\mathbf{Z}(f)|/|\mathbf{Z}_{ref}(f)|$ and phase differences $\Delta\Phi_{zr}(f)=\Phi_z(f)-\Phi_{ref}(f)$.

Another solution in Figure 2b implements the cross-correlation procedure with the signals V_z and V_{ref}. The product r_{zr} of the cross-correlation function (CCF) includes the information about the amplitude and phase of the both signals, and only a single FFT-block is used for producing a cross-power spectrum $|P_{zr}(f)|$ and the spectrum of phase differences $\Delta\Phi_{zr}(f)$. This method is somewhat slower due to the calculations of CCF, but enables to obtain some additional information on declinations of **Z** from $\mathbf{Z}_{ref}$ throughout the CCF (see Figure 3) and has a better noise immunity.

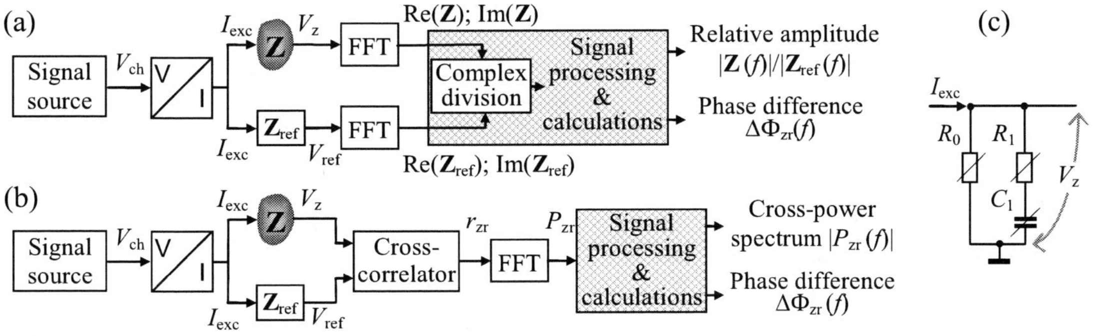

Figure 2. The structures of measurement system: (a) with two separated FFT channels; (b) with cross-correlation calculus; (c) general model of the bioimpedance used in simulations and calculations.

4. Simulation results

The following simulation results are obtained by the model developed in accordance with the circuit in Figure 2b. A simple 3-element RC-circuit in Figure 2c (R_0=1 kΩ, R_1=200 Ω and C_1= 20 nF) represents the reference impedance $\mathbf{Z}_{ref}$, which is approximating of an "unknown" object **Z** as well as.

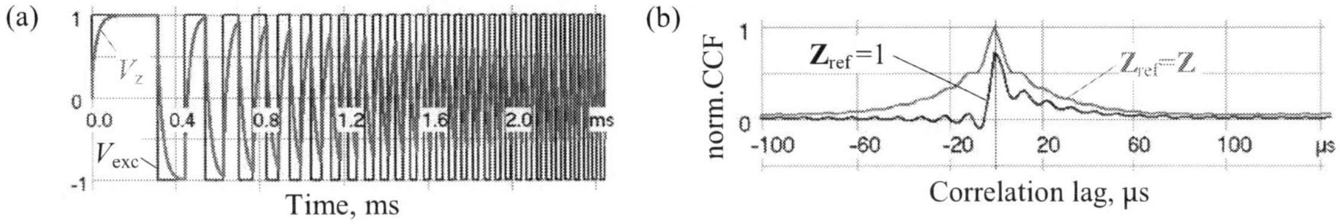

Figure 3. (a) Normalized voltage signals; (b) normalized cross-correlation functions: $\mathbf{Z}_{ref}$=1 and $\mathbf{Z}_{ref}$=**Z**.

Figure 3a shows initial cycles of the binary chirp V_{exc} (f_0=0, f_{fin}=100 kHz, T_{ch}=10 ms) and the respective response V_z to this excitation. Figure 3b shows the CCF in two extreme cases: $\mathbf{Z}=\mathbf{Z}_{ref}$ and $\mathbf{Z}_{ref}$=1.

Figure 4a demonstrates the main result of the proposed measurement method – the possibility to distinguish very small variations of components of **Z** through the spectra $\Delta\Phi_{zr}(f)$. The concurrent simulations proved that the MLS excitation cannot ensure such clearness of the phase spectrum as the binary chirp excitation does – see Figure 4b.

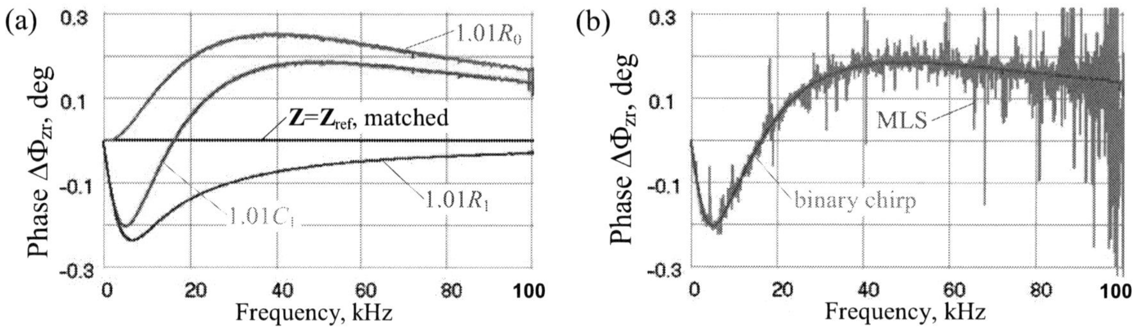

Figure 4. Spectra of phase declinations at the +1% increments of **Z** components in the case of following excitations: (a) binary chirp; (b) binary chirp and pseudo-random MLS at the changing C_1.

5. Conclusion

The proposed method conjoins different means to improve the quality of bioimpedance measurements.

First, the use of binary waveform chirps for excitation simplifies the required hardware and software essentially in the practical developments, but assures the measurement in a wide frequency band at the same time. In addition, simulations pointed out some advantages as independent scaling in both time and frequency domain when compared with another popular binary signal – maximum length sequence (MLS).

Second, by focusing on the phase spectrum, the binary chirp waveform enables to avoid the disturbing impact caused by the higher harmonic components inherent to all the binary excitations.

Third, the using of two-channel measurement system with the reference impedance in one of two channels (designed on the basis of *a priori* knowledge), increases the accuracy of measurement substantially and permits to detect tiny changes in the structure and/or parameters of the unknown bioimpedance to be measured.

Acknowledgments

This research was supported by the European Union through the European Regional Development Fund in frames of the research center CEBE and the competence center ELIKO, partly also by Estonian target financed project SF0140061s12.

References

[1] Grimnes S and Martinsen Ø G 2008 *Bioimpedance and Bioelectricity Basics* (Elsevier-Academic Press).

[2] Gawad S, Sun T, Green N G and Morgan H Impedance spectroscopy using maximum length sequences: application to single cell analysis 2007 *Rev. of Scient. Instruments* **78** 054301-1.

[3] Pollakowski M and Ermert H Chirp signal matching and signal power optimization in pulse echo mode ultrasonic nondestructive testing 1994 *IEEE Trans. Ultrason., Ferroelect. Freq. Contr.*, **41**(5) 655-9.

[4] Trebbels D, Woelki D and Zengerle R High precision phase measurement technique for cell impedance spectroscopy 2010 *J. Phys.: Conf. Ser.* **224** 012159 1-4.

[5] Min M, Paavle T and Ojarand J, Time-frequency analysis of biological matter using short-time chirp excitation 2011 *Proc. Europ. Conf. on Circuit Theory and Design ECCTD'11* 585-8.

[6] Min M, Paavle T, Annus P and Land R 2009 Rectangular wave excitation in wideband bioimpedance spectroscopy *Proc. IEEE 4th Int. Workshop on Medical Measurements and Applications (MeMeA2009)* 268-71.

Optimisation of multisine waveform for bio-impedance spectroscopy

Jaan Ojarand[1], Paul Annus[1,2] and Mart Min[2]

[1]ELIKO Competence Centre, Mäealuse 2/1, 12618 Tallinn, Estonia
[2]Thomas Johann Seebeck Department of Electronics, Tallinn University of Technology, Ehitajate tee 5, 10919 Tallinn, Estonia

E-mail: jaan.ojarand@eliko.ee

Abstract. The multisine excitation is widely used in impedance measurements to retain the advantages of the sine wave, while reducing the measurement time. Adding up sine waves increases the amplitude of the excitation signal, but, for the linearity assumption to be valid, the overall amplitude of the signal needs to be kept low. Thus, the crest factor (CF) of the excitation signal must be minimized. A novel empirical method for minimization of the CF is presented in this paper. As in case of other known methods, the computed CF may be guaranteed to be only a local minimum. However, a systematic variation of initial parameters, which is possible due to the sparing algorithm, ensures the CF value, very close or equal to the global minimum. A brief analysis of the calculation errors and comparison with the results from other sources is provided. In the DSP applications, initial phases of the waveforms must match the grid of sample points if separate signal sources are used to generate the components. An equation for calculation of the indexes of the sample points corresponding to the optimal initial phases is provided, and the time discreteness caused relative error of CF is illustrated.

1. Introduction

Sinusoidal waveforms can be readily measured using ordinary signal processing solutions. Use of the multisine excitation which is the sum of sinusoids, allows us retain the advantages of the sine wave, while reducing the measurement time. A multisine with k frequency components can be expressed as

$$S_{exc}(t) = \sum_{i=1}^{i=k} A_i \cdot \sin(2\pi f_i + \Phi_i) \qquad (1)$$

where A_i is the amplitude of the i-th component, and Φ_i is its initial phase.

Adding up sine waves increases the excitation amplitude, however, for the linearity reasons, the overall amplitude of the signal must be kept low. On the other hand, a better signal-to-noise ratio (SNR) is also desired. To satisfy these contradictory requirements the amplitude extremes for a given energy of the signal should be minimized. The ratio of the peak value of a signal to its root mean square (RMS) level is called as crest-factor (CF). RMS value of a multisine signal is independent of initial phases Φ_i of separate signal components, and thus, the CF minimization problem reduces to finding of the best set of phases, which ensures the minimum peak value of the multisine signal.

Much research has been done in this direction [1-7]. A simple formula proposed by Schroeder [1] produces better CF for the consecutive harmonic frequencies. However, in case of non-consecutive frequency distributions, e.g. with 10 prime harmonics (pos. 3 in table 1) the Shroeder formula gives CF = 2.24, when our optimization method gives CF = 1.48. Our calculation with the same frequency distribution but 1×10^4 random phase sets gave a CF approximately 20% above the minimum value.

Journal of Physics: Conference Series **434** (2013) 012030 doi:10.1088/1742-6596/434/1/012030

An exhaustive search of all possible phase combinations provides the best results, however, it is practical only for spectra having small numbers of harmonics k since the number of calculation steps n increases in power of k-1: $n = (360 / \Delta\Phi)^{k-1}$, where $\Delta\Phi$ is the phase step. More sophisticated iterative methods [2-5] and genetic algorithms (GA) [6] or a hybrid of GA with nonlinear iteration [7] provide a solution for large k.

Minimization of the CF of multisine is not straightforward, it can be more formally stated as the minimization of a non-linear function and many local minima are present in the parameter space during the optimization process. There is no guarantee that iterative algorithms converge to a global minimum. GA has less probability to get stuck in the local minima since it uses a population of candidate solutions [6]. However, the converging to a global minimum is still not guaranteed and exploring various local optima takes more time.

2. A novel empirical method for the minimization of crest factor CF

Examination of amplitude minimums of the multisine signals shows that several equal minimums exist if the combinations of initial phases are changing within the full range from 0 to 360 degrees.

It was found that a phase combination corresponding to the amplitude minimum obtained with large phase steps, will appear also in the next pattern received with smaller phase steps (a fractal phenomenon). These observations led to the hypothesis that the initial phases of the frequency components that produce the minimal amplitude of the sum signal could be found using sequential searches with limited number of phase steps. Computer simulations with different frequency distributions proved that a recursive algorithm with 2 phase steps in each iteration could be used for finding the near-to-optimal set of initial phases of the included sine waves $\Phi_{opt}(i)$. Figure 2 gives a flowchart describing the algorithm.

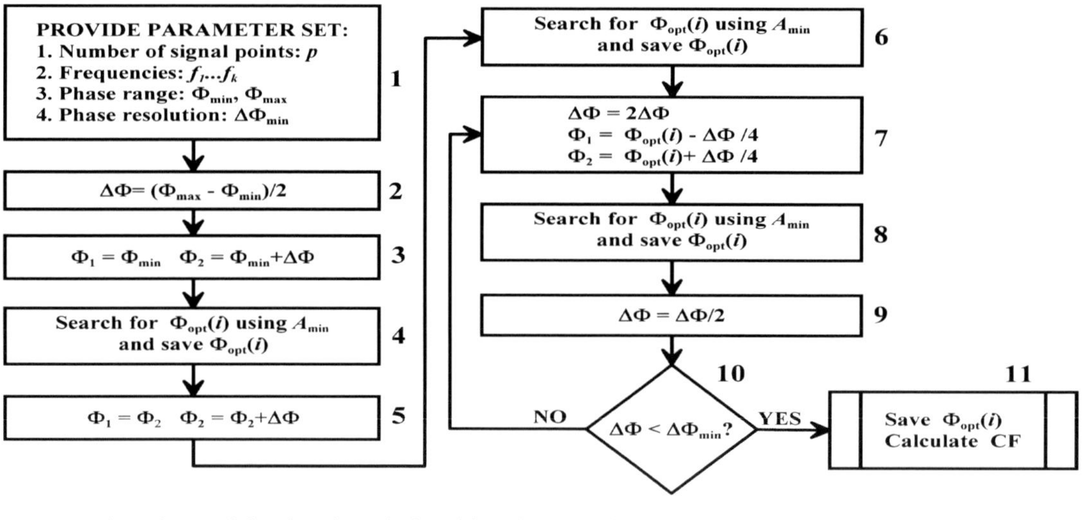

Figure 1. Flowchart of the developed algorithm for calculation of optimal phases of sine waves $\Phi_{opt}(i)$ of a multisine. $A_{min} = \mathrm{Min}\sum_{i=1}^{i=k} |A_i| \sin(2\pi f_i + \Phi_j)$ where $j = 1$ or 2.

As with the other known methods, the computed phases can only be guaranteed to be local optima since the result of the calculations depends also on a frequency distribution and the initial values of the parameters Φ_{min} and Φ_{max}. The performance of the new method was further tested by a variation of these initial parametres. The results of two tests are presented in figure 2. Φ_{min} was fixed to 0° and Φ_{max} was varied with a step of 1°. The tests showed that choosing a proper value for Φ_{max} allows detecting CF_{min} with relative error below 1%. In the case of randomly chosen Φ_{max}, the relative error is typically less than 10 %.

For the new method, the number of calculation steps n^* for searching of optimal initial phases is expressed as

$$n^* = \log_2\left(\frac{|\Phi_{max} - \Phi_{min}|}{\Delta\Phi}\right)2^k + 1 \qquad (2)$$

The results obtained with the new method were compared with other published data (table1).

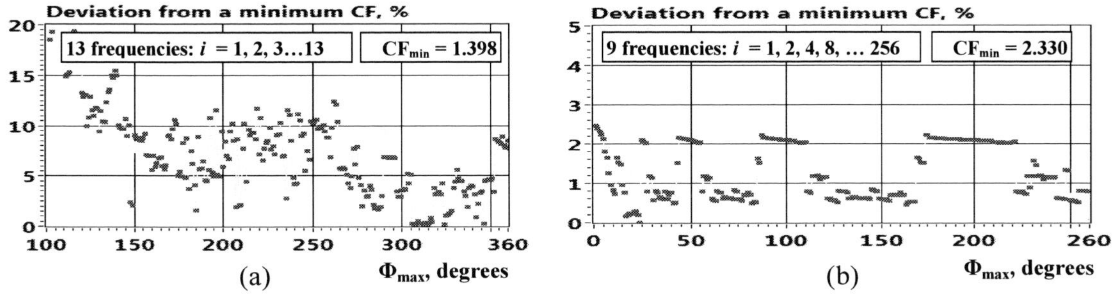

Figure 2. Deviation of the optimized CF from its minimum value for the different frequency distributions of multisine waves (a) and (b).

Table 1. Comparison of calculated CF with reference data, CF_{ref}

Pos.	List of frequency components i	k	CF	CF_{ref}	Reference
1	5, 21, 37, 53, 69, 85, 101	8	1.59	1.64	[7]
2	1, 2, 3, 4, 5...10	10	1.40	1.51	[5]
3	3, 5, 7, 11, 13, 17, 19, 23, 29, 31	10	1.50	1.57	[5]
4	10, 12, 15, 18, 22, 27, 33, 40, 48, 58, 70, 84, 100	13	2.00	1.96	[4]
5	1, 2, 3, 4, 5...15	15	1.40	1.80	[5]
6	3, 5, 7, 11, 13, 17, 19, 23, 29, 31, 37, 41, 43, 47, 43	15	1.65	1.77	[5]
7	1, 2, 3, 4, 5...16	16	1.41	1.42	[4]

In figure 4, the optimization results for some typical frequency distributions are shown. To note in figure 4, the CF of equally distributed multisine components are close to the value of $\sqrt{2}$ if $k > 3$. In case of very sparse distribution, as a decimal logarithmic one, CF tend to the worst case of $\sqrt{2k}$. For all other frequency distributions, the optimized CF falls in between these two limiting values. It was found that the CF of optimized multisine with the binary logarithmic frequency distribution tends to the value of $\pi\sqrt[3]{k}/2\sqrt{2}$.

3. Accuracy of CF calculations and recalculation of optimal phases

Despite the method used, the accuracy of the calculated CF values depends on the sampling rate. It was found that the number of samples per period of the highest frequency component must be at least 10, in order to achieve relative error near 1 %. Close to the Nyquist sampling-rate, the relative error could increase over 20 %. The error depends also on the number of frequency components and their distribution as illustrated in Fig. 5.

Optimized initial phases of the frequency components with some phase range and resolution can be recalculated to another suitable phase range (e.g., from 0 to 360°) and resolution. Figure 6 gives a flowchart describing this algorithm. An example: optimized multisine (pos. 2 in table 1) has CF 1.402 and $\Phi_{opt}(i)$ in the range from -64.00° to 230.37°. After recalculation $\Phi_{opt}(i)$ to the phase range from 0 to 360° with the new phase resolution $\Delta\Phi_n$ = 1°, CF is 1.405 and in case of $\Delta\Phi_n$ = 10° CF increases to 1.459. Initial phases can be also recalculated to the corresponding indices of sample points $I_s(i)$ as

$$I_s(i) = (1 - (\Phi_i / 360)) \times (n_s / i) \qquad (3)$$

where n_s is a total number of sample points. Φ_i must be in the range from 0 to 360°.

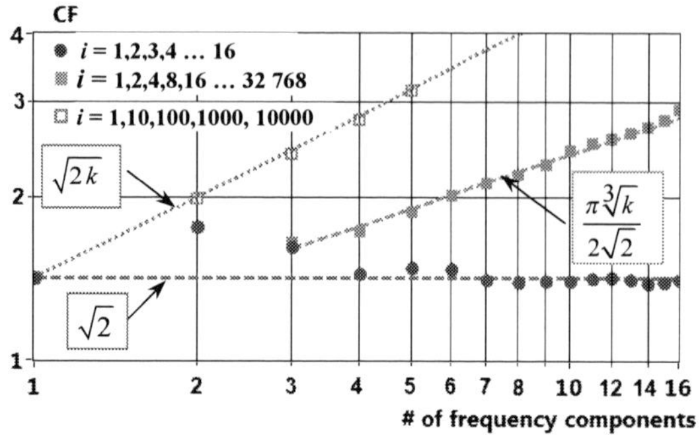

Figure 4. Dependence of the CF on the number of frequency components and their distribution.

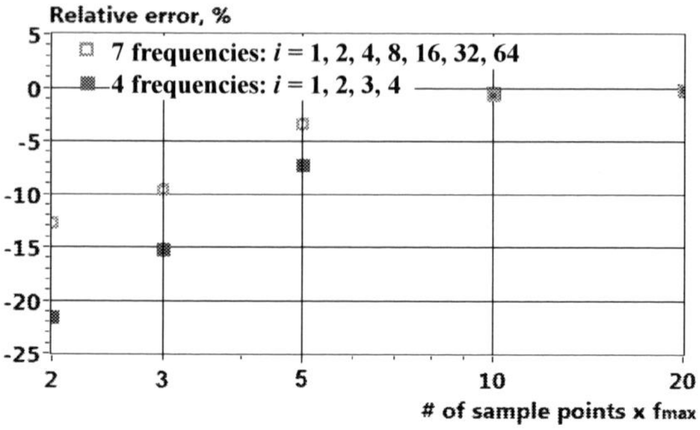

Figure 5. Dependence of the relative error of CF calculation on sampling rate.

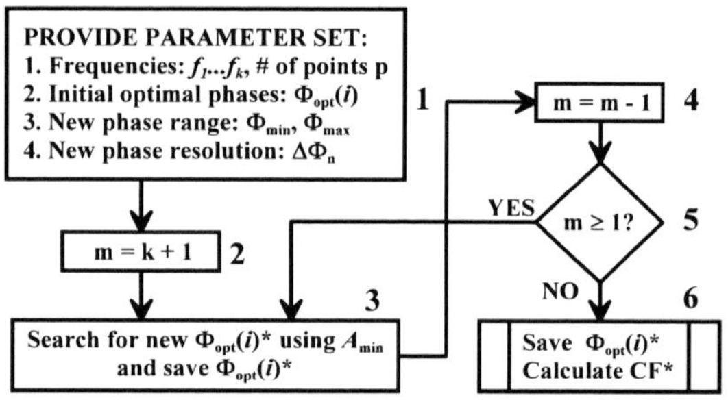

$A_{\min} = \mathrm{Min}\sum_{i=1}^{i=k} |A_i| \sin(2\pi f_i + \Phi_j)$

For $i = m-1$ $\Phi_j = \Phi_{\min}$ to $\Phi_{\max}$ with the step $\Delta\Phi_n$

For $i \neq m-1$ $\Phi_j = \Phi_{opt}(i)$

Note: In some cases repeating calculations from the step 2 using $\Phi_{opt}(i)^*$ may improve CF*

Figure 6. Flowchart of an algorithm for recalculation of optimal phases.

4. Conclusions

The developed new method for the minimization of CF is fast, but gives the same results as the full exhaustive search if the number of frequency components is below 6. For a higher number of components, the results were compared with other published data and found to be similar or better.

An algorithm for recalculating of optimized initial phases to another phase range and resolution is presented, which is practical but highly preserves the previously obtained optimization results (CF_{min}).

5. Acknowledgments

This research was supported by the European Union through the European Regional Development Fund in frames of the research center CEBE and competence center ELIKO, partly also by Estonian target financed project SF0140061s12.

References

[1] Schroeder M 1970 Synthesis of low-peak-factor signals and binary sequences with low autocorrelation *IEEE Trans. Inf. Theory* **16** 85–9

[2] Van den Bos A 1987 A new method for synthesis of low-peak-factor signals IEEE *Trans. Acoust.* **35** 120–2

[3] Van der Ouderaa A, Schoukens J and Renneboog J 1988 Peak factor minimization using a time-frequency domain swapping algorithm *IEEE Trans. Instrum. Meas.* **37** 145–7

[4] Guillaume P, Schoukens J, Pintelon R and Kollar I 1991 Crest-factor minimization using nonlinear Chebyshev approximation methods *IEEE Trans. Instrum. Meas.* **40** 982–9

[5] Rees D, Jones D 1991 Design and application of non-binary low peak factor signals for system dynamic measurement *Proc. Int. Conf. on Control (Edinburgh, UK,* 1991) pp 644–50

[6] Horner A 2000 Low peak amplitudes for wavetable synthesis *IEEE Trans. Speech Audio Process.* **8** 467–70

[7] Ong M S, Kuang Y C, Liam P S and Ooi M P-L 2012 Multisine with optimal phase-plane uniformity for ADC testing *IEEE Trans. Instrum. Meas.* **61** 566–78

Simplified signal processing for impedance spectroscopy with spectrally sparse sequences

P Annus[1, 2], R Land[1, 2], M Reidla[2], J Ojarand[2], Y Mughal[1] and M Min[1, 2]
[1]Thomas Johann Seebeck Department of Electronics, Tallinn University of Technology, Ehitajate tee 5, 10919 Tallinn, Estonia
[2]ELIKO Competence Centre, Mäealuse 2/1, 12618 Tallinn, Estonia

E-mail: paul.annus@elin.ttu.ee

Abstract. Classical method for measurement of the electrical bio-impedance involves excitation with sinusoidal waveform. Sinusoidal excitation at fixed frequency points enables wide variety of signal processing options, most general of them being Fourier transform. Multiplication with two quadrature waveforms at desired frequency could be easily accomplished both in analogue and in digital domains, even simplest quadrature square waves can be considered, which reduces signal processing task in analogue domain to synchronous switching followed by low pass filter, and in digital domain requires only additions. So called spectrally sparse excitation sequences (SSS), which have been recently introduced into bio-impedance measurement domain, are very reasonable choice when simultaneous multifrequency excitation is required. They have many good properties, such as ease of generation and good crest factor compared to similar multisinusoids. Typically, the usage of discrete or fast Fourier transform in signal processing step is considered so far. Usage of simplified methods nevertheless would reduce computational burden, and enable simpler, less costly and less energy hungry signal processing platforms. Accuracy of the measurement with SSS excitation when using different waveforms for quadrature demodulation will be compared in order to evaluate the feasibility of the simplified signal processing. Sigma delta modulated sinusoid (binary signal) is considered to be a good alternative for a synchronous demodulation.

1. Introduction

Complex electrical impedance has been widely used to characterize properties of biological tissues for many years [1]. One of the promising application areas is postoperative monitoring of the healing process. Bioimpedance reflects changes in the tissue state, which can be used to characterize the revivability or resuscitation ability of the tissue. Typically simultaneous measurement at several frequencies is required. Low and high frequency currents flow through the resistive interstitial space between the cells (extracellular component), but only the high frequency current can flow through the electrical capacitances of the insulating cell membranes (intracellular current). Ratio of these currents will reflect the state of the cells, accumulation of fluid in extracellular space, amount of swelling and ultimately warn before massive edema is about to develop. Generally it is desirable to monitor tissue parameters from different tissue locations simultaneously. These multifrequency and multisite measurements may be relatively rare and slow due to the dynamics of the underlying processes, and can be even sequential. Monitoring is warranted during the first few days after the surgical intervention, and possible changes take several minutes to occur. One of the critical tasks is restoration of the spontaneous blood circulation postoperatively. Cardiac activity will cause modulation of the measured bioimpedance due to varying amounts of blood pushed through the tissue. Therefore faster changes in impedance need to be monitored as well, in order to assess whether the blood vessels are connected, and blood flow is restored. Extensive laboratory experiments have been conducted during longer period on isolated tissue samples, such as pigs heart, tongue etc. of pigs, for preclinical investigation of the impedance changes in the biological tissues. Device prototype has been developed for conducting these bioimpedance measurements in laboratory conditions (figure 1). Device has the ability to conduct measurements simultaneously on 16 frequencies in the range from 1 Hz to 100 kHz. Spectrally sparse sequences have been proven to be viable excitation signals for such a device [2]. SSS as two level or binary signal has many good properties, such as ease of generation, and good crest factor compared to similar multisinusoids. They can have tens of almost arbitrarily chosen, spectral lines with tailored magnitudes. The drawback is that there is also certain amount of energy on unwanted spectral lines, on so called snow lines, which extend well beyond the highest measurement frequency of interest. Combination of oversampling and correction is used in order to avoid large measurement errors. Another drawback is that real time calculation of that many discrete Fourier coefficients puts a heavy burden even on the internal resources of the used digital signal processor (DSP) and has been main limiting factor so far. Therefore simplification of the signal processing step is highly desirable.

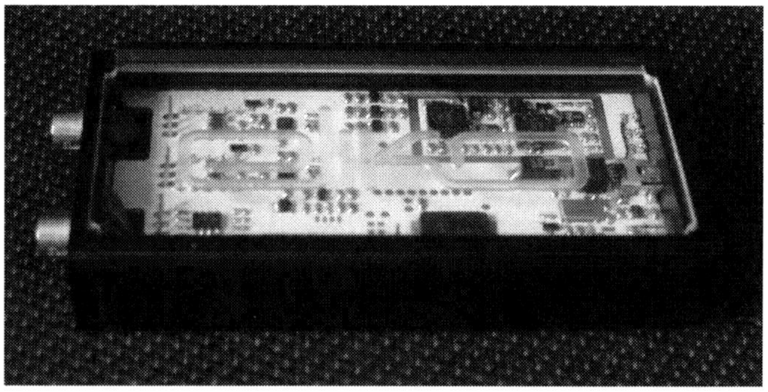

Figure 1. Two channel tissue monitor with spectrally sparse sequence as an excitation signal.

2. Description of simplification steps and the test

One of the possibilities is replacement of the sinusoidal signals used for quadrature demodulation with suitable binary counterparts. In an essence it would enable usage of simple summation (or switching in analogue domain) instead of more complex multiply and accumulate (MAC) operations. Eliminating MAC from the signal processing chain widens choice of possible signal processing platforms. It has been shown that simple low power complex programmable logic device (CPLD) can be used to process acquired signals if only summation is involved [3]. Three of the possible candidates are considered for replacement, and resulting measurement errors are compared (figure 2, table 2). First is the set of simple quadrature square waves, second is the set of shortened square waves (SSW) [3] with shortening angle of 22.5 degrees, and lastly sinusoids modulated with first order sigma delta circuit are considered. Simple first order RC low pass circuit is used with varying corner frequency to make simulations in LabVIEW environment more realistic, since impedance of various biological tissues can be modelled as RC circuitry. Filter corner frequency is swept form 1 Hz to 1.1 kHz, covering wide variety of possible target impedances. SSS, used in simulation, contains ten measurement frequencies (2, 4, 8, 16, 32, 64, 128, 256, 512, and 1024 Hz) with equal magnitudes. Result of the demodulation of the response to the SSS excitation with binary signals is compared against measurements with classical single sinusoidal excitation signal which is demodulated by multiplying it with two quadrature sinusoids.

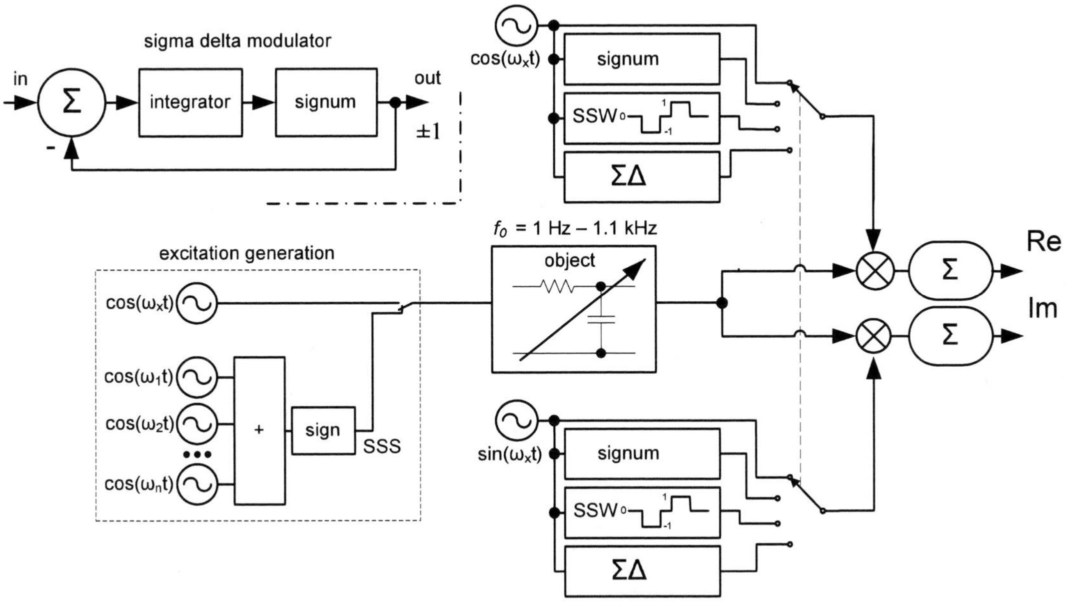

Figure 2. Simplified first order sigma delta modulator (upper left), and block diagram of the experiment with excitation generation, object and demodulation, where f_0 is filter corner frequency, ω_1 - ω_n frequencies of the SSS components, and ω_x is an actual measurement frequency.

Quadrature sinusoids were also used for demodulation of the response to SSS excitation, however resulting errors were very low compared to purely sinusoidal case, and are therefore omitted from analysis. It should be noted that results are not general and depend on chosen excitation sequence, nevertheless similar tests can be conducted for any arbitrary sequence needed and results compared.

3. Results

Since RC circuit is modelled using exponential function certain caution is in order. Measured real and imaginary components of the response signal should be equal exactly when the corner frequency of the sweeping RC filter coincides with the measurement frequency. In reality there is some discrepancy between these frequencies due to limited resolution of the simulation. To assess simulation accuracy this discrepancy is calculated for purely sinusoidal measurement case:

Table 1. Measurement frequency versus frequency when real and imaginary parts coincide in case of sinusoidal excitation and demodulation with sinusoidal signals.

f_{sampling}, kHz	Frequency when real and imaginary parts coincide in Hz									
	Measurement frequency in Hz									
	2	*4*	*8*	*16*	*32*	*64*	*128*	*256*	*512*	*1024*
100	1.99	3.99	7.99	15.97	31.92	63.74	126.9	251.9	495.8	960.8
1000	1.99	3.99	7.99	15.99	31.99	63.99	127.9	255.6	510.3	1017.4

The sampling frequency should exceed maximal filter corner frequency roughly 1000 times according to table 1. Raising it further would make simulation too slow to complete within reasonable time.

Generally relative errors are presented in literature, whereas deviation from the correct value is divided by the correct value. In some cases these errors exceed 50% for real and imaginary parts.

However closer examination shows that it happens when the value of the measurand is very small. In real measurement situation it could mean that it is already buried in the noise. Therefore it seems reasonable to compute full scale errors instead, whereas deviation from the correct value is divided by maximal value of the measurand. Resulting errors are shown in table 2. Also absolute phase errors in degrees are included in italics for better clarity.

Table 2. Demodulation quality with square wave (SQW), shortened square wave (SSW), and sigma delta modulated sinusoid is compared against demodulation of the response to the sinusoidal excitation with pure sinusoids. Errors below 0.001% are denoted by "-". Absolute phase errors are added in italic.

Signal component detected, signal used for detection	Maximal full scale measurement error in % / *absolute phase error*									
	Measurement frequency in Hz									
	2	*4*	*8*	*16*	*32*	*64*	*128*	*256*	*512*	*1024*
Re, SQW	1.42	3.85	3.23	3.48	2.65	2.80	1.95	0.048	0.63	0.47
Im, SQW	3.24	5.00	5.27	2.89	1.79	1.32	0.22	0.85	0.55	1.22
Magnitude, SQW	1.36	3.85	3.23	3.48	2.64	2.78	1.93	0.18	0.39	0.38
Phase, SQW (deg, abs)	*0.96*	*1.55*	*1.63*	*1.00*	*0.61*	*0.55*	*0.2*	*0.25*	*0.29*	*0.48*
Re, SSW	3.81	2.60	2.36	2.52	1.54	0.81	0.85	0.42	0.32	0.27
Im, SSW	1.11	1.79	2.13	1.81	1.77	1.10	0.15	0.29	0.92	0.54
Magnitude, SSW	3.81	2.60	2.36	2.52	1.52	0.78	0.83	0.36	0.05	0.15
Phase, SSW, (deg, abs)	*0.24*	*0.56*	*0.63*	*0.56*	*0.56*	*0.34*	*0.09*	*0.14*	*0.33*	*0.23*
Re, sigma-delta	0.002	0.007	0.006	0.001	-	0.024	-	0.005	0.004	-
Im, sigma-delta	0.005	0.005	0.01	0.02	0.007	0.009	0.005	0.004	0.003	0.007
Magnitude, sigma-delta	0.002	0.007	0.006	0.001	-	0.024	-	0.005	0.004	0.003
Phase, sigma-delta, (deg, abs)	*0.002*	*0.002*	*0.004*	*0.006*	*0.004*	*0.002*	*0.002*	*0.003*	*0.004*	*0.12*

4. Discussion

Even though error calculation results depend on actual SSS used, and on actual object to be measured, some observations are in order. First of all, when comparing measurement results with those of pure sinusoidal excitation and classical Discrete Fourier Transform (DFT), SSS perform almost exactly as well when DFT is applied. Secondly, sufficient oversampling is in order when SSS is used for measurement, due to higher harmonic content. Regarding demodulation with different binary signals it is probably fair to say that sigma delta modulated sinusoids perform almost as well as real sinusoids, with both relative and full scale magnitude errors being below 0.1% and absolute phase error below 0.2 degrees. Even measurement results with simple square waves are within 4% of actual magnitude and 2 degree of real phase. Question remains if the usage of shortened square waves is justified due to little improvement over simple square waves, and added complexity.

5. Acknowledgments

This research was supported by the European Union through the European Regional Development Fund in frames of the research centre CEBE and competence centre ELIKO, partly also by Estonian target financed project SF0140061s12.

References

[1] Kun S, and Peura R A 1994 Tissue ischemia detection using impedance spectroscopy *Proc. IEEW/EMBS 16th Ann. Conf.* pp 868–9

[2] Min M, Ojarand J, Martens O, Paavle T, Land R and Annus P 2012 Binary signals in impedance spectroscopy *Proc. IEEE EMBC 2012* pp 134–7

[3] Paavle T, Annus P, Kuusik A, Land R and Min M 2007 Bioimpedance monitoring with improved accuracy using three-level stimulus *Proc. of ECCTD'07* pp 412–5

Journal of Physics: Conference Series **434** (2013) 012032 doi:10.1088/1742-6596/434/1/012032

Integrated dielectric spectrometer for wideband and high-speed measurements using pseudo-noise codes

J. Sachs[1], M. Helbig[1], M. Kmec[1], R. Herrmann[1], K. Schilling[2], P. Rauschenbach[3]

[1] Ilmenau University of Technology
[2] Ilmsens – TU Ilmenau Service GmbH
[3] MEODAT GmbH
juergen.sachs@tu-ilmenau.de

Abstract. The paper gives a short introduction into some system-theoretic aspects and features of dielectric measurements by periodic wideband signals. Based on these considerations, a related measurement concept will be presented which is well suited for both compact device implementation and monolithic integration. Some implementation examples are shown and first measurement results are presented.

1. Introduction

"Broadband Dielectric Spectroscopy (BDS) is besides Nuclear Magnetic Resonance (NMR) spectroscopy the experimental tool to study molecular dynamics in a broad frequency (10^{-6} Hz to 10^{11} Hz) and temperature range. It has the additional advantage that the sensitivity of the measurements increases with decreasing separation of the electrodes and hence with decreasing amount of sample material." [2]

Seen from the viewpoint of the measurement device, such a setup involves the capturing of weak signals in order to avoid overloading of the test sample. Since recent measurement devices are still quite bulky, they have to be connected with the measurement cells via cables. They introduce an additional parasitic impedance and a signal transformation affecting, e.g., the wanted impedance of the measurement cell. In order to remove these uncertainties, a calibration of the measurement setup may be performed. However, every variation of the setup due to cable bending (leading to variations of propagation time and triboelectric current generation), temperature variation or temperature gradient etc. will degrade the calibration and hence it will reduce the reliability of the measurement. In order to alleviate the susceptibility of dielectric measurements to such influences and to break the restraints of the 50 Ω environment for RF-measurements, the actual measurement plane has to be moved as close as possible towards the electrodes of the measurement cell. This is best done by monolithic integration of the critical device components.

Furthermore, the field exposure of the device or material under test (DUT, MUT) and the measurement speed may be critical issues if small volumes and a plenty of items must be continuously evaluated. This requires sounding signals of low crest factor and a wide instantaneous bandwidth. Binary pseudo-noise (PN) codes (in particular the M-Sequence) represent a signal which meets these requirements. They may be generated by a simple and stable approach providing a large fractional bandwidth. In the future, this will open the way to join micro-fluidic devices and wideband measurement electronics on a single chip (Lab on a Chip).

In what follows, we shortly introduce some system-theoretic aspects and features of dielectric measurements by periodic wideband signals. Based on these considerations, a related measurement concept is presented which is well suited for compact device implementation and monolithic integration, as well. Some implementation examples – restricted to the RF-domain for the sake of shortness – are shown and first measurement results are presented.

2. Dielectric measurement based on signals of large fractional bandwidth

The polarization behavior and conductivity of a material under test (MUT) are usually summarized by the frequency dependent complex permittivity $\underline{\varepsilon}(f)=\varepsilon_0(\varepsilon_r'(f)-j\varepsilon_r''(f))$ (ε_0- permittivity of vacuum, $\varepsilon_r', \varepsilon_r''$ - relative dielectric constant of the MUT - real and imaginary part; complex quantities are denoted by underlined symbols). The determination of the permittivity is based on specimen-holders or measurement setups which represent either one- or two-port systems. A summary of different measurement configurations is given in [1]. Depending on the operational frequencies, the behavior of these devices is typically described by impedance $(\underline{Z})$ or admittance $(\underline{Y})$ functions and scattering parameters $(\underline{S})$. To abstract from the different options, we write the relations between the measured signals and the characteristic functions of the test setup as:

$$\underline{P}(f)=\underline{Q}(f)\,\underline{R}(f) \tag{1}$$

in which $\underline{Q}=\underline{Q}(f,\underline{\varepsilon}(f),\mathbf{\Theta})$ stands for $\underline{Y},\underline{Z},\underline{S}$-parameters. $\underline{Q}$ may be a scalar quantity or a matrix whose entries depend on frequency f, the geometric structure of the test setup (symbolized by the vector $\mathbf{\Theta}$ containing geometric parameters; corresponding relations can be found in [1, 3]) and the permittivity $\underline{\varepsilon}(f)$ of the MUT. $\underline{P}$ and $\underline{R}$ represent the measured quantities (i.e. voltage, current or normalized waves) who may be either a scalar or a column vector. In order to gain $\underline{\varepsilon}(f)$, we have to calculate $\underline{Q}(f)$ from the measurements first. This is simply done by:

$$\underline{Q}(f)=\underline{P}(f)/\underline{R}(f) \tag{2}$$

Relation (2) extends to a related matrix operation in the case of a two port test setup (see chapter 2.5.4 in [4]). The complex notation in (1) and (2) assumes the measurement of signal magnitudes and phases. Since this is only feasible for sine waves, both relations cannot be immediately applied for arbitrary wideband signals. Instead of (1), we have to write a convolution

$$p(t)=\int_{-\infty}^{\infty} q(\tau)\,r(t-\tau)\,d\tau=q(t)*r(t) \tag{3}$$

in which $p(t)$ and $r(t)$ represent the measured wideband signals of arbitrary shape and $q(t)$ is the impulse response function of the test setup containing the wanted information about $\underline{\varepsilon}(f)$ since $\underline{Q}(f)=\mathfrak{F}\{q(t)\}$ (with $\mathfrak{F}\{\ \}$ denoting Fourier Transform). Due to the limited recording time T_R in the case of a real measurement, (3) actually becomes:

$$p(t)=\int_{-T_R/2}^{T_R/2} q(\tau)\,r(t-\tau)\,d\tau=\left[\operatorname{rect}\left(\frac{t}{T_R}\right)q(t)\right]*r(t) \xrightarrow{\mathfrak{F}\{\}} \underline{P}(f)=\left[\underline{Q}(f)*\operatorname{sinc}(f\,T_R)\right]\underline{R}(f) \tag{4}$$

Hence, the wanted $\underline{Q}(f)$ is affected by a sinc-function. As shown in [4], its influence can be eliminated by choosing periodic wideband signals of period T_0 for which $T_R=n\,T_0;\quad n\in\mathbb{N}$ holds.

In order to estimate $\underline{Q}(f)$, we subject the measured signals to a Fourier Transform and apply (2). However, it becomes ill-conditioned at higher frequencies due to the lack of spectral components at the upper end of the stimulus spectrum. Unfortunately, this spectral gap has to be accepted since we have to enforce the Nyquist theorem by anti-aliasing filters which will never have abrupt edges. By regularization (see [4]), the ill-conditioned problem may be suppressed up to a certain degree.

3. Pseudo noise measurement device

The PN-device as depicted in Figure 1 exploits the above introduced wideband approach for dielectric measurements in frequency and time domain. A stable clock generator pushes a binary shift register which provides the PN-sequence (M-sequence). This signal –having a comb spectrum – stimulates the MUT. Bandwidth and spectral line spacing are fixed by the clock rate and the shift register length. In contrast to the classical sine wave approaches, the measurement may run quite fast since all spectral lines appear in parallel. A sampling receiver captures the measurement signal. They are processed to give either $\underline{Q}(f)$ or $q(t)$. In subsequent steps, one has to extract the wanted information about the MUT. The low pass-filters in Figure 1 are aimed to keep the Nyquist sampling theorem and the

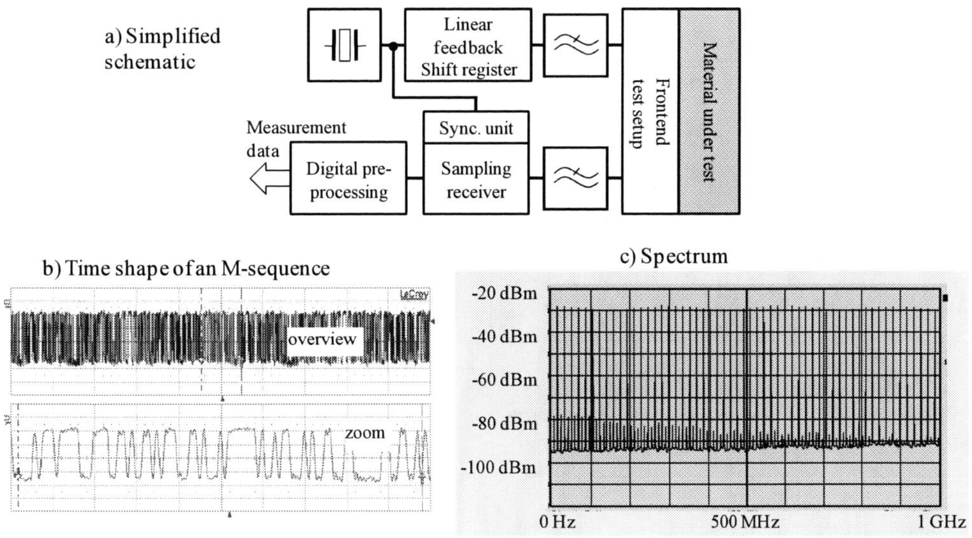

Figure 1: Simplified schematic of a PN-device for one-port measurements and typical appearance of a PN-signal in time and frequency domain

frontend is either a directional bridge or an auto-balancing bridge depending on the considered frequency band. For two-port measurements, one needs a multi-channel version of the PN-device. The interested reader is addressed to [4] for further discussions of the device functioning and performance.

4. Implemented devices and measurement examples

For the sake of shortness, we only refer to two RF-examples even if the measurement method is not restricted to that domain only. Figure 2 shows the structure and a photo of an integrated RF-frontend. It is a key component of a small device for reflection factor measurements as depicted in Figure 3. Due to its small volume and weight, a coaxial probe can be immediately connected to the device.

The second example refers to a transmission line measurement and data evaluation in time domain. Figure 4 depicts the measurement schematic und shows some results. We measured the transmitted impulse response function of a coplanar waveguide which was in contact with a liquid under test. The liquid was running through the test fixture in portions (assigned by capital letters A ... F). For the sake of demonstration, we simply used saltwater with minor variations in salt concentration, fresh and distilled water (all of same temperature). As expected, the measurements have shown that mainly the peak value of the pulse was affected by the different salt concentrations. In other experiments, the temperature of fresh water was slightly changed. Now, pulse position, pulse width and magnitude have changed due to permittivity variation. A rough estimation has shown that temperature variations of water are traceable down to $5 \cdot 10^{-3}$ °C with the applied prototype device.

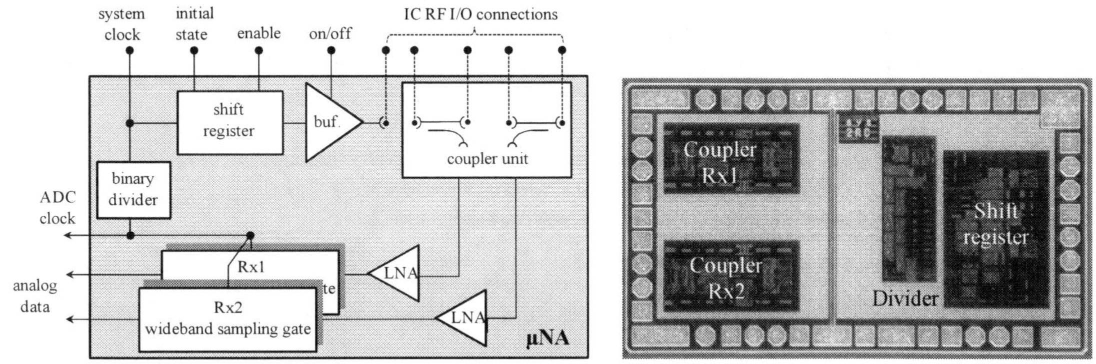

Figure 2: Structure of the RF-part and microphotograph of an integrated PN-sequence device

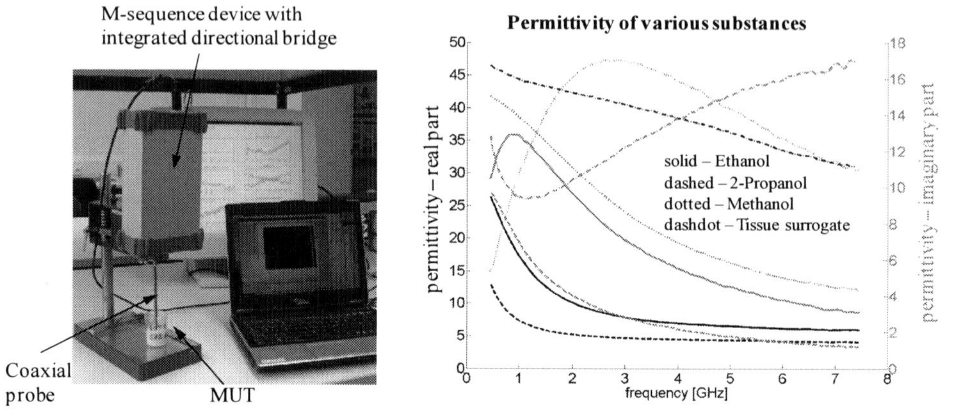

Figure 3: Example of permittivity measurement by a coaxial probe

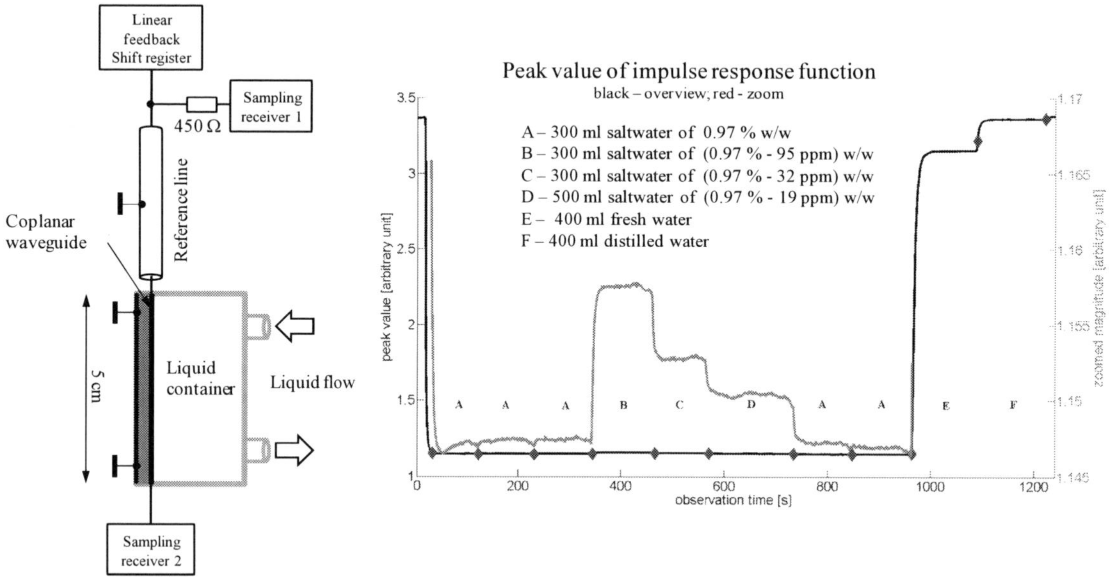

Figure 4: Continuous time domain measurement for tracking of permittivity variations.

5. Conclusion

Dielectric spectroscopy based on M-sequence signals represents an interesting alternative to the classical sine wave approaches. The devices promote monolithic integration of key components and permit the construction of small and light weight equipment for precise and continuous observation.

References

[1] Kremer F, Serghei A, Sangoro J R, Tress M, and Mapesa E U 2009 Broadband Dielectric Spectroscopy in nano-(bio)-physics *IEEE Conference on Electrical Insulation and Dielectric Phenomena, CEIDP '09*, pp. 1-6.

[2] Baker-Jarvis J, Janzic M D, Riddle B F, Johnk R T, Kabos P, Holloway C L, Geyer R G, and Grosvenor C A 2005 Measuring the Permittivity and Permeability of Lossy Materials: Solids, Liquids, Metals, Building Materials, and Negative-Index Materials *NIST Technical Note 1536.*

[3] Kupfer K 2005 *Electromagnetic aquametry: electromagnetic wave interaction with water and moist substances*. (Berlin, New York: Springer)

[4] Sachs J 2012 *Handbook of Ultra-Wideband Short-Range Sensing - Theory, Sensors, Applications*. (Berlin: Wiley-VCH)

Contactless Investigations of Yeast Cell Cultivation in the 7 GHz and 240 GHz Ranges

J Wessel[1], K Schmalz[1], G Gastrock[2], B P Cahill[2], C Meliani[1]

[1]IHP, Im Technologiepark 25, 15236 Frankfurt (Oder), Germany

[2]IBA, Institute for Bioprocessing and Analytical Measurement Techniques, Rosenhof, 37308 Heilbad Heiligenstadt, Germany

wessel@ihp-microelectronics.com

Abstract. Using a microfluidic system based on PTFE tubes, experimental results of contactless and label-free characterization techniques of yeast cell cultivation are presented. The PTFE tube has an inner diameter of 0.5 mm resulting in a sample volume of 2 µl for 1 cm sample length. Two approaches (at frequencies around 7 GHz and 240 GHz) are presented and compared in terms of sensitivity and applicability. These frequency bands are particularly interesting to gain information on the permittivity of yeast cells in Glucose solution. Measurements from 240 GHz to 300 GHz were conducted with a continuous wave spectrometer from Toptica. At 7 GHz band, measurements have been performed using a rat-race based characterizing system realized on a printed circuit board. The conducted experiments demonstrate that by selecting the phase as characterization parameter, the presented contactless and label-free techniques are suitable for cell cultivation monitoring in a PTFE pipe based microfluidic system.

1. Introduction

The continuing trend in microbiology to conduct cell cultivation with highly parallel techniques and decreasing culture volumes creates an urgent need for efficient and reliable tools to provide automated cell cultivation monitoring. The segmented flow architecture presented in [1] allows for performing high throughput cultivation in pipe based bioreactors. In addition, cultivation methods based on parallelism are more commonly applied, which increases the need for effective contactless and label-free techniques to characterize the cultivation progress. These techniques mainly use the fact that the increase of number of cells within a single compartment evokes a change in the average permittivity, which is detectable using microwave spectroscopy. The permittivity of disperse substances has been actively studied and detailed models for disperse media can be found in [2]. This work demonstrates two techniques being well suited for contactless and label-free characterization of cell cultivation. Other approaches to investigate the relative permittivity of liquids often require the liquid under test to be in direct contact with the sensing coplanar waveguide structure. Considering the described pipe based microfluidic system, a much higher sensitivity is demanded, since the electric field needs to penetrate the wall thickness of the PTFE channel. Therefore, a differential method, presented in [3], was modified and put into operation, resulting in a phase sensitive biosensor working at 7 GHz. Measurements beyond 200 GHz were conducted with a continuous wave (CW) Terahertz spectrometer

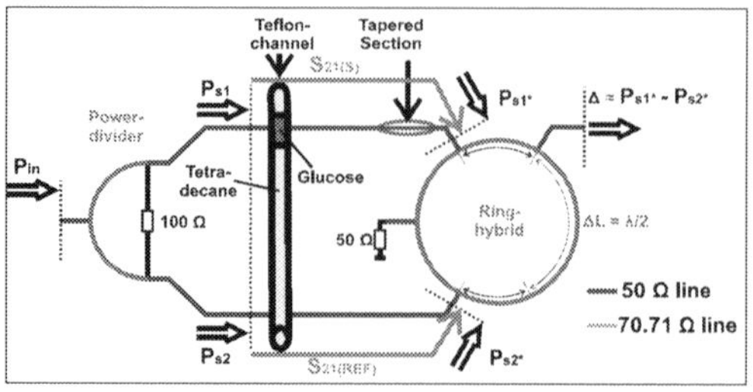

Fig. 1 Concept of the 7 GHz biosensor

to investigate the possibilities and limitations of monitoring yeast cultivation in the sub-terahertz range. Sec. 2 presents the function and experimental setup of the 7 GHz Biosensor followed by a discussion of the simulated results. Sec. 3 introduces the continuous wave Terahertz spectrometer and describes the conducted experiments. The measurement results are presented in Sec. 4 and finally, Sec. 5 discusses both methods.

2. 7 GHz Biosensor

To investigate cell cultivation at comparatively low frequencies, a biosensor structure working in the 7 GHz range has been designed and put into operation.

2.1. Concept of the 7 GHz Biosensor

The 7 GHz Biosensor consists of passive microwave structures and follows a differential principal. The operation concept is shown in Fig 1. The input signal is fed to a Wilkinson power divider. The divider splits the signal up into two ideally identical signals to a reference path and a sensing path, each consisting of microstrip lines. These lines are crossed by a PTFE tube building the microfluidic channel. The PTFE tube contains the sample that is to be investigated with the Biosensor. The microstrip line used as the reference path ($S_{21(ref)}$ in fig. 1), is covered by a section of the tube, containing tetradecane, which is used to confine the liquid solution to be examined. Moreover, tetradecane serves as a reference fluid. The sensing path is covered by a section of the PTFE tube containing the sample of interest, in this case a glucose yeast mixture. To compare the propagation characteristics of the two paths, the signals labeled

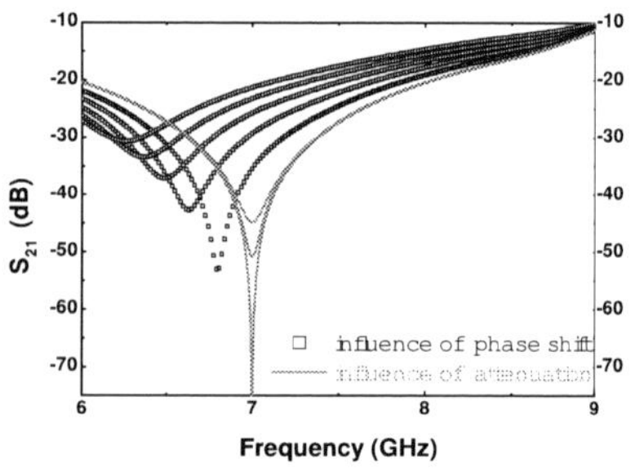

Fig. 2 Numerical Simulation of the Biosensor's transfer function S21 for different attenuations (red curves) and phase shifts (black curves) on the sensing path.

$S_{21(S)}$ and $S_{21(REF)}$ are fed to a ring hybrid. This coupler provides the interference of $S_{21(S)}$ and $S_{21(REF)}$ with 180 ° phase shift for the designated centre frequency at its output. A change in the permittivity of the sample alters the signal $S_{21(S)}$. This variation can be observed at the hybrids output. Theoretical investigations on the transfer function of the designed biosensor show that the shift in frequency of the observed notch in fig. 2 and thus the level shift of the signal at a fixed frequency are mainly caused by a phase variation in the sensing path ($S_{21(S)}$ in Fig. 1). Fig. 2 presents the numerically simulated transmission ($S_{21(S)}$ in Fig. 1) of the biosensor. Attenuation and phase shift have been artificially introduced in the sensing path: the phase shift in the propagation path of $S_{21(S)}$ is swept from 0 to 10 ° and the attenuation from 0 to 10 dB. One can see that the main parameter that changes the frequency of the observed notch is obviously the phase shift

2.2. Experimental Setup

For the conducted experiments, the described sensor structure has been realized on a Rogers 3003 printed circuit board using SMA connectors for measurements. The PTFE tube, being part of a microfluidic system, is pressed on top by PVC jaws.

3. CW Spectrometer system

To investigate the effects at frequencies beyond 200 GHz, a continuous wave Terahertz spectrometer from Toptica [4] was used.

3.1. Principle of the CW System

The continuous wave spectrometer makes use of GaAs-photomixers to convert the beat signal of two near-infrared lasers into tunable terahertz radiation. The resulting signal is then guided by an interferometer, propagates through the sample under test (SUT) and is subsequently detected by a second photomixer. The measured parameter is the photo current in the receiver photomixer, which is a measure of the electric field of the terahertz wave. It is $I_{PH} \propto E_{THz} \cdot \cos\Delta\varphi = E_{THz} \cdot \cos 2\pi\Delta L f/c$, where E_{THz} is the amplitude of the terahertz electric field, f the terahertz frequency, c the speed of light in vacuum and ΔL basically the path length difference between transmitted and received signal of the spectrometer, resulting in a phase shift $\Delta\varphi$. [5] gives a detailed description. As the equation shows, the signal amplitude can directly be derived from the measured photo current. Using measurements with a reference sample, for example a yeast glucose mixture that has been in cultivation for a certain time, a relative phase can be extracted from the gathered data.

3.2. Experimental Setup

To investigate the sample of interest with the CW spectrometer, it needs to be exposed to the THz beam, what is accomplished by a sample holder fixing the PTFE tube. Since the thickness of the tube (0.55 mm in fig. 5) is almost equal to tube interior diameter (0.5 mm in fig. 5), a considerable amount of the radiated power is transmitted through the PTFE walls, which reduces measurement sensitivity. To investigate the propagation behavior of the liquid exclusively, additional measurements have been performed using an epoxy resin aperture, which covers the sidewalls of the tube. The setup is sketched in Fig. 5. Since all the THz radiation, which is detected at the second photo mixer propagated exclusively trough the sample, the phase shift due to cell cultivation is expected to be higher than in the case without the epoxy resin aperture.

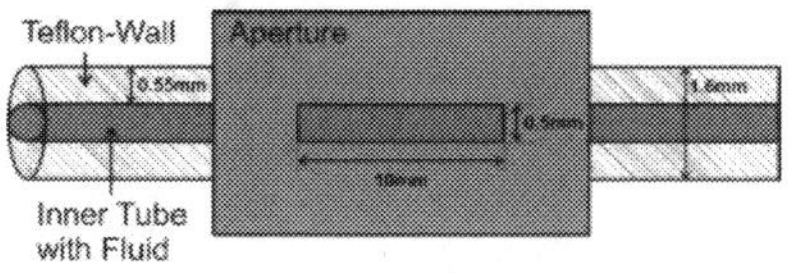

Fig. 5 Aperture to ensure exclusive transmission through the liquid under test

4. Measurement Results

4.1. 7 GHZ Measurements

To experimentally determine the transfer characteristic of the microwave biosensor, scattering parameters were measured with a standard network analyzer. From the measured transmission, the shift in the notch frequency (in fig. 6) over time was derived. The result is depicted in Fig. 6. As it is shown, the biosensor exhibits a 125 MHz frequency shift of the transmission notch after 20 hours of cultivation. After that time, considering a cell doubling time of 90 min [6], one can assume no considerable increase in number of cells for the considered sample volumes. Comparing the measured results and the simulated ones from Sec. 2.1, one can say that the observed behavior in the measured transmission is mainly caused by a change in phase in the sensing paths transmission $S_{21(S)}$. According to the performed ADS simulations, the frequency shift of 125 MHz corresponds to a phase change in $S_{21(S)}$ of 1.2 °.Thanks to the rat-race based characterization procedure, this change is emphasized in order to achieve relevant measurement sensitivity.

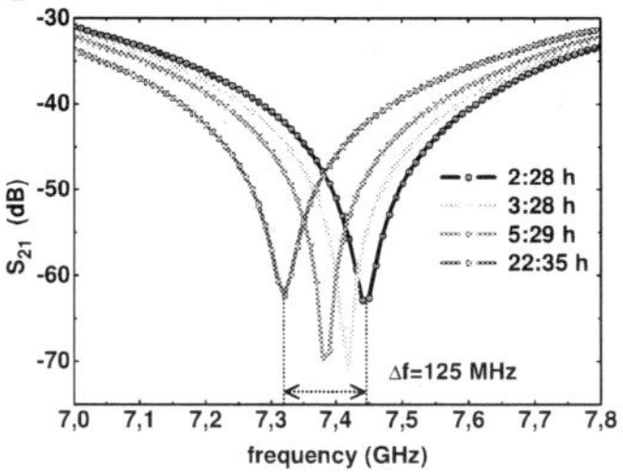

Fig. 6 Frequency Change of S_{21} Notch Versus Time (2:28 h up to 22:35 h cultivation time)

4.2. CW Measurements

In [7] it has been shown that the evaluation of the amplitude of the photo current is not sufficient to distinguish between yeast samples at different cultivation times. Here the phase needs to be exploited

(a) (b)

Fig. 7 Change in phase versus cultivation time at 240 GHz, (a) without epoxy aperture, (b) with epoxy aperture

for proper monitoring of the cultivation state. Fig. 7 (a) displays the change in phase over cultivation duration for a Glucose yeast solution at 240 GHz. As the figure depicts, a change of 46° within 16 h is observed. As already mentioned in sec. 3.2, since a part of the radiated power is transmitted through the sidewalls of the tube, which decreases the effect on the phase of the signal, the sensitivity of the measurement can be emphasized by using the aperture described in Sec. 3.2 assuring that all radiation that is reaching the receiving photo mixer was propagating through the sample under test exclusively and not through the walls of the tube. For this modified configuration, the relative phase change over time is depicted in Fig. 7 (b). It is demonstrated that a much higher change in phase is achieved by using the aperture depicted in Fig. 5. Within 7 h, a change of 120° has been observed.

5. Conclusion

Investigations on the feasibility of characterization techniques of yeast cell cultivation at 7 GHz and 240 GHz have been performed. The amplitude information can be used to differentiate between samples exhibiting large difference in relative permittivity. However, to monitor yeast cell cultivation, one has to be able to differentiate between glucose with and without yeast. Thus it is necessary to detect comparatively small variations in permittivity. For this case, it is mandatory to exploit the phase information. This leads to the conclusion that mainly the change of the relative permittivity of the mixture evoked by progress of a yeast culture can be detected by the demonstrated techniques and hence be used for monitoring cell cultivation. Owing to its differential principle, the presented 7 GHz biosensor exhibits a high sensitivity of $\Delta f_m/t_c$=125 MHz/20h, enabling characterization of cell cultivation in pipe based microfluidic systems. The high sensitivity at 240 GHz of $\Delta\varphi/t_c$=2.9°/h along with the advantage of potentially compact sensor structures renders an exploitation of this frequency range extremely attractive for biosensing applications. Further experiments revealed that this sensitivity can be increased significantly to $\Delta\varphi/t_c \approx 17$°/h by using the demonstrated aperture.

References

[1] K. Martin et al., "Generation of larger numbers of separated microbial populations by cultivation in segmented-flow microdevices," Lab Chip, 2003, 3, pp. 202–207.

[2] C. Persch, "Messung von Dielektrizitätskonstanten im Bereich von 0.2 GHz bis 6 GHz und deren Bedeutung für die Mikrowellenerwärmung von Lebensmitteln," Dissertation, Universität Fridericiana Karlsruhe, 1997.

[3] C. Song and P. Wang, "A radio frequency device for measurement of minute dielectric property changes in microfluidic channels," Appl. Phys. Lett. 94, 023901 (2009).

[4] http://www.toptica.com/products/terahertz_generation/lasers_and_photomixers_for_cw_terahertz_generation.html, "cw Packages: Lasers and Photomixers for Frequency-Domain Terahertz Spectroscopy," TOPTICA Photonics AG, 1998-2012, called september 9, 2012

[5] A. Roggenbuck et al., "Coherent broadband continuous-wave terahertzspectroscopy on solid-state samples," New Journal of Physics,12, April 2010, pp 043017 (13pp)

[6] R. Lohaus, "Simulation der Hefezelle(Saccharomyces cerevisiae) und ihres Lebenszyklus," mbi Technical Report, 2002

[7] J. Wessel et al., "Contactless Characterization of Yeast Cell Cultivation at 7 GHz and 240 GHz," Biowireless 2013, Austin, USA, pp 70-72

Journal of Physics: Conference Series **434** (2013) 012034 doi:10.1088/1742-6596/434/1/012034

High frequency fluidic and microfluidic sensors for contactless dielectric and in vitro cell culture measurement applications

T Nacke, A Barthel, B P Cahill, M Meister, Y Zaikou

Institut für Bioprozess- und Analysenmesstechnik e.V., Rosenhof, D 37308 Heilbad Heiligenstadt

E-mail: thomas.nacke@iba-heiligenstadt.de

Abstract. There is a widespread need for highly-sensitive robust sensors that operate without direct contact to the fluid for analysis of fluids in bioprocess technology. Measuring the variation of dielectric properties (conductivity and permittivity) in the microwave frequency band can be used as an approach to investigate biological and chemical matter and processes such as, cell growth, cell metabolism and the concentration of large aqueous based molecules. In comparison to measurement at lower frequencies, DC conductivity (σ) effects on material properties (permittivity ε) can be neglected with increasing of the frequency. This presentation describes a high frequency sensor, which combines detection in macro- or microfluidic networks with quick and precise analysis. It is composed of a fluidic channel placed contactless between a micro-strip line waveguide combined with resonant properties.

1. Introduction

The use of flexibly constructed lab-on-a-chip diagnostics systems with many interchangeable modules will allow a cost-effective implementation of analytical functions across a wide application spectrum. Through the use of small microelectronic systems and miniaturised sensor modules, biological interaction mechanisms can be used to implement easy-to-use medical diagnostic systems, for process control in the food industry and for environmental monitoring. Microfluidic systems find various applications: (i) classical chemical/biochemical analysis, (ii) microreactors for chemical synthesis, (iii) drug development, (iv) diagnostics, (v) DNA amplification, (vi) cell biology, (vii) tissue engineering and many more. There have been a huge number of publications describing the application, construction and challenges of microfluidic systems in the last 15 years [1, 2].

Contactless measurement by capacitive coupling of the excitation signal avoids many of the electrode effects associated with direct electrode contact [3]. At RF frequencies, contactless electrodes operate within a very narrow band and have non-linear characteristics within the plateau phase of the ß-dispersion and the influence of medium conductivity changes directly influences the measured capacitance. Contactless measurements at microwave frequencies determine the sum of polarizable contents. The integrity of the cell membrane and in turn the cell's viability cannot be assessed at frequencies greater than 50 MHz. In any case, the growth characteristics of batch cell cultures can be ascertained by measuring the influence of metabolites on the permittivity and conductivity of the medium. These properties enable the development of miniaturized sensors based on high-frequency analysis for fluidic and microfluidic applications.

2. Materials and Methods

Microwave sensors measure the interaction of matter with high frequency electromagnetic electric fields. The interaction of microwave radiation with polar materials, for instance, water, depends on how the polarized molecules orient themselves to the applied field. If the polarity of the field reverses, the molecules reorient themselves. The phase difference of the molecules with regard to the applied high-frequency electrical field is described macroscopically by the complex relative permittivity.
The relative permittivity expresses how much greater the capacitance of a capacitor would be if air ($\varepsilon_r = 1$) was replaced by a dielectric material of identical thickness. The real part of the complex relative permittivity ε_r‘ depends on the polarisability of the material. The imaginary part of the complex relative permittivity (ε_r‘‘) depends on the dielectric losses through absorption of energy during the reversal of polarity and on Ohmic losses. The frequency-dependent conductivity σ‘ is related to the imaginary part of the complex relative permittivity. The direct current (DC) conductivity σ_{DC} must be taken into account when calculating the permittivity of higher conductivity media, such as seawater [4] or culture medium. Above 100 MHz, the influence of the moisture content and conductivity (ionic strength) of the object under test can be measured by separating the complex permittivity into its real and imaginary parts (1):

$$\varepsilon_r(\omega) = \varepsilon_r'(\omega) - j\varepsilon_r''(\omega) - j\frac{\sigma_{DC}}{\varepsilon_0 \omega} \quad (1)$$

Measurements with microwave sensors are particularly well suited for measurement application in non-destructive testing because microwaves can penetrate deep into the material without any direct contact and can even be transmitted through the material. For example the permittivity of water differs strongly from that of other liquids and aqueous mixtures ($\varepsilon_{H2O} \sim 80.5$, alcohol with $\varepsilon_{C2H5OH} \sim 24.1$, cellulose with $\varepsilon_{Cellulose} \sim 3.05$ at 1 GHz frequency and 25°C temperature), and this property highlights the great potential microwave sensors have for precise and fast applications in medicine, biotechnology and food technology. A variety of applicator designs have been described with regard to solving the basic electromagnetics. Concerning the coupling of applicator and object under test, the general types of applicators are: a) strip line applicators, b) stray field applicators, c) radiation field applicators, d) resonators and e) hybrid forms of the previously listed applicators. It is suggested that readers refer to these resources for a complete description of the relevant theory and practical approaches regarding the interaction of microwaves with dielectric materials [5, 6].

The compact sensor consists of a flow-through chamber; tubing can be reproducibly placed between strip-line electrodes that act as the sensor element (Figure 1). The tubing can be easily placed in the sensor and replacement is uncomplicated. Media can be pumped through the tubing so that the variation of the material under test is measured by the MW-Sensor. The strip-line waveguide is made from two copper cantilevers of 36 mm length and 3 mm width. The cantilevers were separated from each other by 3 mm. The strip-line waveguide was placed inside an aluminium housing (51 x 26 mm) and connected directly to the soldering terminals of an SMA connector.

Fig. 2 shows schematically how the frequency response of a resonator can be described by the width at half maximum, W, and the area, A, where the reflection coefficient (S_{11} expressed in dB) is presented as a function of frequency for liquids with varying water content. These electrical properties are the basis of numerous calculations for the determination of the complex permittivity of mixtures or as a calibration function for determining the composition of material compounds. Sometimes the exact estimation of the resonant frequency and the width at half maximum was only possible at reduced accuracy. For this reason, equation 2 was developed, to improve the calculation of the area underneath a resonant curve (in this equation SI is a reflection index).

$$SI_{(S11)} = \frac{A_x}{f_0 - f_x} \quad (2)$$

Further detail of this measurement set up (sensor design and data processing) can be found in [7].

Figure 1. Schematic of the strip line resonator

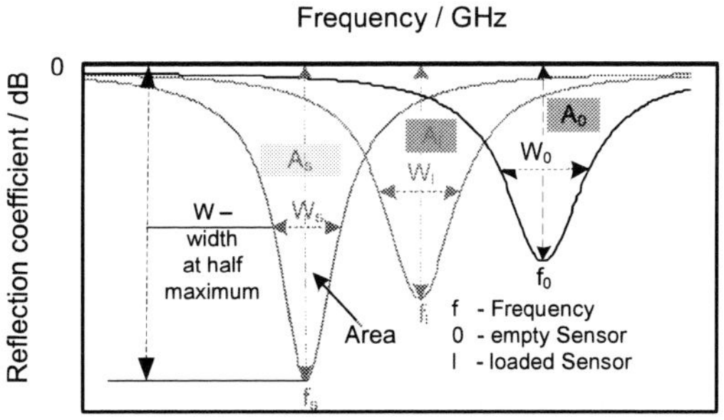

Figure 2. Frequency response of a resonator, measured reflection coefficient, basics of parameter estimation

3. Experiment and Results

The performance of the sensor was tested using several liquids with well-known dielectric and conductive properties: air, ethanol, deionized water, salt solution with different ionic concentrations and mixtures (Fig: 3, Fig 4). A special application was the determination of the ethanol concentration from a condensate during ethanol fermentation (Fig. 5) and the growth of s. cerevisiae in growth medium BM001B (SY-LAB) inoculated with 2.6×10^5 cells per ml. Measurements where performed in an incubator at a controlled temperature of 25°C. The complex reflection coefficient S_{11} was measured with a network analyzer (Anritsu MS MS4644A).

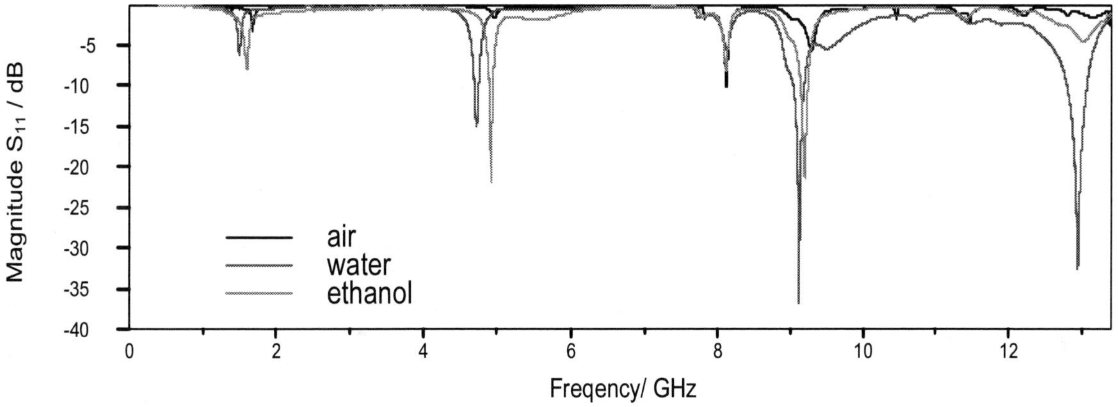

Figure 3. Wide band frequency response of strip line sensor figure 1 with pure media

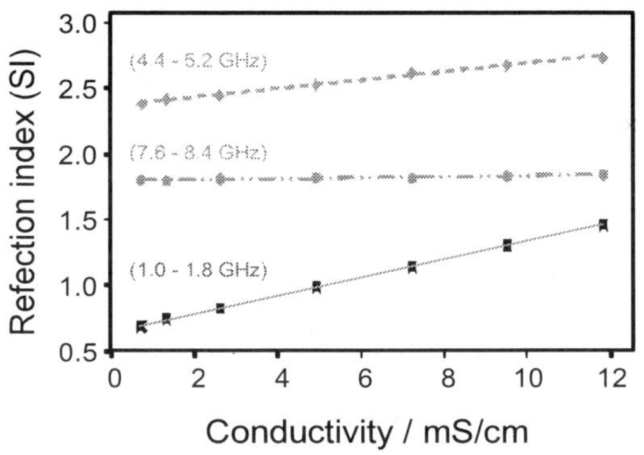

Figure 4. Reflection index calculated with formula 2 as a function of the ethanol-water mixture with different salt concentrations

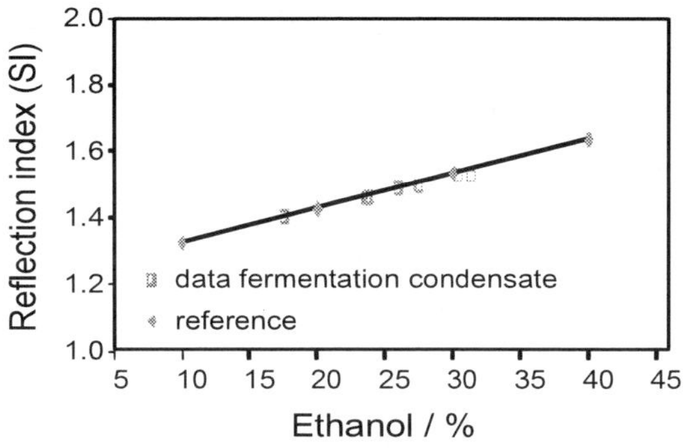

Figure 5. Reflection index SI resonance frequencies between 1.5 – 1.9 GHz as a function of the ethanol/ water dilution

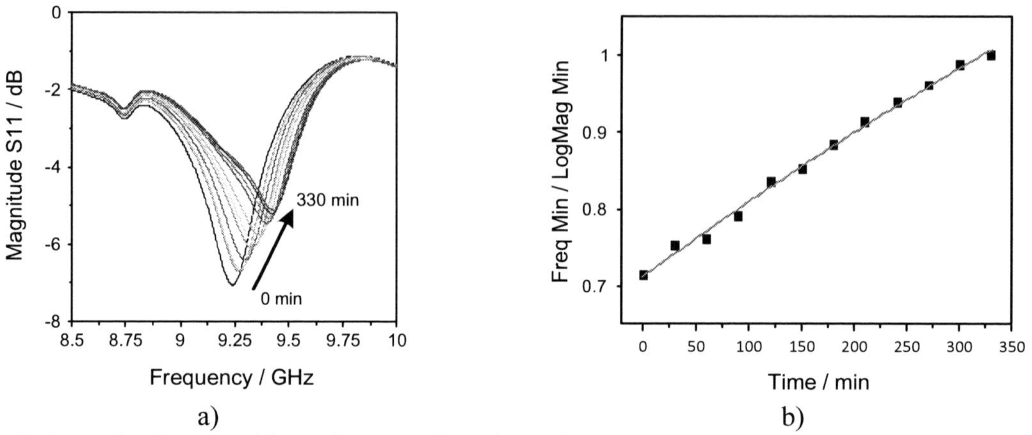

Figure 5. Growth of S. cerevisiae microorganisms in a stirred culture medium as a batch, fermentation time 5 h, a) shows the spectra and b) shows the time duration of the exponential grown phase

4. Conclusion and Future Work

The contactless nature of these sensors enables the cost-effective production of microwave sensors that can avoid sample contamination. Numerical simulation with high-frequency full wave electromagnetic software is necessary for optimization of the development of these sensors. Reproducible measurements require high precision in the mechanical production of the geometry of each sensor: obviously microfabrication offers such specifications.The next step will be the determination of the variation of conductivity and dielectric properties of a cell culture in a droplet application (volume >100 nl) in a segmented flow system. The number of cells, consumption of nutrients and cell metabolism directly influence the passive dielectric properties.

Together with reasonably priced high frequency electronic circuits (for example mini-circuits delivered high frequency synthesizers and Analog Deveces hole RF Gain-Phase Detector with bandwidths up to 2.7 GHz and the design and manufacturing of a scalar MW Device is also much cheaper than a laboratory system) this techniques opens up the possibility of automation simplification and miniaturization of routine application and in a wide variety of sample pretreatments..

References

[1] Gao D, Liu H, Jiang Y and Lin J M 2012 Recent developments in microfluidic devices for in vitro cell culture for cell-biology research *TrAC Trends in Analytical Chemistry.* ***35***, 150-164

[2] Wu J and Gu M 2011 Microfluidic sensing: state of the art fabrication and detection techniques. *Journal of Biomedical Optics.* 2011, *16*, 13

[3] Cahill B P, Land R, Nacke T, Min M and Beckmann D 2011 Contactless Sensing of the Conductivity of Aqueous Droplets in Segmented Flow *Sensors and Actuators B: Chemical* **159**(1): 286-293

[4] Stogryn A and Desargant G 1985 The dielectric properties of brine in sea ice at microwave frequencies." Antennas and Propagation *IEEE Transactions on* Antennas and Propagation **33**(5): 523-532

[5] Kupfer K 2005 Electromagnetic Aquametry, Electromagnetic Wave Interaction with Moist Substances. Springer Berlin Heidelberg NewYork, ISBN 3-540-22222-7

[6] Chen L, Ong C K, Neo C P, Varadan V V and Varadan V K 2004 Microwave electronics: measurement and materials characterization, John Wiley & Sons Inc., ISBN 0-470-84492-2

[7] Nacke T, Barthel A, Frense D, Meister M and Cahill B P 2013 Application of High Frequency Sensors for Contactless Monitoring in Disposable Bioreactors *Chemie Ingenieur Technik* **85**(1-2): 179-185

Detection of person presence and its activity in the bathtub

Adam Bujnowski, Arkadiusz Palinski, Piotr Koscinski, Lukasz Skalski, Anna Skurczynska, Jerzy Wtorek

Biomedical Engineering Department, Gdansk University of Technology, Narutowicza 11/12 , 80-233 Gdansk, Poland

E-mail: bujnows@biomed.eti.pg.gda.pl

Abstract. A practical application of a bioimpedance technique for a detection of a bathing person is presented in the paper. It addresses the possibility of supervising people in the bathtub without voiding of their intimacy. The measurement system installed in a fiber-glass or a plastic bathtub is able to detect a presence of the bathing person, to estimate its activity and thus to detect potentially dangerous events. In the paper a principle of measurement, working prototype and measurements are presented. The proposed method can be useful for supporting and supervising bathing of elders, partially disabled or people with some health state risk during the bath and living alone.

1. Introduction

The bathroom is a very important place in every house for hygiene and maintaining of the health. It is also the place where one can expect maintenance of his (her) intimacy. However, due to the tasks performed and the type of furniture it can be also one of the most dangerous places in the house. Wet floor and hard furniture might be dangerous especially for elders with some mobility dysfunctions. There are reports from social care employees of elders that could not manage to leave the bathtub itself and thus were kept staying in the water for prolonged period (eg. a whole night). There are also reports of sudden when bathing [3]. In such situations the supervision of bathing person without voiding of their intimacy could be invaluable assistance. The bioimpedance technique seems to be a well suited for a detection of biological cells presence in solutions. Detection of a person presence in the bathtub can be thus regarded as a problem of detection of living cells in the water.

In the frequency range between 1 kHz and 10 MHz impedance of a water (with salts and detergents) exposes mainly resistive character. In contradiction, for the living cells there is a characteristic magnitude of the impedance drop at certain frequency which is related to polarization phenomena. In this frequency range a characteristic phase angle shift is also present.

Presence of the measuring electrodes and associated connecting wires introduce stray capacitances. They limit the resistive range to below 1 MHz. When a human body is immersed in a water character (frequency dependence) of the measured impedance is changed. It is known that living cells exhibit, aside the resistive, capacitive properties. Moreover, electrical properties of the living cells are also dispersive characteristic impedance change around specific frequency and it is associated with a significant change of phase angle. The principle of operation of a developed system is based on measuring the impedance using set of electrodes placed at the bottom of the bathtub. Changes of the impedance modulus measured for the selected frequency are monitored and used for evaluating the bathing person activity. Variance of modulus value is utilized to estimate movement of the bathing person.

2. Materials and Methods

The concept of bathing person monitoring relies on the impedance measurement. Thus, the plastic (or acrylic) bathtub is equipped with three electrodes placed on the bottom of the bathtub. Each electrode is compound one, i.e. it contains both, the current and voltage parts. The impedance is measured between selected pair of the compound electrodes, giving two independent measurements.

2.1. Sensitivity of measurements

Assume bounded domain with two regions: one of the background conductivity of σ_1 and another one of a perturbation conductivity of σ_2. Following the Gesselovitz sensitivity relationship we obtain [5]:

$$\Delta Z = \int_V -\Delta\sigma \frac{\delta\phi(\sigma)}{I_{1-2}} \cdot \frac{\Delta\Psi(\sigma + \Delta\sigma)}{I_{3-4}} dv$$

where the V denotes the volume of considered domain Ω , the σ and $\Delta\sigma$ are the conductivity, and conductivity change of the domain respectively, ϕ is the potential within Ω caused by current I_{1-2} issued between ports $1-2$ and ψ is the hypothetical potential within Ω caused by hypothetical current I_{3-4} applied between ports 1 and 2.

In general conductivity is non-uniform and it depends on spatial conductivity distribution in relation to the measurement ports and in case of dispersive materials on the frequency.

In the considered problem - detection of presence of the human in the bath - the measured impedance change depends on the frequency, conductivity of the bathing fluid and position of the person in relation to the measurement electrodes.

Considering the frequency range of $2kHz - 100kHz$ the bathing fluid exhibits constant, frequency independent and real (no imaginary part) conductivity. For the human body the modulus of the impedance is strongly frequency-dependent value in this frequency range. It is related to the dispersion phenomena, mainly involved by a presence of the living cells (cell membranes). Thus analysing the impedance spectrum of bathing fluid it is possible to distinguish between bathtub filled only with fluid from the bath with the person immersed in it. Under certain circumstances it is possible to detect presence of bathing person by measuring constantly impedance on selected frequency.

2.2. The measurement system

The measurement system was constructed basing on AD5933 (Analog Devices Ltd) integrated circuit. The AD5933 is a complete impedance analyser including DDS with sine-table stored in ROM, output amplifier, input I/U converter and a complete digital synchronous demodulator using the 1024-bit real and imaginary register. The AD5933 is controlled by means of the IIC using ATMega32 (Atmel) microcontroller. The AD5933 utilizes a two-electrode method. It creates a major problem for using it in the considered application. In a case of the bioimpedance measurement using two-electrode method leads to false readings as the electrode impedance is included in the readings.

To meet the measurement requirements an input circuit of AD5933 has been enhanced with additional one. It utilizes Howlands voltage controlled current source, an instrumentation amplifier, additional filters and polarization circuitry [4]. However, an application of AD5933

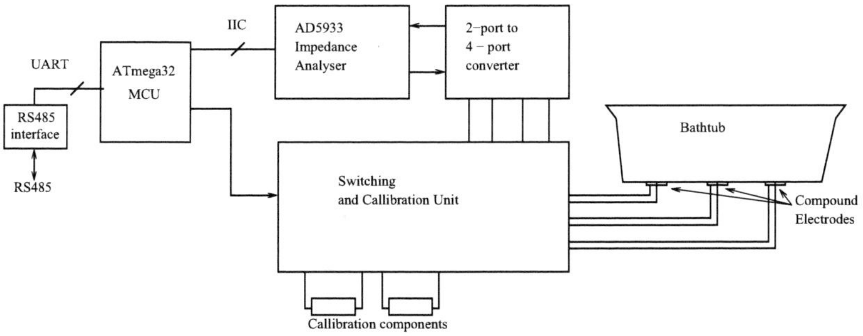

Figure 1. Block diagram of the measurement system

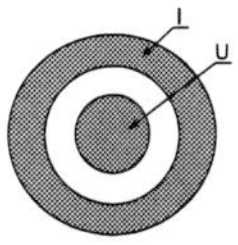

Figure 2. Measurement electrode shape

in a four-electrode technique involves a development of a calibration procedure. To improve performance of the measurement system the calibration resistors are included in the circuit. Moreover, the calibration procedure must be performed for each individual frequency, selected from the prescribed frequency table. The calibration procedure requires to determine three parameters: magnitude scale factor, offset value, and the phase correction angle. It has been developed and tested using the reference components which were integrated with the measurement circuit.

3. Results

The measurement system has been designed, and the prototype of the measurement board has been developed as it is shown in figure 3. An acrylic bathtub has been equipped with the measurement electrodes (figure 4). Electrodes are made of silver-plated brass and glued on the bottom of the bathtub. The measurement system is a part of broader one containing displays and servomotors used for bathtub filling and flushing [1, 2]. Thus, the impedance meter has serial RS485 interface for data transmission. Tests of the developed system were performed. However, instead of two frequency measurements a spectroscopic ones were utilized. A volunteer has been asked for taking the bath using the developed system. He has been asked to perform several exercises (movements) during the test. We used a casual liquid soap for bathing and in a second test we used a bathing salt. After performing the calibration the bathtub was filled with a warm tap water. Impedance of the water has been measured . When the person entered the bathtub impedance measurements have been performed again (see figure 5).

Additionally volunteer was asked to stop moving for certain period, and then to start moving. Instead os spectroscopy, such activity could be monitored using continous single frequency measurements. Exemplary result is shown in the figure 6.

During set of measurements the performance of the measurement system was checked. Usually for the same conditions the measurement error is less than 1%. A special caution must be taken when selecting frequency of the measurement signal. It appears that for a certain frequency measurement errors are greater than for other. Fortunately, a selection of an adjacent frequency (e.g. 1Hz higher) can solve the problem.

4. Conclusion

We have proposed the monitoring method of the bathing person. It is based on electrical impedance measurement without using camera. We have designed ant tested the performance of the measurement system. The information gained can be assigned to several categories e.g. no body in the bathtub, body present in the bathtub and it is moving, body present , no movement detected. Such information is used for detecting of dangerous situations when taking the bath,

Figure 3. Prototype of the measurement system with the self-calibration

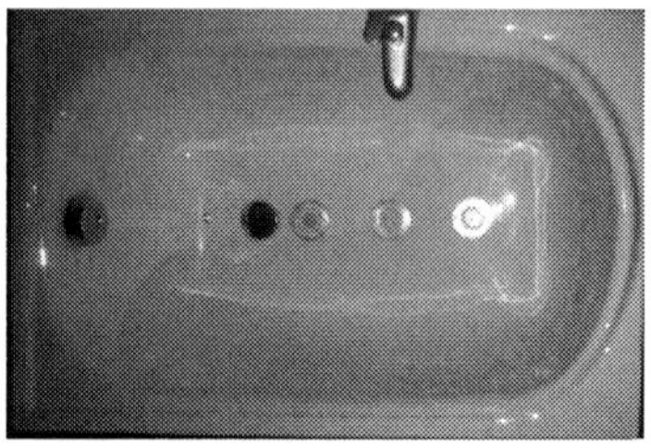

Figure 4. Modified bathtub

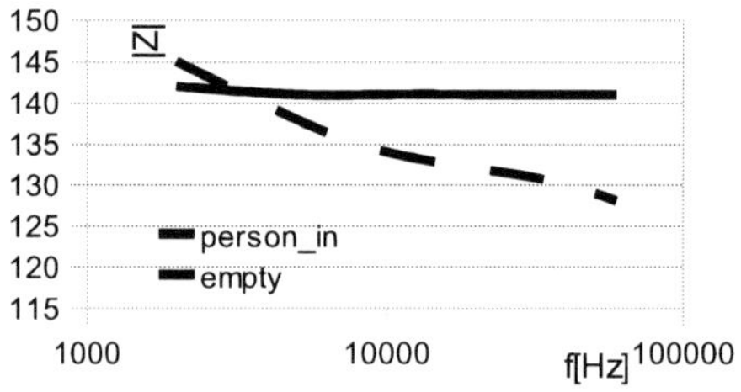

Figure 5. Comparison of the bath impedance with and without person

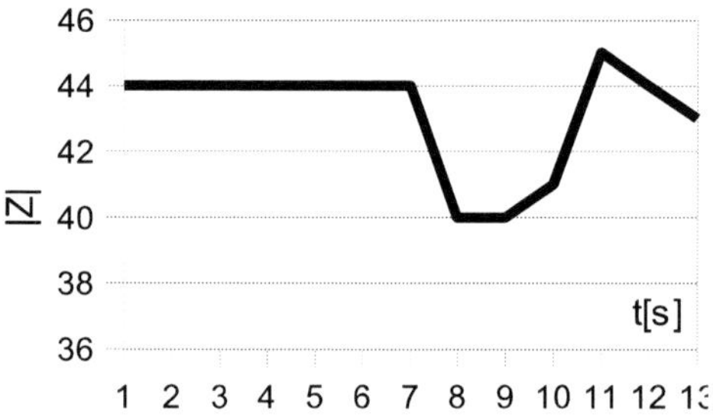

Figure 6. Continous monitoring of person activity - modulus of the impedance measured at 30kHz in 1s time interval. Until 7 sec volunteer was relaxed, after started to move

and in connection to other systems (alarms, water flushing system, etc) can improve safety of the bathing person without violating her/his intimacy and privacy.

5. Acknowledgements
This work has been partially supported by European Regional Development Fund concerning the project: UDA-POIG.01.03.01-22-139/09-00 - "Home assistance for elders and disabled - DOMESTIC", Innovative Economy 2007-2013, National Cohesion Strategy.

References

[1] Bujnowski A, Wtorek J and Palinski A 2011 Intelligent bathroom, *Proceedings of Federated Conference on Computer Science and Information Systems - FedCSIS 2011* Szczecin, Poland pp 381-386
[2] Bujnowski A, Skalski L and Wtorek J, Monitoring of a Bathing Person 2012 *J. Med. Imaging Health Inf.* **2** pp 27-34
[3] Chibaa T, Yamauchia M, Nishidaa N, Kanekob T, Yoshizakib K and Yoshiok N 2005 Risk factors of sudden death in the Japanese hot bath in the senior population *Forensic Science International* **149** pp 151-158
[4] Ferreira J, Seoane F, Ansede A and Bragos R 2010 AD5933-based spectrometer for electrical bioimpedance applications *J.Phys., Conference series* **244** No. 1
[5] Gesselowitz D An Application of electrocardiographic lead theory to impedance plethysmography 1997 *IEEE Trans. Biomed. Eng.* **18** pp 38-41

Initial Systolic Time Interval (ISTI) as a Predictor of Intradialytic Hypotension (IDH)

JD Biesheuvel[1], MG Vervloet[2], RM Verdaasdonk[1], JH Meijer[1]

[1] Department of Physics and Medical Technology, VU University Medical Center, Amsterdam, the Netherland
[2] Department of Nephrology, VU University Medical Center, Amsterdam, the Netherlands

E-mail: jh.meijer@vumc.nl

Abstract. In haemodialysis treatment the clearance and volume control by the kidneys of a patient are partially replaced by intermittent haemodialysis. Because this artificial process is performed on a limited time scale, unphysiological imbalances in the fluid compartments of the body occur, that can lead to intradialytic hypotensions (IDH). An IDH endangers the efficacy of the haemodialysis session and is associated with dismal clinical endpoints, including mortality. A diagnostic method that predicts the occurrence of these drops in blood pressure could facilitate timely measures for the prevention of IDH. The present study investigates whether the Initial Systolic Time Interval (ISTI) can provide such a diagnostic method. The ISTI is defined as the time difference between the R-peak in the electrocardiogram (ECG) and the C-wave in the impedance cardiogram (ICG) and is considered to be a non-invasive assessment of the time delay between the electrical and mechanical activity of the heart. This time delay has previously been found to depend on autonomic nervous function as well as preload of the heart. Therefore, it can be expected that ISTI may predict an imminent IDH caused by a low circulating blood volume. This ongoing observational clinical study investigates the relationship between changes in ISTI and subsequent drops in blood pressure during haemodialysis. A registration of a complicated dialysis showed a significant correlation between a drop in blood pressure, a decrease in relative blood volume and a substantial increase in ISTI. An uncomplicated dialysis, in which also a considerable amount of fluid was removed, showed no correlations. Both, blood pressure and ISTI remained stable. In conclusion, the preliminary results of the present study show a substantial response of ISTI to haemodynamic instability, indicating an application in optimization and individualisation of the dialysis process.

1. Introduction

Haemodialysis (HD) is a therapy that supports or replaces kidney function in the case of renal failure. Although haemodialysis is a life saving therapy for end-stage renal patients it entails various side effects [1], and the patient population shows a high morbidity rate [2]. Especially intradialytic hypotension (IDH) is a persistent problem for about 20-25% of the HD patients. IDH is defined as an abrupt (over 10 to 15 min) decrement in blood pressure accompanied by symptoms such as: abdominal discomfort, yawning, nausea, vomiting, muscle cramps, dizziness, fainting, or a decline of blood pressure requiring a nursing intervention [3].

Journal of Physics: Conference Series **434** (2013) 012036 doi:10.1088/1742-6596/434/1/012036

IDH originates primarily from a decrease in blood volume due to fluid removal by HD and a delayed refill of fluid from other parts of the body. The body can be considered to consist of three compartments that contain fluid: the intravascular compartment containing blood, the interstitial space surrounding cells and the intracellular space inside cells (see figure 1). Fluid removed by HD is withdrawn from the intravascular compartment which should be complemented by fluid from the interstitial and the intracellular space. However, in some circumstances this process is not fast enough, or is counteracted by osmotic fluid transport from the hypotonic blood compartment to the hypertonic interstitial and intracellular space, which can result in an IDH.

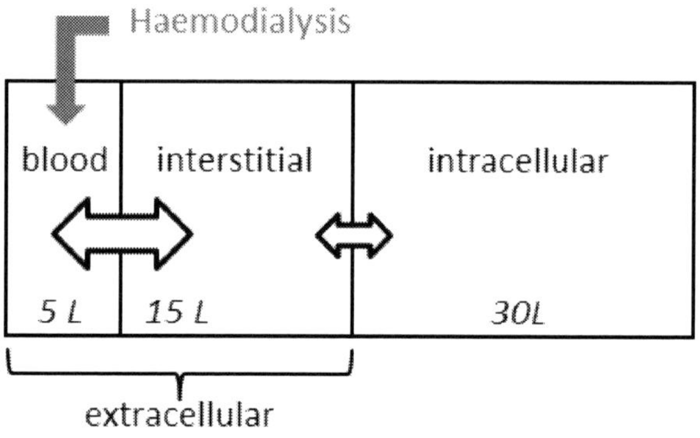

Figure 1. The three compartment model of the human body. Haemodialysis intervenes with the intravascular compartment only. The arrows represent fluid transport, whereby the size of the errors corresponds to the degree of transport.

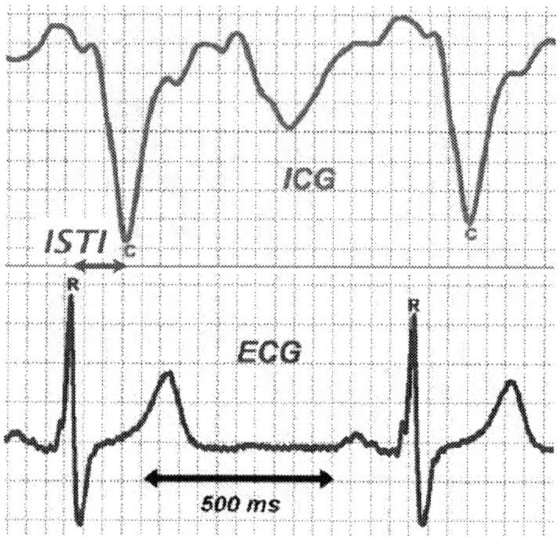

Figure 2. A typical example of simultaneous registration of an impedance cardiogram (ICG) and an electrocardiogram (ECG). The time difference between the R-point in the ECG and the C-point in the ICG is defined as the Initial Systolic Time Interval (ISTI).

The use of impedance cardiography (ICG) to study haemodynamic stability has been reported before by Meijer et al. in patients, admitted to the intensive care unit after cardiac surgery [4]. In these patients, fluid was administered intravenously in order to expand the circulating volume of the intravascular compartment. This situation can be viewed as a mirror image of haemodialysis.

Meijer et al. showed that an interesting aspect of the ICG-signal can be found in the time-relationships, when the moments of occurrence of specific waves in the ICG are compared to specific points in the electrocardiogram (ECG) [5]. They introduced the Initial Systolic Time Interval (ISTI), where the ISTI was defined as the time interval between the R-point in the ECG and the C-point in the ICG (see figure 2).

Assuming that ISTI depends on preload of the heart by way of the Frank-Starling mechanism and on autonomic nervous control of the heart, it can be expected that ISTI is influenced by an imminent IDH during haemodialysis and that ISTI may be used as a predictor of IDH. Moreover, previous research showed an inverse relationship between ISTI and cardiac output [4] and a decrease of ISTI during a Valsalva manoeuvre [6]. The present study investigates whether ISTI can be used to evaluate and predict hypotensive periods during dialysis, and the present paper reports on preliminary results obtained from patients undergoing haemodialysis.

2. Method

ICG recordings were made using a four-electrode system attached to the left side of the body described in detail by Meijer et al. [5]. The outer two electrodes applied a small alternating (64 kHz) electrical current i of 0.3 mA (r.m.s.) through the upper part of the body. The inner two electrodes continuously measured the subsequent electrical voltage difference $V(t)$ over the region of the heart, from which the electrical impedance *Z(t)* was computed continuously. The ECG signal was derived simultaneously from the two inner electrodes. Recordings were made using an impedance cardiograph designed and built at the Department of Physics and Medical Technology, VU University Medical Center, Amsterdam, the Netherlands. The signals were AD-converted using an ADInstruments PowerLab 8SP and processed on a personal computer using ADInstruments Chart (version 5.1.1) for Windows.

The study was performed on the Nephrology department of the VU University Medical Center. Continuous registrations of ECG and ICG were made during standard dialysis sessions. Also blood pressure (BP) and the relative blood volume (RBV) were registered by the dialysis machine (Nikkiso DBB-07). Recordings of 1 minute with a time interval of 15 minutes of these parameters were made in a resting state (lying or sitting) and processed. All patients gave their written informed consent prior to participation in the study.

3. Preliminary results

Figure 3 shows examples of a complicated dialysis (A) and one of an uncomplicated dialysis (B).

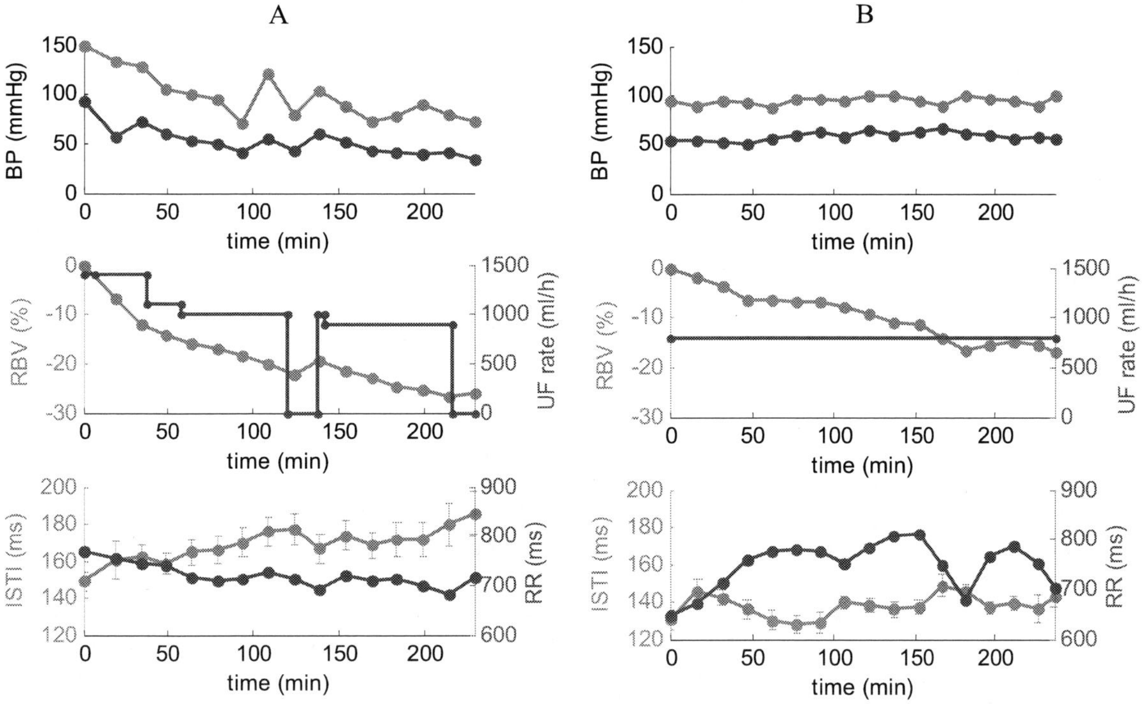

Figure 3. Two typical examples of a complicated dialysis (A) and an uncomplicated dialysis (B). The top graph shows both the systolic (red) and diastolic (blue) blood pressure. The middle graph shows the relative blood volume as a percentage of the initial blood volume (left) and the ultrafiltration rate (right). The bottom graph shows the mean ISTI ± SD (left) and the RR-interval of the ECG (right).

The registration of the complicated dialysis was taken from a 61-years old woman (162 cm, 90 kg), during a normal haemodialysis session of 4 hours. Due to dizziness, the session was limited to 3.45 hours in which 3510 ml of fluid was extracted by ultrafiltration. In order to prevent a severe hypotension the ultrafiltration rate was adjusted after 120 minutes of haemodialysis. During this dialysis session, the systolic blood pressure (SBP) decreased from 149 mmHg at the start to 72 mmHg at the end. Parallel, ISTI showed an increase from 149 ms to 186 ms. Regression analysis showed a strong significant correlation between ISTI and SBP ($p < 0.005$) and between ISTI and RBV ($p < 0.001$), shown in respectively figure 4 and 5.

A registration of an uncomplicated dialysis is shown in figure 3B. This registration was taken from a 50-years old male (176 cm, 105 kg) during a normal dialysis session of 4 hours in which 3100 ml of fluid was extracted. During this uncomplicated dialysis session SBP remained stable at 95 mmHg while the ultrafiltration rate was high and RBV decreased. ISTI and RR also remained stable. Regression analysis showed no significant correlation between ISTI and SBP ($p > 0.05$) and between ISTI and RBV ($p > 0.05$).

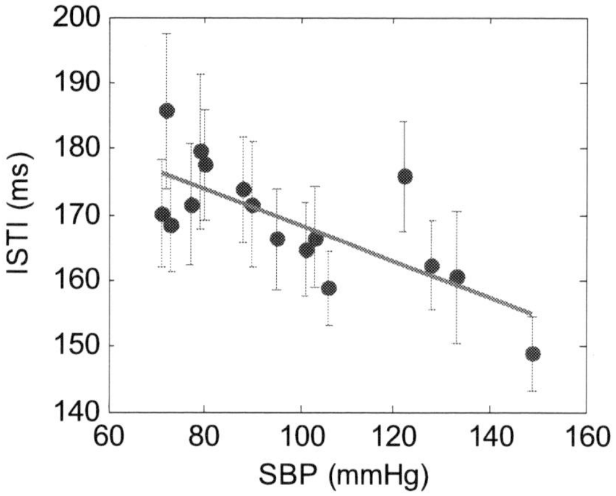

Figure 4. Relationship between ISTI and systolic blood pressure (SBP) during the complicated dialysis session ($r = 0.736$, $p < 0.005$). Data are means ± SD.

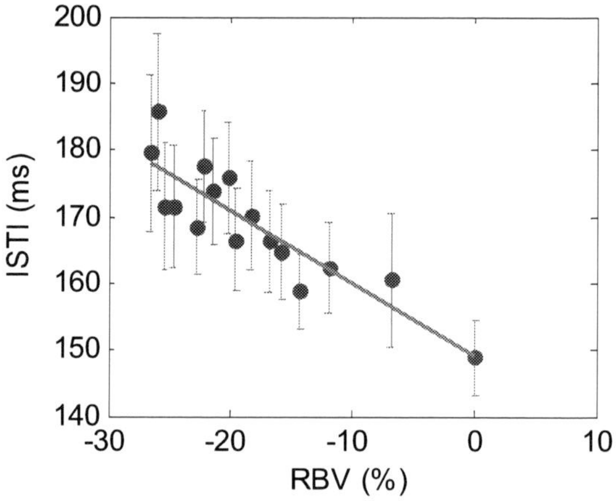

Figure 5. Relationship between ISTI and relative blood volume (RBV) during the complicated dialysis session ($r = 0.887$, $p < 0.001$). Data are means ± SD.

4. Discussion and conclusion

The present study focuses on the evaluation of the use of ISTI to predict IDHs. So far, not sufficient registrations of an IDH were made to draw decisive conclusions. The preliminary results are consistent with the expectation that in the case of an IDH the ISTI increases and in the case of an uncomplicated dialysis the ISTI remains stable. The major difference between the two examples is that during the complicated dialysis session an IDH occurs and during the uncomplicated session not, while a similar amount of fluid was removed. Although the cause of an IDH is not fully understood, the extraction rate of fluid from the intravascular compartment in relation to the refill by fluid from the other body compartments is of importance. Further, the degree of overfilling may play an important role. An overfilled patient has a stable, high preload and an operating point of the heart on the upper part of the Frank-Starling curve. Because ISTI depends on the preload of the heart, ISTI is expected not to change in an overfilled person during haemodialysis, which is consistent with the observations.

In conclusion, the preliminary results of the present study show a substantial response of ISTI to haemodynamic instability, indicating an application in optimization and individualisation of the dialysis process.

References

[1] Eknoyan G 1984 *N. Engl. J. Med.* **311** 915-7
[2] Rayner H C et al. 2004 *Nephrol. Dial. Transplant.* **19** 108-20
[3] Dheenan S and Henrich W L 2001 *Kidney Int.* **59** 1175-81
[4] Meijer J H, Smorenberg A, Lust E J, Verdaasdonk R M and Groeneveld A J 2010 *J. Electr. Bioimp.* **1** 80-3
[5] Meijer J H, Boesveldt S, Elbertse E and Berendse H W 2008 *Physiol. Meas.* **29** S383-91
[6] Meijer J H, Hoekstra F, Habers E, Verdaasdonk R M and Janssen T W 2010 *J. Phys.: Conf. Ser.* **224** 012119

Journal of Physics: Conference Series **434** (2013) 012037 doi:10.1088/1742-6596/434/1/012037

Bioimpedance-based respiratory gating method for oncologic positron emission tomography (PET) imaging with first clinical results

T Koivumäki[1,2], M Vauhkonen[2], J Teuho[3,4], M Teräs[4] and M A Hakulinen[1]

[1]Diagnostic Imaging Centre, Kuopio University Hospital, Kuopio, Finland
[2]Department of Applied Physics, University of Eastern Finland, Kuopio, Finland
[3]GE Healthcare, Helsinki, Finland
[4]Turku PET Centre, University of Turku and Turku University Hospital, Turku, Finland

E-mail: tuomas.koivumaki@kuh.fi

Abstract. Respiratory motion may cause significant image artefacts in positron emission tomography/computed tomography (PET/CT) imaging. This study introduces a new bioimpedance-based gating method for minimizing respiratory artefacts. The method was studied in 12 oncologic patients by evaluating the following three parameters: maximum metabolic activity of radiopharmaceutical accumulations, the size of these targets as well as their target-to-background ratio. The bioimpedance-gated images were compared with non-gated images and images that were gated with a reference method, chest wall motion monitoring by infrared camera. The bioimpedance method showed clear improvement as increased metabolic activity and decreased target volume compared to non-gated images and produced consistent results with the reference method. Thus, the method may have great potential in the future of respiratory gating in nuclear medicine imaging.

1. Introduction

Positron emission tomography (PET) combined with computed tomography (CT) imaging is an important tool in oncology. Studies with fluorine 18 ([^{18}F]) –based radiopharmaceuticals can be used for cancer staging and treatment response evaluation, for example [1]. Respiratory motion is a significant error source in PET/CT studies of thoracic region. Due to the long imaging time of PET, which is typically minutes, respiratory motion leads to image blurring, which may result in overestimation of target's size and underestimation of its metabolic activity [1].

At present, there are many studies on minimizing motion artefacts by synchronizing, or gating, the image reconstruction with a simultaneously recorded respiratory signal. The respiratory signal is most often derived from thoracic or abdominal chest wall motion, which is typically monitored by infrared camera or pressure belt [1, 2]. However, respiratory gating has been adopted slowly in clinical routine. One reason for this may be that the methods have been considered complicated and cumbersome to use.

Our aim is to develop a bioimpedance-based respiratory gating method, which is as accurate as previously introduced respiratory gating methods but easier and faster to use. Previously, an optimized four-electrode bioimpedance measurement configuration was determined horizontally on anterior upper thorax using sensitivity analysis and volunteer measurements [3]. Further, high correlation and

Journal of Physics: Conference Series **434** (2013) 012037 doi:10.1088/1742-6596/434/1/012037

very good agreement was confirmed between respiratory bioimpedance measurements and reference spirometer measurements in healthy volunteers [4]. The aim of this paper is to compare developed bioimpedance gating method with a widely studied reference method and to show that the results are consistent between the methods.

2. Methods

Twelve oncologic patients (4 women, 8 men) were recruited to this study in Turku PET Centre. The patients' age was 46-77 (mean 65) years and their body mass index varied in the range of 21.3-38.5 (mean 26.3) kg/m^2. The gating study was conducted alongside the clinical PET/CT study, to which the patients were referred. Written informed consent was obtained from all of the patients and the study was approved by the ethical committee of Kuopio University Hospital (Dnro 90/2011).

The PET/CT imaging was performed with a high-resolution Discovery 690 PET/CT scanner (GE Medical Systems, Milwaukee, USA) and [^{18}F]-labeled radiopharmaceuticals ([^{18}F]-fluorodeoxy-glucose, [^{18}F]-FDG or [^{18}F]-fluorodihydroxyphenylalanine, [^{18}F]-FDOPA). The bioimpedance measurement was carried out with EBI100C bioimpedance amplifier (Biopac, Goleta, USA) and processed in MATLAB R2007B (MathWorks Inc., Natick, USA). The processing of bioimpedance measurement consisted of the removal of cardiac component from the band between 0.6-20 Hz using 5th order Butterworth filters and smoothing of the remaining respiratory signal with moving average filter of 100 ms. If a clearly non-physiological drift was observed in the signal, a 50 s moving average filter was additionally applied for detrending. Real-time Position Management (RPM) system (Varian, Palo Alto, USA) was used as reference gating system in the study. The gated image reconstruction was performed with GE Research Gating Tool (RGT) based on breathing depth. Thus, only PET data during end-expiration, which was considered the longest and most stable phase of respiration, was included in the image reconstruction. In this study, end-expiration was determined as the lowest fifth of the respiratory amplitude of each cycle individually (figure 1). Due to excluding large part of the PET data in the gating process, the PET imaging time was increased to 10 minutes compared to 2 minutes used in the non-gated study. An averaged CT over the whole respiratory cycle was used for the attenuation correction of the PET data.

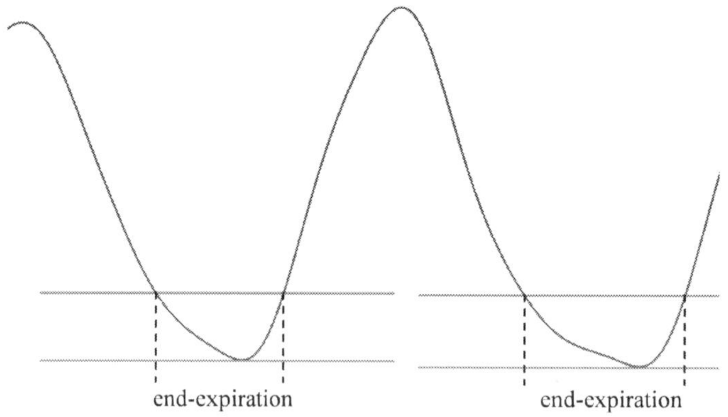

Figure 1. Selection of end-expiratory periods from the respiratory bioimpedance measurement for the reconstruction of PET image. Horizontal lines illustrate the amplitude limits (the lowest fifth of the respiratory amplitude) that determine the boundaries of end-expiratory periods in time. End-expiratory periods are depicted by dashed vertical lines.

Three image sets of each patient were created in the study: standard non-gated, bioimpedance-gated and RPM-gated. The bioimpedance gating method was evaluated by studying the difference and correlation of three parameters between bioimpedance- and RPM-gated images as well as the change of these parameters compared to non-gated images. The studied parameters were maximum metabolic activity of the target, observed target size and target-to-background ratio. The parameters were determined with standard clinical AW Workstation 4.5 (GE Medical Systems, Milwaukee, USA).The targets were delineated utilizing the volume of interest (VOI) tool. Maximum metabolic activity (g/ml) was determined as the maximum voxel value in the target. The target size was determined as the volume of the target (cm^3), where the activity of the target was 50-70 % of its maximum activity. The threshold varied slightly between the patients in order to achieve reliable delineation of the target in all image sets. Target-to-background ratio was determined as the ratio of mean metabolic activity in the observed target volume and mean of background activity within VOI, which was determined on the flank of the patient with no specific uptake of the radiopharmaceutical.

3. Results

The radiopharmaceutical uptake of the target is more visible in both gated images compared to the non-gated image (figure 2). The maximum metabolic activity was determined in altogether 15 targets in 11 patients. The targets of one patient were excluded from the analysis, since no averaged attenuation correction CT was available. The maximum metabolic activity changed -6.3 - +122.4 % (mean 20.5 ± 31.5 %) from non-gated to bioimpedance-gated images, whereas with the RPM-gated images the corresponding range was 1.6-104.1 % (mean 20.6 ± 26.1 %). On average, the maximum metabolic activity between bioimpedance and RPM gating differed 3.8 ± 3.8 % (of the RPM-based measurement). The correlation of maximum metabolic activity measurements of gated images was very high ($R^2 = 0.978$) (figure 3a).

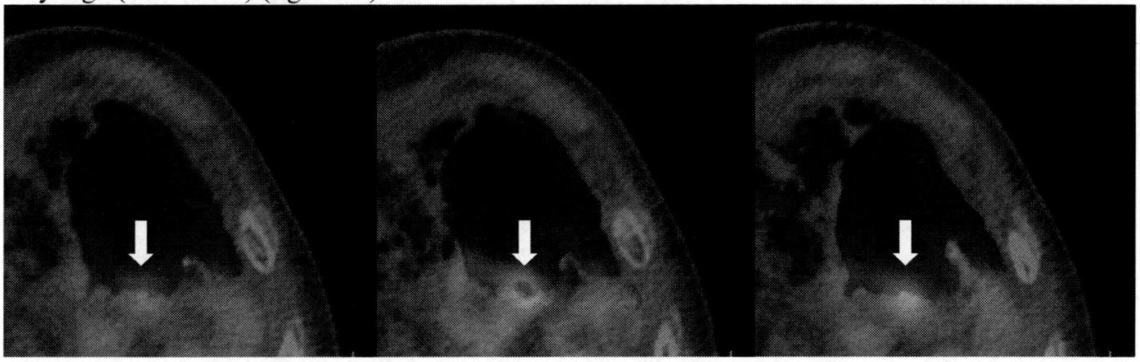

a) b) c)

Figure 2. The radiopharmaceutical accumulation of the target (pointed by the arrow) in non-gated (a), bioimpedance-gated (b) and RPM-gated (c) transaxial image. The PET image is fused with the CT image. In each case (a, b, c) the slice with the most distinguishable target is shown.

The target size was observed from 13 targets in 9 patients. Two additional patients were excluded in this analysis, due to the complicated form of the target, for which it was not possible to delineate the volume of the target reliably. Compared to non-gated images, the target size changed -62.0 - +16.4 % (mean -18.0 ± 25.5 %) in bioimpedance-gated images and -63.9 - +27.4 % (mean -16.0 ± 25.6 %) in RPM-gated images having the average difference of 8.2 ± 5.7 % between gated images. Furthermore, a very good correlation ($R^2 = 0.954$) was observed between the gating methods (figure 3b).

Target-to-background ratio was determined from the same targets as the size, since the observed target size essentially affects the mean metabolic activity value, which is used in calculating the target-to-background ratio. Target-to-background ratio increased from the non-gated to bioimpedance- and RPM-gated images by -4.8 - +132.3 % (mean 22.7 ± 36.9 %) and 0.0-116.1 % (mean 21.1 ± 32.4 %), respectively. The mean difference between the gating methods was 2.4 ± 2.8 % and the measurements correlated excellently ($R^2 = 0.994$) with each other (figure 3c).

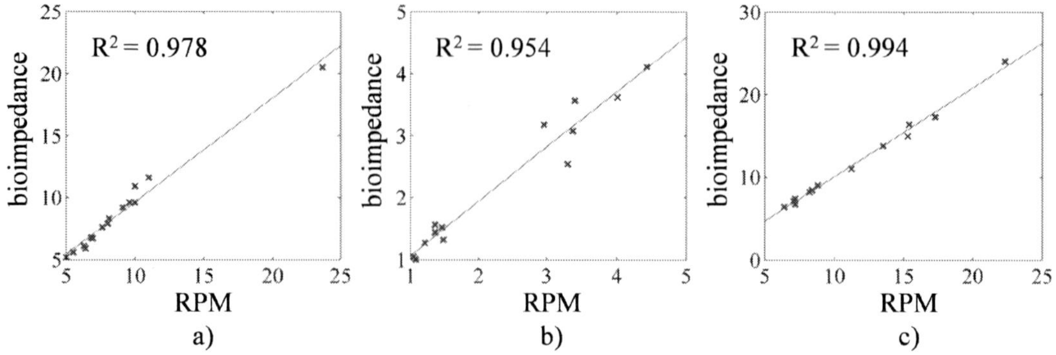

Figure 3. Correlation plots of measured maximum metabolic activity of the target in g/ml (a), target size in cm^3 (b) and target-to-background ratio (c).

4. Discussion and Conclusion

As expected, both gating methods compensate effectively the overestimation of target size and the underestimation of the metabolic activity of the target. The increase of target-to-background ratio was also expected. In addition, the measurements of all three parameters are consistent between bioimpedance and RPM gating and have a very high correlation. The mean differences between the bioimpedance- and RPM-gated measurements are small in all studied parameters, though the range of measurement results is considerably wide. Comparing the means of the measured change in parameters between non-gated and gated image sets, bioimpedance gating has a slightly stronger effect on parameters than the RPM method. However, more patient studies are required for a definite conclusion on the relation between bioimpedance and RPM gating.

The results of this study show that bioimpedance-based respiratory gating produces consistent results in oncologic PET/CT studies with the widely used RPM method. Moreover, the bioimpedance method is very straightforward and fast for the technologists to use. Thus, the method may have great potential in the future of respiratory gating in nuclear medicine imaging.

Acknowledgments

The study was conducted within the Finnish Center of Excellence in Molecular Imaging in Cardiovascular and Metabolic Research supported by the Academy of Finland, University of Turku, Turku University Hospital and Åbo Akademi University. This work was funded by Academy of Finland (International Doctoral Programme in Biomedical Engineering and Medical Physics), Kuopio University Hospital (EVO, project 5031345), Oskar Öflunds Stiftelse and Instrumentarium Science Foundation.

References

[1] Nehmeh S A, Erdi Y E, Clifton C L, Rosenzweig K E, Schoder H, Larson S M, Macapinlac H A, Squire O D and Humm J L A reference 2002 Effects of respiratory gating on quantifying PET images of lung cancer *J. Nucl. Med.* **43** 876-81

[2] Bundschuh R A, Martínez-Möller A, Essler M, Martínez M-J, Nekolla S G, Ziegler S I and Schwaiger M 2007 Postacquisition detection of tumor motion in the upper abdomen using list-mode PET data: a feasibility study *J. Nucl. Med.* **48** 758-63

[3] Koivumäki T, Vauhkonen M, Kuikka J T and Hakulinen M A 2011 Optimizing bioimpedance measurement configuration for dual-gated nuclear medicine imaging: a sensitivity study *Med. Biol. Eng. Comput.* **49** 783-91

[4] Koivumäki T, Vauhkonen M, Kuikka J T and Hakulinen M A 2012 Bioimpedance-based measurement method for simultaneous acquisition of respiratory and cardiac gating signals *Physiol. Meas.* **33** 1323-34

Journal of Physics: Conference Series **434** (2013) 012038 doi:10.1088/1742-6596/434/1/012038

Possibilities of electrical impedance tomography in gynecology

Trokhanova O V[1], Chijova Y A[1], Okhapkin M B[1], Korjenevsky A V[2] and Tuykin T S[2]

[1]Yaroslavl State Medical Academy, Yaroslavl, Russia; [2]Kotel'nikov Institute of Radioengineering and Electronics of the RAS, Moscow, Russia

trokhanova@yandex.ru

Abstract. The paper describes results of comprehensive EIT diagnostics of mammary glands and cervix. The data were obtained from examinations of 170 patients by EIT system MEM (multi-frequency electrical impedance mammograph) and EIT system GIT (gynecological impedance tomograph). Mutual dependence is discussed.

1. Introduction

Electrical impedance tomography (EIT) is a method of imaging the spatial distribution of the electrical impedance provided by measurements taken from the object surface with further solving an inverse problem.

A mammary gland is an integral part of the female reproductive system, and its diseases, which are generally benign, are consequences of general processes in the organism and reproductive organs of a woman. It is important to note that to consider pathologies of genitals and mammary glands separately is a basic mistake. When planning treatment of patients with gynecological pathology it is necessary to consider the condition of genitals, as well as to conduct a complex survey of mammary glands. Multi-frequency electrical impedance mammography system "MEM" (MEM) [1, 2] with 256 electrodes has successfully been used for several years only for mammary gland diagnosis.

Recently impedance measuring has been applied for detection of cervical intraepithelial neoplasia [3]. Nowadays there is also a possibility to visualize tissues of a cervix uterus by Gynecological Impedance Tomorgraph "GIT" (GIT) [4]. Combination of both EIT investigations (MEM and GIT) enables to improve quality of women's health diagnostics. Possible scenarios of such a combination and preliminary clinical results of a complex assessment and diagnostics of the reproductive system will be presented in the paper.

2. Materials and methods

The paper presents an integrated study of the breast and the neck of the womb of 170 women between the age of 21 and 50 who had never been tested and had no complaints concerning their reproductive organs.

Multi-frequency electrical impedance tomography at a frequency of 50 kHz and 10 kHz was used for mammary glands and gynecological diagnostics. Both MEM and GIT are 3D electrical impedance tomography systems, which have a planar array of 256 and 48 measuring electrodes correspondingly. While MEM device has a regular and orthogonal electrode array with diameter 12-cm [1, 2], GIT device has a non-regular, non-orthogonal electrode array with diameter 30-mm [4]. GIT non-regular electrode placement enables to increase covering and spatial density of electrodes on a small area of 33 mm diameter. To reduce spurious couplings and wiring complexity, the measuring electronics (analogue front-end, microcontroller and communication unit) of the MEM and GIT are embedded inside the corresponding probes. This is quite challenging in case of the GIT system as the available space is only 30 mm in diameter and 20 mm in height inside the vaginal probe. 3D back-projection algorithms are used for the image reconstruction.

There were also other methods used for the cervix and breast examinations: history taking, physical examination, palpation, breast ultrasound, X-ray mammography, a visual inspection of cervix by mirrors, extended colposcopy, target biopsy of cervix, Papanikolau`s smear method. The obtained results of the study were statistically processed.

3. Results

Considering the mammary gland examination by electrical impedance mammography, five groups were formed [5]: norm, mastopathy, cyst, fibroadenomas and cancer.

In the first group (norm) 53 women (31.2%) out of 170 showed no changes in the mammary glands: a clear electrical impedance anatomy, correspondence of the electrical impedance image according to the age type of mammary gland structure, absence of deformations of the image contours, distinct architectonics of internal structures, absence of focal masses, absence of any difference in images depending on the patient position and the side of scanning, symmetry of the graph of frequency distribution of the conductivity. 82 women (48.2%) showed signs of mastopathy of different types: well-defined undistorted contour of the mammary gland, absence of the internal structure displacement, abnormalities of the image architectonics due to the change of ratio of the mammary gland tissues, resulting into mismatch of the electrical impedance (juvenile, reproductive, premenopausal, postmenopausal) type of image and the age, increase of hyper-impedance areas on the images due to fibrous changes of the adipose tissue, fibrosis of the duct walls and the Cupper ligaments, increase of the mean electrical conductivity zones due to hyperplasia of glandular tissue at adenosis. 21 women (12.4%) were identified to have an isolated cyst of a mammary gland: hypoimpedance formations with rather blurred contours, featuring electrical conductivity of 0.65 – 0.94 conventional units. 12 women (7%) had fibroadenomas of the mammary gland: isoimpedance formations with rather blurred contours, featuring electrical conductivity of 0.50 – 0.64 conventional units. 2 women (1.2%) were diagnosed to have cancer of the mammary glands: hypoimpedance areas with high electrical conductivity above 0.95 conventional units. Additionally there are observed: deformation of contours, impaired architectonics, and displacement of internal structures and difference in conductivity depending on the side of scanning.

All breast pathologies, founded by electrical impedance mammography method were confirmed by conventional techniques (ultrasound and X-ray mammography).

Preliminary results of the EIT investigation with GIT device were presented in [6]. In the current report extended results are enlisted. According to the new results patients were divided into 4 groups under their pathologies, which were diagnosed by conventional methods (exploration by mirrors, an extended colposcopy and cytology). Thus, the 4 groups were named: "in norm", "inflammation changes", "CIN I", "CIN II&CIN III".

Group "Inflammation changes" combines nonspecific changes of the neck of the womb: erosion, leucoplakia, polyps, warts, inflammation. Group "CIN I" means cervical intraepithelial neoplasia grade I. It is characterized by a low level of squamous intraepithelial lesion. Group "CIN II and CIN III" combines cervical intraepithelial neoplasia of grade II and III. A cytological examination reveals a high degree of squamous intraepithelial lesion and an extended colposcopy diagnoses suspicion of an invasion.

The results of electrical impedance imaging of cervix (typical for a group) are shown in figure 1. Mutual correlated statistics by diagnostics of the cervix and mammary glands are presented in table 1.

Table. 1. Pathology of the neck of the womb in various diseases of the mammary glands.

Pathology of cervix / Path. mammary glands	In norm	Inflammation changes	CIN I	CIN II, CIN III
The absence of disease (n=53)	32 (60.4%)	14 (26.4%)	3 (5.7%)	4 (7.5%)
The mastopathy of different type (n=82)	37 (45.2%)	28 (34.1%)	11 (13.4%)	6 (7.3%)
Isolated cycts (n=21)	15 (71.4%)	4 (19.1%)	2 (9.5%)	-
Fibroadenoma (n=12)	3 (25%)	3 (25%)	4 (33.3%)	2(16.7%)
Cancer (n=2)	1(50%)	1(50%)	-	-

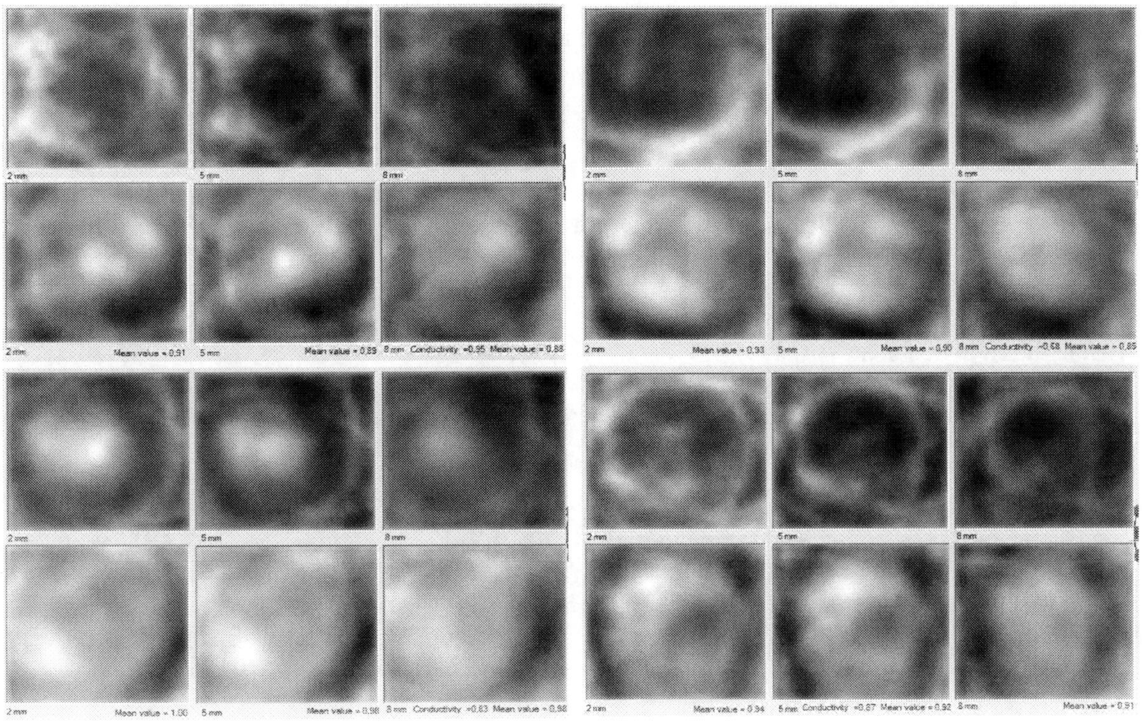

Figure. 1. Electrical impedance tomograms of the neck of the womb. 4 shots (patients). Each shot has two frequency images (upper line of 10 kHz, lower line of 50 kHz) and three slices (2mm, 5mm, 8mm depth in row from left to right). Left up: norm, 41 years old; right up: with different inflammation changes, 25 years old; left down: with CIN I, 31years old; right down: with CIN II and CIN III, 25 years old.

The characteristics of the average cervix electrical conductivity for the groups (in normal state and 3 pathology types) for two frequencies are given in table 2.

Table 2. Average value electrical conductivity at a frequency of 50 kHz and 10 kHz for 4 patient groups. The average values by group ± standard deviation are presented.

Depth & freq. / Pathology	1 level (2 mm)		2 level (5 mm)		3 level (8 mm)	
	50 kHz	10 kHz	50 kHz	10 kHz	50 kHz	10 kHz
The absence of disease (n=88)	0.89±0.06	0.40±0.1	0.85±0.07	0.36±0.09	0.83±0.09	0.30±0.05
Inflammation changes (n=50)	0.91±0.04	0.47±0.1	0.88±0.05	0.37±0.1	0.86±0.05	0.35±0.1
CIN I (n=20)	0.89±0.04	0.41±0.02	0.85±0.04	0.3±0.02	0.83±0.04	0.28±0.02
CIN II, CIN III (n=12)	1.01±0.1	0.38±0.05	1.0±0.1	0.26±0.05	1.0±0.1	0.23±0.04

Analysis of the results showed that the average electrical conductivity decreases statistically significantly ($\rho<0.05$) when scan depth increases on both frequencies scanning in all groups, except for the group "CIN II and CIN III", in which there was no statistically significant ($\rho>0.05$) dependence of the average electrical conductivity on scan depth at a frequency of 50 kHz only. Group "CIN II and CIN III" average electrical conductivity obtained at a frequency of 50 kHz (for any scan depth) is statistically significantly higher ($\rho<0.05$) than average electrical conductivity of all the other groups. The average electrical conductivity at a frequency of 50 kHz is statistically significantly ($\rho<0.05$) higher than at 10 kHz for all the groups.

4. Conclusions

The results showed possibility to distinguish pathology by EIT investigation for mammary glands previously [1, 2, 5] but now this has been demonstrated for cervix uteri. Thus, the method of electrical impedance tomography can be used as a method of screening for both mammary glands and cervix.

The method of electrical impedance tomography for cervix diagnostics is especially effective to identify higher degree of squamous intraepithelial lesion and suspicion of invasion. This will allow choosing the right treatment.

A comprehensive examination of the reproductive system by the method of electrical impedance tomography will improve the quality of diagnostics and the formation of risk groups for both types: mammary gland cancer and cervix uteri cancer.

References

[1] Cherepenin V, Karpov A, Korjenevsky A, Kornienko V, Kultiasov Yu, Ochapkin M, Trochanova O and Meister D 2002 Three-dimensional EIT imaging of breast tissues: system design and clinical testing *IEEE Trans. Medical Imaging* **21** 662 – 7

[2] Trokhanova O V, Okhapkin M B and Korjenevsky A V 2008 Dual-frequency electrical impedance mammography for the diagnosis of non-malignant breast disease *Physiol. Meas.* **29** S331 – 44

[3] Brown B H, Milnes P, Abdul S, Tidy J A 2005 Detection of cervical intraepithelial neoplasia using impedance spectroscopy: a prospective study *BJOG* **112** 802 – 6

[4] Cherepenin V A, Gulyaev Yu V, Korjenevsky A V, Sapetsky S A and Tuykin T S 2012 An electrical impedance tomography system for gynecological application GIT with a tiny electrode array *Physiol. Meas.* **33** 849 – 62

[5] Trokhanova O V, Okhapkin M B, Korjenevsky A V, Kornienko V N and Cherepenin V A 2009 Diagnostic possibilities of the electrical impedance mammography method *Biomeditsinskaya Radioelektronika* No 2 66-77

[6] Trokhanova O V, Chijova Yu A, Okhapkin M B, Korjenevsky A V and Tuykin T S 2010 Using of electrical impedance tomography for diagnostics of the cervix uteri diseases *J. Phys.: Conf. Ser.* **224** 012068

Journal of Physics: Conference Series **434** (2013) 012039 doi:10.1088/1742-6596/434/1/012039

Synchronous absolute EIT in three thoracic planes at different gravity levels

G Hahn, A Just, J Dittmar, K H Fromm and M Quintel

Department of Anaesthesiology, Emergency and Intensive Care Medicine, University Medical Centre Göttingen, Robert-Koch-Str. 40, D-37075 Göttingen, Germany

E-mail: ghahn@gwdg.de

Abstract. The validity of absolute Electrical Impedance Tomography (a-EIT) for assessment of local lung volume has been investigated far less than the well evaluated ventilation monitoring by functional EIT (f-EIT). To achieve progress in a-EIT we investigated 10 healthy volunteers in an upright sitting position by using a-EIT at normal gravity (1 g), weightlessness (0 g) and approx. double gravity (1.8 g) during parabolic flight manoeuvres. Lung resistivity in three thoracic planes was determined by a-EIT using a multiple-plane synchronised Goe-MF II EIT system. Tomograms of resistivity at end-expiration in normal spontaneous breathing were reconstructed by a modified SIRT algorithm. Local lung resistivity was determined separately for both lungs. The respective resistivity values at 1 g and 1.8 g before and after weightlessness show an almost reversible behaviour along the sequence of gravity changes with a tendency to be lower after occurrence of weightlessness. The results reveal not only the expected varying resistivity of lung tissue in cranio-caudal direction but also a clear difference in these cranio-caudal stratifications of local lung volume between the left and right lung. The resolution and stability of absolute EIT seem to be valid and expressive for future investigations of unilateral lung volume under different physiological and pathological conditions.

1. Introduction

Aeration and ventilation of the lung are not homogeneously distributed. The relations of these properties have been investigated formerly by elaborate methods, e.g. [1], [2]. In the healthy lung the cranio-caudal distribution is expected to be in accordance to the lung model of West [3], [4]. That means the upper lung regions will contain more air than the lower ones due to the gravity dependent gradient of the transpulmonary pressure. We investigated this distribution of aeration in the human lungs by extending the usage of normal CE-certified EIT devices (Goe-MF II type) to synchronous operation in three thoracic planes. We thereby avoided any hardware modifications that would have resulted in a long-lasting new CE-certification process for human application. The resistivity values of lung tissue determined by absolute EIT (a-EIT) were the target as a measure for the quantification of lung tissue aeration. In this study the feasibility and meaningfulness of this approach using a-EIT was tested and compared to the lung model of West by changing the level of gravity.

2. Methods

2.1. Realisation of synchronous multiple plane operation of the Goe-MF II EIT systems

Three EIT systems of the Goe-MF II type were used for measurements in three thoracic planes. The CE-certified EIT systems (for use on humans) are controlled by the bidirectional Universal Serial Bus

(USB) which is per se poorly real-time capable. Due to the non-deterministic occurrence of latencies in the operation of multiple EIT systems which are controlled exclusively by USB the following concept was applied to ensure sufficient synchronous operation of the EIT devices: A timestamp is sent from the host computer to all clients (EIT-systems) and the difference in time between each EIT system and the host computer is continuously determined by a sliding average. The value of this deviation is used to speed up or slow down the internal clock in each client which is controlled by a voltage controlled crystal oscillator (VCXO). The entire design is basically realized as a digital proportional-integral (PI) controller in each EIT system. It is important to achieve a nearly identical relation in time between the three EIT systems whereas the absolute difference in time between the host computer and the EIT systems is not of relevance. After an initialisation phase the entire arrangement is locked in a fixed phase relationship between the attached EIT systems. Therefore a synchronised control of the interlaced measurement procedure in all of the three EIT systems is possible in a very flexible manner. The software ensures this state of operation and displays quantitatively the remaining tiny differences of the timing jitter.

2.2. EIT measurements and data evaluation
We investigated 10 healthy volunteers (7 male and 3 female, mean body weight 76.3 kg ± 14.5 kg SD, age between 19 and 60 years) in an upright sitting position by using a-EIT under conditions of changing gravity between weightlessness (0 g) and approx. double gravity (1.8 g) during parabolic flight manoeuvres. Parabolic flight manoeuvres were performed by an Airbus A 300 operated by NOVESPACE (France) during the 18th Parabolic Flight Campain of the German Aerospace Center (DLR). Each of the parabolas results in five consecutive gravity levels of 20 to 25 s duration in the following sequence: normal gravity (1 g before), hyper-gravity (1.8 g before), weightlessness (0 g), hyper-gravity (1.8 g after) and normal gravity (1 g after). Measurements of lung resistivity by a-EIT were carried out synchronously in three thoracic planes. The middle plane was located at the level of the xiphoid process and the cranial and caudal plane 5 to 6 cm above and below the middle plane. A multiple-plane synchronised Goe-MF II EIT system (see above) was used. The total frame rate was 6.6 Hz (interlaced operation of the three systems) and the injection current 5 mA_{eff} at 50 kHz for all measurements. Lung tissue resistivity, which is considered to be linearly correlated to local lung volume (air content) [5-9], was quantified by a-EIT at end-expiration in normal spontaneous breathing. The absolute tomograms were reconstructed by a modified SIRT algorithm [10], [11] whereas resistivity was quantified for the total lung and separately for both lungs. The location and number of the pixels (regions of interest [ROIs] for a-EIT) that usually represent the areas of the lungs in healthy volunteers were determined by f-EIT. They were used as predefined and fixed ROIs for all volunteers. Statistical analysis was performed applying a paired t-test for the differences between the right and left lung and the effect of gravity was tested using a one way ANOVA for repeated measures.

3. Results

3.1. Methodological results
After switch-on, a short synchronisation phase was needed to achieve a stable and synchronous operation of the three EIT systems. The application of the described synchronisation concept resulted in a timing jitter of always less than 100 μs between the three EIT systems. The applicability of the experimental approach and setup under the conditions of changing gravity during parabolic flights was successfully demonstrated. Based on the synchronous raw data from three different thoracic planes it was possible to calculate the intrathoracic stratification of resistivity values in cranio-caudal direction under the dynamic conditions of spontaneous breathing. Also a clear separation of the resistivity values of lung tissue - being mainly dependent on lung aeration - between the right and the left lung was achieved by the applied SIRT algorithm.

3.2. Physiological results

Figure 1 displays the resistivity values in the three thoracic planes of the total lung (left panel), the left lung (middle panel) and the right lung (right panel) (mean ± SE) in the course of the different gravity levels as they occur during a parabolic flight manoeuvre. The resistivity at 0 g is always lower than that at 1 g and 1.8 g in all planes. A significant influence ($p<0.05$) of gravity compared to 0 g is denoted by *. Stratification from higher resistivities (top plane) to lower ones (bottom plane) is observed in both lungs whereas it is more pronounced in the right lung. For all gravity levels the resistivity of the right lung in the bottom plane (black line, bot) is significantly ($p<0.05$) lower than that of the left lung. In the middle (mid) and top (top) plane a higher variation of individual resistivities is observed compared to the bottom plane.

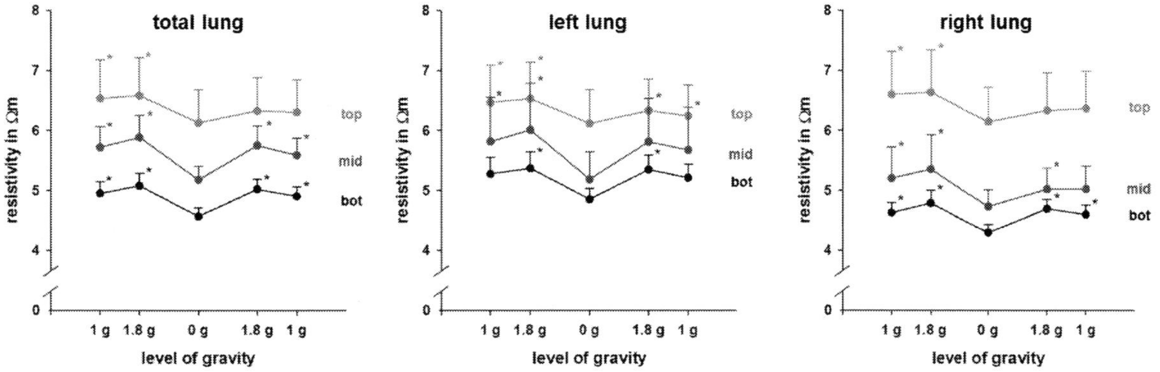

Figure 1 Resistivity of the total, left and right lungs in three thoracic planes in the course of parabolic flight manoeuvres (mean ± SE). A significant influence ($p<0.05$) of gravity compared to 0 g is denoted by *.

The respective resistivity values at 1 g and 1.8 g before and after weightlessness were averaged for each of the planes since they show an almost reversible behavior along the sequence of gravity changes with a tendency to be lower after occurrence of weightlessness. In detail, the corresponding values for the cranial, middle and caudal plane in the right lung are: 6.48 Ωm, 5.36 Ωm and 4.61 Ωm at 1 g; 6.48 Ωm, 5.50 Ωm and 4.74 Ωm at 1.8 g and 6.14 Ωm, 4.93 Ωm and 4.29 Ωm at 0 g. In the left lung the respective resistivities are: 6.35 Ωm, 5.95 Ωm and 5.25 Ωm at 1 g; 6.42 Ωm, 6.13 Ωm and 5.36 Ωm at 1.8 g and 6.11 Ωm, 5.43 Ωm and 4.85 Ωm at 0 g.

4. Discussion

The concept of synchronising existing CE-certified single plane EIT systems via USB is modular and extendible to a useful number of EIT systems operating in parallel. It offers more flexibility by firmware and software adaptation than a simple extension to a multiple plane operation by multiplexing the analog front end of an EIT system. This concept allows also current injection in one plane and a simultaneous measurement of the resulting potential differences in all of the other planes. But this option has not been used in the current study.

The resistivity range (cf. figure 1) we found *in vivo* at expiratory level for the human lungs at 50 kHz fits well to the value of 5.9 Ωm reported by [12] for deflated pig lungs (interpolated between the *in vivo* measurements at 10 kHz and 100 kHz from figure 7 in [12]). Direct measurements *in vivo* on human lung tissue have not been performed to our knowledge for ethical reasons. Since resistivity of lung tissue is mainly dependent on aeration resulting in stretching the surrounding tissue of the pulmonary alveoli [13], blood and fluid content will probably play a minor role. But a shift in blood or fluid content into the lungs may be one of the reasons for the tendency of lower values after weightlessness. The results of stratification of aeration determined by a-EIT for the total lung at end-expiration are in full agreement with those predicted by the lung model of West [2]. a-EIT used in this study is more easily applicable than other approaches that were formerly used to investigate local aeration of the lungs, which is an evident advantage of EIT that may be of interest for future space missions. The physiological plausible results of the total lung are considered to be an indirect validation of the whole methodological setup for synchronous multiple plane a-EIT based on the Goe-MF II hardware and the applied SIRT algorithm. The resolution and stability of a-EIT seem to be valid and expressive.

Looking in more detail at the results of the left and right lungs a different level of cranio-caudal stratification in lung tissue aeration is observed. This is also the case at weightlessness which implies an inherent inhomogeneity of lung aeration.

This study was supported by the German Aerospace Center, DLR, on behalf of the Federal Ministry of Economics and Technology, BMWi, project funding reference number 50WB0915.

References

[1] Bates D V, Kaneko K, Henderson J A, Milic-Emili J, Anthonisen N R, Dollfuss R, Dolovich, M 1966 Recent experimental and clinical experience in studies of regional lung function *Scandinavian journal of respiratory diseases.* **62** 15-29

[2] Milic-Emili J, Henderson J A, Dolovich M B, Trop D and Kaneko K 1966 Regional distribution of inspired gas in the lung *J. Appl. Physiol.* **21**(3) 749-59

[3] West J B 1962 Regional differences in gas exchange in the lung of erect man *J. Appl. Physiol.* **17** 893-8

[4] West J B 1966 Distribution of pulmonary blood flow and ventilation measured with radioactive gases *Scandinavian journal of respiratory diseases* **62** 9-13.

[5] Frerichs I, Hinz J, Herrmann P, Weisser G, Hahn G, Dudykevych T, Quintel M and Hellige G 2002 Detection of local lung air content by electrical impedance tomography compared with electron beam CT *J. appl. physiol.* **93** 660-6

[6] Hahn G, Sipinkova I, Baisch F and Hellige G 1995 Changes in the thoracic impedance distribution under different ventilatory conditions *Physiol. Meas.* **16** A161-73

[7] Hahn G, Just A, Dudykevych T, Frerichs I, Hinz J, Quintel M and Hellige G 2006 Imaging pathologic pulmonary air and fluid accumulation by functional and absolute EIT *Physiol. Meas.* **27** S187-98

[8] Hinz J, Hahn G, Neumann P, Sydow M, Mohrenweiser P, Hellige G and Burchardi H 2003 End-expiratory lung impedance change enables bedside monitoring of end-expiratory lung volume change *Intensive care medicine* **29** 37-43

[9] Victorino J A, Borges J B, Okamoto V N, Matos G F, Tucci M R, Caramez M P, Tanaka H, Sipmann F S, Santos D C, Barbas C S, Carvalho C R and Amato M B 2004 Imbalances in regional lung ventilation: a validation study on electrical impedance tomography *Am. J. Respir. Crit. Care. Med.* **169** 791-800

[10] Just A, Küpper T, Kürschner D and Jacobs F 1998 Core imaging by electrical resistivity tomography *Proc. 4th SEGJ Int. Symp.* 231-7

[11] Just A, Ruecker C, Hahn G, Hellige G 2007 Sensitivity of two different reconstruction algorithms to body shape and electrode position erroers in absolute EIT *IFMBE Proceedings* 17 (Graz, A, 29 August - 2 Spetember 2007) ed H Scharfetter and R Merwa (Berlin Heidelberg New York: Springer) pp 404-7

[12] Gabriel C, Peyman A and Grant E H 2009 Electrical conductivity of tissue at frequencies below 1 MHz *Physics in medicine and biology* **54** 4863-78

[13] Moerer O, Hahn G, Quintel M 2011 Lung impedance measurements to monitor alveolar ventilation *Current opinion in critical care* **17(3)** 260-7

Does thorax EIT image analysis depend on the image reconstruction method?

Zhanqi Zhao[1,4], Inéz Frerichs[2], Sven Pulletz[3], Ullrich Müller-Lisse[4], Knut Möller[1]

[1]Institute of Technical Medicine, Furtwangen University, VS-Schwenningen, Germany

[2]Department of Anaesthesiology and Intensive Care Medicine, University Medical Center Schleswig-Holstein, Campus Kiel, Kiel, Germany

[3]Department of Anaesthesiology and Intensive Care Medicine, Medical Center Osnabrück, Osnabrück, Germany

[4]Department of Radiology, University of Munich, Munich, Germany

zhanqi.zhao@hs-furtwangen.de

Abstract. Different methods were proposed to analyze the resulting images of electrical impedance tomography (EIT) measurements during ventilation. The aim of our study was to examine if the analysis methods based on back-projection deliver the same results when applied on images based on other reconstruction algorithms. Seven mechanically ventilated patients with ARDS were examined by EIT. The thorax contours were determined from the routine CT images. EIT raw data was reconstructed offline with (1) filtered back-projection with circular forward model (BP_C); (2) GREIT reconstruction method with circular forward model ($GREIT_C$) and (3) GREIT with individual thorax geometry ($GREIT_T$). Three parameters were calculated on the resulting images: linearity, global ventilation distribution and regional ventilation distribution. The results of linearity test are 5.03±2.45, 4.66±2.25 and 5.32±2.30 for BP_C, $GREIT_C$ and $GREIT_T$, respectively (median ±interquartile range). The differences among the three methods are not significant ($p=0.93$, Kruskal-Wallis test). The proportions of ventilation in the right lung are 0.58±0.17, 0.59±0.20 and 0.59±0.25 for BP_C, $GREIT_C$ and $GREIT_T$, respectively ($p=0.98$). The differences of the GI index based on different reconstruction methods (0.53±0.16, 0.51±0.25 and 0.54±0.16 for BP_C, $GREIT_C$ and $GREIT_T$, respectively) are also not significant ($p=0.93$). We conclude that the parameters developed for images generated with $GREIT_T$ are comparable with filtered back-projection and $GREIT_C$.

1. Introduction

Electrical impedance tomography (EIT) is a non-invasive, radiation-free imaging technique that can be used at the bedside for monitoring regional lung ventilation and aeration distribution [11]. EIT measures the electrical potentials at the chest wall surface and detects air volume changes based on the phenomenon that changes in regional air content modify the electrical impedance of lung tissue [4].

The reliability of EIT has already been confirmed by comparison with different conventional methods [6, 9, 10]. Unfortunately, due to the limited number of electrodes the resolution of the reconstructed EIT images is relative low compared to computed tomography (CT). Another problem that prevents wide acceptance of EIT in the clinical environment is the difficult interpretation of the results. Because of the ill-posed problem, the electrical impedance within the thorax cannot be perfectly reconstructed. Up to now, most EIT devices idealize the thorax to a 2D circular forward model to simplify the image reconstruction. The shape distortion of the lungs and thorax impair the quality of the EIT images [8].

Recently, the so-called GREIT (Graz Consensus Reconstruction Algorithm for EIT) was developed [1]. With the help of extensible software based EIDORS [2], EIT raw data can be reconstructed for individual thorax geometry, which improves the EIT images [13].

Since the EIT measurements contain large amount of information, different parameters were proposed to analyze the EIT images [5, 7, 12, 14]. Most of them were based on the images reconstructed with filtered back-projection and circular forward model. The aim of our study was to examine if the analysis parameters based on back-projection are still valid when they were applied on images reconstructed with GREIT.

2. Methods

A total of seven mechanically ventilated patients with acute lung injury were studied (4 male, 3 female; age 60±8 years; height 177±13 cm; weight 81±12 kg; mean±SD). The patients were ventilated under pressure controlled mode with individual positive end-expiratory pressure (PEEP). Immediately precediung the examination, PEEP was temporarily reduced to zero end-expiratory pressure without changing any other ventilation settings for one minute. Then a stepwise increase in airway pressure to 35 cmH_2O was induced for one minute followed by a stepwise decrease to the initial zero end-expiratory pressure (figure 1). EIT data were acquired during the whole measurement period at a rate of 25 images/s (Goe-MF II device, CareFusion, Höchberg, Germany). The study was approved by the local ethics committee. Written informed consent was obtained from all patients prior to the study.

The individual thorax contours were determined from routine computed tomography (CT) images (Brilliance CT 64 channel scanner, Philips, Andover, USA) based on a threshold value of >800 Hounsfield units as well as dilation and erosion filtering. The CT analysis was performed in the fifth intercostal space, i.e. in the same cross-sectional plane as EIT. EIT raw data was reconstructed offline with (1) filtered back-projection with circular forward model (BP_C) [3]; (2) GREIT reconstruction method with circular forward model ($GREIT_C$) and (3) GREIT with individual thorax geometry ($GREIT_T$). Three parameters calculated from the resulting images based on these reconstruction methods were examined: (1) linearity, i.e. the ratio of tidal and deep inflation impedance changes obtained during the final five breaths preceding the stepwise increase in airway pressure and between the beginning and end of the 1-min long inflation (figure 1, D_2/D_1); (2) global ventilation distribution, indicated by the tidal impedance changes in the right and left lungs (figure 2, $D_{Right}/(D_{Right}+D_{Left})$); (3) regional ventilation distribution, indicated by the global inhomogeneity (GI) index [14].

Data analysis was performed using MATLAB 7.2 (The MathWorks Inc., Natick, MA, USA). For non-normally distributed data, results were expressed as median value and interquartile range. Kruskal-Wallis test was applied to assess the differences among reconstruction methods for each parameter. A P value < 0.05 was considered statistically significant.

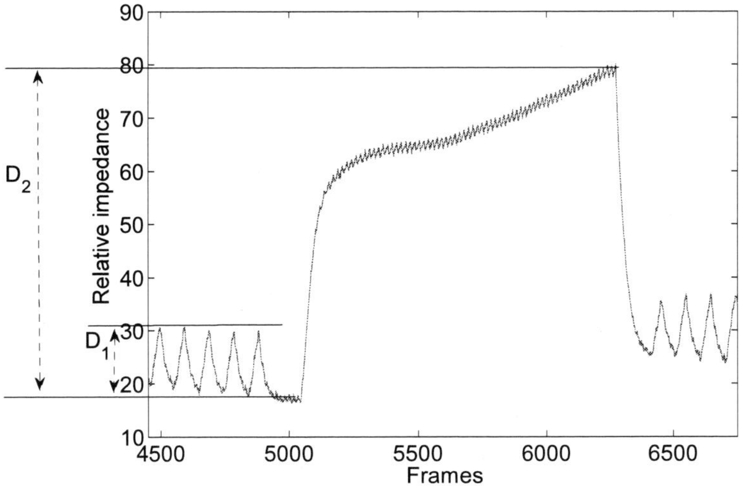

Figure 1. A typical total relative impedance (arbitrary unit) curve of one patient during the EIT examination. During the initial phase, the patient was temporarily ventilated at zero end-expiratory pressure followed by an increase in airway pressure to 35 cmH_2O. D_1 represents the mean tidal impedance change during zero end-expiratory pressure. D_2 denotes the impedance change caused by the large airway pressure step. The impedance curve is based on $GREIT_C$. Similar curves are obtained with other reconstruction methods (BP_C, $GREIT_T$).

3. Results

Table 1 summarizes the parameters derived for images with different reconstruction methods and the results of Kruskal-Wallis test. Although the parameter values vary among patients, no significant differences are found in any of these three parameters among different reconstruction methods.

Table 1. Parameters obtained using different reconstruction methods.

	Linearity D_2/D_1	Global distribution $D_{Right}/(D_{Right}+D_{Left})$	Regional distribution The GI index
BP_C	5.03 (2.45)	0.58 (0.17)	0.53 (0.16)
$GREIT_C$	4.66 (2.25)	0.59 (0.20)	0.51 (0.25)
$GREIT_T$	5.32 (2.30)	0.59 (0.25)	0.54 (0.16)
P value	0.93	0.98	0.93

Data are expressed as median values and interquartile ranges. P values are calculated with Kruskal-Wallis test.

4. Discussion and conclusion

In the present study, we chose three representative parameters for EIT image analysis and examined their variation in different reconstruction methods. P values in Kruskal-Wallis test close to 1 suggest that the methods used for EIT image reconstruction do not influence the results of image analysis. The parameters developed for images generated with $GREIT_T$ are comparable with filtered back-projection and $GREIT_C$. Since the former algorithm is adapted to individual thorax geometry it improves the EIT image quality and may accelerate the acceptance of EIT in routine clinical use compared with older image reconstruction methods. The variation of parameters among different patients emphasizes the importance of individual monitoring and strategy optimization in mechanical ventilation therapy.

Acknowledgment

This work was partially supported by the Deutsche Forschungsgemeinschaft (DFG, # MO 579/1-2 PAR) and by Bundesministerium für Bildung und Forschung (BMBF, FKZ 01|B10002D, WiMVENT).

Journal of Physics: Conference Series **434** (2013) 012040 doi:10.1088/1742-6596/434/1/012040

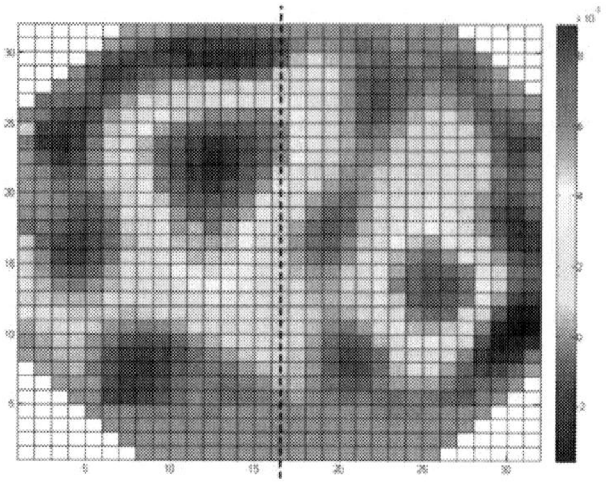

Figure 2. Functional EIT of the same patient in figure 1 during ventilation at zero end-expiratory pressure, reconstructed with $GREIT_C$. Colour bar indicates the magnitude of relative impedance values in each pixel (arbitrary unit). Dashed line divides the image into identical halves. Note that EIT images have the same orientation as CT images.

References

[1] Adler A, Arnold JH, Bayford R, *et al.* 2009 GREIT: a unified approach to 2D linear EIT reconstruction of lung images *Physiol Meas* **30** S35-55

[2] Adler A and Lionheart WR 2006 Uses and abuses of EIDORS: an extensible software base for EIT *Physiol Meas* **27** S25-42

[3] Barber DC 1989 A review of image reconstruction techniques for electrical impedance tomography *Med Phys* **16** 162-9

[4] Brown BH, Barber DC, Morice AH and Leathard AD 1994 Cardiac and respiratory related electrical impedance changes in the human thorax *IEEE Trans Biomed Eng* **41** 729-34

[5] Frerichs I, Hahn G, Golisch W, *et al.* 1998 Monitoring perioperative changes in distribution of pulmonary ventilation by functional electrical impedance tomography *Acta Anaesthesiol Scand* **42** 721-6

[6] Frerichs I, Hinz J, Herrmann P, *et al.* 2002 Regional lung perfusion as determined by electrical impedance tomography in comparison with electron beam CT imaging *IEEE Trans Med Imaging* **21** 646-52

[7] Grant CA, Fraser JF, Dunster KR and Schibler A 2009 The assessment of regional lung mechanics with electrical impedance tomography: a pilot study during recruitment manoeuvres *Intensive Care Med* **35** 166-70

[8] Grychtol B, Lionheart WR, Wolf GK, *et al.* 2011 The importance of shape: thorax models for GREIT In *Conf. EIT 2011*: Bath, UK

[9] Hinz J, Neumann P, Dudykevych T, *et al.* 2003 Regional ventilation by electrical impedance tomography: a comparison with ventilation scintigraphy in pigs *Chest* **124** 314-22

[10] Marquis F, Coulombe N, Costa R, *et al.* 2006 Electrical impedance tomography's correlation to lung volume is not influenced by anthropometric parameters *J Clin Monit Comput* **20** 201-7

[11] Putensen C, Wrigge H and Zinserling J 2007 Electrical impedance tomography guided ventilation therapy *Curr Opin Crit Care* **13** 344-50

[12] Wrigge H, Zinserling J, Muders T, *et al.* 2008 Electrical impedance tomography compared with thoracic computed tomography during a slow inflation maneuver in experimental models of lung injury *Crit Care Med* **36** 903-9

[13] Zhao Z, Pulletz S, Frerichs I and Möller K 2012 EIT image reconstruction with individual thorax geometry *Biological and Medical Systems* **8 Part 1** Aug 29-31, Budapest, Hungary

[14] Zhao Z, Steinmann D, Frerichs I, et al. 2009 Evaluation of an electrical impedance tomography-based global inhomogeneity index for pulmonary ventilation distribution *Intensive Care Med* **35** 1900-1906

Detection and localization of internal haemorrhaging using electrical bioimpedance

J Morse, M Fenech[1]

550 Cumberland Street, Ottawa, ON, Canada K1N 6N5

Email: *MFenech@UOttawa.ca*

Abstract. Electrical bioimpedance is an effective measuring tool to provide quick, non-invasive, real-time results which will be applied to the detection of internal haemorrhaging. Experiments were performed on female Fancy Rats weighing 333±44g, and 10mL of porcine blood was injected abdominally over 3 minutes. Data was collected using an 8x8 needle electrode array at 5 kHz, and 95 kHz and sent to the *BioParHom Z-Flow*. A strong correlation was found between the electrode paths crossing directly through the blood injection site, showing a decrease of about -0.17±0.1Ω/mL for the 5 kHz frequency. This correlation allows us to quickly detect internal haemorrhaging and also localize it with the current path set-up in the electrode array.

1. Introduction

In emergency settings, rapid real-time detection and localization of internal bleeding can be crucial to preserve the welfare of patients. The majority of patients who die from multiple injuries, die unnecessarily due to undetected internal bleeding [1]. Internal bleeding is the cause of 2/3 of all the deaths from multiple injuries in emergency patients. This gives rise to a need for rapid detection that can also estimate the rate of bleeding. On top of this, any added mobility would be an advantage to allow for on-site or in transit diagnosis.

Internal haemorrhaging is the dangerous bleeding found inside a major cavity. Its presence causes localized high pressure because of the volume of blood entering the cavity which puts undue strain on the nearby organs. It also causes hypoxia since the blood is not staying in the vessels and entering the organs, but rather pooling in the cavity which can cause irreparable organ damage. Internal haemorrhaging is commonly found in patients already suffering from other medical concerns.

CT Scans are capable of detecting and locating internal bleeding, however current detection methods lack the ability to estimate bleeding rates in patients. CT Scans are expensive, large and used as a diagnostic tool for many other medical issues. This creates a medical environment with few CT Scans in large hospitals with large wait lists. As a result of these current drawbacks, an inexpensive method capable of detecting, locating and estimating the rate of internal bleeding in real-time is believed to be a useful addition to clinical situations. The aim of this research is to outline the possibility of using electrical bioimpedance as a method to further a clinician's ability to accurately diagnose and assess internal bleeding.

Electrical bioimpedance is a method of detecting resistance changes *in vivo*. Using a mild electric current at a specific frequency, the resistance and capacitance can be measured to determine the constituents found along the current path. Using this *in vivo*, we can probe through the skin without causing any harm and measure internal components of the body.

[1] To whom any correspondence should be addressed.

Published under licence by IOP Publishing Ltd

The addition of a pool of blood found in a cavity affects the overall resistance of the cavity. The resistance of blood, primarily that of the plasma is lower than that of tissue. Blood has an average resistance of 191Ω for 1cc of blood at 87,000 cycles per second [2]. For skeletal muscle, the resistance is close to 500Ω per cm [3]. With this in mind, the more blood pooling near an electrode the lower its resistance since the electrical current will have more paths with less resistance.

2. Materials and Methods

Blood samples were taken from food grade pigs and anticoagulated with Ethylenediaminetetraacetic acid (EDTA) at 2mg/ml with 8ml/l of diluted water to porcine blood. The blood was injected into dead feeder female Fancy Rats weighing 379.5±61g. A rat was determined to be an appropriate phantom since it had the same blood to tissue ratio as a human of 7% [4] [5].

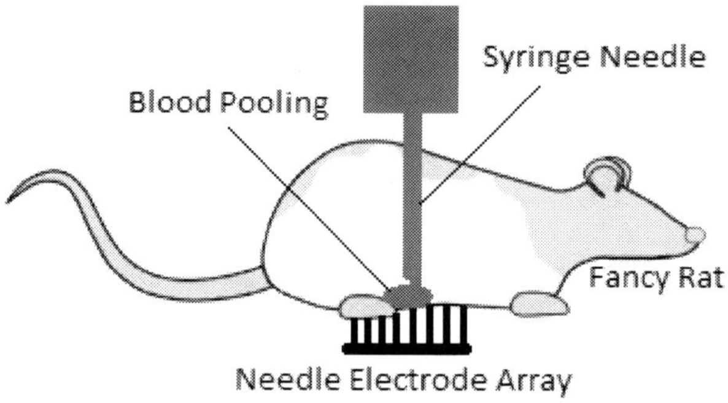

Figure 1 Schematic showing the location of electrode array, blood pooling and the syringe.

Blood was injected dorsally into the abdominal cavity as seen in Figure 1. The experiment was run for 3 minutes, while 10mL of porcine blood was injected at a rate of 3.33mL/min. The amount equalled a potentially fatal amount of bleeding found in the human body of roughly 24mL/kg [6]. The site of the injected blood was centered over the electrode array at a depth of 0.5cm from the skin.

The measurements were made using an 8x8 needle electrode array built to 5cm^2, taking inputs from a current source (*BioParHom Z-Flow*) which is sent through a multiplexer (*Arduino Mega2560*). Four electrodes along the electrode path were used, the electrodes on either end alternate inputting the current while the middle two electrodes measure the signal. The initial set-up was run at 5 kHz for the first 8 rats, the second 9 rats were done at 95 kHz. The current was alternated between various electrode paths in the array to systematically work through a grid. The electrode array was placed *in vitro* on a 5cm^2 shaved spot on the abdominal cavity. The signal alternated between 8 current paths as seen in Figure 2, taking 1 measurement along each path before changing. Each path received 25 measurements per minute.

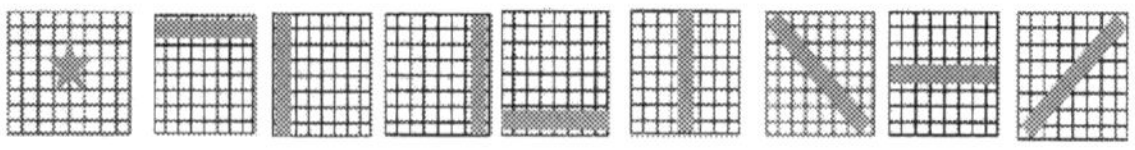

Figure 2 Schematic showing the location of the blood pooling as a star on the electrode array as well as the 8 different current paths.

3. Results and Discussion

For electrical current in the human body, there are several factors which must be considered. The first is the desired path of the current. It has been stated in various papers that the cell membrane acts as a capacitor [2] [3] [7]. At higher frequencies, current is capable of passing through biological tissue with less resistance. This is due to the capacitive effect of the cell membrane. Above 150 kHz, the current can pass through the cell membrane with little resistance. Below 5 kHz, the current cannot pass the capacitive cell membrane. With these values, we would expect a higher resistance found in the 5 kHz tests compared to the 95 kHz tests. Due to this higher initial resistance, any extra current paths with a low resistance would have a much more prominent effect. For the 5 kHz tests, we expect to see a larger effect on resistance change than at 95 kHz when in the presence of blood pooling.

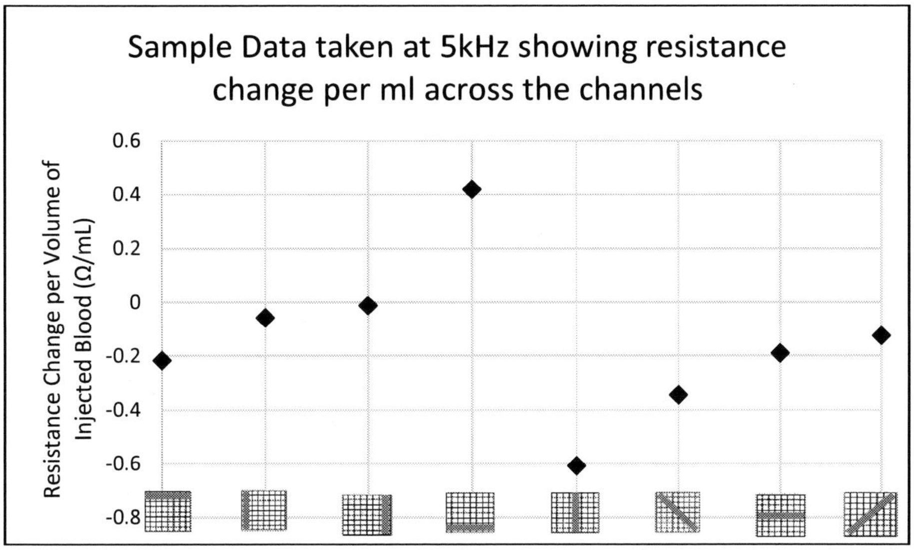

Figure 3 Sample data from a rat at 5 kHz highlighting the drop in resistance across the electrode paths passing through the blood pooling.

The above plot shows the data points acquired for a single rat over the course of the test. The last four data points are those that cross through the location of the blood pooling, and show a marked decrease in resistance per ml of blood.

Averaging the data points, we get a trend across all the rats tested. The comparison of the trends of the two frequencies show a large difference in the variability of the mean resistance drops. This change in resistance can be seen in Figure 4, with a drop in resistance per ml injected for the last four current paths which cross through the location of the blood pooling. For the 95 kHz values, there is little change as the current already has many paths to go through since it can cross cell membranes with less impedance.

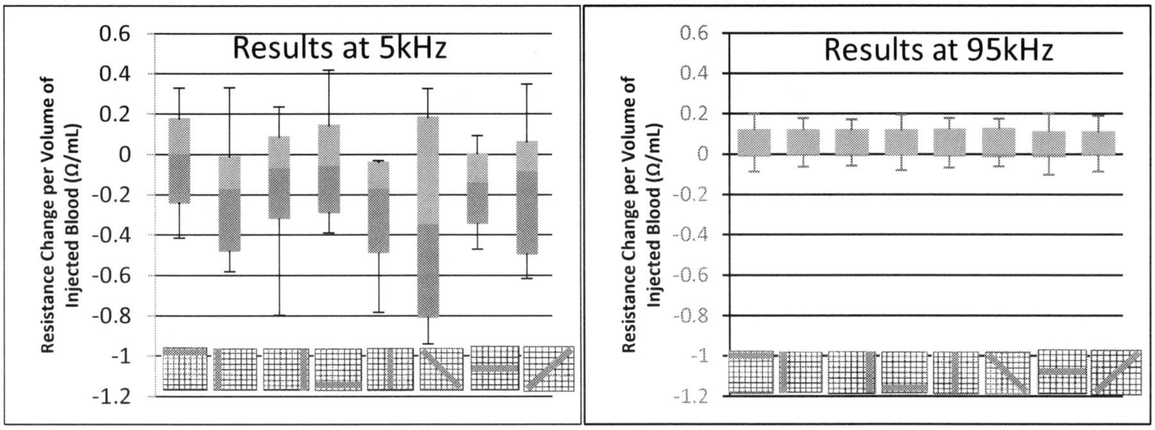

Figure 4 Resistance change per volume of blood injected across current paths in a rat at 5 kHz and 95 kHz.

The change in variability between the two tests shows that at lower frequencies a more magnified change in resistance can be detected. This is in part due to the added importance of the blood plasma at the lower frequency. The results give rise to the possibility of finding an optimal frequency to measure resistance changes in the body. As a result, more research will be conducted to find that optimal frequency using impedance spectroscopy.

Combining the average resistance drop and overlaying it onto the electrode array area, we can find the location of the highest rate of blood accumulation. For the 5 kHz frequency, we find that the blood pooling occurs in the bottom right corner of the electrode shown in Figure 5.

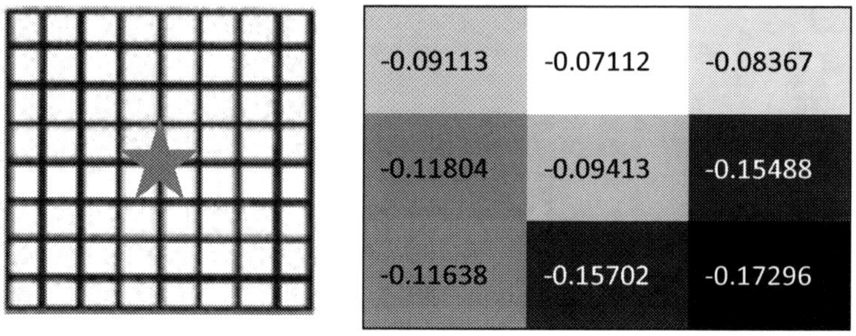

Figure 5 Estimated location of the blood pooling and measured results for the 5 kHz experiments. Measured values in average Ω/ml.

For locating the internal bleeding under the electrode array, a 60% drop in resistance per ml isolated where the blood was pooling. This gives a clear and visible indication as to where to investigate for the source of the bleeding.

4. Conclusion

A set up using a *Z-Flow* with and 8x8 electrode array at 5 kHz was able to detect and localize blood pooling. Potential for this detection technique has been demonstrated by the initial research. A 60% greater decrease in resistance per ml was able to indicate the location of the blood pooling. Further work is required to refine the applied frequency to maximize the detection of blood pooling.

References

[1] Sadleir R, Fox R, van Kann F and Attikiouzel Y 1992 Estimating volumes of intra-abdominal blood using electrical impedance imaging *Eng. Med. Biol. Soc.: Conf. Ser.* **14** 1750-1

[2] Fricke H and Morse S 1925 The electric resistance and capacity of blood for frequencies between 800 and 4 1/2 million cycles *J. Gen. Physiol.* **9(2)** 153-67

[3] Sperelakis N and Hoshiko T 1961 Electrical Impedance of Cardiac Muscle *Circ. Res.* **9(6)** 1280

[4] Cameron J, Skofronick J and Grant R 1999 Physics of the Body *Medical Physics Publishing*

[5] Lee H and Blaufox M 1995 Blood Volume in the Rat *J. Nucl. Med.* **26(1)** 72-6

[6] Krantz T, Cai Y, Lauritsen T, Warberg J and Secher N 2000 Accurate monitoring of blood loss: thoracic electrical during hemorrhage in the pig *Acta Anaesthesiol. Scand.* **44(5)** 598-604

[7] McAdams E and Jossinet J 1995 Tissue impedance: a historical overview *Physiol. Meas.* **16(3A)** A1

Assessment of cerebral blood flow autoregulation (CBF AR) with rheoencephalography (REG): studies in animals

Djordje Popovic[1], Michael Bodo[2], Frederick Pearce[2], Stephen van Albert[2], Alison Garcia[2], Tim Settle[2], Rocco Armonda[3]

[1] Advanced Brain Monitoring, Inc., Carlsbad, CA, USA; [2] Walter Reed Army Institute of Research (WRAIR), Silver Spring, MD, USA; [3] Walter Reed National Military Medical Center (NMMC), Bethesda, MD, USA

Email: popovic@usc.edu

Abstract. The ability of cerebral vasculature to regulate cerebral blood flow (CBF) in the face of changes in arterial blood pressure (SAP) or intracranial pressure (ICP) is an important guard against secondary ischemia in acute brain injuries, and official guidelines recommend that therapeutic decisions be guided by continuous monitoring of CBF autoregulation (AR). The common method for CBF AR monitoring, which rests on real-time derivation of the correlation coefficient (PRx) between slow oscillations in SAP and ICP is, however, rarely used in clinical practice because it requires invasive ICP measurements. This study investigated whether the correlation coefficient between SAP and the pulsatile component of the non-invasive transcranial bioimpedance signal (rheoencephalography, REG) could be used to assess the state and lower limit of CBF AR. The results from pigs and rhesus macaques affirm the utility of REG; however, additional animal and clinical studies are warranted to assess selectivity of automatic REG-based evaluation of CBF AR.

1. Introduction

Brain injuries caused by vascular incidents or a physical trauma are a major public health concern worldwide [1]. In the United States alone, stroke and traumatic brain injury (TBI) annually account for 1.1 million hospitalizations, 200,000 deaths and $133 billion in medical costs [2, 3]. Moreover, 15 million Americans are permanently disabled because of the acquired brain damage [2]. Unfortunately, contemporary medicine can do little to remedy the direct damage caused by the primary vascular or traumatic insult, and the treatment focuses on maintenance of adequate cerebral blood flow (CBF) and prevention of the secondary ischemia in the hours and days following the primary injury.

Cerebral blood flow depends on systemic arterial blood pressure (SAP), intracranial pressure (ICP), and the cerebral vasculature, which continuously adjusts diameters of vessels in response to changes in cerebral perfusion pressure (CPP = SAP - ICP) [4]. The autoregulation (AR) of CBF keeps it constant for a wide range of CPP and protects the brain during transient episodes of hypo- or hypertension (**Figure** 1). Impairment of autoregulation, which often accompanies TBI or stroke, narrows the 'safe' range of CPP, and increases the risk of secondary ischemia even if the standards of care are strictly observed [4 - 6]. Treatment guidelines, therefore, recommend continuous monitoring of the state of CBF autoregulation and consideration of this information when making therapeutic decisions [7].

CBF AR is continuously assessed by correlating spontaneous slow (<0.05Hz) oscillations of SAP or CPP with related changes in cerebral blood volume (CBV): a negative correlation indicates

Published under licence by IOP Publishing Ltd

Journal of Physics: Conference Series **434** (2013) 012042 doi:10.1088/1742-6596/434/1/012042

preserved autoregulation, whereas positive values suggest its impairment or loss [8]. As direct measurement of CBV with radiological methods is cumbersome, CBV fluctuations are usually assessed indirectly, through related changes in ICP. The correlation coefficient between SAP and ICP – called pressure-reactivity index (PRx) – has been established in clinical studies as a sensitive metric of CBF AR [4 – 6]. Its long-term monitoring (>30 minutes) allows for determination of the limits of autoregulation, defined as the range of SAP/CPP for which PRx is negative (**Figure** 2): this is an important piece of information, as SAP/CPP should, by all means, be kept within this range. Autoregulation is, unfortunately, rarely monitored in clinical practice because invasive ICP probes are not routinely implanted to patients with cerebrovascular pathologies. Noninvasive transcranial bioimpedance monitoring (rheoencephalography, REG) might, however, substitute ICP measurements, as REG also captures slow oscillations of cerebral intra-arterial volume caused by MAP changes. This study investigated whether the correlation coefficient between MAP and REG (called REGx) could be used to assess the reactivity of cerebral vasculature and determine the lower limit of CBF AR.

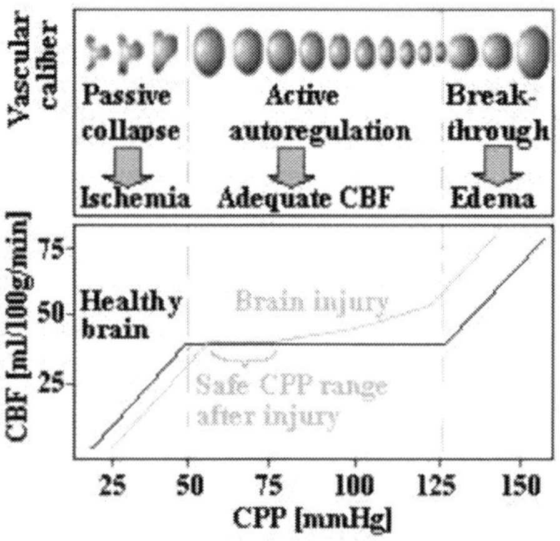

Figure 1. Schematic relationship between the cerebral perfusion pressure (CPP) and blood flow (CBF) in a healthy individual and a patient with an acute TBI or stroke.

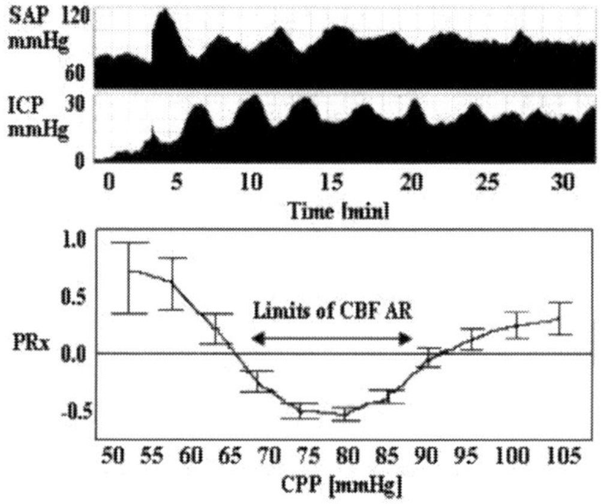

Figure 2. Detection of the limits of CBF AR from the pressure-reactivity index (PRx). In this example. AR is preserved for CPP between 65 - 90 mmHg (modified from [11]).

2. Materials and methods

REG-based estimates of the state and lower limit of CBF AR were compared with the method based on the pressure-reactivity index (PRx) and visual assessment by an expert in anesthetized pigs (N=13) and rhesus macaques (N=8) during the terminal phase of experimental exsanguination. Exsanguination is chosen as a model for CBF AR assessment as it causes SAP to gradually lower and cerebral arteries to maximally dilate until the autoregulation is exhausted. REG and ICP were measured concurrently (and *invasively*) in pigs, with the goal to establish direct correspondence between the REG- and ICP-based assessments of CBF AR. The study on monkeys aimed, on the other hand, to assess the utility of non-invasive REG recordings for evaluation of CBF AR.

SAP was measured with a micro-tip pressure transducer (Millar Instruments, Houston, TX) inserted into femoral artery and connected to the Digi-Med Pressure Analyzer, (Micro-Med, Louisville, KY). ICP in pigs was recorded with an intraparenchymal transducer (ICP Express; Codman, Raynham, MA). REG was acquired with intracerebral needle electrodes in pigs and surface (noninvasive) electrodes in macaques using bipolar systems (OTE Galileo, Italy and Cerberus, Quintlab, Hungary). All signals were digitized at 200Hz. SAP and ICP were averaged using a 10-second window, and REG was differentiated, rectified, integrated and time-averaged in order to remove fluctuations and derive the trend signals (tSAP, tICP, tREG) that capture the aforementioned slow oscillations [8 - 10].

Automated PRx- and REGx-based assessment of autoregulation was performed in MATLAB R2008 software (Natick, MA). PRx [10] and REGx [11] were computed by correlating tSAP with the tICP and tREG signals, respectively, using a 5-minute window that was slid along the signals in 1-minute steps. In accord with [10], the automated ICP and REG-based methods qualified CBF AR in each recording as impaired (-) if PRx or REGx were above 0.2 during at least half of the recording time. Distribution of the PRx and REGx indices were analyzed as a function of the averaged SAP signal (as depicted in Figure 2), and the lower limit of CBF AR was determined as the lowest SAP level for which the mean value of pertinent correlation index was still negative. The expert used the WRAIR DataLyser software package to view the signals, assess the state of CBF AR, and determine its lower limit if CBF AR was considered active (+).

3. Results

1.1. Invasive REG vs. ICP-based and expert assessment in pigs

The PRx- and REGx-based classifications of the CBF-AR status were identical in all 13 animals (**Table** 1), and there was a close agreement between the PRx- and REG-based estimates of the lower limit of CBF AR, when the latter was determinable (n= 7; PRx-based: 58.5 ± 6.1 mmHg; REGx-based: 54.3 ± 6.4 mmHg). The pooled point-to-point correlation between the REGx and PRx values was significant (Spearman r = 0.44, p<0.001), but the observed degree of variability of the short-term (5-minute) estimates of both indices warrants additional studies with the focus on selectivity of this approach under various conditions. The automated evaluation of the CBF AR status agreed with the expert's assessment in all but one animal. REGx-based estimates of the lower limit of CBF AR were, however, lower than the values determined by the expert (61.4 ± 10.5 mmHg, p=0.11).

Table 1. CBF AR evaluation in pigs.

Pig No.	State of CBF AR			Lower limit [mmHg]		
	Exp	PRx	REGx	Exp	PRx	REGx
03	+	+	+	68	63	63
04	+	+	+	54	68	58
08	+	+	+	79	53	58
09	+	+	+	51	n/d[a)]	48
12	+	+	+	56	58	58
13	+	+	+	60	57	48
10	+	+	+	n/d[b)]	n/d[b)]	n/d[b)]
06	-	+	+	n/a[c)]	52	47
11	-	-	-	Cannot be determined if CBF AR is lost		
07	-	-	-			
01	-	-	-			
02	-	-	-			
05	-	-	-			

Table 2. CBF AR evaluation in macaques.

Monkey No.	State of AR		Lower limit	
	Exp	REGx	Exp	REGx
02	+	+	50	58
05	+	+	70	62
07	+	+	55	62
08	+	+	35	38
01	-	-	Cannot be determined if AR is lost	
03	-	-		
04	-	-		
06	-	-		

a) Not determinable as 0.0 < PRx <0.2 throughout most of the record.
b) PRx and REGx <0 throughout the record.
c) Expert assessment not attempted.

1.2. Non-invasive REG vs. expert assessment in monkeys

There was a perfect agreement of the REG-based and expert classifications of the state of CBF AR (**Table** 2), and a close agreement between the expert and REGx-based estimates of the lower limit of AR, when the latter was determinable (n = 4; expert: 52.5 ± 14.4 mmHg; REG: 55.0 ± 11.9 mmHg).

4. Conclusion

The results obtained in pigs concur to the findings (in rats) that amplitude changes in the intracranial REG signal reflect changes in cerebral vascular resistance [9], and further strengthen the hypothesis that the REGx index could substitute the PRx index in the assessment of CBF AR [11]. The results from monkeys represent, on the other hand, the first demonstration that surface REG recordings and the respective *noninvasive* correlation index can be used for the assessment of the state and limits of CBF AR. This is important for affirmation of the method in the light of the decades-old debate about the relevance of extracranial impedance measurements in evaluation of intracranial phenomena [12]. Additional studies are, however, warranted in animal models, human volunteers and clinical populations to quantify sensitivity and selectivity of the automated, surface REG-based evaluation of CBF AR during physiological CBF manipulations, hemorrhage, brain blast injury and stroke.

5. Acknowledgments

This work has been partially funded by a grant from the US Department of Defense to Advanced Brain Monitoring (#W81-XWH-10-C-0061). The authors are grateful to members of the WRAIR Departments of Resuscitative and Veterinary Medicine for their assistance.

Disclaimer: Material has been reviewed by the Walter Reed Army Institute of Research. There is no objection to its presentation and/or publication. The opinions or assertions contained herein are the private views of the author, and are not to be construed as official, or as reflecting true views of the Department of the Army or the Department of Defense. Research was conducted in compliance with the Animal Welfare Act and other federal statutes and regulations relating to animals and experiments involving animals and adheres to principles stated in the Guide for the Care and Use of Laboratory Animals, NRC Publication (1996 and 2011 editions).

References

[1] World Health Organization 2012 *World Health Statistics 2012* (Geneva, Switzerland: WHO)

[2] Langlois J A, Rutland-Brown W and Thomas K E 2006 *Traumatic brain injury in the United States: emergency department visits, hospitalizations, and deaths* National Center for Injury Prevention and Control (Atlanta, Georgia: US Department of Health & Human Services)

[3] National Stroke Association 2010 *Stoke Facts Sheet* (Cenntenial, Colorado: NSA)

[4] Lassen N A 1974 Control of cerebral circulation in health and disease. *Circ. Res.* **34** 749 – 60

[5] Steiner L A, Czosnyka M, Piechnik S K, Smielewski P, Chatfield D, Menon D and Pickard J D 2002 Continuous monitoring of cerebrovascular pressure reactivity allows determination of optimal cerebral perfusion pressure in patients with TBI *Crit. Care Med.* **30** 733 – 8

[6] Lang E and Chestnut R 2000 A bedside method for investigating the integrity and critical thresholds of cerebral pressure autoregulation in severe TBI patients *Brit. J. Neurosurg.* **14** 117 – 26

[7] Brain Trauma Foundation, American Association of Neurological Surgeons, Congress of Neurological Surgeons, Joint Section on Neurotrauma and Critical Care, AANS/CNS 2007 Guidelines for the management of severe traumatic brain injury, IX: cerebral perfusion thresholds *J. Neurotrauma* **24** 59 – 64

[8] Van Beek A H E A, Claassen J A H R, Rikkert M G M and Jansen R W M M 2008 Cerebral autoregulation: an overview of current concepts and methodology with special focus on the elderly *J. Cereb Blood Flow Metab* **28** 1071 – 85

[9] Bodo M 2010 Studies in rheoencephalography *J. Electr. Bioimp.* **1**18 - 40

[10] Czosnyka M, Smielewski P, Kirkpatrick F, Laing R, Menon D and Picard J D 1997 Continuous assessment of the cerebrovascular reactivity in head injury *Neurosurgery* **41** 11-7

[11] Brady K M, Mytar J O, Kibler K K, Easley R B, Koehler R C, Czosnyka M, Smielewski P, Zweifel C, Bodo M, Pearce F J and Armonda R A 2010 Monitoring cerebrovascular pressure reactivity with REG *J. Phys.: Conf. Ser.* **224** 012089

[12] Perez J J, Guijarro E and Sancho J 2005 Spatiotemporal pattern of the extracranial component of the rheoencephalographic signal *Phys. Meas.* **26** 925 – 38

Measurement of localized tissue water – clinical application of bioimpedance spectroscopy in wound management.

LC Ward[1], K Sharpe[2], D Edgar[2,3], V Finlay[3] and F Wood[2,3]

[1] School of Chemistry and Molecular Biosciences, The University of Queensland, Brisbane, QLD 4072, Australia
[2] Burn Injury Research Unit, University of Western Australia, Australia
[3] Fiona Wood (Burn Research) Foundation of WA, Perth, Australia

l.ward@uq.edu.au

Abstract. Wound healing is a complex process which can be impeded by the presence of accumulated cell fluid or oedema. A simple and convenient method for the assessment of wound oedema would aid improvement in patient care. In this proof of concept study we investigated whether bioelectrical impedance spectroscopy has the potential to provide such a tool. A number of important observations were made. Firstly, the method was highly reproducible and data can be obtained from electrodes located at different positions around the region of interest; important given the highly variable topography of surface wounds, e.g. burns. Secondly, the method was highly sensitive with the potential to detect changes of as little as 20 μl in extracellular fluid. Thirdly the relative changes in R0, R∞ and Ri following sub-cutaneous injections of saline were consistent with redistribution of water from the extracellular to intracellular space and /or removal from the local area as may occur during wound healing.

1. Introduction

The burden of chronic and acute wound care on the hospital system is immense and conditions involving these lesions are increasing in prevalence [1]. Wound healing is a complex but well described physiological process, which is common to all lesions and injuries in spite of their origin. Factors known to impede wound healing have been identified, amongst which is oedema. Oedema, which represents an expansion of the tissue's extracellular fluid compartment, impairs the healing process by increasing the diffusion distance between the nutrient source and the cell, compromising cell function and cellular waste removal, leading to cell death and a toxic wound environment. Bioimpedance spectroscopy (BIS) is a safe, rapid and non-invasive technique that has been used for the objective assessment of whole body and tissue fluid compartments. The technique lends itself to the assessment of conditions involving fluctuations of fluid within and between the intra-and extracellular fluid compartments, and could be useful in the assessment of periwound oedema. A preliminary report has suggested that it may also be useful for the assessment of epithelialization that occurs during wound healing [2].

We report here the results of studies, designed as a proof of concept, to demonstrate the potential and feasibility of the technique as a wound area-specific, non-invasive means of wound oedema monitoring.

Published under licence by IOP Publishing Ltd

Journal of Physics: Conference Series **434** (2013) 012043 doi:10.1088/1742-6596/434/1/012043

2. Experimental

2.1 Assessment of the feasibility of localized impedance measurements. Bioimpedance measurements were made across a 60-mm diameter region of the dorsal surface of the forearm mid-way between the head of the ulna and the olecranon using an SFB7 (ImpediMed, Brisbane) bioimpedance spectrometer. Current drive electrodes (Ag-AgCl, Kendall Care 610) were located at the base of fingers and on the foot; voltage sense electrodes were located at 45° intervals around the circumference of the measurement area. Impedance was measured, in triplicate, between all combinations of electrode pairs at 256 discrete current frequencies (4-1000 kHz) across a body region and data analysed according to the Cole model using Bioimp version 4.8.0.0. to determine resistance at zero frequency (R0), infinite frequency ($R\infty$) and Ri, the intracellular resistance. Data were obtained for six healthy control subjects.

2.2 Assessment of the ability of impedance measurements to detect localized changes in extracellular fluid. A 21-gauge needle was inserted subcutaneously into the forearm of a participant at the midpoint between sense electrodes located 40 mm apart along the midline and was left in situ for the duration of the study. Current drive electrodes were located on the hand and foot. Impedance measurements were repeated prior to and following injection of a 1 ml bolus of normal saline. Further injections (1 ml) to a total volume of 5 ml were made 3, 5, 7, 10 minutes from baseline. Impedance was recorded, in triplicate as described above, following each injection. The study concluded with continued monitoring of impedance during resorption of injected fluid over the subsequent 50 minutes.

3. Results

3.1 Assessment of the feasibility of localized impedance measurements.

Impedance measurements were highly reproducible, with mean a coefficient of variation of 2.3%. Figure 1 represents a map of the measured resistances (R0) between pairs of electrodes for a representative subject.

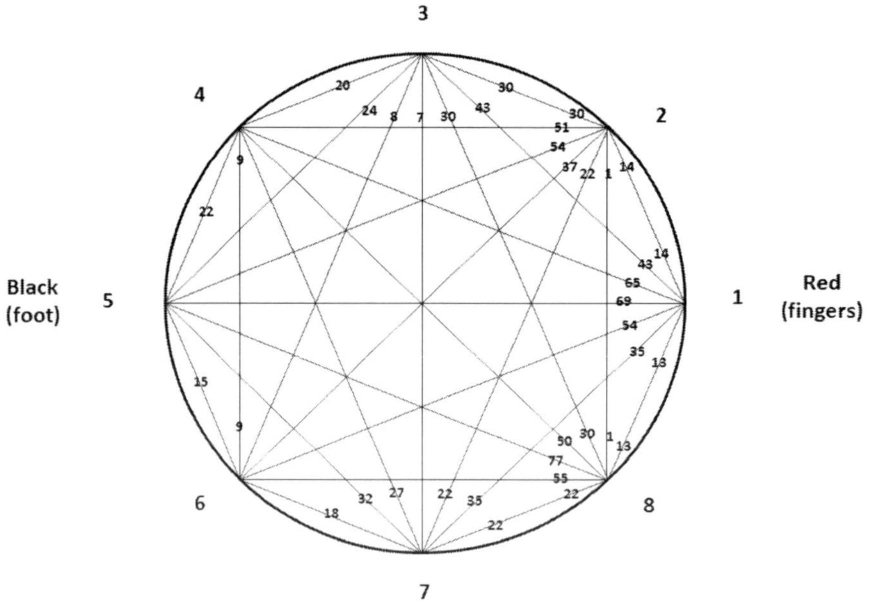

Figure 1. Map of resistances measured between surface electrodes located on the dorsal surface of the forearm. Voltage sense electrodes (numbered 1 to 8) were located at 45° intervals around the circumference of a circle 60 mm in diameter; current drive electrodes were located on the hand and foot. Lines and numbers within the circle indicate the measured resistance (ohm) between the pair of electrodes connected by the line.

Resistance measurements ranged from 1 to approximately 80 ohm between adjacent or diametrically opposite electrode pairs respectively and between different subjects. Impedance measurements were achievable for all electrode combinations with the exception of electrodes located orthogonal to the direction of current flow in some subjects. The sum of impedances for each pair of adjacent electrodes equalled (within ± 2%) that of equatorially placed electrodes along the line of current flow. For example in Figure 1, the sum of resistances between electrode positions 1-8-7-6-5 is 68 ohm compared to 69 ohm for 1-5 directly. In this participant, electrode positions for pairs 2-8, 4-6 and 3-7, although aimed to be orthogonal to the direction of current flow (red-black), this was not fully achieved as resistances of 1, 9 and 7 ohm respectively were recorded.

3.2 Assessment of the ability of impedance measurements to detect localized changes in extracellular fluid. The change in R0, R∞ and Ri are presented in Figures 2a,b,c respectively. The times of injection of each bolus (1 ml) of saline are indicated by the vertical lines.

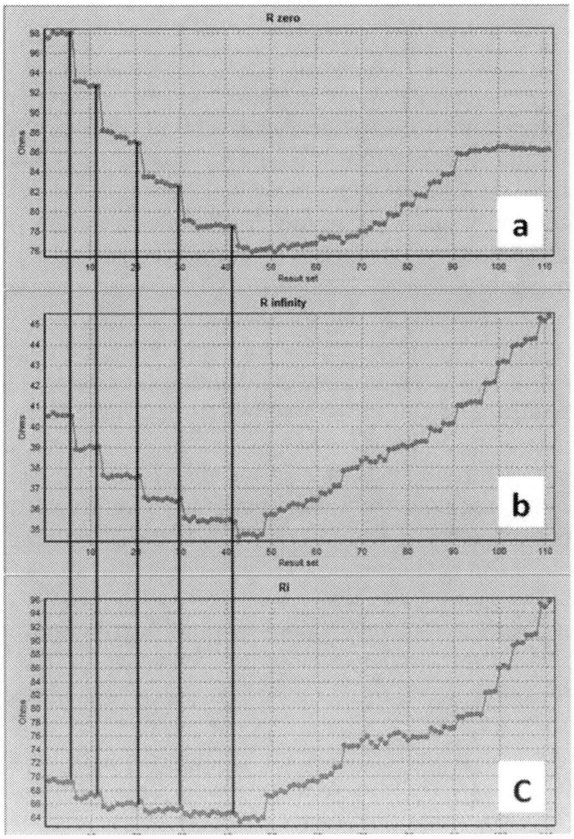

Figure 2. Change in tissue resistance during subcutaneous injection and resorption of saline.
Panel a: Change in R0 (inversely proportional to change in extracellular water volume).
Panel b: Change in R∞ (inversely proportional to change in total tissue water volume).
Panel c: Change in Ri (inversely proportional to change in intracellular water volume).
Vertical lines indicate time of injection of 1 ml of saline.

The relationship between resistance and fluid volume is described by the following

$$R \propto \frac{l}{V} \qquad (1)$$

where l is the distance between the measurement electrodes, R is resistance and V is the volume of fluid. R0 decreased by 4.9 ohm with the first subcutaneous injection of 1 ml of saline. The drop in resistance decreased progressively with each injection to 2.1 ohm by the fifth injection. The corresponding decreases in R∞ were 1.6 and 0.7 ohm respectively and the decreases in Ri were 2.2 and 0.9 ohm respectively.

Following the last injection of saline (at result data set 45) resistance progressively of all three fluid compartments increased as fluid was resorbed and redistributed between the intra and extracellular compartments. Notably resistance of no compartment returned to baseline values by 50 minutes post-injection.

4. Discussion

It is clear from these preliminary data that impedance measurements can track in a quantitative manner accumulation of fluid and its subsequent resorption in a localized tissue region. A number of important observations can be made from this pilot study. Firstly, the method appears to be robust with respect to electrode positioning. Wounds are frequently not regular in shape and it will not always be possible to locate electrodes such that current drive and sense electrodes are in-line.

Nevertheless, it is clear from the positioning study that an impedance will be measurable unless sense electrodes are perfectly orthogonal to the direction of current flow, suggesting that electrodes may be located at clinically acceptable locations even of not electrically ideal. Secondly, the method is potentially very sensitive. The impedance device used in this pilot study measured resistance with a resolution of 0.1 ohm. We were able to detect a change of approximately 5 ohm in R0 per ml of saline injected. This implies an ability potentially to detect changes in fluid volume of as little as 20 μl. Thirdly, during the period of injection the resistance of the intracellular compartment decreased to a much smaller and diminishing extent suggesting that the injected fluid was being predominantly sequestered in the extracellular space. Fourthly, the rate of increase in R0 following the last injection was much slower than that observed for the decrease in resistance during injections and approached a plateau after approximately 35 minutes. Conversely, the increase in R_∞, an index of both extra-and intracellular fluid showed a continued increase for as long as monitoring was continued (50 min). These observations are consistent with redistribution of water from the extracellular to intracellular space and/or removal from the local area.

Acute and chronic wounds are the feature of many illnesses including diabetes, venous stasis and burns. Pathological fluid accumulation around the site of wound is known to impair reparation, even in volumes imperceptible to the human eye. Whilst there are numerous medical interventions designed to reduce oedema, assessing their efficacy has been difficult on account of the relative paucity of simple and reliable tools to measure swelling. This pilot study shows that focal bioimpedance spectroscopy has great promise as a painless, non-invasive and potentially quantitative tool for the measurement of localized oedema; particularly, if as suggested by this study, it is capable of perceiving changes in tissue fluid volume of as little as a ml or potentially much less.

The instrument used in this trial was designed primarily for body composition assessment although it has been applied in variety of clinical settings, notably assessment of lymphedema. Its use here is a novel application of this technology and, hence, care must be taken in the interpretation of the results. The electrode arrangement described in this study should not be construed as representing the optimal arrangement. The region spanned by the measurement electrodes was 40 mm but clearly wounds are highly variable in shape and size and further studies are required to replicate the above findings for different electrode spacings and orientations. Wounds are also of variable depth and hence volume. The volume of tissue defined by the impedance measurement in these studies is unknown.

The present study measured acute changes in fluid only; the measurement of fluid distribution in the tissues long-term brings its own challenges. In order to detect any true changes in fluid accumulation and distribution in a particular location, both the configuration and placement of the electrodes must be replicated exactly at each measurement occasion; raising issues of both inconsistency and inconvenience. Preliminary studies in burns patients have underscored these difficulties. These studies continue, seeking not only to test the validity of this method for the purpose of wound monitoring, but also looking at ways of overcoming any practical issues in its implementation. Although much further investigation is required, this single study demonstrates the potential of bioimpedance spectroscopy for wound monitoring represents an exciting step toward objective evaluation of wound changes and their management.

References

[1] Mulligan S, Prentice J and Scott L 2011 WoundsWest Wound Prevalence Survey 2011 State-wide Overview Report. Ambulatory Care Services, Department of Health, Perth, Western Australia.

[2] Lukaski H C and Moore M M 2012 Bioelectrical impedance assessment of wound healing *J. Diabetes Sci. Technol.* 6 209-12

Journal of Physics: Conference Series **434** (2013) 012044 doi:10.1088/1742-6596/434/1/012044

Screening for sarcopenia in obesity

R Sammarco, M Marra, C Montagnese, E De Rosa, M Onufrio, V Amato, L Santarpia, C De Caprio, F Contaldo and F Pasanisi.
Department of Clinical Medicine and Surgery, Federico II University, Naples, Italy.
e-mail: marra@unina.it

Abstract. Sarcopenia is a progressive and generalized loss of skeletal muscle mass (SM) and function which can also be found in obese adults. The aim of the study was to evaluate the prevalence of sarcopenia in 1 245 obese women (18 – 67 years, weight 114.7±24.5 kg; BMI 44.1±9.2; fat mass 49.0±6.2%) from Southern Italy. Body composition was evaluated by bioimpedance analysis (BIA) and SM calculated by using Janssen's equation; therefore the sex-specific cut-off points of percentage skeletal muscle index were used. The whole population was divided in two age groups: A (18-40 years; n. 808; weight 115.4±23.5 kg; BMI 43.8±8.8 kg/m^2) and B (41-67 years; n. 438; weight 113.4±26.3 kg; BMI 44.8±9.9 kg/m^2). In all the sample there was 2.7% moderate and 0.6% severe sarcopenia; in group A, 1.9% moderate and 0.6% severe sarcopenia whilst in group B 4.3% moderate and 0.7% severe sarcopenia. The results of our study suggest that, based on a screening examination by BIA, moderate/severe sarcopenia can be detected in an unselected middle-aged obese population. Further studies are required to clarify the diagnosis with functional tests.

1. Introduction

In 1989 the term "sarcopenia" (Greek *sarx* or flesh + *penia* or loss) was suggested by Rosemberg to describe the loss of skeletal muscle mass due to ageing [1].
Sarcopenia describes a progressive and generalized loss of skeletal muscle mass and strength below a critical threshold, associated with reduced quality of life and increased risk of disability and death [2,3]. Moreover, ageing and physical disability are related to an increase of fat mass and, in particular, of visceral fat, an important risk factor for metabolic and cardiovascular diseases [4].
In 1996 Sarcopenic Obesity (SO) was defined for the first time by Heber et al. as the combination of reduced fat free mass evaluated by bioimpedance analysis (BIA) and fat mass excess, expressed as body weight percentage [5].
In Southern Italy a high prevalence of obesity, associated with low physical activity, is observed [6] and requires different levels of care, as defined in a recent consensus document [7].
The growing older population and the increasing incidence of obesity in western societies require accurate studies to evaluate the real prevalence of sarcopenia and its related risks.
Aim of the study was to evaluate prevalence and characteristics of sarcopenia in a group of female obese outpatients, by performing a bioimpedance analysis (BIA) as screening evaluation and using different methods for muscle mass evaluation, according to present literature.

2. Methods

One thousand and two hundred-fortyfive, 18-to-67 year-old, female obese (Body Mass Index BMI ≥ 30 kg/m^2) outpatients were evaluated in the Obesity Outpatient Clinic of the Department of Clinical Medicine and Surgery of Federico II University in Naples, Italy. All sample was divided in two age groups: A) age 18-40 years; B) 41-67 years.
Anamnesis, in particular as concerning weight history, was collected. Complete clinical exam was performed and blood samples for lab test evaluation were collected in order to check the presence/absence of associated metabolic abnormalities.
Body composition and Skeletal Muscle Mass (SM) were evaluated by BIA with a Human IM Plus II (DS Medica – Milano_Italy). Measurements were performed on the non dominant body side, at room temperature (22 - 24 °C), after urination and having been supine for 20'. A standard tetrapolar technique was used by sensor electrodes on the anterior surface of wrist and ankle and injector electrodes on the dorsal surface of hand and foot.

2.1. Definition of Sarcopenia

The presence of sarcopenia in our population was assessed by using the cut-off points of skeletal muscle mass index (SMI%) reported in Janssen et al. 2002 study [8]. Skeletal muscle mass (SM) was calculated by using the BIA equation by Janssen et al. 2000 [9]:

$$SM\ (kg) = [(h^2/\ BIA\ resistance \times 0.401) + (gender \times 3.825) + (age \times 0.071)] + 5.102$$

where height (h) is expressed in cm, BIA resistance in ohms; for gender: men =1 and women =0; age in years.

2.2. SMI% cut-off points

As in Janssen's study [8], absolute skeletal muscle mass was converted into percentage of skeletal muscle mass. We defined this index SMI% = SM/total body mass × 100.
In Janssen's 2002 study, data from young adults (3116 men and 3298 women aged 18 - 39 years), selected among the population of NHANES III study [9, 10], were used as reference data to define cut-off values for normal skeletal muscle mass and sarcopenia. In details, subjects were considered to have a normal SM if their SMI% was greater than –1 Standard Deviation (SD) of the sex-specific mean for young adults (men: SMI% > 37.0; women: SMI% > 27.6). Class I sarcopenia was considered to be present in subjects whose SMI% was between –1 and –2 SD (men: 37.0 ≥ SMI% > 31.5; women: 27.6 ≥ SMI% > 22.1); class II sarcopenia was present if SMI% was below –2 SD (men: SMI% ≤ 31.5; women: SMI% ≤ 22.1). Data were computerized and analyzed by using SPSS 14.0 software. Mean and standard deviation of all the variables were calculated. Prevalence of sarcopenia was evaluated in our population according to the criteria described in the literature. T-Student test for unpaired data was used to evaluate the differences between groups.

3. Results

Table 1 shows the main characteristics of our 1245 obese patients population. Height and Fat Mass (%) were significantly higher in group A than in group B, while BMI was significantly lower; weight, Fat Mass (kg) and Fat Free Mass (kg) didn't result significantly different in the two groups (Table 1).
Skeletal muscle mass (SM), calculated by Janssen et al. BIA equation [9], was 34.0 ± 3.8 kg in group A and 32.8 ± 4.3 kg in group B ($p = 0.001$). SMI% was 30.2 ± 4.3 % in group A and 29.6 ± 4.8 % in group B ($p = 0.036$) (table 1).
According to SMI% cut-off points by Janssen et al. 2002, in all sample the prevalence of moderate sarcopenia was 2.7%, while the prevalence of severe sarcopenia was 0.6%; in group A 1.9% presented moderate sarcopenia and 0.6% had severe sarcopenia, whereas in group B prevalence was 4.3% and 0.7% respectively.

Table 1. Antropometric parameters, body composition, SM, SMI% and of 1245 female obese patients.

	Group A (n. 808)		Group B (n = 438)		
	Mean	SD	Mean	SD	*p*
Age	28,9	7,0	48,9	5,9	< 0,001
Weight (kg)	115,4	23,5	113,9	26,2	n.s
Height (cm)	162,3	6,2	159,1	6,6	< 0,001
BMI (kg/m^2)	43,8	8,8	45,0	9,9	0,035
Fat free mass (kg)	57,5	9,4	57,8	10,3	n.s.
Fat mass (Kg)	57,9	16,5	56,1	19	n.s.
Fat mass (%)	49,5	5,7	48,3	7	0,002
SM (kg)	34,0	3,8	32,8	4,3	< 0,001
SMI %	30,1	4,3	29,6	4,8	0,036

3. Conclusions

Most data reported in literature deal with the evaluation of sarcopenia and sarcopenic obesity in older people with eterogeneous BMI values, whilst information about the prevalence of sarcopenia in obese adults are limited. Our study highlights the presence of a high risk of sarcopenia in obese patients even at a mean age of fifty.

In conclusion, this study suggests that sarcopenic obesity is not only a "geriatric" syndrome; indeed, based on a screening by BIA, some degree of sarcopenia can be detected in an outpatient obese population. Further studies are required to better define the diagnosis with functional tests.

In the future, the screening for sarcopenic obesity could be useful in all age groups in order to obtain an early diagnosis and to develop prevention strategies for this cause of physical disability in obese subjects.

References

[1] Rosenberg 1989 *Am J Clin Nutr* **50** 1231-3

[2] Cruz-Jentoft AJ, Baeyens JP, Bauer JM, Boirie Y, Cederholm T, Landi F, Martin FC, Michel JP, Rolland Y, Schneider SM, Topinková E, Vandewoude M and Zamboni M 2010 *Age Ageing* **39** 412-23

[3] Janssen I, Baumgartner RN, Ross R and Roubenoff R 2004 *Am J Epidemiol* **159** 413-21

[4] Riechman SE, Schoen RE, Weissfeld JL, Thaete FL and Kriska AM 2002 *Obes Res.* **10** 1065-73

[5] Heber D, Ingles S, Ashley JM, Maxwell MH, Lyons RF and Elashoff RM 1996 *Am J Clin Nutr.* **64** S472-S477

[6] Finelli C, Gallipoli P, Celentano E, Cacace G, Saldalamacchia G, De Caprio C, Contaldo F and Pasanisi F 2006 *Nutr Metab Cardiovasc Dis.* **16** 168-73

[7] Donini LM, Cuzzolaro M, Spera G, Basso N, Bollea MR 2010 *Eat Weight Disord.* **15** 1-31
[8] Janssen I, Heymsfield SB and Ross R 2002 *J Am Geriatr Soc* **50** 889-96
[9] Janssen I, Heymsfield SB, Baumgartner RN and Ross R 2000 *J Appl Physiol* **89** 465-471
[10] NHANES III Reference Manuals and Report, 1996

Journal of Physics: Conference Series **434** (2013) 012045 doi:10.1088/1742-6596/434/1/012045

Study of the Dynamics of Transcephalic Cerebral Impedance Data during Cardio-Vascular Surgery

S.R Atefi [1], F Seoane[1,2] and K Lindecrantz[1,3]

[1]School of Technology and Health, Royal Institute of Technology, 141 52 Huddinge, Sweden
[2]School of Engineering, University of Borås, 501 90, Borås, Sweden
[3]Department of Clinical Science, Intervention and Technology, Karolinska Institute, 14186 Stockholm, Sweden

E-mail: atefi@kth.se

Abstract. Postoperative neurological deficits are one of the risks associated with cardio vascular surgery, necessitating development of new techniques for cerebral monitoring. In this study an experimental observation regarding the dynamics of transcephalic Electrical Bioimpedance (EBI) in patients undergoing cardiac surgery with and without extracorporeal circulation (ECC) was conducted to investigate the potential use of electrical Bioimpedance for cerebral monitoring in cardio vascular surgery. Tetrapolar transcephalic EBI measurements at single frequency of 50 kHz were recorded prior to and during cardio vascular surgery. The obtained results show that the transcephalic impedance decreases in both groups of patients as operation starts, however slight differences in these two groups were also observed with the cerebral impedance reduction in patients having no ECC being less common and not as pronounced as in the ECC group. Changes in the cerebral impedance were in agreement with changes of haematocrit and temperature. The origin of EBI changes is still unexplained however these results encourage us to continue investigating the application of electrical bioimpedance cerebral monitoring clinically.

1.Introduction

Coronary artery disease (CAD) is rated as the first cause of death in world[1]. Many patients with CAD require coronary artery bypass surgery in order to reduce angina and the risk of death. However major neurological and neurophysiological deficits have been reported after cardiac surgery for cardiopulmonary bypass [2]. Different etiologic mechanisms have been reported to be responsible for these deficits [3-5]. One common problem associated with cardio vascular surgery is hypo-perfusion, but irrespective of the origin of the problem a reliable and easy to use method for monitoring the brain during a cardiac bypass surgery would be advantageous. EBI could be one candidate for such purpose.

1.1. Electrical bioimpedance cerebral monitoring (EBCM)

Electrical bioimpedance (EBI) is a safe, non-invasive method currently clinically available for different applications like skin cancer screening[6], impedance pneumography or body composition assessment [7]. It has been shown that following an incident of neurological damage like stroke or

Journal of Physics: Conference Series **434** (2013) 012045 doi:10.1088/1742-6596/434/1/012045

cerebral hypoxia the electrical properties of the cerebral tissue changes as a result of changes in the tissue structure and composition which is detectable by electrical bioimpedance[8-11].

1.2. Monitoring hypo-perfusion
Hypo-perfusion reduces the blood flow at first changing the electrical properties of cerebral, if hypo-perfusion extends over time cerebral hypoxia or even ischemia may occur, thus modifying further the changes in the electrical properties of brain tissue.

Under the assumption that EBI measurements could be used to monitor the status of the brain during cardiopulmonary bypass (CPB), in this paper the dynamics of EBI measurements from two groups of patients undergoing cardio vascular surgery with and without ECC have been analysed together with other physiological variables to study the possibility of using cerebral impedance from EBI recordings.

2.Materials and Methods
Tetrapolar EBI measurements were recorded from two groups of patients undergoing cardio vascular surgery. Cardiovascular surgery can be done with or without the use of heart-lung-machine, extracorporal circulation (ECC). This study includes seven (n=7) patients with ECC and six (n=6) patients without ECC. It has been done with an ethical approval from the regional ethical review board of Gothenburg.

2.1 EBI recordings
Transcephalic EBI measurements were recorded at the single frequency of 50 kHz. Recordings started at least 30 minutes prior to start of the cardio vascular surgery and continued for the whole operation.

2.2 Haematocrit and temperature recordings
Together with EBI and cerebral impedance measurements haematocrit (HCT) and temperature were also recorded from all patients.

3. Results

3.1 EBI dynamics with and without ECC
In figure 1, it is possible to observe that after a relatively stable baseline recorded for 20-30 minutes, the recorded cerebral EBI decreases during surgery for the seven patients undergoing the cardio vascular surgery with ECC. For the Non-ECC cases it can be seen in figure 2, that in half of the patients the changes are very small, less than 1 Ohm, while in the other 3 impedance reduction from baseline is as large as the changes seen in the ECC cases.

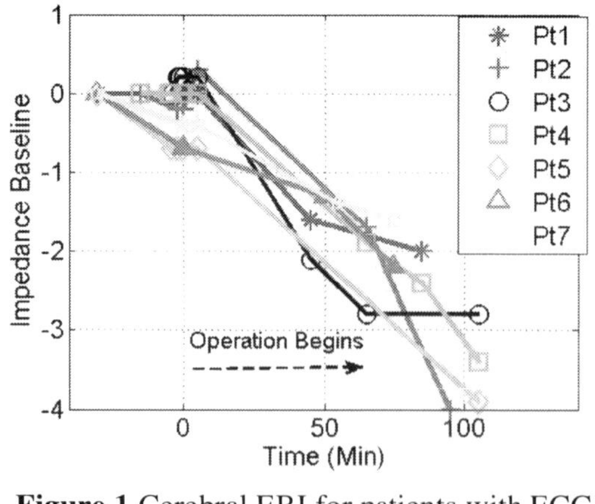

Figure 1.Cerebral EBI for patients with ECC

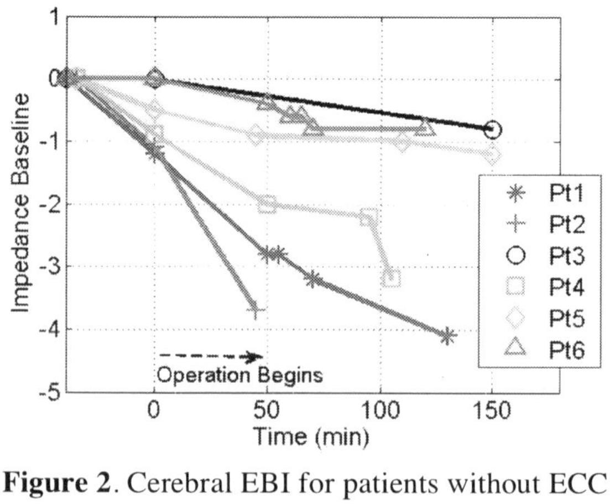

Figure 2. Cerebral EBI for patients without ECC

Table 1. Changes in the hematocrit value in each patient before the operation and during the operation (with ECC)

Patient Nr	1	2	3	4	5	6	7	Mean
HCT_{PRE_OP} - HCT_{PER_OP}	27	43	61	30	5	31	41	34

Table 2. Changes in the hematocrit value in each patient before the operation and during the operation (without ECC)

Patient Nr	1	2	3	4	5	6	Mean
HCT_{PRE_OP} - HCT_{PER_OP}	4	7	8	14	16	19	11.3

3.1 Haematocrit and temperature

In both groups the pre-operation haematocrit level HCT_{PRE_OP} was very similar with a mean group value of 115 for the ECC group and 124 for the non-ECC group. While in the non-ECC group the mean HCT level decreased slightly to 113, in the ECC group the level decreased down to 81. In Table 1 and 2, the differences between HCT levels before the surgery and during the surgery for each patient are listed. As it can be seen, the HCT level of patients with ECC exhibit higher changes than to those without ECC. Haematocrit decreases as the operation starts.

Figure 3, shows the mean value of temperature for each group. It is observed that in the ECC cases the temperature varies between 35.5 and 37, whereas for the non-ECC group the mean value remains relatively stable throughout the surgery.

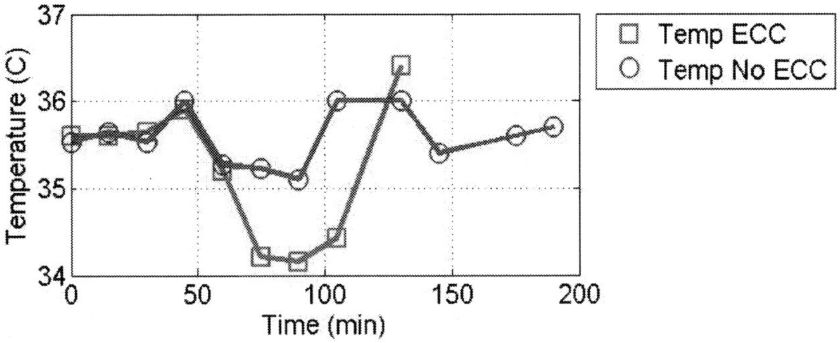

Figure 3. Temperature over time for patients undergoing the operation with and without ECC

4. Discussion

The results indicate that the cerebral impedance drops as the operation starts. The reduction in the cerebral impedance can be easily observed not only on patients having ECC but it can also be seen in half of the patients not having ECC. A general reduction on the haematocrit level has been also reported for both cases, but more pronounced in the ECC patients. Since Geddes and Saddler have shown that resistivity is exponentially related to haematocrit [12], the reported reduction might be partially responsible for the observed reduction in the measured cerebral impedance.

A temperature increase over 2.5 ° Celsius is seen in the ECC group. Since the resistivity of ionic conductors depends on temperature and changes of around 2%/°C on EBI can be expected as reported in [13] it is very likely that the temperature change also contributes to the change of cerebral EBI.

5. Conclusion

For EBCM to become a monitoring tool for prompt detection of Brain damage during Cardiovascular surgery, it is required to establish a reference baseline for comparison and if the baseline value changes due to external factors of the surgery like haematocrit levels or temperature the reference value must be compensated properly to keep the possibility of performing a fair comparison between EBI recording that would allow changes produced by the hypo-perfusion.

References

[1] Mackay J and Mensah G A 2004 The Atlas of Heart Disease and Stroke (Geneva: Myriad Ed. Ltd for WHO) p 112

[2] Shaw P J, Bates D, Cartlidge N E, French J M, Heaviside D, Julian D G and Shaw D A 1987 Neurologic and neuropsychological morbidity following major surgery: comparison of coronary artery bypass and peripheral vascular surgery *Stroke* **18** 700-7

[3] Baker A J, Naser B, Benaroia M and Mazer C D 1995 Cerebral microemboli during coronary artery bypass using different cardioplegia techniques *Ann. Thorac. Surg.* **59** 1187-91

[4] Brooker R F, Brown W R, Moody D M, Hammon J W, Jr., Reboussin D M, Deal D D, Ghazi-Birry H S and Stump D A 1998 Cardiotomy suction: a major source of brain lipid emboli during cardiopulmonary bypass *The Annals of thoracic surgery* **65** 1651-5

[5] Stockard J J, Bickford R G and Schauble J F 1973 Pressure-dependent cerebral ischemia during cardiopulmonary bypass *Neurology* **23** 521-9

[6] Aberg P, Nicander I, Hansson J, Geladi P, Holmgren U and Ollmar S 2004 Skin cancer identification using multifrequency electrical impedance-a potential screening tool *IEEE transactions on bio-medical engineering* **51** 2097-102

[7] Kushner R F and Schoeller D A 1986 Estimation of total body water by bioelectrical impedance analysis *The American journal of clinical nutrition* **44** 417-24

[8] Atefi S R, Seoane F and Lindecrantz K 2012 Electrical Bioimpedance cerebral monitoring. Preliminary results from measurements on stroke patients. In: *Engineering in Medicine and Biology Society (EMBC), 34th Annual International Conference of the IEEE,* pp 126-9

[9] Bonmassar G and Iwaki S 2004 The Shape of Electrical Impedance Spectroscopy (EIS) is altered in Stroke Patients. In: *Engineering in Medicine and Biology Society (EMBC), 26th Annual International Conference of the IEEE,* pp 3443-6

[10] Liu L X, Dong W, Ji X, Chen L, Chen L, He W and Jia J 2006 A new method of noninvasive brain-edema monitoring in stroke: cerebral electrical impedance measurement *Neurological research* **28** 31-7

[11] Liu L X, Dong W W, Wang J, Wu Q, He W and Jia Y J 2005 The role of noninvasive monitoring of cerebral electrical impedance in stroke *Acta Neurochir Suppl.* **95** 137-40

[12] Geddes L A and Sadler C 1973 The specific resistance of blood at body temperature *Medical & biological engineering* **11** 336-9

[13] Lingwood B E, Dunster K R, Healy G N and Colditz P B 2004 Effect of cooling and re-warming on cerebral and whole body electrical impedance *Phys. Meas.* **25** 413-20

Journal of Physics: Conference Series **434** (2013) 012046 doi:10.1088/1742-6596/434/1/012046

Estimated preejection period (PEP) based on the detection of the R-wave and dZ/dt-min peaks in ECG and ICG

René van Lien[1], Nienke M Schutte[1], Jan H Meijer[2], Eco J C de Geus[1]

[1] Department of Biological Psychology, VU University Amsterdam
[2] Department of Physics and Medical Technology, VU University Medical Center, Amsterdam, the Netherlands

E-mail: R.van.lien@vu.nl

Abstract. The validity of estimating the PEP from a fixed value for the Q-wave onset to the R-wave peak (QR) interval and from the R-wave peak to the dZ/dt-min peak (ISTI) interval is evaluated. Ninety-one subjects participated in a laboratory experiment in which a variety of physical and mental stressors were presented and 31 further subjects participated in a sequence of structured ambulatory activities in which large variation in posture and physical activity was induced. PEP, QR interval, and ISTI were scored. Across the diverse laboratory and ambulatory conditions the QR interval could be approximated by a fixed interval of 40 ms but 95% confidence intervals were large (25 to 54 ms). Multilevel analysis showed that 79% to 81% of the within and between-subject variation in the RB interval could be predicted by the ISTI. However, the optimal intercept and slope values varied significantly across subjects and study setting. Bland-Altman plots revealed a large discrepancy between the estimated PEP and the actual PEP based on the Q-wave onset and B-point. It is concluded that the estimated PEP can be a useful tool but cannot replace the actual PEP to index cardiac sympathetic control.

1. Introduction

The preejection period (PEP) is currently the measure of choice to monitor changes in cardiac sympathetic activity non-invasively. Under conditions of stable preload and afterload, changes in PEP reflect changes in contractility [1] which are influenced by sympathetic but not parasympathetic activity. PEP can be obtained from the thoracic impedance cardiogram (ICG) and electrocardiogram (ECG) [2] and is defined as the interval from the onset of left ventricular depolarization, reflected by the Q-wave onset in the ECG, to the opening of the aortic valve, reflected by the B-point in the ICG [2, 3, 4]. Figure 1 shows the ECG and ICG signals with the relevant landmarks.

PEP is usually scored from the ICG waveform after ensemble averaging over multiple beats, time locked to the R-wave peak to reduce noise. However, substantial problems in detection of the Q-wave onset and B-point remain [2, 3, 5]. Current practice is to estimate the Q-wave onset by subtracting a fixed value of 48 ms from the time of the R-wave peak [2, 6]. The validity of this practice has not been verified to date. To assist in the detection of the B-point in the ICG the dZ/dt-min peak can be exploited. The dZ/dt-min peak (also called C-point or Z-point) can be detected with much more fidelity than the B-point. Changes in cardiac contractility, the main concept that PEP aims to asses, are reflected in the time it takes the left ventricle to build up sufficient force to open the aortic valve (B-point) but also in the time it takes to reach peak ventricular ejection (dZ/dt-min). Previously, two groups have reported that the time interval between the R-wave peak and the dZ/dt-min, called the Initial Systolic Time Interval (ISTI), is a significant predictor of both the R-wave peak to B-point (RB) interval as well as the actual PEP [3, 7]. This suggests that an adequate estimation of the PEP could be achieved by the detection of the two most prominent features in the ECG or ICG, the R-wave peak and the dZ/dt-min peak respectively. An estimated PEP can be defined as the sum of a fixed QR interval and an estimated RB interval derived from the ISTI by a regression relationship. As this bypasses the detection of the Q-wave onset and the B-point, PEP estimation could be achieved in automated way.

Journal of Physics: Conference Series **434** (2013) 012046 doi:10.1088/1742-6596/434/1/012046

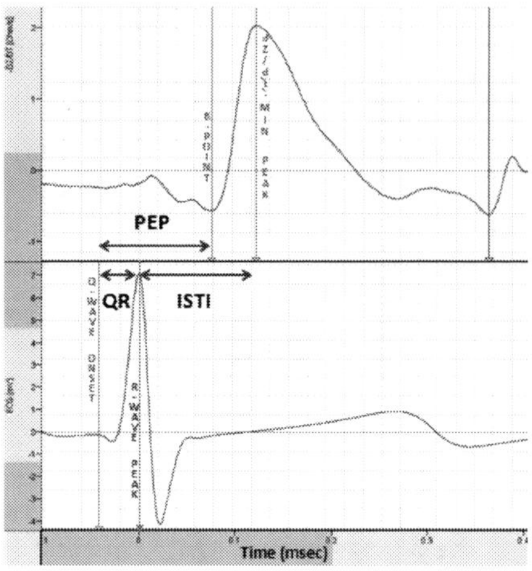

Figure 1. The impedance cardiogram (ICG, top) and the electrocardiogram (ECG, bottom) with the four landmarks defining the PEP (Q-wave onset to B-point) and the ISTI (R-peak to dZ/dt-min peak). The figure shows the R-wave peak locked ensemble averaged ICG and ECG, which is used for analysis rather than single beats, in order to reduce noise.

The current paper reports on two studies that evaluated the estimated and actual PEP in subjects undergoing a wide variety of stress manipulations in a laboratory setting, and in subjects who underwent a protocol resembling daily activities with a large variance in posture and physical activity. Firstly, it was explored whether the QR interval can be approximated by a fixed interval even in very diverse laboratory and ambulatory conditions. Secondly, multilevel regression analysis was used to derive an equation to estimate the RB interval from the ISTI in both laboratory and ambulatory recordings. It was tested whether this analysis resulted in a set of fixed regression coefficients that can be applied to all subjects as was suggested by Lozano et al. [3]. Finally, Bland and Altman plots were used to investigate whether the estimated PEP from the R-wave and dZ/dt-min peaks and fixed QR interval adequately predicts the actual PEP across this set of subjects and a wide range of conditions.

2. Methods

2.1. Subjects
In the laboratory study 91 young, healthy volunteers (20 male, 71 female) with a mean age of 21.7 years (SD = 3.2) and a mean Body Mass Index (BMI) of 22.2 (SD = 2.9) participated. The participants in the ambulatory study were 31 young, healthy volunteers (11 male, 20 female) having a mean age of 22.0 (SD = 1.9) and a mean BMI of 23.4 (SD = 4.3). Participants in both studies did not report any psychiatric diseases or cardiovascular problems and none were using cardio-active medication (e.g. antihypertensives). All gave their written, informed consent prior to participation.

2.2. General Procedures
Laboratory study: The subjects were seated in front of a 19" monitor in a dimly lighted, electrically-shielded, sound-attenuated cabin. The experimental session commenced with general instructions and a brief period of rest. To induce variation in SNS activity, various experimental physical and mental stressors were presented in a fixed order. Ambulatory study: A sequence of protocolized, supervised ambulatory activities was performed in the lab, outdoors, and in a sports centre to create variations in posture and intensity of physical and mental activity in close resembling normal daily activities.

2.3. Physiological recordings
ECG and ICG leads were attached to the participants using pregelled Ag/AgCl spot electrodes. The electrodes were attached as described by Goedhart et al. [8]. The ECG and ICG signals in the laboratory study were continuously recorded at a sample rate of 1000HZ with use of the ECG100C and NICO1000C modules of the BioPac data-acquisition system (BioPac systems INC, Santa Barbara, CA). The electrocardiogram (ECG) and the impedance cardiogram (ICG) of the ambulatory study were recorded continuously with the VU-AMS5fs device (VU University, Amsterdam).

2.4. *Signal Analyses and Data Reduction*

The mean QR interval was computed per condition across all valid beats in the laboratory data. For detection of the Q-wave onset in the ambulatory data, the ECG was imported into the VU-AMS5fs software. For each experimental condition the interbeat interval (IBI, ms) was scored from the R-wave peaks. Poor ECG signal fragments were removed. The mean Q-wave onset for each ambulatory condition was visually scored. The procedures to score the ISTI were identical for both studies: After obtaining the IBI time series, visual scoring from all valid beats was used to mark the B-point and the dZ/dt-min peak (see figure 1 for an example). The actual PEP was computed as the interval from the Q-wave onset in the ECG to the B-point in the ICG signal. The ISTI was computed as the time interval between the R-wave peak and the dZ/dt-min peak [7].

2.5. *Statistical Analyses*

The grand averaged QR interval across all subjects and all laboratory and ambulatory conditions was used to compute a weighted average QR across both studies to be used as a fixed QR interval. To test whether the QR interval can be approximated by a fixed interval the actual QR values were compared to the grand weighted average to determine the absolute agreement and its error values across both studies. Multilevel regression was applied to establish the optimal regression equation to predict the RB interval from the ISTI for both studies separately. Previous research suggested that a fixed single equation can be used for all subjects [3], which would mean that models with a random slope and intercept should not provide a significant better fit than a model with a fixed intercept and slope. Finally, the estimated PEP was computed for each individual in every condition in both studies by summing the weighted grand averaged QR interval across both studies to the RB interval estimated from the ISTI using the slope and intercept parameters from the best fitting model of the laboratory study, with the ambulatory data acting as the confirmatory set. A Bland-Altman analysis was used to test the absolute agreement between the estimated PEP and the actual PEP for both studies.

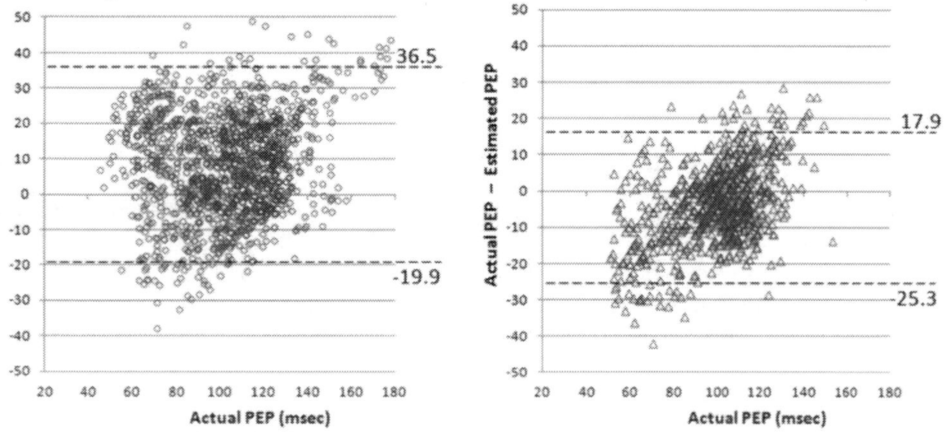

Figure 2. Bland-Altman plots of the difference between the actual PEP based on the ECG Q-wave onset and the ICG B-point and the estimated PEP based on the ECG R-wave and ICG dZ/dt-min peaks. The difference is plotted as a function of the absolute value of the actual PEP for the laboratory study (left) and the ambulatory study (right). Dotted lines represent the 95% confidence intervals around the mean difference.

3. Results

3.1. *QR interval*

The mean QR interval across all conditions was 39 ms (sd = 8 ms) in the laboratory study and 42 ms (sd = 6 ms) in the ambulatory study. To obtain the best possible value for a fixed QR interval the weighted grand average of both studies of ((39 * 91) + (42 * 31))/122 = 40 ms was computed. The 95% confidence interval (-14.5 ms – +14.5 ms) was found to be substantial.

3.2. Estimation of the R-wave peak to B-point interval using the ISTI

The multilevel analysis showed that the best model to describe the relationship between the RB interval and the ISTI in the laboratory study was the extended linear model with a random intercept and a random slope, RB = -46 + (0.9 * ISTI) ms. The model explained 79 % of the total variance in the RB interval. The same was true for the ambulatory study where the linear model with a random intercept and a random slope explained 81 % of the variance in RB. However, the slope and intercept were significantly different from the laboratory study: RB = -15 + (0.7*ISTI) ms.

3.3. Absolute agreement between actual PEP and estimated PEP

Figure 2 depicts the crucial test whether the estimated PEP adequately reflects the actual PEP across a wide range of laboratory and ambulatory conditions. PEP was estimated as: 40 + (-46 + (0.9 * ISTI)). The mean difference between the actual PEP and the estimated PEP in the laboratory study was +8 ms, and in the ambulatory study -4 ms The 95% confidence intervals were very large, ranging from -20 to +36 ms for the laboratory study and from -25 to +18 ms in the ambulatory study.

4. Discussion and conclusions

Two studies in different samples and settings found substantial discrepancies between the estimated PEP and the actual PEP based on the Q-wave onset and B-point. In the laboratory study, at least 84% of the differences between the estimated PEP and the actual PEP exceeded 3.5 ms, which is the mean reactivity found to two often used tasks in laboratory stress studies. About half of the error is due to the use of a fixed QR interval. The mean value of QR interval across both settings was 40 ms which is in agreement with previous studies. For instance, Goldberger and Bhargava [9] reported a QR interval of around 37 ms at rest which decreased by a few ms during exercise. Apart from the QR interval, the estimation of the RB interval from the ISTI also strongly contributed to the difference between estimated and actual PEP. These findings are in contrast to previous reports showing the ISTI to be a significant predictor of both the R-wave peak to B-point (RB) interval as well as the actual PEP [3]. Specially, Lozano et. al [3] found that the equation $RB = -31.59 + (1.233 * ISTI) + (0.0032 * ISTI^2)$ accounted for 95% of the variance in the actual RB interval during rest, a mental arithmetic task and a speech preparation task. The present study, with a wider range of conditions, shows that a single regression equation relating ISTI to RB is not adequate.

It is concluded that for valid PEP scoring the detection of the Q-wave onset and B-point remains mandatory. PEP estimated from the R-wave and dZ/dt-min peaks should not be used to replace the actual PEP, but could be a useful in helping to locate the Q-wave and B-points. Focusing the detection algorithms in a window around these expected locations can assist automated detection and reduce the effort of visual inspection.

References

[1] Newlin D B and Levenson R W 1979 *Psychophysiology* **16** 546-53
[2] Willemsen G H, de Geus E J, Klaver C H, van Doornen L J and Carroll D 1996 *Psychophysiology* **33** 184-93
[3] Lozano D L, Norman G, Knox D, Wood B L, Miller B D, Emery C F and Berntson G G 2007 *Psychophysiology* **44** 113-9
[4] Sherwood A, Allen M T, Fahrenberg J, Kelsey R M, Lovallo W R and van Doornen L J 1990 *Psychophysiology* **27** 1-23
[5] Berntson G G, Lozano D L, Chen Y J and Cacioppo J T 2004 *Psychophysiology* **41** 333-7
[6] Brydon L, O'Donnell K, Wright C E, Wawrzyniak A J, Wardle J and Steptoe A 2008 *Obesity* **16** 2642-7
[7] Meijer J H, Boesveldt S, Elbertse E and Berendse H W 2008 *Physiol. Meas.* **29** S383-91
[8] Goedhart A D, Kupper N, Willemsen G, Boomsma D I and de Geus E J 2006 *Biol. Psychol.* **72** 110-7
[9] Goldberger A L and Bhargava V 1983 *J. App. Physiol.* **54** 1083-8

Ventilation and Heart Rate Monitoring in Drivers using a Contactless Electrical Bioimpedance System

R Macías[1], M A García[1], J Ramos[1], R Bragós[1] and M Fernández[1]

[1]Department of Electronic Engineering, Universitat Politécnica de Catalunya, Campus Nord, Edifici C-4, 08034, Barcelona, Spain

E-mail: raul.macias@upc.edu

Abstract. Nowadays, the road safety is one of the most important priorities in the automotive industry. Many times, this safety is jeopardized because of driving under inappropriate states, e.g. drowsiness, drugs and/or alcohol. Therefore several systems for monitoring the behavior of subjects during driving are researched. In this paper, a device based on a contactless electrical bioimpedance system is shown. Using the four-wire technique, this system is capable of obtaining the heart rate and the ventilation of the driver through multiple textile electrodes. These textile electrodes are placed on the car seat and the steering wheel. Moreover, it is also reported several measurements done in a controlled environment, i.e. a test room where there are no artifacts due to the car vibrations or the road state. In the mentioned measurements, the system response can be observed depending on several parameters such as the placement of the electrodes or the number of clothing layers worn by the driver.

1. Introduction

According to [1], crash statistics suggest that about 10-20% of all crashes might be sleep or fatigue related. For that reason, detecting drowsiness or fatigue states has received increasing attention recently. Research efforts on that field can be classified into three main branches. First one is based on driving performances i.e. unintended lane departures, steerings and brakes. The second one is related to camera systems that allow detect the percentage of eye closure (PERCLOS), head movements and blinks [2]. Finally, the last one is based on monitoring biomedical signals, e.g. EEG, ECG or heart rate variability (HRV) [3].

Therefore, in this paper a set of measurements realized using a system based on the last group is shown. This system, through measuring the bioimpedance using the 4-wire technique, is capable of monitoring the ventilation, and also heart rate, of a driver. It is worth mentioning that the measurements were done in a laboratory environment where there are no interferences caused by the state of the road or the car engine vibrations.

2. System

The bioimpedance system designed for monitoring the ventilation consists of several textile electrodes and two main blocks: an analog front-end (AFE) and a generator-demodulator block, see figure 1. It is worth mentioning that all the system is controlled by LabView and, the generator-demodulator block consists of the PXI System by National Instruments. Briefly, a sinusoidal signal with a frequency of 62.5 kHz is generated using one of the PXI modules, NI

PXI-5422. Later, this generated signal is injected to the AFE block. In the injection stage of the AFE, the voltage signal coming from the PXI is converted to a current signal by a second generation current conveyor (CCII) [4]. By this way, the injected signal to the driver by the textile electrodes connected to HC and LC in figure 1 is a current driving instead of a voltage driving. In the voltage sensing stage, a differential to single ended voltage conversion is done by a wideband differential amplifier. However, before this conversion, the voltage difference between the pair of sensing electrodes, HP and LP in figure 1 is measured by a pair of high-impedance buffers. Finally, the output signal of the AFE is acquired using other PXI module, NI PXI-5122, and an In Phase-Quadrature (IQ) demodulation is done by LabView. It is worth mentioning that there is also a Common-Mode Feed-Back (CMFB) stage in the AFE which is useful to reduce the effect of high electrode mismatch.

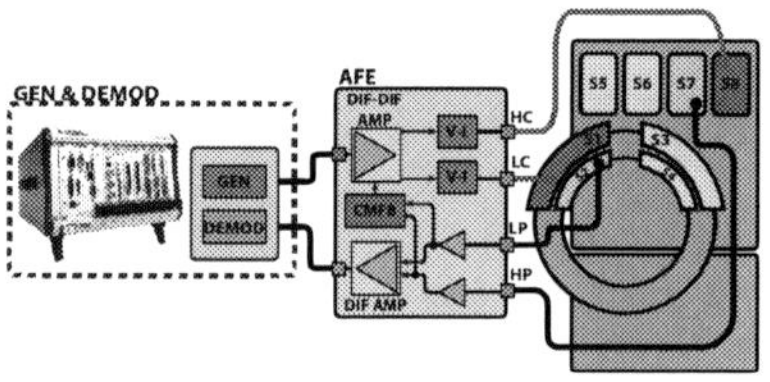

Figure 1. A block diagram of the system. It consists of a PXI system, an AFE and several textile electrodes attached to the steering wheel and the car seat.

3. Measurements

3.1. Comparison to a Reference Signal

To check the proper work of the system, several volunteers are monitored not only by the designed system but also by a commercial one made by BIOPAC Systems. This commercial device acquires the ventilation signal using a piezoresistive thoracic band. Therefore, the designed system works properly if the measured signal fits to the signal related to the thoracic band, i.e. for the same period of time, the exhalation-inhalation ratio is the same in both signals.

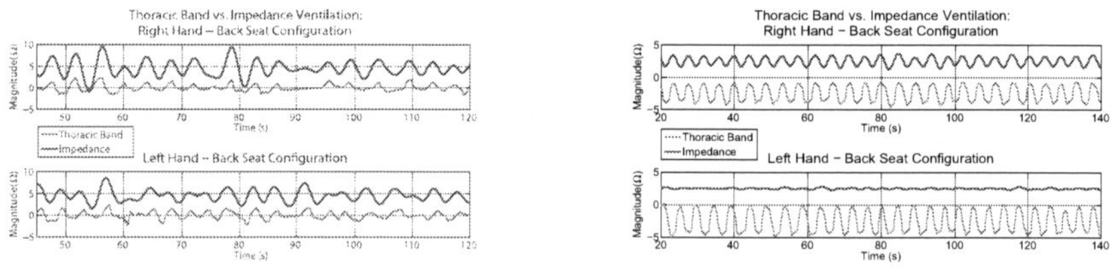

(a) Volunteer 1 (b) Volunteer 2

Figure 2. Comparison between the Thoracic Band and the Bioimpedance Device for two volunteers. (Top) Configuration where both pairs of driving-sensing electrodes are in the right side of the body. (Bottom) Configuration where a pair of electrodes is in the left hand and the other pair is in the right side of the back. In both plots, the upper line is related to the bioimpedance device and the bottom one comes from the thoracic band.

As observed in figure 2, for each volunteer two different configurations are tested. On one hand, in the upper graphs, all four electrodes are placed at the right side, i.e. whereas S1 and S5 in figure 1 are the driving electrodes, S2 and S6 are the sensing electrodes. On the other hand, in the bottom graphs whereas the electrodes on the back seat remain at the same point (S5 and S6), both electrodes on the steering wheel are moved to the left side (S3 and S4 in figure 1).

Note that for all cases except for the bottom graph of the second volunteer, both signals, from the bioimpedance device and from the thoracic band, match up and show a normal respiration rate, i.e. between 11 and 20 breaths per minute in adults and in normal conditions. The special case can be due to the lack of contact between any textrode and the volunteer. Due to this fact, it is important to check the system according others parameters such as the placement of electrodes or the dependencies on clothing.

3.2. Configuration of Electrodes

In this second group of measurements, to check the system whereas the placement of the electrodes is changed, the same test is done. That test consists of a two-minute monitoring and, around the last 30 seconds, five deep breathing are taken. Therefore changing the place of driving and sensing electrodes, three configurations are tested: Steering Wheel-Steering Wheel Configuration (SW-SW), Steering Wheel-Back Seat Configuration (SW-BS) and Back Seat-Back Seat Configuration (BS-BS).

Briefly, in SW-SW configuration, whereas in figure 1 S1 and S3 are the driving electrodes, S2 and S4 are the sensing ones. In SW-BS configuration, S3 and S5 are the driving electrodes and S4 and S6 are the sensing ones. Finally, in BS-BS configuration, whereas S5 and S8 are the pair of driving electrodes, S6 and S7 are the pair of sensing.

In figures 3(a), 3(b) and 4(a), three signals are plotted. The middle one is the raw signal measured by the bioimpedance device and without processing. This signal is based on two components: a low-frequency component, between 0.1 Hz and 0.3 Hz, and a high-frequency component, over 1 Hz, that are related to the ventilation and the cardiac rhythm, respectively. To show better the component related to the heartbeat, figure 4(b) corresponds to the zoomed-in signal of the figure 4(a) in a one-minute period. Note that whereas in figures 3(a) and 4(b) the heartbeat component can be observed, in figure 3(b) that is not possible. Therefore, to measure the signal related to heartbeat, to be in contact directly to a hand seems required.

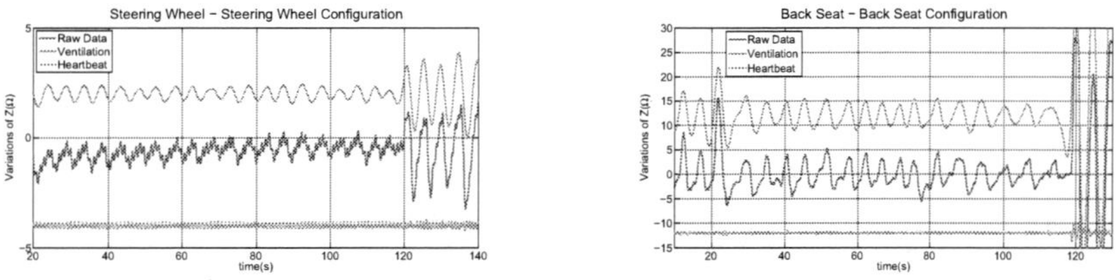

(a) SW - SW Configuration (b) BS - BS Configuration

Figure 3. Two configurations of electrodes: (a) Steering Wheel - Steering Wheel Configuration. (b) Back Seat - Back Seat Configuration. In both cases, (Middle) The raw data. (Top) The ventilation signal. (Bottom) The signal related to the cardiac rhythm.

3.3. Influence of the Thickness of Clothing

In the last group of measurements, dependencies on clothing are checked. Applying the same test explained above and the SW-BS configuration, a subject is monitored wearing several clothing. So, whereas in figure 4(a) the volunteer wears a thin 100% cotton T-shirt, in figures 5(a) and 5(b), the volunteer wears, over the same T-shirt as before, a thin 100% polyester jacket and a thick sweater, respectively.

Note that the respiration rate is different in the three figures. Meanwhile in figures 5(a) and 5(b), the respiration rate is around 8 breaths per minute, in figure 4(a) a normal rate of 11 breaths per minute can be observed. Therefore, there seems to be a correlation between clothing and the measured signal. Depending on the clothing, the system could not work properly because of some inhalations or exhalations cannot be monitored.

Journal of Physics: Conference Series **434** (2013) 012047 doi:10.1088/1742-6596/434/1/012047

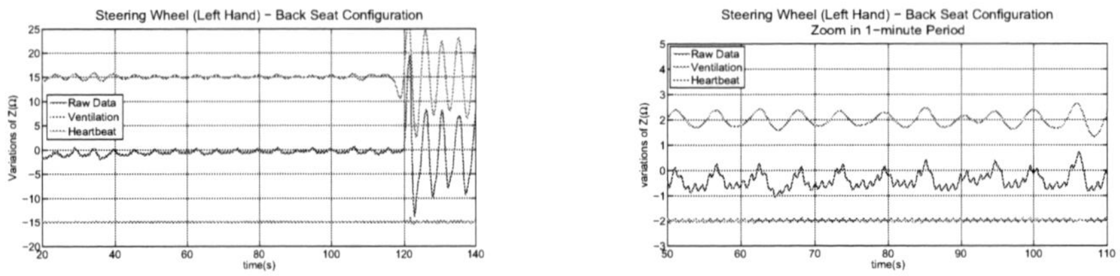

(a) SW - BS Configuration (b) Zoom in a one-minute period

Figure 4. Steering Wheel - Back Seat configuration of electrodes. (b) is a zoom in a one-minute period to the signal shown in (a). In both cases, (Middle) The raw data. (Top) The ventilation signal. (Bottom) The signal related to the cardiac rhythm.

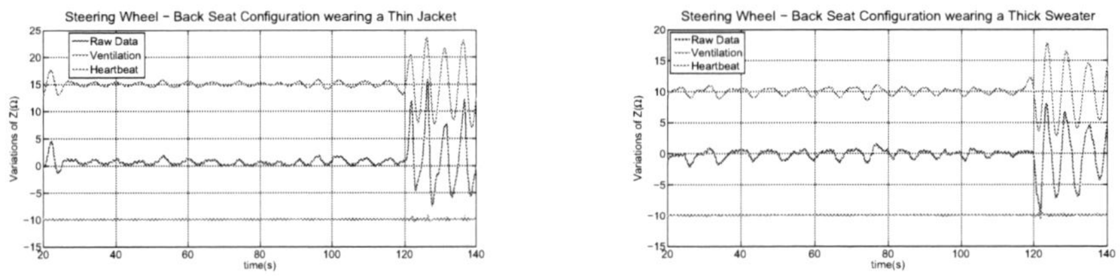

(a) Wearing a thin Jacket (b) Wearing a thick sweater

Figure 5. Measurements wearing several clothes in a SW-BS configuration. In (a), the subject is wearing a thin jacket made of polyester (100%) over a thin T-shirt made of cotton (100%). In (b), the subject is wearing a thick sweater made of woolen (33%), polyester (27%), acrylic (27%) and polyurethane (13%) over a thin T-shirt made of cotton (100%).

4. Conclusions

Basically, in this paper is shown that monitoring the ventilation, and heart rate, using a bioimpedance system with textile electrodes placed over the clothes is possible. To check this, the designed system and a commercial device are compared and, as shown in this paper, both signals match up. However, it is worth mentioning that the system has been only tested in a simulation environment. It is to be expected that in a real environment, the lack of contact with the textile electrodes will appear because of the vibrations and the state of the road. Furthermore, the system could not work properly at an injected frequency of 62.5 kHz due to the high capacitive behaviour of the interface between clothing and textile electrodes. Thus, although testing the system in a real environment or with a higher injected frequency is required, the results obtained in a controlled environment are encouraging.

References

[1] Anund A, Kecklund G, Peters B and Åkerstedt T 2008 Driver sleepiness and individual differences in preferences for countermeasures *J. Sleep. Res.* **17** 16–22
[2] Sommer D and Golz M 2010 Evaluation of PERCLOS based current fatigue monitoring technologies *Proc. of the 32th Annual Int. Conf. of the IEEE EMBS* (Buenos Aires) 4456–9
[3] Michail E, Kokonozi A, Chouvarda I and Maglaveras N 2008 EEG and HRV markers of sleepiness and loss of control during car driving *Proc. of the 30th Annual Int. Conf. of the IEEE EMBS* (Vancouver) 2566–9
[4] Bragós R, Rosell J and Riu P 1994 A wide-band AC-coupled current source for electrical impedance tomography *Physiol. Meas.* **15** A91–9

Non-invasive method for the aortic blood pressure waveform estimation using the measured radial EBI

Andrei Krivoshei[1,2], Jürgen Lamp[2,3], Mart Min[1], Tiina Uuetoa[2,4], Hasso Uuetoa[1,5] and Paul Annus[1,2]

[1] Tallinn University of Technology, Estonia
[2] ELIKO Competence Center, Estonia
[3] JR Medical OY, Estonia
[4] East-Tallinn Central Hospital, Estonia
[5] Sahlgrenska University Hospital, Sweden

E-mail: andrei.krivoshei@gmail.com

Abstract. The paper presents a method for the Central Aortic Pressure (CAP) waveform estimation from the measured radial Electrical Bio-Impedance (EBI). The method proposed here is a non-invasive and health-safe approach to estimate the cardiovascular system parameters, such as the Augmentation Index (AI). Reconstruction of the CAP curve from the EBI data is provided by spectral domain transfer functions (TF), found on the bases of data analysis. Clinical experiments were carried out on 30 patients in the Center of Cardiology of East-Tallinn Central Hospital during coronary angiography on patients in age of 43 to 80 years. The quality and reliability of the method was tested by comparing the evaluated augmentation indices obtained from the invasively measured CAP data and from the reconstructed curve. The correlation coefficient $r = 0.89$ was calculated in the range of AI_{CAP} values from 5 to 28. Comparing to the traditional tonometry based method, the developed one is more convenient to use and it allows long-term monitoring of the AI, what is not possible with tonometry probes.

1. Introduction

Increasing mortality in developed countries, caused by cardiovascular disease, and mainly by atherosclerosis and/or hypertension, is a major concern in modern medicine nowadays. Thereupon, the situation requires developing of novel, non-invasive and reliable monitoring and diagnosing tools capable to recognize a heart disease on its early progress stages.

Along with other parameters, the Central Aortic Pressure (CAP) and the Augmentation Index (AI) have been indicated in several studies as important cardiovascular risk markers [1, 2]. It is well known, that the CAP curve is formed from the sum of the direct aortic pressure wave, just pumped out from the heart, and from the traveling wave reflected from peripheral arteries. A quickly arrived reflected wave will produce an increase in the CAP, and significant rise of which can be recognized as a risk marker. Thus estimating the contribution of a reflected wave to the CAP gives important information about condition of arteries and patient health risk.

Non-invasive estimation of the AI and the CAP is usually done by a well-known applanation tonometry, when a peripheral blood pressure is measured on a radial artery and then calibrated by the pressure values obtained using the brachial cuff. After that both, the CAP and the AI values, are estimated from the measured radial pressure and generalized transfer function [3 - 6].

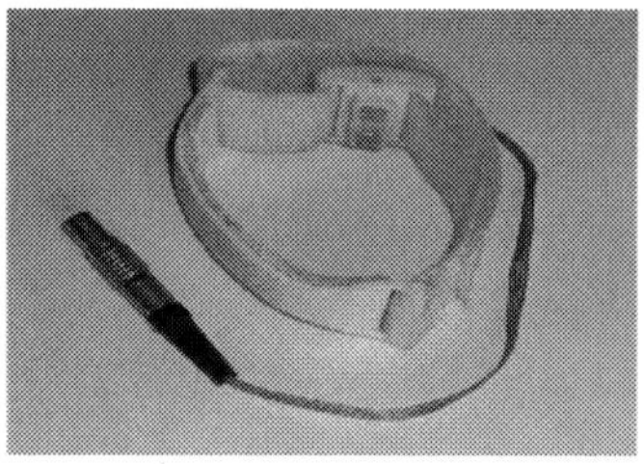

Figure 1 Experimental EBI sensor.

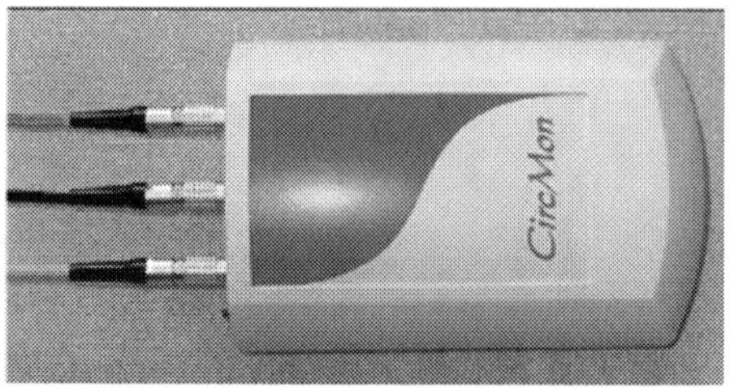

Figure 2 View of the impedance cardiograph – circulation monitor "Circmon BT101".

In our research we show that AI can be estimated from a radial electrical bio-impedance (EBI) as well as from radial blood pressure curve. Comparing to the traditional tonometry based method, presented one is more convenient to use and it allows long-term monitoring of the AI, what is not possible with tonometry probes.

2. Method

Clinical experiments were carried out on 30 patients during coronary angiography in the Center of Cardiology of East-Tallinn Central Hospital. After that, the found out spectral domain transfer function TF was used to estimate the CAP waveform from the noninvasively measured radial EBI. All patients in age of 43 to 80 years were informed about the purpose and threats of an investigation and they all have signed consents. The investigation was carried in according to the decision of the Ethics Committee of the National Institute for Health Development.

The EBI was measured using the bracelet-type holder and specially designed tetra-polar sensor (figure 1), placed on a wrist directly onto a radial artery in close region to a scaphoid bone. The electrodes were made of non-toxic and non-irritable silver compound. The distance between measuring electrodes of 5 mm was used to limit the volume of the tissue under the measurement and thus to avoid any possible influence from surrounding veins and arteries. The distance of 14 mm was used between the current injecting electrodes, which were placed on the both sides from the voltage measuring electrodes. The frequency of the injecting current was 125 kHz and the amplitude 0.2 mA.

The measuring procedure, on the whole, was carried out using the JR Medical's CE-certified wireless multichannel impedance cardiograph – circulation monitor "Circmon BT101" (figure 2), which allows simultaneous measurement of 6 impedance channels and one ECG channel in its default configuration. For the proposed investigation, two channels of the cardiograph were used for the EBI and for the ECG signal measurement, and another one was adapted to simultaneous acquiring of invasive CAP data using the PVB's XTRANS sensor.

The block diagram of the BT101 is shown in the figure 3 and presented in more details in [7]. The major advantage of the BT101 is the significant increase in the precision of the EBI measurements, which has been achieved by using the carrier compensation technique [7] provided by the DDS2, the compensation signal which is subtracted from the modulated carrier in the summing element in figure 3. Both, the amplitude and phase of the compensating signal are estimated before the main EBI

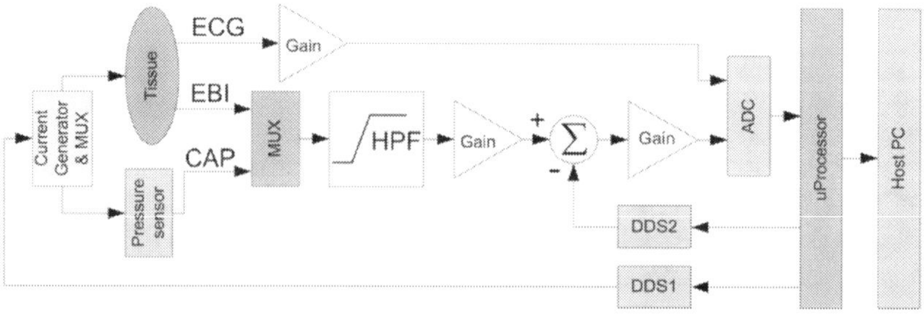

Figure 3 A simplified block diagram of the impedance measurement device BT101.

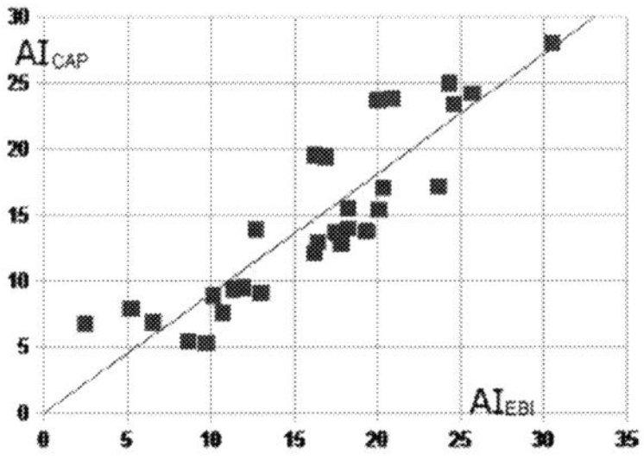

Figure 4 Correlation between of AI_{CAP} and AI_{EBI}. $r = 0{,}89$.

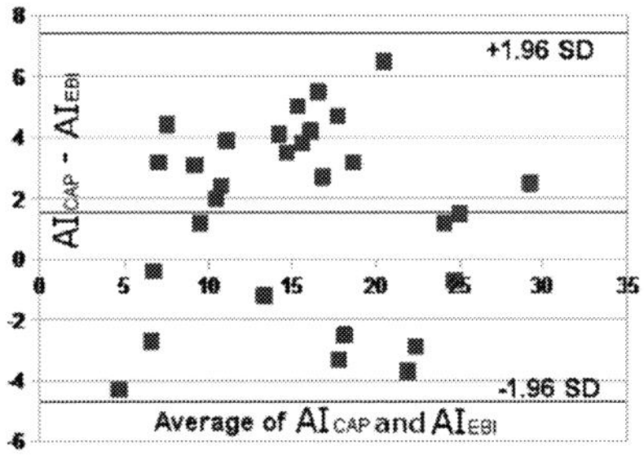

Figure 5 Bland-Altman plot of experimental data. All the data remain in +/-1,96 STD area.

detection procedure. As a result, the large basal impedance signal (carrier) is compensated and the base band EBI signal begins from the DC component without any distortions usually caused by the low pass filtering in traditional impedance cardiographs.

The data with 5 ms sampling intervals were transmitted via Bluetooth™ link to a personal computer in which the CircMon™ program runs for performing the initial data processing.

3. Transfer function (TF) estimation technique

To estimate the wanted transfer function TF, the initial cardiac periods of EBI and CAP were taken between the beginnings of two systolic intervals. The periods have zero mean value and are normalized by theirs standard deviations. To normalize the relative time positions, the periods are aligned in time domain by shifting the fronts of EBI and CAP waveforms of the systolic interval to the middle position of signal periods. The circular shifting of signal arrays was used for this purpose.

The transfer functions TF between the EBI and CAP spectra (1) are estimated by dividing (2) the complex spectra of CAP by the EBI spectra for each patient's data individually.

$$\begin{aligned} S_{EBI}(j\omega) &= \mathrm{DFT}(\mathrm{EBI}(t)) \\ S_{CAP}(j\omega) &= \mathrm{DFT}(\mathrm{CAP}(t)) \end{aligned} \tag{1}$$

$$\mathrm{TF}(j\omega) = S_{CAP}(j\omega)/S_{EBI}(j\omega) \tag{2}$$

Further, the estimated transfer functions (TF) are scaled (stretched or squeezed) in spectral domain by corresponding heart rate (HR) values. And, after ensemble averaging of all the found individual transfer functions the generic TF was found.

4. Results

The on-line operating algorithm for the CAP waveform reconstruction and AI value estimation from the measured radial EBI waveform is implemented as a DLL function, programmed in the NI LabVIEW environment.

During experimentation, the EBI signal periods are sent individually into the DLL, where the generic TF, found on the analysis stage, is scaled back (squeezed or stretched) by corresponding HR value and multiplied by the harmonic spectrum of the received EBI signal period. The reconstructed CAP signal spectrum is transformed back into time domain and returned from the DLL into the main CircMon™ program together with the estimated AI value (3) for displaying and further diagnosing (see figure 6).

The AI values for original CAP (AI_{CAP}) and for reconstructed one (AI_{EBI}) are calculated as

$$AI_{\{CAP|EBI\}} = (\mathrm{CAP}_B(t) - \mathrm{CAP}_C(t))/(\mathrm{CAP}_B(t) - \mathrm{CAP}_A(t)) \cdot 100[\%], \tag{3}$$

where $\mathrm{CAP}_n(t)$, $n = \{A, B, C\}$ are pressure values of the CAP curves at points A, B, C in figure 6.

The quality and reliability of the method were tested by comparing two AI values, AI_{CAP} and AI_{EBI}

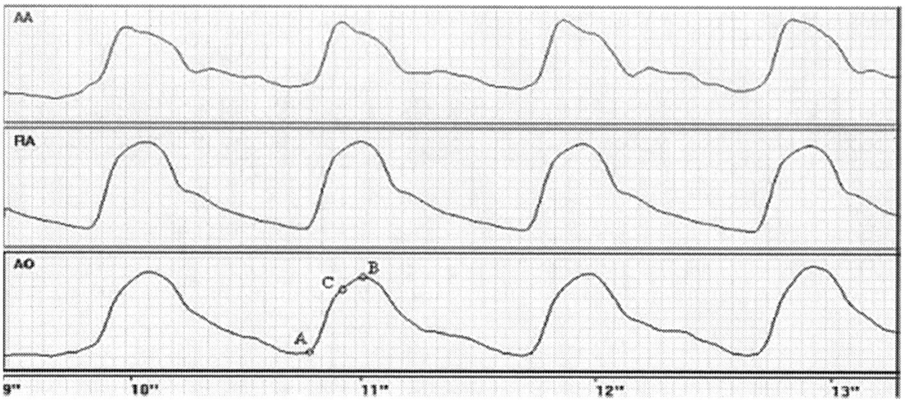

Figure 6 The measured and reconstructed waveforms as the outputs of the CircMon^{TM} program: AA – radial artery bio-impedance signal, RA – invasively measured CAP waveform, AO – reconstructed CAP waveform.

– those based on the invasively measured CAP data (channel RA in figure 6) and those obtained from the reconstructed CAP curve (channel AO in figure 6). A correlation coefficient for all 30 patients is $r = 0.89$ in the range of 5 to 28 of the AI_{CAP} values (see figure 4 and figure 5).

5. Discussions and conclusions

As shown above, the EBI measurement at radial artery can give a new non-invasive and easy-to-use reliable and operator independent method for estimating the beat-by-beat CAP curve continuously. This is helpful for cardiovascular disease risk assessment in clinical practice. Moreover, further development of the method can open new frontiers for dynamic CAP analysis, e.g., during exercising or mental stress testing of patients.

Acknowledgment

The research was supported by the European Union through the European Regional Development Fund in frames of the research center CEBE and the competence center ELIKO, partly also by the Estonian target financed project SF0140061s12.

References

[1] Avolio A P, Butlin M and Walsh A 2010 Arterial blood pressure measurement and pulse wave analysis – their role in enhancing cardiovascular assessment *Physiol. Meas.* **31** R1–R47
[2] Nürnberger J, Keflioglu-Scheiber A et al 2002 Augmentation index is associated with cardiovascular risk *J. Hypertens.* **20** 2407-14
[3] Karamanoglu M,O'Rourke M F, Avolio AP, Kelly RP 1993 An analysis of the relationship between central aortic and perópheral upper limb pressure waves in man *European Heart J.* 1993 **14** 160-7
[4] O'Rourke M F 1993 Method for ascertaining the pressure pulse and related parameters in the ascending aorta from the contour of the pressure pulse in the peripheral arteries. *US Patent 5,265,011*
[5] O'Rourke M F 2000 Non-invasive determination of aortic flow velocity waveforms *US Patent 6,010,457*
[6] Hahn J-O, Reinsner A T and Asada H H 2010 Estimating aortic blood pressure from non-invasive extrimity blood pressure. *US Patent Application 2010/0016736 A1*
[7] Annus P, Lamp J, Min M and Paavle T 2005 Design of a Bioimpedance Measurement System Using Direct Carrier Compensation *Proc. of the European Conference on Circuit Theory and Design Aug.29-Sept.02 2005 Cork Ireland* vol III pp 23 – 6

Determination of time delay between ventricles contraction using impedance measurements

M Lewandowska, A Poliński and J Wtorek
Biomedical Engineering Department, Gdansk University of Technology, Narutowicza 11/12, 80-233 Gdansk-Wrzeszcz, Poland

E-mail: apoli@biomed.eti.pg.gda.pl

Abstract. The paper presents a novel approach to assessment of ventricular dyssynchrony basing on multichannel electrical impedance measurements. Using a proper placement of electrodes, the sensitivity approach allows estimating time difference between chambers contraction from over determined nonlinear system of equations. The theoretical considerations which include Finite Element Method simulations were verified using measurements on healthy 28 year's old woman. The nonlinear least squares method was applied to obtain a time difference between heart chambers contraction. The obtained value was in a good agreement with theoretical values found in literature.

1. Introduction

Congestive heart failure (CHF) is nowadays one of the most frequent heart diseases in developed countries. Very promising treatment for patients with CHF is cardiac resynchronization therapy (CRT). Current criteria for CRT candidacy, as well as evaluation of the therapy effectiveness are not sufficient, as still 20-30% patients take no benefit from CRT [1]. The most popular marker of dyssynchrony, is based on electrocardiographic examination [2]-[3]. However, electrical dyssynchrony is not always linked to mechanical dyssynchrony and vice-versa [4], therefore examinations that can evaluate also mechanical activity of the heart are needed. The most popular supporting examination in CRT is echocardiography. Despite its many advantages correct diagnosis is strongly dependent on the chosen biomarker depicting pathology physiological substrate as well as physician interpretation of performed examination [5].

Our proposed method is a multichannel bioimpedance measurement. This technique has been introduced for the evaluation of cardiac function in the late 1930s [6] and is still used for stroke volume assessment [7]. Mechanical activity of the heart causes changes of internal organs volumes and placement, and changes of electrical properties of tissues (due to different blood contents). Thus, such measurements can be used to improve the evaluation of the mechanical work of heart. The measured impedance change, ΔZ, invoked by a conductivity change $\Delta\sigma$ can be evaluated using the relationship proposed by Geselowitz [8]

$$\Delta Z = -\int_V \Delta\sigma(x,y,z)\frac{\nabla\phi(\sigma,x,y,z)}{I_\phi}\cdot\frac{\nabla\psi(\sigma+\Delta\sigma,x,y,z)}{I_\psi}dv\,, \tag{1}$$

where ϕ represents the potential associated with current I_ϕ flowing between the "current" electrodes, ψ is the potential associated with hypothetical current I_ψ flowing between the "voltage" electrodes after the occurrence of the conductivity change $\Delta\sigma$; V is the volume of the region of changed conductivity, and a dot stands for the scalar product. The local conductivity change is thus weighted by the scalar product of two gradients of potential normalized by corresponding currents I_ϕ and I_ψ. The gradient has to be calculated twice, before and after the conductivity change. The integral is calculated only in the region where the conductivity change, $\Delta\sigma$, is not equal to zero.

Using Geselowitz relationship the impedance change can be view as volume (in which $\Delta\sigma \neq 0$) multiplied by some constant

$$\Delta Z = aV\ , \tag{2}$$

where

$$a = \frac{-\int_V \Delta\sigma(x,y,z)\frac{\nabla\phi(\sigma,x,y,z)}{I_\phi}\cdot\frac{\nabla\psi(\sigma+\Delta\sigma,x,y,z)}{I_\psi}dv}{V}\ . \tag{3}$$

Assume that a body is divided into regions of different conductivity change, and conductivity change inside each of such regions is constant. Assuming that the potential gradients are constant in regions (which is poor near the boundaries of regions) one can approximate the coefficients a_i i=1,…,P, where P is the number of different regions

$$a_i = \Delta\sigma_i(x,y,z)\frac{\nabla\phi_i(\sigma,x,y,z)}{I_\phi}\cdot\frac{\nabla\psi_i(\sigma+\Delta\sigma,x,y,z)}{I_\psi}\ . \tag{4}$$

In fact it is enough that integral of the scalar product of potential gradients do not depend on conductivity change inside each region (local variations in potential can cancel each other).

In the resynchronization the main interest is into mechanical work of the heart. The blood volume changes inside the chest and blood movement caused impedance changes measured by impedance technique. If we consider only the systole the impedance changes will be caused mainly by blood volume changes in heart chambers and partially by blood flowing in aorta and lungs. However there are only two sources of the blood volume changes: right and left heart chamber. Thus our approach assumes that measured impedance change during systole can be approximated by some (unknown) coefficients and blood volume change in right and left heart chamber. Under the assumption of (almost) constant value of coefficients a_L and a_R the following system of equations is obtained

$$\Delta Z(t_k) = a_L \Delta V_L(t_k) + a_R \Delta V_R(t_k) \tag{5}$$

where t_k is time, k=1,…,N, $\Delta Z(t_k)$ is measured impedance change at time t_k, $\Delta V_L(t_k)$ is left chamber volume change at time t_k, $\Delta V_R(t_k)$ is right chamber volume change at time t_k. The constant value of coefficients a_L and a_R can be achieved by proper electrode configuration (so potential gradient almost do not vary with conductivity change). Such configuration also ensures that the influence of aorta and lungs can be neglected. Small changes in the coefficients' values can be treated as noise and removed by our method.

2. Material and Method

Basing on the periodicity and shape of measured impedance change signal, the source signals (blood volume changes) were approximated by sinusoidal signals. Although such approach suggest single period approximation due to the variations in RR interval in ECG another problem appears: that time of systole and diastole are no necessary equal. Thus two cases were considered. In the first case only the systole time was considered basing on ECG and impedance cardiogram (ΔZ) analysis. The analysis begun from the R point in ECG signal and the end time of analysis was valve closing time detected from ΔZ signal. In the second case whole impedance cardiogram between two successive R points in ECG was considered. For estimation of two chambers, two measurement channels are required at least. Thus the system of $2N$ equations is obtained

$$\begin{cases} \Delta Z_1(t_k) = a_{L,1}\Delta V_L(t_k) + a_{R,1}\Delta V_R(t_k) \\ \Delta Z_2(t_k) = a_{L,2}\Delta V_L(t_k) + a_{R,2}\Delta V_R(t_k) \end{cases} \tag{6}$$

According to our assumptions the blood volume change can be approximated by

$$\Delta V_i(t) = A_i \sin(2\pi f_i t + \phi_i), \quad i = L, R. \tag{7}$$

Since it is impossible to distinguish between magnitude of blood volume change signal A and mixing coefficients a_i the measured signal was normalized, as so we assume that A=1. Hence the problem is reduced to finding four unknown coefficients $a_{L,1}$, $a_{L,2}$ $a_{R,1}$ $a_{R,2}$, two phases ϕ_L, ϕ_R and two frequencies of the signal f_L and f_R. The working frequency of each chamber should be the same or almost the same due to physiological requirements. This mean that the number of parameters reduces to five (f_L=f_R). However weighted sum of two sinusoidal signals produces sinusoidal signal which phase depend not only on the phases of the source signals but also on weights. Hence our problem has no unique solution. However, there are two sources with normalized amplitudes thus it can be assumed that

$$a_{L,i} + a_{R,i} = 1, \quad i = 1, 2. \tag{8}$$

This assumption is valid if the phases of two signals are equal. Thus we have $2N$+2 equations and only 5 unknowns. We use nonlinear least squares to solve this over-determined system of nonlinear equations. The advantage of least squares approach is that each equation has not to be satisfied accurately. Thus the strength of the last condition can be weighted

$$M(a_{L,i} + a_{R,i}) = M, \tag{9}$$

where M>0 was chosen on the base of N i.e. sampling frequency and length of time interval analysis. Our final system of equations consists of $2N$ equations from equation (6) with additional constrain (two equations – one from each measurement channel) generated by equation (9). Two values of M were considered during analysis, M=N/10 and M=N/2.

Data for analysis have been collected using a specially developed measurement system [9]. It consisted of a sinusoidal, 40 kHz, current source $I(\omega)$ and two identical measurement channels. The current was passing between electrodes I_+ and I_- while two pairs of electrodes $\{V_{1+},V_{1-}\}$ and $\{V_{2+},V_{2-}\}$ formed two measurement leads (figure 1). The sampling frequency was equal to 1 kHz. The measurements were done on healthy 28 years old woman.

3. Results

Example of the approximation results for M=N/2 for systole time is shown in figure 2. The obtained time difference between chamber contractions was equal to 35 ms while time delay between approximated signals was 21 ms. The approximations of measured signals were very good. RMS approximation error for the first measurement channel was equal to 0.0521 for systole period and 0.2428 for whole RR period analysis, while for the second measurement channel 0.0720 and 0.1689 respectively.

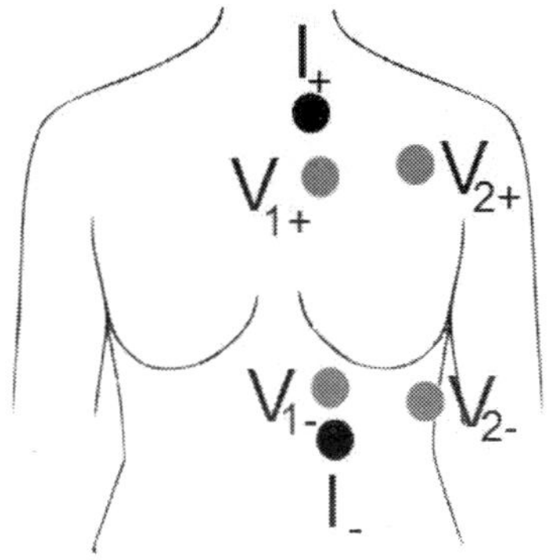

Figure 1. Measurement electrodes configuration.

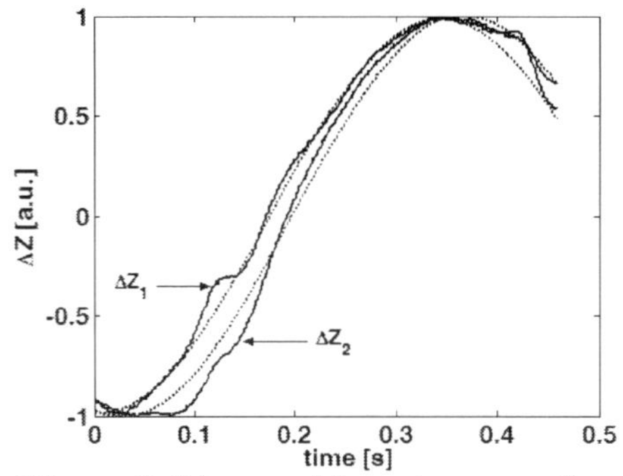

Figure 2. Measured impedance cardiograms (solid lines) and theirs approximations (dotted lines).

4. Discussion

The approximation of whole RR period does not give good results. The shape of this signal was too far from sinusoidal one, so for further analysis only systole interval was taken into account.

The obtained approximations of measured impedance cardiograms during systole time by sinusoidal signals were very good. The proposed algorithm does not allow indicating which signal comes from each chamber. This is due to symmetry of the problem (changing indexes L and R in equations produce the same approximation but different interpretation of sources). Also the indicated by our method domination chamber will depend not only on position of electrode but also on initial condition for least squares approximation (relation between a coefficients). This means that we can obtain time difference between chambers contraction without indicating the sequence. To do this anatomical placement of electrodes is required and support from FEM studies to estimate influence of each chamber, which allows setting proper initial conditions.

The weighting factor M should be chosen with care since even good approximation and small time difference for two chambers may be not correct. Its value should depend on difference of time contraction between chambers also, since the equation (9) is not exactly satisfied in general case. Better results were obtained for $M=N/2$ than for $M=N/10$.

The electrode should be placed according to the anatomy of the patient to ensure that chambers will dominate measured signals, and the placement should enable to dominate of each chamber for each measurement channel. However the last condition is not necessary to obtain proper results, but it can improve the quality of reconstruction of source signals.

Measurement system was verified by interchanging the measurement channels. Obtained results were the same. Approximated time difference (35 ms) was in good agreement with value given in literature for healthy people (about 20 ms). However, further studies include Doppler imaging as a reference method are under preparation.

5. Conclusions

The obtained results of time difference coincide with theoretical ones. Thus, the presented approach is a very promising in measuring mechanical dyssynchrony of heart chambers. The proposed method allows beat-to-beat evaluation of ventricles' synchrony. However, further studies including experimental one are needed.

References

[1] Boriani G, Diemberger I, Biffi M, Martignani C, Valzania C, Ziacchi M, Bertini M, Specchia S, Grigioni F and Rapezzi C 2006 *J. Interv. Card. Electrophysiol.* **17** 215-24
[2] Gras D, Leclercq C, Tang A S, Bucknall C, Luttikhuis H O and Kirstein-Pedersen A 2002 *Eur. J. Heart Fail.* **4(3)** 311-20
[3] Dickstein K, Vardas P E, Auricchio A, Daubert J C, Linde C, McMurray J, Ponikowski P, Priori S G, Sutton R and van Veldhuisen D J 2010 *Eur. J. Heart Fail.* **12(11)** 1143-53
[4] Hawkins N M, Petrie M C, MacDonald M R, Hogg K J and McMurray J J V 2006 *Eur. Heart J.* **27(11)** 1270-81
[5] Lafitte S, Reant P, Serri K and Roudaut R 2008 *Echocardiography* **25 (9)** 1040-6
[6] Yang F and Patterson R P 2008 *Ann. Biomed. Eng.* **36 (5)** 762-8
[7] Siebert J, Wtorek J, Rogowski J 1999 *Ann. N. Y. Acad. Sci.* **873** 182-90
[8] Geselowitz D B 1971 *IEEE Trans. Biomed. Eng.* **18** 38-41
[9] Lewandowska M, Wtorek J, Bujnowski A and Mierzejewski L 2010 *J. Phys.: Conf. Ser.*, **224** 012162

Journal of Physics: Conference Series **434** (2013) 012050 doi:10.1088/1742-6596/434/1/012050

Pulse Wave Velocity and Cardiac Output vs. Heart Rate in Patients with an Implanted Pacemaker Based on Electric Impedance Method Measurement

Ladislav Soukup[1], Vlastimil Vondra[2], Ivo Viščor[2], Pavel Jurák[2], Josef Halámek[2]

[1]International Clinical Research Center, Brno, Czech Republic
[2]Institute of Scientific Instruments AS CR, Brno, Czech Republic

E-mail: vond@isibrno.cz

Abstract
The methods and device for estimation of cardiac output and measurement of pulse wave velocity simultaneously is presented here. The beat-to-beat cardiac output as well as pulse wave velocity measurement is based on application of electrical impedance method on the thorax and calf. The results are demonstrated in a study of 24 subjects. The dependence of pulse wave velocity and cardiac output on heart rate during rest in patients with an implanted pacemaker was evaluated. The heart rate was changed by pacemaker programming while neither exercise nor drugs were applied. The most important result is that the pulse wave velocity, cardiac output and blood pressure do not depend significantly on heart rate, while the stroke volume is reciprocal proportionally to the heart rate.

1. Introduction

Pulse wave velocity (PWV) and cardiac output (CO) are essentially parameters for appraisal of the cardiovascular system. PWV is a marker of arterial stiffness and atherosclerosis. Structural changes are one of factors, by which arterial stiffness can be affected. They involve the composition of the arterial wall [1]. In several studies estimations of dependence of PWV on heart rate were proved to be different, e.g. [3], [4], [5]. Similarly, it is well known that the magnitude of the stroke volume (SV) is, among other things, dependent on heart rate (HR) [1], [2]. Monitoring of the CO and PWV during rest with respect to HR in patient with an implanted pacemaker could be important for deepening the knowledge about hemodynamics of cardiovascular system. Likewise it could help to basic setting of pacemakers. Heart rate was changed by pacemaker programming while neither exercise nor drugs were applied. The bioimpedance method appears to be an appropriate method, because it is absolutely non-invasive and unpretentious for patients and does not require any special knowledge of medical personal. Our goal is to prove results in [3], [4] and [5] on our set of patients. At the same time we want to test using of the bioimpedance method for PWV measurement by the device designed in the ISI Brno [6] in clinical research.

2. Material and methods

The present study was performed in 44 volunteers, who had an implanted pacemaker. Pacemaker implantation was performed at least one year before measurement and the patients were investigated during a regular check-up at the clinic. Within experiment, subjects were in the supine position on the examination bed in a wide-awake state. Each measurement consists of at least eight intervals of 6 minutes each according to the sequence (with HR controlled by the pacemaker): HR=80bpm;

HR=120bpm; HR=80bpm; HR=100bpm; HR=80bpm; HR=100bpm; HR=80bpm; HR=120bpm.The following biological signals were recorded continuously: ECG (6 or 12 leads), blood pressure by Finapres-2300 (Ohmeda), phonocardiogram (PCG), bioimpedance of thorax and calf. At the beginning and the end of the measurement, the blood pressure from the cuff on the patient's left arm was recorded manually. The analogue signals were sampled at 500 Hz, and converted to digital form by a 16-bit Analogue-to-Digital Converter; data was stored in binary files for subsequent analyses. Two current sources were used for measurement of bioimpedance, see Figure 1. The separation of channels was accomplished by the frequency offset of 2 kHz (i.e. 50 kHz and 52 kHz) [6].

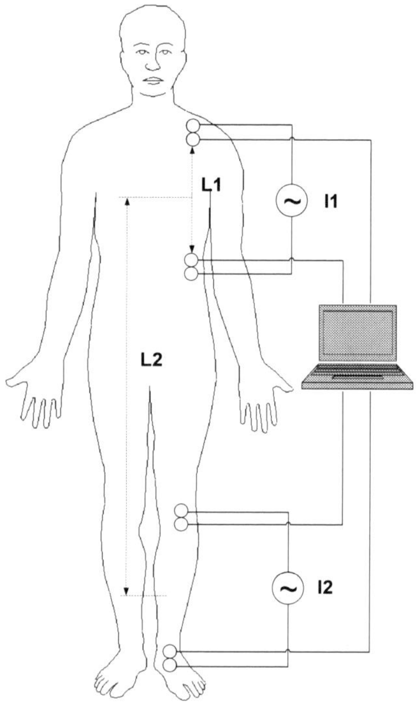

Figure 1. Bioimpedance setup

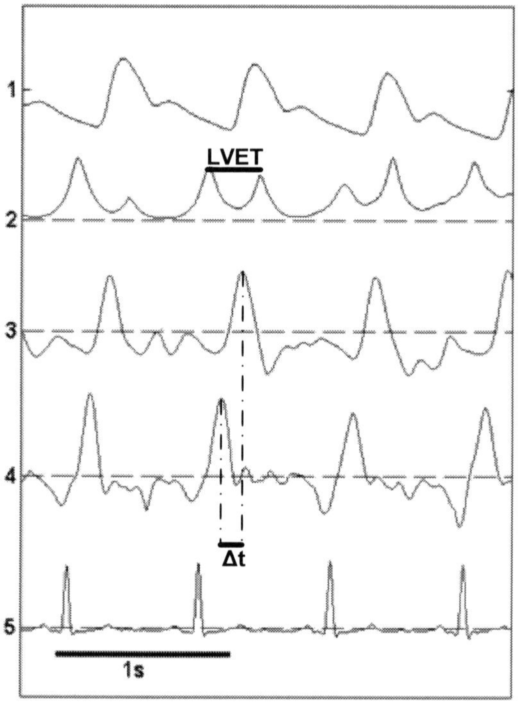

Figure 2. Example of measured and computed signals, from top: 1. BP, 2. energy of the heart sounds, 3. $-dZ_{calf}/dt$, 4. $-dZ_{thorax}/dt$, 5. ECG

For the estimation of stroke volume (SV), the Sramek-Bernstein equation was applied [2], [7]. The left ventricular ejection time (LVET) was calculated from energy of the PCG by filtering and subsequent detection of the heart sounds, see Figure 2. [8]. Cardiac output (CO) was calculated by multiplying SV and HR.

Pulse wave velocity (PWV) was determined using the following formula: PWV=L2/Δt, where: L2 is distance between the centres of voltage electrodes on thorax and calf (see Figure 1.), Δt is time delay between maximum of derivative impedance signals from thorax ($-dZ_{thorax}/dt$) and calf ($-dZ_{calf}/dt$), see Figure 2.

3. Results

Only those subjects, for whom it was possible to securely evaluate all parameters on HR=80, 100 and 120 bpm were chosen for statistical analyses. The statistical group consisted of 24 subjects, average age 74,6±9,4 years (20 men and 4 women). Reasons for elimination were: signals interruption or corruption, a lot of artefacts, extrasystoles, etc. Every selected subject for further evaluation had to have all measured signals with enough quality on all HR. Following hemodynamic parameters were computed from available signals: SV, LVET, CO, PWV, systolic blood pressure (SBP), diastolic blood pressure (DBP), pulse pressure (PP) and mean arterial pressure (MAP). The following normalization was done for a comparison between individual parameters in the whole group. Parameters for every subject were normalized individually so that its mean value over all HR was taken and subsequently, the corresponding parameter was normalized to this value. These mean relative changes (± standard deviation) and p-value are displayed in Table 1.

Table 1 Relative changes±SD and p-value

	HR=80 bpm		**HR=100 bpm**		**HR=120 bpm**	
Parameters	change±SD	p-value	change±SD	p-value	change±SD	p-value
SV	121,7±4,5	6,1e-005	98,6±4,9	3,1e-005	81,0±4,2	3,1e-005
LVET	107,6±3,5	1,9e-006	100,2±2,7	1,9e-006	92,1±4,1	1,9e-006
CO	100,2±4,1	6,1e-005	100,5±3,4	3,1e-005	99,3±4,3	3,1e-005
PWV	101,1±3,7	6,1e-005	100,1±3,7	3,1e-005	98,2±5,0	3,1e-005
SBP	103,3±15,3	1,9e-006	99,7±3,3	1,9e-006	97,0±13,8	1,9e-006
DBP	98,5±15,3	1,9e-006	100,0±3,4	1,9e-006	101,5±14,0	1,9e-006
PP	106,8±6,2	3,1e-005	100,3±4,6	1,5e-006	93,4±6,3	1,5e-006
MAP	100,3±14,7	1,9e-006	100,1±2,7	1,9e-006	99,6±14,0	1,9e-006

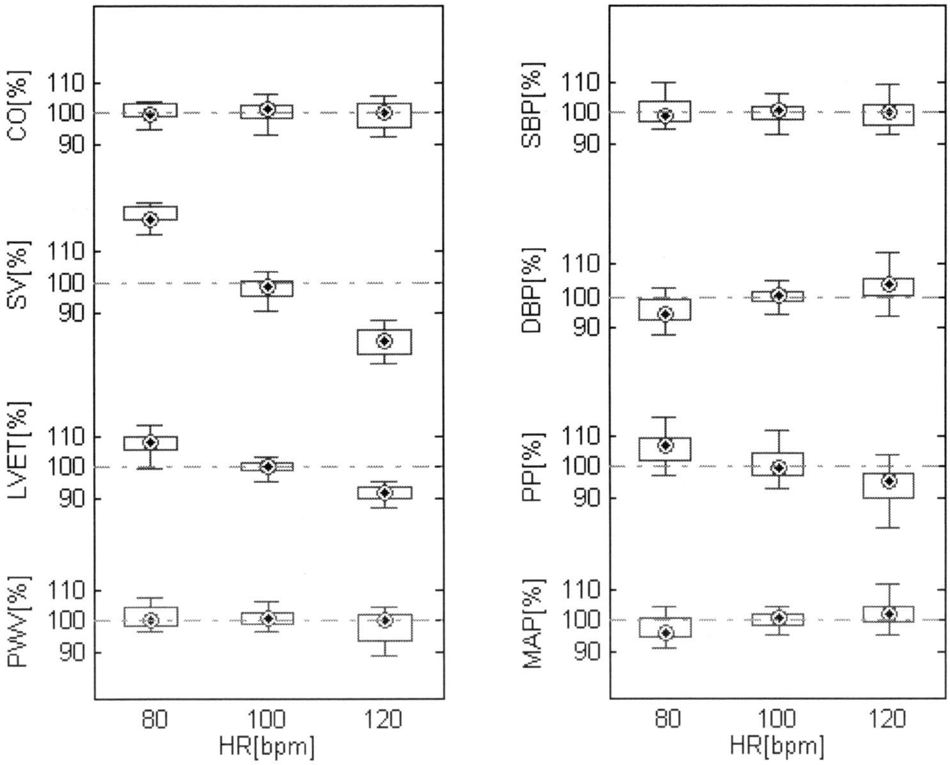

Figure 3. Graphic representation of relative changes in following hemodynamic parameters (CO, SV, LVET, PWV, SBP, DBP, PP and MAP)

A two-sided sign test was used to determine whether the parameters come from a continuous distribution with zero median. The p-value<0,05 was detected as significant. Figure 3. shows the relative changes in box plot, where a chain line is 100% (mean of absolute value).

4. Discussion and conclusion

The presented study investigates behaviour of cardiovascular system in dependence on heart rate. A group of 44 subjects was examined by the bioimpedance method on the thorax and on the calf. At the same time the continuous BP, PCG and ECG signals were registered. On the basis of available signals chosen hemodynamic parameters (CO, SV, LVET, PWV, SBP, DBP, PP and MAP) were evaluated. This study was aimed on pulse wave velocity and the cardiac output (or stroke volume) primarily. For our results group of 24 subjects (from 44 subjects, who was measured) were carefully chosen. Other 20 subjects were eliminated due to errors and/or artefacts in measured signals (often due to estimation of CO), which were mainly caused by the health state and age of the subject. This selection was done to obtain as reliable results as possible. The HR was changed by the programming of pacemaker, without of exercise, drugs, etc. The bioimpedance method was used to determine CO and PWV. The obtained parameters were statistically processed. It was found that CO and PWV do not change significantly on HR at paced patient during rest (see Table 1 and Figure 3.). There is 2,9% decrease in PWV and 1,2% decrease in CO across whole HR interval (80-120bpm). It follows that the CO and PWV do not significantly depend on HR at rest in patients with an implanted pacemakers. It was also found that the bioimpedance method is possible and robust method for PWV evaluation.

5. Acknowledgement

The study was supported by project PRECEDENCE (CZ.1.05/3.1.00/13.0292).

References

[1] Sramek BB, Valenta J and Klimes F 1995 Biomechanics of the Cardiovascular System (1st ed. Prague-Irvine: Czech Technical University Press) 214-31

[2] Leondes C.T 2007 Biomechanical systems technology (River Edge, NJ, USA: World Scientific) 49-88

[3] Park B J, Lee H R, Shim J Y, Lee J H, Jung D H and Lee Y J 2010 Association between resting heart rate and arterial stiffness in Korean adults *Archives of Cardiovascular Diseases* **103** 246-52

[4] Lantelme P, Mestre P, Lievre M, Gressard A and Milon H 2002 Heart rate: an important confounder of pulse wave velocity assessment *Hypertension* **39** 1083–87

[5] Albaladejo P, Copie X, Boutouyrie P, Laloux, B, Déclère A D, Smulyan, H, and Bénétos A 2001 Heart rate, arterial stiffness, and wave reflections in paced patients *Hypertension* **38** 949-52

[6] Viščor I, Vondra V, and Halámek J 2007 Two-channel high dynamic range bioimpedance monitor for cardiography *13th International Conference on Electrical Bioimpedance and the 8th Conference on Electrical Impedance Tomography* **17** 225-28

[7] Bernstein D P and Lemmens H J 2005 Stroke volume equation for impedance cardiography. *Medical and Biological Engineering and Computing* **43** 443-50

[8] Wang X, Li Y, Sun C and Liu C 2009 Detection of the First and Second Heart Sound Using Heart Sound Energy *Biomedical Engineering and Informatics* **2** 1-4

[9] Cotter G, Schachner A, Sasson L, Dekel H, and Moshkovitz Y 2006 Impedance cardiography revisited *Physiological measurement* **27** 817-27

[10] Nitzan M, Khanokh B and Slovik Y 2001 The difference in pulse transit time to the toe and finger measured by photoplethysmography *Physiological measurement* **23** 85-93

[11] Solà J, Vetter R, Renevey P, Chételat O, Sartori C and Rimoldi S F 2009 Parametric estimation of pulse arrival time: a robust approach to pulse wave velocity *Physiological Measurement* **30** 603-15

Journal of Physics: Conference Series **434** (2013) 012051 doi:10.1088/1742-6596/434/1/012051

The new criterion for cardiac resynchronization therapy treatment assessed by two channels impedance cardiography

K Peczalski[1, 2], T Palko[1], D Wojciechowski[2, 3], Z Dunajski[1], M Kowalewski[2].
[1] Institute of Metrology and Biomedical Engineering Warsaw Technical University, Warsaw
[2] Stopczyk Centre of Cardiology Wolski Hospital, Warsaw
[3]IBIB PAS Warsaw

E-mail: k.peczalski@mchtr.pw.edu.pl

Abstract. The cardiac resynchronization therapy is an effective treatment for systolic failure patients. Independent electrical stimulation of left and right ventricle corrects mechanical ventricular dyssynchrony. About 30-40% treated patients do not respond to therapy. In order to improve clinical outcome authors propose the two channels impedance cardiography for assessment of ventricular dyssynchrony. The proposed method is intended for validation of patients diagnosis and optimization of pacemaker settings for cardiac resynchronization therapy. The preliminary study has showed that bichannel impedance cardiography is a promising tool for assessment of ventricular dyssynchrony..

1. Introduction

The cardiac resynchronization therapy (CRT) is a common treatment for systolic heart failure patients. The method is based on resynchronization of improper timing of the heart conductive system by multipoints pacing. The independent electrical stimulation of left ventricle (LV) and right ventricle (RV) corrects the mechanical dyssynchrony (MD) and increases stroke volume of the heart. The common accepted criteria for CRT are: ejection fraction (EF) less than 35%, New York Heart Association (NYHA) class III or IV and QRS complex duration (QRSd) longer than 120 msec [1, 2]. One can see that the only criterion related to MD is QRSd which describes sum of depolarization potentials of LV and RV. The delay between systole of right and left side of the heart, triggered by depolarization potentials, is indicated by extended duration of QRS. Two above mentioned criteria confirm the symptoms of heart failure in the patient.

The clinical outcome of CRT is limited by fact that about 30% of treated patients do not respond to therapy [1]. No response to CRT seems to be caused by improper selection of patients for resynchronization pacing. Echocardiography-guided methods (USG) have been most commonly employed for measurement of MD. According to published data and authors clinical results the echo indices of MD are difficult to collect, appear to be discordant, and fail to predict clinical outcomes [3, 4]. There are two possible explanations for the poor accuracy of USG. The first explanation is related to how the blood flow velocity is detected with Doppler sonography. Doppler sonography is used to measure blood flow velocity by the Doppler frequency shift of echoes from red blood cells. The angle between the ultrasound beam and the blood flow seems to be a significant source of measurement uncertainty since the Doppler shift for given velocity of blood is proportional to the cosine of the angle α. Maximum Doppler frequency shift occurs when blood is flowing directly towards or away from the ultrasound beam. Thus, scanning as close as possible to 0° will increase sensitivity and minimize ambiguity in the flow direction. The crucial limitation of the cardiosonography technique is the dependence of the measurements on the angle between the ultrasound beam and the direction of blood flow. The resulting error in blood flow velocity measurements is unavoidable and varies from one procedure to another. The second explanation is related to the technique of estimation of MD by means of USG.

The measurement procedure consists of three steps:

1. measurement of the pre ejection period (PEP) between the Q wave of QRS complex in the reference ECG record and the onset of ejection through pulmonary valve for RV.
2. measurement of PEP between Q wave of QRS complex in reference ECG record and onset of the ejection through the aortal valve for LV.
3. the MD is calculated by PEP(LV) minus PEP(RV).

The global uncertainty of the measurement method of MD depends on following components:

- the uncertainty due to the indirect mode of measurement (subtraction of two separate measurements gives bigger error),
- the uncertainty caused by possible changes of the patient's heart rhythm due to time differences between consecutive measurements,
- the uncertainty of the blood velocity measurement due to the inconsistency of the angle between the ultrasound beam and the blood stream direction and consequently poor repeatability of the results (as described above).

These arguments indicate inaccuracies of the USG methods in the assessment of the MD in systolic heart failure patients. Therefore the new more accurate method is desirable for improving the clinical outcomes of CRT.

2. The proposed measurement method

We propose to apply two channel impedance cardiography (IC) for the assessment of MD in order to improve the fidelity of selection of systolic heart failure patients for the CRT treatment [5]. This method has following advantages to USG:

- measurements for LV and RV are simultaneously executed on the same beat of the heart,
- direct measurement of MD avoids uncertainty related to subtraction of two measurements curried out at different times,
- absence of the error due to inconsistent angle of the probe application (the angle between beam and blood flow).

In addition the measurement procedure is simple, less time consuming and does not require a highly trained operator.

The system consists of two channels for current tetrapolar impedance measurement method. Different frequencies of alternating application current (AC) current are applied for each channel in order to avoid interference between them. The parameters of AC are: first channel 1 mA, 63 kHz and second channel 1 mA, 95 kHz. The terapolar method was applied to reduce influence of contact impedance (electrode – skin of patients) on the measured signal.

The lead system was dedicated to collecting flow data in the right pulmonary artery and aorta. The flow in the right pulmonary artery was assessed by two pairs of leads located to the right of the sternum in the second intercostals area. The distance between signal electrodes was in the range of 6-8 cm depending on the patient. The flow in upper aorta was assessed also by two pair of leads located longitudinally to the sternum (higher in first intercostals area and second on the level of xiphisternum). In this case the distance between signal electrodes was in range of 12 – 18 cm.

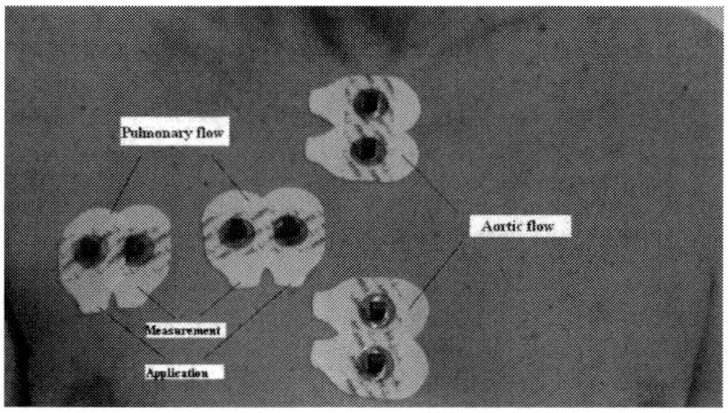

Figure 1. proposed biIC lead system

The MD interval was calculated directly from the distance between the onset of aortic flow and the onset of pulmonary flow.

3. Preliminary clinical study

The preliminary clinical study was performed upon approval of the Ethical Committee.

1. The 60 years old volunteer in I class of NYHA with a QRS duration of less than 120 milliseconds and an EF higher than 35%, underwent two channels IC examination. The MD was found to be zero (fig. 2). No additional echo study was done,
2. Five patients treated by CRT were examined by IC and cardiosonography for MD assessment. The criteria for resynchronization pacing were fulfilled for all of them. The differences (in ms) between results achieved for DDD and CRT pacing for IC and USG are as follows:

   - first patient CRT – 10, DDD – 20,
   - second patient CRT – 0, DDD – 20,
   - third patient CRT – 0, DDD – 80,
   - fourth patient CRT – 30, DDD – 50,
   - fifth patient CRT – 70, DDD – 30 (unstable rhythm).

The goal of further optimization of MD was to get the delay between pulmonary and aortic onsets close to zero. The result of adjustment of MD in range 0 – 20 ms was terminated the optimization procedure in all examined patients. The tailoring of MD for each patient by USG was postponed due to the time needed for achieving sufficient results. However, exactly the same results for both methods were observed in two patients. The switch from the CRT pacing mode (after optimization) to DDD (pacing mode supporting right atrium and right ventricle of the heart) caused the significant increase of MD in four of five patients.

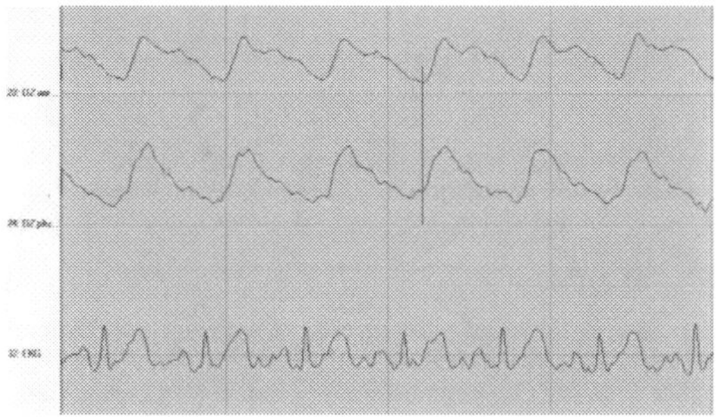

Figure 2. MD in healthy volunteer

4. Conclusions

1. Preliminary clinical studies have showed that this new technique has potential for more accurate diagnosis using parallel measurements of pulmonary and aortal flow with two channels Impedance Cardiography,
2. The proposed new method needs to be tested in a larger group of patients in order to statistically validate its accuracy for mechanical dyssynchrony measurements.

5. References

[1] Cazeau S, Alonso C, Jauvert G, Lazarus A, Ritter P 2004 *Europace* **14 (suppl 5)** 42
[2] Donahue T, Niazi I, Leon A, Stucky M, Herrmann K 2011 *J Cardiovasc Transl Res* **4** 30
[3] Galderisi M, Cattane F, Mondillo S 2007 *Cardiovascular Ultrasound* **5** 28
[4] Barold S S, Ilercil A, Herweg B. 2008 *Europace* **10 (suppl 3)** 88
[5] Palko T 1984 *Post Fiz Med (in polish)* **19** 3

XV Int. Conf. on Electrical Bio-Impedance & XIV Conf. on Electrical Impedance Tomography IOP Publishing
Journal of Physics: Conference Series **434** (2013) 012052 doi:10.1088/1742-6596/434/1/012052

Optimal configuration of an electrode array for measuring ventricles' contraction

M Lewandowska[1], A Poliński[1], B Truyen[2] and J Wtorek[1]

[1]Biomedical Engineering Department, Gdansk University of Technology, Narutowicza 11/12, 80-233 Gdansk-Wrzeszcz, Poland, [2]Vrije Universiteit Brussel, Belgium

E-mail: maglew@biomed.eti.pg.gda.pl

Abstract. An influence of an electrode-array configuration on an impedance signal composition for a fixed spatial distribution of its sources is examined in the paper. The Finite Element Method and Geselowitz relationship were used for examining three different electrode-arrays. A sensitivity approach was used to evaluate each configuration assuming that localization of the signal source is known. A conductivity change, thus the source of the impedance signal was considered as two hemispheres covered by a shell.

1. Introduction

Diagnostics of a heart is a relatively complicated task due to different phenomena involved, i.e. electrical, biochemical, mechanical, etc. Some cardiac diseases manifest in the form of ventricles' dysfunction, for example as a dyssynchrony. This dysfunction, among others, can be manifested as a distortion of the time relationship between the ventricles contractions. Thus, it is very important to have reliable technique which allows examining activity of each cardiac ventricle separately. Potentially, an impedance technique can be used in this application [1]-[3]. It is due to the fact that the impedance technique is also sensitive to a change of examined object's volume. However, an essential problem when using the impedance technique is to separate signal associated with the cardiac ventricles contraction from other ones. Then, the cardiac component of the signal has to be decomposed into two components, each corresponding to the other ventricle. The problem is not trivial as almost all conductivity changes located in the thorax are the consequence of ventricles' contraction. However, it follows from our previous studies that remarkably good results can be obtained when the total impedance signal contains mainly ventricles' components [4]. It can be to some extent, at least theoretically, achieved by using the appropriate electrode-array. Thus, performing at least two-channel measurements and using the appropriate electrode-arrays it should be possible to decompose of the impedance signal into two components arising from different ventricle. The impedance measurement technique is sensitive to conductivity changes undergoing throughout a whole volume of the object to which the electrodes are attached. However, the sensitivity is unevenly distributed in the object. In effect, a contribution of the selected conductivity changes is enhanced in the impedance signal when comparing it to the others. The main aim of the paper is to examine of the spatial properties of the impedance technique for the selected configurations of the electrode-arrays and to choose the optimal one for an anticipated application. It was achieved by means of a Geselowitz's relationship [5] and Finite Element Method (FEM).

2. Method

The change of impedance involved by conductivity variation is described by the relationship proposed by Geselowitz [4]

$$\Delta Z = -\int_V \Delta\sigma(x,y,z)\frac{\nabla\phi(\sigma,x,y,z)}{I_\phi}\cdot\frac{\nabla\psi(\sigma+\Delta\sigma,x,y,z)}{I_\psi}dv, \tag{1}$$

where ϕ represents the potential associated with current I_ϕ flowing between the "current" electrodes, ψ is the potential associated with hypothetical current I_ψ flowing between the "voltage" electrodes after the occurrence of the conductivity change $\Delta\sigma$; V is the volume of the region of changed conductivity, and a dot stands for the scalar product. Basing on this approach the spatial sensitivity is defined as

$$S = \frac{\nabla\phi(\sigma,x,y,z)}{I_\phi}\cdot\frac{\nabla\psi(\sigma+\Delta\sigma,x,y,z)}{I_\psi}. \tag{2}$$

The relationship (2) allows determining the contribution to the total signal, ΔZ, from each subregion assuming that the sensitivity S is known.

The numerical analyses were conducted using FEM and the relationship (2). The analyzed model is presented in figure 1. Cardiac ventricles were simulated as a two hemispheres surrounded by a shell. Two conductivity distributions were considered: homogenous, where $\sigma_1 = \sigma_2 = \sigma_3 = 0.1$ [S/m] and inhomogeneous, where $\sigma_1 = 0.1$ [S/m], $\sigma_2 = 0.22$ [S/m] and $\sigma_3 = 0.5$ [S/m].

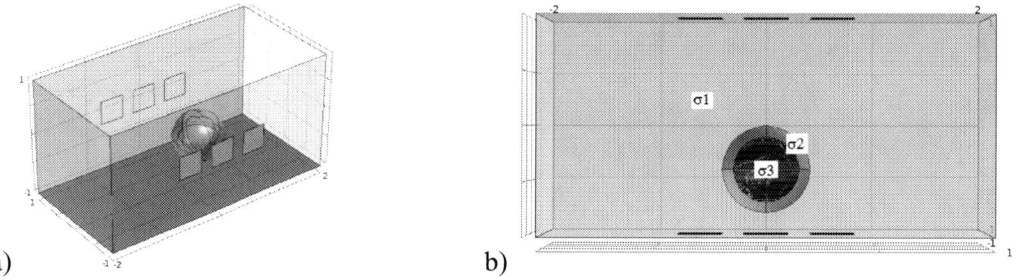

a) b)

Figure 1. Numerical model of volume conductor a) 3D view, b) top view.

A few different configurations of the electrode-arrays were examined (figure 2). A symmetrical electrode-array for two-channel measurements was composed of current electrodes, I_- - I_+, and two pairs of voltage electrodes $V1_-$, $V1_+$, and $V2_-$, $V2_+$ (figure 2a). In a practical application the distances between electrodes would be different. A localization of the current and voltage electrodes could be the same when using compound electrodes (figure 2b) and the electrode-array (for one-channel measurement) is modelled as a two electrode one. The voltage and current electrodes can interleave and distances between them can also be different (figure 2c).

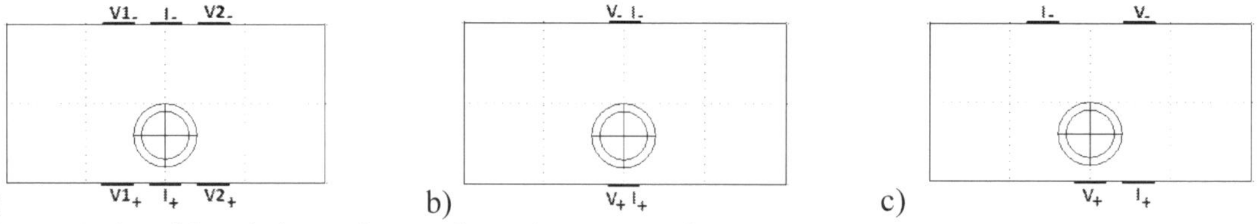

a) b) c)

Figure 2. Considered electrodes configurations: a), c) four-electrode impedance technique, b) two-electrode technique.

3. Results

Examples of the obtained sensitivity distributions are respectively shown in the figures 3, 4, 5 and 6. For comparison the sensitivity distributions obtained for uniform objects are also included (figures 3a, 4a, 5a and 6a). Figures 3b and 3c show the sensitivity distributions for different placement of V2+ electrode when a conductivity of the sphere is different from the surrounding one.

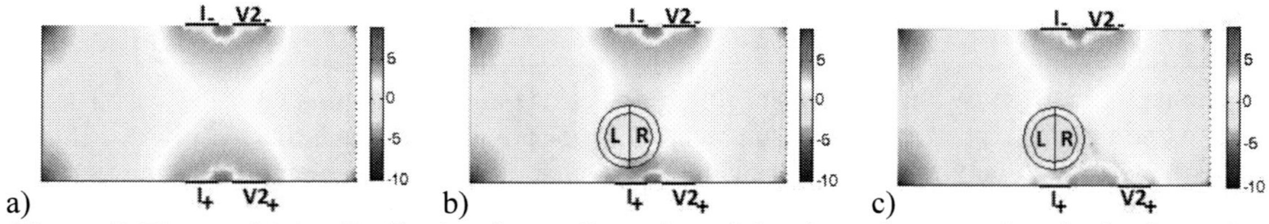

Figure 3. The sensitivity distribution for configuration of the electrode-array given in the figure 2a.

Figures 4b and 4c show the sensitivity distributions for two different positions of the sphere when a conductivity distribution outside the sphere is different and homogeneous.

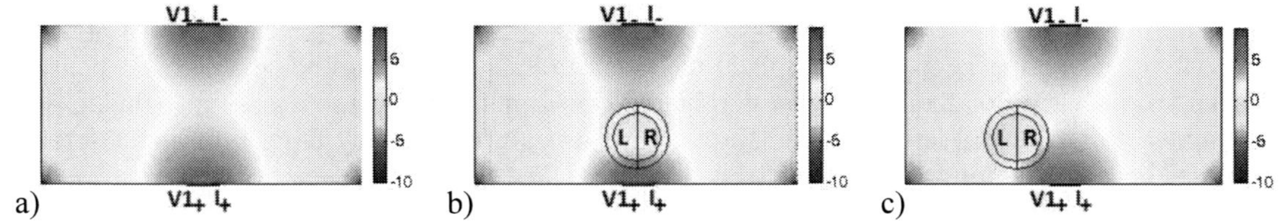

Figure 4. The sensitivity distribution obtained for configuration of electrode-array given in figure 2b.

Figures 5b and 6b show sensitivity distribution for different placement of V1+ electrode for inhomogeneous conductivity distribution when voltage and current electrodes interleave.

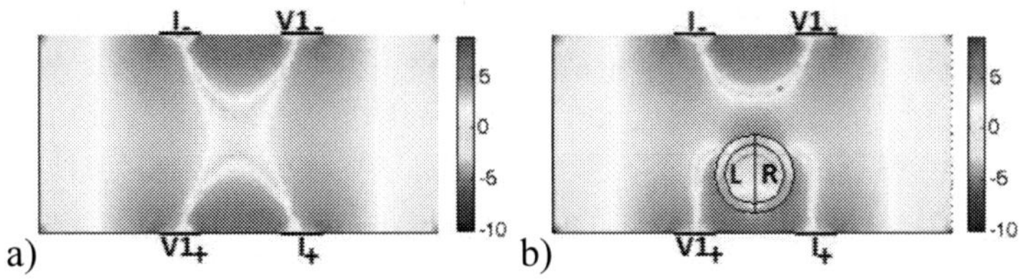

Figure 5. The sensitivity distribution obtained for configuration of electrode-array given in figure 2c.

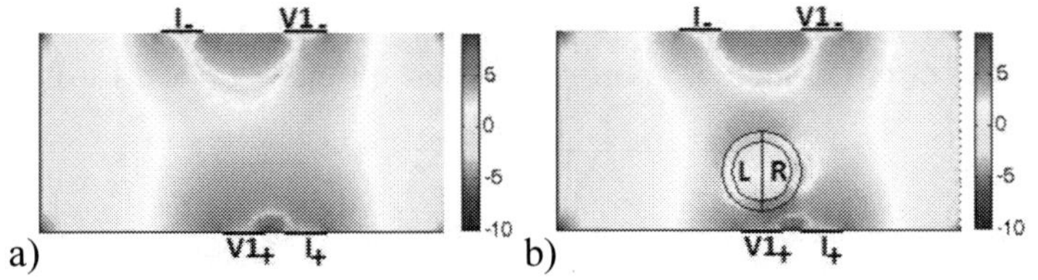

Figure 6. The sensitivity distribution obtained for configuration of electrode-array given in figure 2c.

Table 1 contains the sensitivity assigned to each hemisphere (it was obtained by integration of the sensitivity distribution) related to the total value of the sensitivity (the sensitivity integrated over the whole volume of the model).

Table 1. The relative sensitivity values.

Electrode configuration	Sensitivity for the left hemisphere [%]	Sensitivity for the right hemisphere [%]
2 (figure 3b)	1.000	1.604
3 (figure 4b)	0.727	0.725
4 (figure 4c)	0.093	0.315
5 (figure 5b)	-10.791	-10.798
6 (figure 6b)	2.098	2.516

4. Discussion

The obtained results show that the choice of the electrode-array configuration depends on many factors, e.g. on a goal of performed measurement. It is not possible to create the sensitivity different from zero and located only in desired sub-region. However, it is, to some extent, possible to change relations between contributions (the sensitivity integrated over the sub-region) from different sub-regions. Then, the signal processing methods can be applied to acquire the desired information.

The first examined configuration (figure 2a, figure 3) allows for the best separation of the signals from left and right hemisphere. The obtained sensitivity distribution is relatively uniform. However, it strongly depends on distance between electrodes (figure 3b, c). The main disadvantage of this configuration is relatively low sensitivity compare to the surrounding. It means that if a large conductivity change appears outside the sphere then measurement signal can be heavily disturbed, i.e. the signal from outside of the heart may dominate. The same may appear if a small conductivity change appears in relatively large volume (for example lungs perfusions by blood). In that case time analysis can be used to diminish errors due to such factors.

The second configuration (figure 2b, figure 3c, d) shows that reduction of electrodes number gives large sensitivity near the electrodes. This may cause that measured signal is much smaller and more susceptible to measurement noise. Changing position of the electrodes compare to sphere position changes the influence of left and right hemisphere. The great advantage of such solution is possibility of increasing the influence of the signal from one of the hemispheres.

The configuration from figure 2c is presented to show that there are configurations, which give more powerful signal from the desired region in comparison to other regions, but are not necessary better when other factors are taken into account. The disadvantage of such configuration is the non-uniform sensitivity distribution in the region of interest. This may cause problems in measurement interpretation.

5. Conclusions

The obtained results show that the electrode configuration should be chosen depending on the goal of performed measurement. It is possible to optimize electrode-array configuration in order to obtain a desired property of the sensitivity distribution, e.g. uniformity in prescribed region, or the highest sensitivity in comparison to other sources. Further studies should be performed using more realistic models.

References

[1] Yang F and Patterson R P 2008 A simulation study on the effect of thoracic conductivity inhomogeneities on sensitivity distributions *Ann. Biomed. Eng.* **36 (5)** 762-8

[2] Lewandowska M, Wtorek J, Bujnowski A and Mierzejewski L 2010 Analysis of thoracic regional impedance changes using PCA approach *J. Phys.: Conf. Ser.*, **224** 012162

[3] Wtorek J 2000 Relations between components of impedance cardiogram analyzed by means of finite element model and sensitivity theorem *Ann. Biomed. Eng.* **28** 1352-61

[4] Wtorek J and Poliński A 1995 Examination of impedance cardiography properties-FEM model studies *Biomed. Sci. Instrum.* **31** 77-82

[5] Geselowitz D B 1971 An application of electrocardiographic lead theory to impedance plethysmography *IEEE Trans. Biomed. Eng.* **18** 38-41

Journal of Physics: Conference Series **434** (2013) 012053 doi:10.1088/1742-6596/434/1/012053

Diagnostic criteria for mass lesions differentiating in electrical impedance mammography

Karpov A, Korotkova M
Clinical Hospital #9
Yaroslavl, Russia

e-mail: kb9@mail.ru

Abstract. The purpose of this research was to determine the diagnostic criteria for differentiating volumetric lesions in the mammary gland in electrical impedance mammography. The research was carried out utilizing the electrical impedance computer mammograph «MEIK v.5.6»®, which enables to acquire images of 3-D conductivity distribution layers within mamma's tissues up to 5 cm depth . The weighted reciprocal projection method was employed to reconstruct the 3-D electric conductivity distribution of the examined organ. The results of 3,710 electrical impedance examinations were analyzed. The analysis of a volumetric lesion included assessment of its shape, contour, internal electrical structure and changes of the surrounding tissues. Moreover, mammary gland status was evaluated with the help of comparative and age-related electrical conductivity curves. The diagnostic chart is provided. Each criterion is measured in points. Using the numerical score for evaluation of mass and non-volumetric lesions within the mammary gland in electrical impedance mammography allowed comparing this information to BI-RADS categories developed by American College of Radiology experts. The article is illustrated with electrical impedance mammograms and tables.

1. Introduction

Electrical impedance mammography is a relatively new method for breast diseases diagnostics (2, 4). In a short time period diagnostic criteria for early breast cancer detection were developed (3, 5), as well as the method for assessment and evaluation of electrical impedance images (6). Percentile curves of age-related electrical conductivity were built up for the mammary gland status evaluation (1). Currently, there emerged a necessity of a standardized description of volumetric lesions and developing of a system for volumetric lesions description in the mammary gland to be in line with ACR BI-RADS categories. The purpose of this research was to determine the diagnostic criteria for differentiating volumetric lesions in the mammary gland in electrical impedance mammography. Moreover, importance was attached to the assessment of the breast in case of non-volumetric lesions too.

2. Materials and Methods

The research was carried out utilizing the electrical impedance computer mammograph «MEIK v.5.6»®, which enables to acquire images of 3-D conductivity distribution layers within mamma's tissues up to 5 cm depth. The weighted reciprocal projection method was employed to reconstruct the 3-D electric conductivity distribution of the examined organ (2, 4). At the same time, the mammary glands were examined with «Ultrasonix SP» ultrasound scanner. The results of 3,710 electrical impedance examinations were analyzed. Visual and quantitative assessment of images including the calculation of the mean electrical conductivity index (IC) / electrical potential was performed.

3. Result

Journal of Physics: Conference Series **434** (2013) 012053 doi:10.1088/1742-6596/434/1/012053

Volumetric lesion is a dimensional lesion, detected in several scan planes. The analysis of volumetric lesion images included assessment of their shape, contour, internal electrical structure and changes of the surrounding tissues. Moreover, mammary gland status was evaluated using the comparative and age-related electrical conductivity curves.

A. Volumetric lesion.
1. *Volumetric lesion contours assessment in accordance with the terms of BI-RADS ACR:round, oval, lobular, urregular.*
For example, round – a lesion of a spherical, circular or spherical shape (Fig. 1). The size is determined by lesion's diameter. Typical example – a cyst.

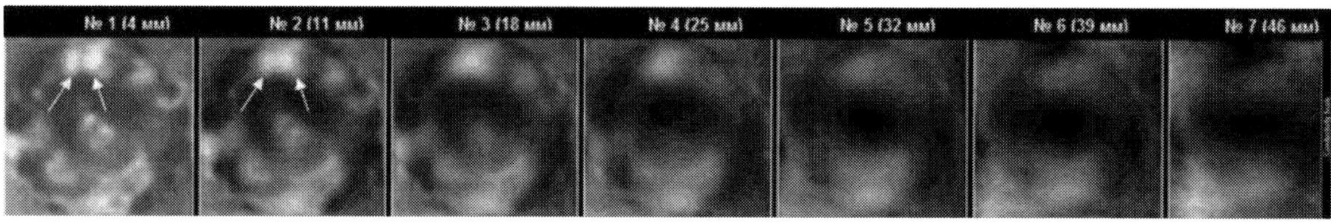

Fig. 1. EIM. Seven scan planes. In the upper segment at 12 on the clock dial, there can be observed two round hypoimpedance lesions (indicated with arrows).

2. *Assessment of volumetric lesion contours in accordance with the terms of BI-RADS ACR: sharp, vague, infiltrated.*
For example, infiltrated - the contours of a lesion are clearly distinguishable and are characterized by hyperimpedance (Fig. 2). Typical example – breast cancer.

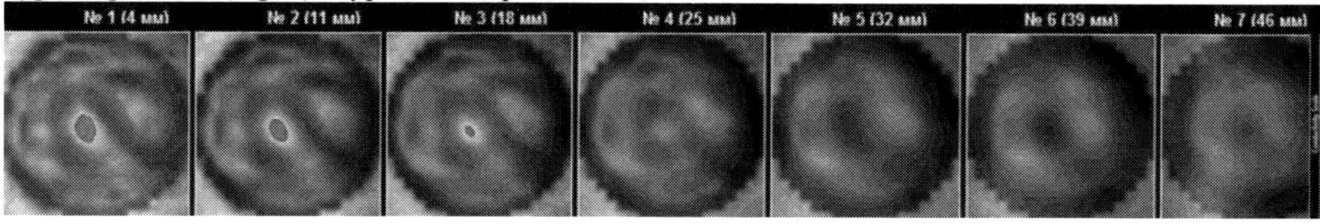

Fig. 2. EIM. Seven scan planes. The obovoid lesion is visualized at 3 on the clock dial (indicated

3. *Assessment of the internal electrical structure of a volumetric lesion in accordance with the terms of BI-RADS ACR: hyperimpedance, isoimpedance, hypoimpedance, animpedance.*
Taken separately, the internal electrical structure of the lesion is not the criterion for judging on its possible malignancy. However, it is an important characteristic, especially in combination with other evaluation criteria. The increase of electrical conductivity/potential is correlated with an increase of malignancy probability.
For example, animpedance - electrical conductivity/potential value of the lesion is considerably higher than that of the surrounding tissue of the breast (Fig. 3). Typical example – breast cancer.

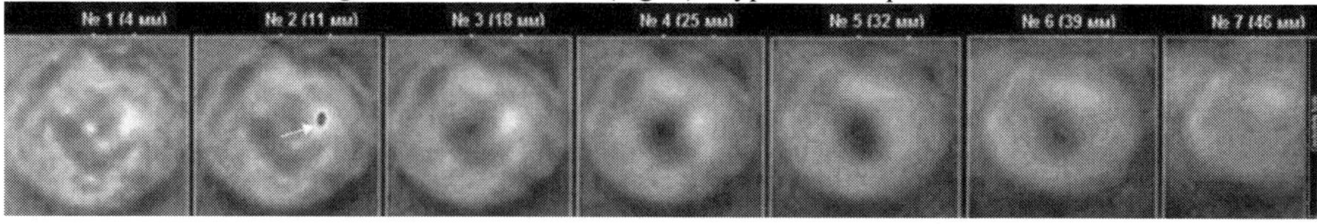

Fig. 3. EIM. Seven scan planes. In the outer segment of the left mammary gland, at 3 o'clock position there is observed an an-impedance area, which is highlighted with red in the second scan plane, less than 10 mm in size (indicated with arrows).

4. *Assessment of volumetric lesion influence on the surrounding tissue in accordance with the terms of BI-RADS ACR: skin thickening, skin extrusion or retraction, skin or nipple infiltration, alterations of the breast anatomy.*

For example, skin thickening – a pronounced unilateral hyperimpedance lesion either focal or located around the mammary gland (Fig. 4). Typical example – inflammatory breast carcinoma.

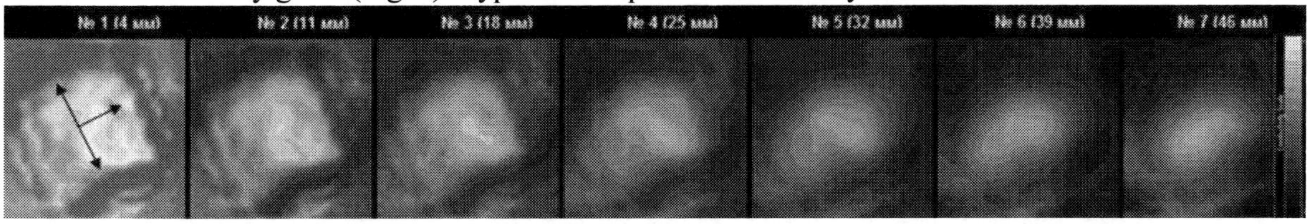

Fig. 4. EIM. Seven scan planes. A pronounced hyperimpedance contour around the mammary gland (indicated with arrows)

B. Comparative electrical conductivity.

Comparative conductivity is the alteration of electrical conductivity of one breast with respect to the other (Fig. 5). Typical examples – breast cancer, mastitis.

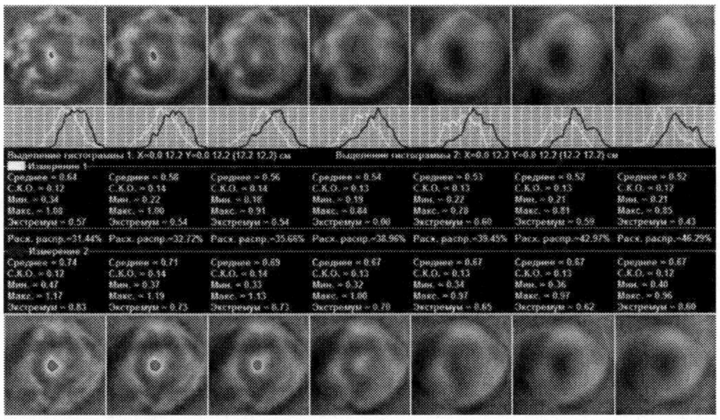

Fig. 5. Upper row – EIM. Seven scan planes. Breast cancer. Bottom row - EIM. Seven scan planes. Healthy gland. The second row shows the divergence between the histograms of electrical conductivity distribution of the affected and healthy mammary glands.

4. Conclusion

Diagnostic chart was created, in which each criterion was measured in points (Table 1).

Table 1. Diagnostic criteria for differentiating mass lesions in electrical impedance mammography (EIM).

Diagnostic criteria	Electrical impedance points
Shape	
• round, oval	1
• lobular, irregular	2
Contour	
• no	0
• sharp	1
• hyperimpedance, indistinct	2
Surrounding tissues	
• preserved	0
• structure alteration/displacement	1
• thickening/extrusion/retraction	2
Internal electrical structure	
• hyperimpedance (IC<x-2 std)	0

Journal of Physics: Conference Series **434** (2013) 012053 doi:10.1088/1742-6596/434/1/012053

• isoimpedance (IC =x±2 std)	1
• hypoimpedance (IC>x+2 std)	2
• animpedance (IC>x+3 std)	3
Comparative electrical conductivity	
• divergence between histograms < 20%	0
• divergence between histograms 20- 30%	1
• divergence between histograms 30-40%	2
• divergence between histograms > 40%	3

Using the numerical score for evaluation of volumetric and non-volumetric lesions within the mammary gland in electrical impedance mammography allowed comparing this information to BI-RADS ACR categories.

EIM	ACR
Common scale	*BI-RADS-categories*
no score	BI-RADS 0 poor image
0-1	BI-RADS 1 lesions are not defined – routine mammography
2-3	BI-RADS 2 benign tumours – routine mammography
4	BI-RADS 3 probably benign findings – in 6 months
5-7	BI-RADS 4 suspicious abnormality - biopsy
> 8	BI-RADS 5 highly suggestive of malignancy – treatment/biopsy

EIM numerical score allows for standardizing the description of volumetric lesions and for the usage of patient monitoring algorithm, developed by the American College of Radiology when conducting electrical impedance mammography. Thus, it may contribute to the compatibility of electrical impedance mammography with the well-established diagnostic methods and the uniformity of volumetric lesions differentiation by various methods.

References

[1] Breast imaging reporting and data system (BIRADS) 2003 *American College of Radiology*

[2] Cherepenin V, Karpov A, Korjenevsky A, Kornienko V, Mazaletskaya A, Mazurov D and Meister D 2001 A 3D electrical impedance tomography (EIT) system for breast cancer detection *Physiological Measurement* **22** 9

[3] Dunaeva O, Gerasimov D, Karpov A, Machin M, Tchayev A, Tsofin Yu and Tsyplyonkov V 2009 Using Backprojection Algorithm for 3D Image Reconstruction in EIT *World Congress on Medical Physics and Biomedical Engineering* **25** 288

[4] Electrical Impedance Tomography 2005 Edited by Holder D 167

[5] Karpov A, Korotkova M, Mumtazuddin Ahmed M, Myachin M and Tsofin Yu 2011 Seminar on electrical impedance potential mammography Yaroslavl

[6] Korotkova M and Karpov A 2007 Procedure for assessment of the mammary gland electrical impedance images *XIII international conference on electrical bio-impedance* **17** 699

Bioimpedance for the spot measurement of tissue density

E S Dylke[1], L C Ward[2], C Stannard[1], A Leigh[1], S L Kilbreath[1]

[1] Faculty of Health Sciences, University of Sydney, Sydney, Australia
[2] School of Chemistry and Molecular Biosciences, The University of Queensland, Brisbane, Australia

Email: l.ward@uq.edu.au

Abstract. Long-standing lymphoedema is characterised by tissues changes which are currently not detectable using bioimpedance spectroscopy. It has been suggested that a combination of bipolar and tetrapolar measurements may be used to detect these tissues changes for a single site in the transverse direction. This was technique was trialled in a group of control participants with no history of lymphoedema or recent upper limb trauma. Repeated spot measurements were done without removal of electrodes to determine biological variability as well as with removal of electrodes to determine technical reproducibility. The inter-limb spot ratio of the controls was then compared to that of a number of women previously diagnosed with secondary lymphoedema in the forearm. Biological variability was not found to greatly influence repeated measures but only moderate technical reliability was found despite excellent co-efficient of variation for the majority of the measurements. A difference was seen between those with more severe swelling and the controls. This novel technique shows promise in detecting tissue changes associated with long-standing lymphoedema.

1. Introduction

Secondary lymphoedema can occur after a disruption to or obstruction of the lymphatic system [1]. Early stages of the condition are characterised by increases in extra-cellular fluid leading to soft pitting swelling of the limb. Bioimpedance analysis (BIA) has been shown to be more sensitive to changes in these stages allowing treatment to be started earlier than traditional circumference or volume measures [2]. Long standing lymphoedema though, is associated with tissue composition changes where the fluid has been predominantly replaced by a fatty, fibrotic tissue [3]. Conventional tetrapolar BIA measurements are not useful in diagnosis or monitoring at this time [4]. It has been suggested that a combination of bipolar and tetrapolar bioimpedance measurements could be used to overcome this deficiency allowing changes in the tissue to be detected at a single site in the transverse direction (spot measurements) rather than impedance longitudinally along a limb [5].

In order to detect possible tissue changes, a series of six bipolar and tetrapolar bioimpedance measurements could be performed at each location of interest bilaterally [5]. When bipolar impedance is measured using skin electrodes, the measured impedance is that of the body tissues both parallel and perpendicular (transverse) to the longitudinal axis (direction of current flow). In contrast, a tetrapolar electrode arrangement measures the longitudinal component only. The transverse impedance at a single site can then be calculated from the combination of these measurements by simple vector algebra. The ratio of the spot measure resistance of the dominant or affected arm to the non-dominant or unaffected arm is then determined.

Published under licence by IOP Publishing Ltd

Journal of Physics: Conference Series **434** (2013) 012054 doi:10.1088/1742-6596/434/1/012054

The aim of this study was to determine the reliability of this method of spot bioimpedance measurements and to compare measurements in control participants with those from women with secondary lymphoedema.

2. Methods

2.1 Participants: Fifteen healthy females without a history of upper limb lymphoedema or recent upper limb trauma were recruited from the staff and student bodies at the University of Sydney. Five women with upper limb lymphoedema secondary to treatment for cancer were recruited as part of a larger study. Participant details are provided in table 1. Ethical approval was obtained from the University of Sydney's Medical Human Research Ethics Committees and all participants provided written informed consent.

Table 1: Participant characteristics

	Control group (n=15)	Lymphoedema group (n=5)
Age (y)*	22.9 (3.2)	61.4 (3.8)
Height*	1.62 (0.07)	1.63 (0.05)
Weight*	56.3 (7.4)	69.0 (12.9)
BMI*	21.4 (2.2)	26.0 (4.9)
Dominance (R:L)	14:1	5:0
Affected side (R:L)		2:3
Inter-arm circumference difference (cm) ^		2 to 8

* mean (SD), ^ range

2.2 Protocol: Three studies were undertaken to examine the utility of the measurement technique. For all studies, participants were positioned in supine and the skin on their hands, forearms and feet was thoroughly cleaned with an alcohol wipe. Bilaterally, a point 15 cm proximal from the ulnar styloid was then marked on the dorsal forearm. An electrode was placed at this location as well as on standard BIA assessment locations: the 3rd metacarpal phalange joint; in line with the ulnar styloid; between the maleoli on the front of the leg and between the 2nd and 3rd metatarsal phalange joints. Care was taken to ensure good adherence between the electrode and the skin. Measurements were completed using a BIM4 single frequency (50 kHz) impedance meter (SEAC-ImpediMed, Brisbane).

Study one examined the magnitude of subject variability. The series of six bipolar and tetrapolar measurements, as described previously [6], were completed bilaterally, three times in immediate succession without removal of the electrodes followed by a fourth measurement after a period of 15 minutes of rest in five controls. Study two examined the technical reliability of the measurement. The spot measurement technique was again completed three times in succession but all electrodes were removed and replacement before each measurement in a larger control sample which included all subjects from study one (n = 15). Finally, study three compared the data from the controls to those of women with secondary lymphoedema secondary (n = 5). For all of lymphedema participants, the location at 15 cm proximal to the ulnar styloid represented their area of self-reported worst swelling.

For each step of the series of bipolar and tetrapolar bioimpedance measurements, the resistance (R), impedance (Z) and reactance (Xc) was recorded and entered into a spreadsheet which, using Pythagoras' theory, calculated the R, Z and Xc for the specific location tested (in this case, 15 cm proximal from the ulnar styloid). The inter-limb spot ratio of this point was determined by dividing the resistance on the dominant side by that found for the non-dominant side. For the lymphoedema group, the affected side was compared to the unaffected side.

2.3 Data Analysis: Cronbach's alpha, a coefficient of reliability,[7] was determined for study one and two to establish the reliability of the inter-limb spot ratios found at each measurement time point. For study 2, the coefficient of variation (CV) for R, Z and Xc for each participant was examined. Finally, a comparison between the control ratios found in study two and that of the lymphoedema group were done using non-parametric independent test as the ratios were not normally distributed.

3. Results

The participants exhibited little biological variability (Cronbach's alpha = 0.816) between the measurements when the electrodes were not removed. While the technical reliability of the method was only moderate (Cronbach's alpha = 0.668), the co-efficient of variation for each of the measure was mostly excellent with less than 5% variability for found in 93% of the individual measurements, implying that the reduction in technical reliability may have been caused by errors with individual measurements and not by the technique itself.

The control group was significantly younger and lighter than the lymphoedema group (both $p < 0.01$). All participants had been diagnosed with lymphedema at least 8 months previously (range 8 to 96 months). Three women with secondary lymphoedema had more severe inter-limb differences (>3 cm) at the location of measurement and their inter-limb spot ratio did not overlap with the control group while those with less severe inter-limb differences did (Figure 1). However there was a significant different between the lymphedema group and the controls (Mann Whitney U test $p < 0.001$).

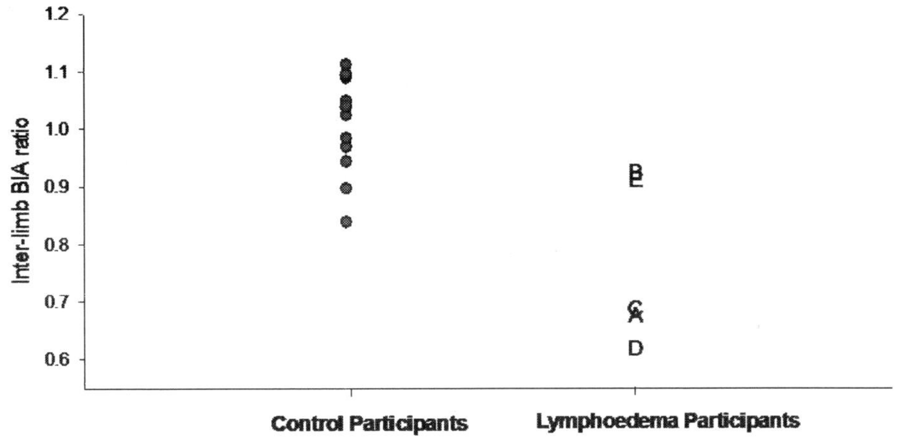

Figure 1: Inter-limb spot ratio of the controls compared to 5 women previously diagnosed with lymphoedema

4. Discussion

This novel technique was found to have acceptable reliability when the electrodes were not removed between measurements. Although the reliability decreased with removal and replacement of the electrodes between measurements, the vast majority of these measurements still exhibited minimal variability. Clinically, therefore, the emphasis when using this technique in its present form, requires stringent attention to the placement and adherence of the electrodes as lack of complete electrode contact in bipolar measurements, such as slight lifting of a corner of an electrode, will affect the measured impedance between the skin and the electrode. This alters the overall measurement increasing error and decreasing the technical reliability, which could be clinically relevant. Furthermore, its present form, the technique is time consuming as a series of six measurements on each arm are necessary before the overall spot impedance can be calculated. Nonetheless, the difference found between those with and without long standing lymphoedema is encouraging and suggests the method's ability to detect changes in tissue composition.

Further research comparing the method with techniques such as ultrasound or MRI that can directly assess tissue composition are required to confirm these observations. Further determination of its sensitivity in detecting early changes in tissue composition, particularly in women with mild lymphoedema, could assist with treatment decisions. Currently most treatment techniques attempt to mobilise the stagnant fluid in a limb with lymphoedema. If tissue changes have occurred, these techniques are no longer appropriate. Spot measurement of tissue composition could, therefore, play an important role in improving treatments efficacy.

In conclusion, spot BIA shows promise in detecting the tissue changes that occur with long-standing lymphoedema. Further research is necessary to confirm its utility and benefit in treatment decisions.

References

[1] Murdaca G, Cagnati P, Gulli R, Spano F, Puppo F, Campisi C and Boccardo F 2012 *American Journal of Medicine* **125** 134-40
[2] Ward L C 2006 *Lymphatic Research & Biology* **4** 51-6
[3] Brorson H, Ohlin K, Olsson G and Karlsson M K 2009 *Lymphatic Research and Biology* **7** 3-10
[4] Shah C and Vicini F A 2011 *Int J Radiat Oncol Biol Phys* **81** 907-14
[5] Cornish B H, Ward L C, Thomas B J and Bunce I H 1998 *Applied Radiation and Isotopes* **49** 651-2
[6] McCullagh W A and Ward L C 2004 *Proceedings of the XII International Conference on Electrical Bioimpedance* **1** 259-61
[7] Tavakol M and Dennick R 2011 *International Journal of Medical Education* **2** 53-5

Journal of Physics: Conference Series **434** (2013) 012055 doi:10.1088/1742-6596/434/1/012055

The measurement of peripheral blood volume reactions to tilt test by the electrical impedance technique after exercise in athletes

AA Melnikov[1], SG Popov[1], DV Nikolaev[2], AD Vikulov[1]

[1] State Pedagogical University, 108 Respublikanskaya St. , 150000, Yaroslavl, Russia
[2] Scientific Research Centre 'Medass', 12 Chistoprudny Blvd., 101000, Moscow

Email: meln1974@yandex.ru; dvn@medass.ru

Abstract. We have investigated the distribution of peripheral blood volumes in different regions of the body in response to the tilt-test in endurance trained athletes after aerobic exercise. Distribution of peripheral blood volumes (ml/beat) simultaneously in six regions of the body (two legs, two hands, abdomen, neck and ECG) was assessed in response to the tilt-test using the impedance method (the impedance change rate (dZ/dT). Before and after exercise session cardiac stroke (CSV) and blood volumes in legs, arms and neck were higher in athletes both in lying and standing positions. Before exercise the increase of heart rate and the decrease of a neck blood volume in response to tilting was lower ($p < 0.05$) but the decrease of leg blood volumes was higher ($p < 0.001$) in athletes. The reactions in arms and abdomen blood volumes were similar. Also, the neck blood volumes as percentage of CSV (%/CSV) did not change in the control but increased in athletes ($p < 0.05$) in response to the tilt test. After (10 min recovery) the aerobic bicycle exercise (mean HR=156±8 beat/min, duration 30 min) blood volumes in neck and arms in response to the tilting were reduced equally, but abdomen ($p < 0.05$) and leg blood volumes ($p < 0.001$) were lowered more significantly in athletes. The neck blood flow (%/CSV) did not change in athletes but decreased in control ($p < 0.01$), which was offset by higher tachycardia in response to tilt-test in controls after exercise. The data demonstrate greater orthostatic tolerance in athletes both before and after exercise during fatigue which is due to effective distribution of blood flows aimed at maintaining cerebral blood flow.

1. Introduction

The assessment of blood volumes simultaneously entering main peripheral regions of the human body is important for the orthostatic instability understanding and presents a rather difficult task. Peripheral blood volumes can be measured by non-invasive technique of electrical impedance on a beat by beat basis in various peripheral regions of the body. This impedance technique has been successfully used for analysis of cardiac stroke volume [1] and lower limb blood flow [2]. We suggest that the impedance method can be applied for assessment of peripheral blood volume reactions in response to ortostasis. The aim of our study was to evaluate blood volume reactions in the peripheral regions of the body in response to the tilt test in endurance-trained male athletes prior and after exercise. We hypothesized that in athletes with high aerobic fitness blood volume reactions will be decreased in the neck vessels and increased in the lower limbs and/or in the abdomen region and that these reactions in athletes after exercise will be less deteriorated.

2. Subjects and methods

Twelve healthy young (22.4±5.5 years) male racing skiers composed the athlete group. All athletes had 4-8 years of total sport experience and 10-15 hours exercise per week for last 2 months. The control group included healthy aged-matched (19.4±1.6) sedentary men (n=23).

2.1. Impedance measures of blood volumes

Measures were made using "Medass" (Russia) impedance analyzer. In this analyzer a current generated with a frequency of 1000 Hz is distributed into 8 channels at 125 Hz. It allows simultaneous monitoring of impedance change rates (dZ/dT) in six regions of the body: two legs, two hands, abdomen, neck, and ECG. For analysis of impedance changes in these peripheral regions five pairs of ECG-electrodes (one is outer current and other is inner potential) are applied to the wrists of both hands, to the ankles of lower limbs and to the temples of head. To assess dZ/dT in proper leads the current and potential electrodes were automatically switched in desired combination, thereby providing required work leads. For example, to measure the impedance of the right hand a current circuit was created where current electrodes are located on the right wrist and the right temple of head, and the potential electrodes are located on the right arm and right leg. Thus, a common part for both circuits is the right hand. The ECG is monitored using seventh channel of the analyzer from a first lead. Cardiac stroke volume was measured by the standard impedance cardiographic method [1]. For calculation of blood volume (in ml) values in neck (BVneck), abdomen (BVabdomen), arms (BVarm), legs (BVleg) and cardiac stroke volume (CSV) additional anthropometric measurements (cm) were used: circumferences of neck, trunk at the levels of the xyphoid process, groin, waist, wrist (minimum) and shoulder (maximum), ankle (minimum) and hip (maximum); the lengths from jugular notch of the neck to the temple and the xyphoid process; shoulder width; the lengths of hands from wrist to clavicle, the lengths of legs from ankle to pelvis. The left and right measurements in legs and arms were averaged. Every calculations were made as an average over 60 heart beats. Blood volumes (BV) in all regions were calculated using Kubicek's equation [1]. We used absolute values of blood volumes in cardiac and peripheral regions because the total anthropometric characteristics and statistical significance of differences of indexed BV between the groups did not change.

2.2. Protocol

All measures of blood volume reactions in response to the head up tilt test (hand-made tilt table with a footboard) were made twice at baseline state before and 10 min after aerobic exercise session. The measurements were performed after a 5 min rest period in a lying position and after 3 min standing period in an orthostatic position. Brachial arterial blood pressure was measured using the Omron 907 sphygmomanometer. After baseline measurements were completed subjects performed an exercise session on a bicycle ergometer (Kettler FX-1). The initial 25 Wt exercise load (70 rev/min) increased stepwise by 25 Wt every 1 min until the heart rate was 140 beat/min. Then operating power remained constant for 30 min. The postexercise measurements were performed in 10 minutes after end of exercise.

2.3. Statistics

In figures data are expressed as mean ±0,95 confidence interval and as M±Std Dev in the text. The analysis of variance for repeated measures was used to compare the cardiovascular reactions to orthostatic stress between controls and athletes and for determination of exercise effects on BV. Post-hoc comparisons were made using the Newman-Keuls criterion. Comparisons of blood volumes between groups in corresponding positions were made using unpaired t-test.

3. Results

Athletes had a higher working load at 140 b/s heart rate (205.4±24.7 Wt vs 117.0±18.2 Wt, $p<0.001$) and performed a higher total work (1907±344 kJ vs 979±176 kJ, $p<0.001$) compared to controls. But subjective exercise intensities as indicated mean HR (156±7 b/min in controls and 156±8 b/min in athletes, p=0.98) during exercise session were equal in both groups. As a result of a higher absolute intensity body weight decrease was higher in athletes (p<0.001). Thus during postexercise tilt-testing athletes were more dehydrated.

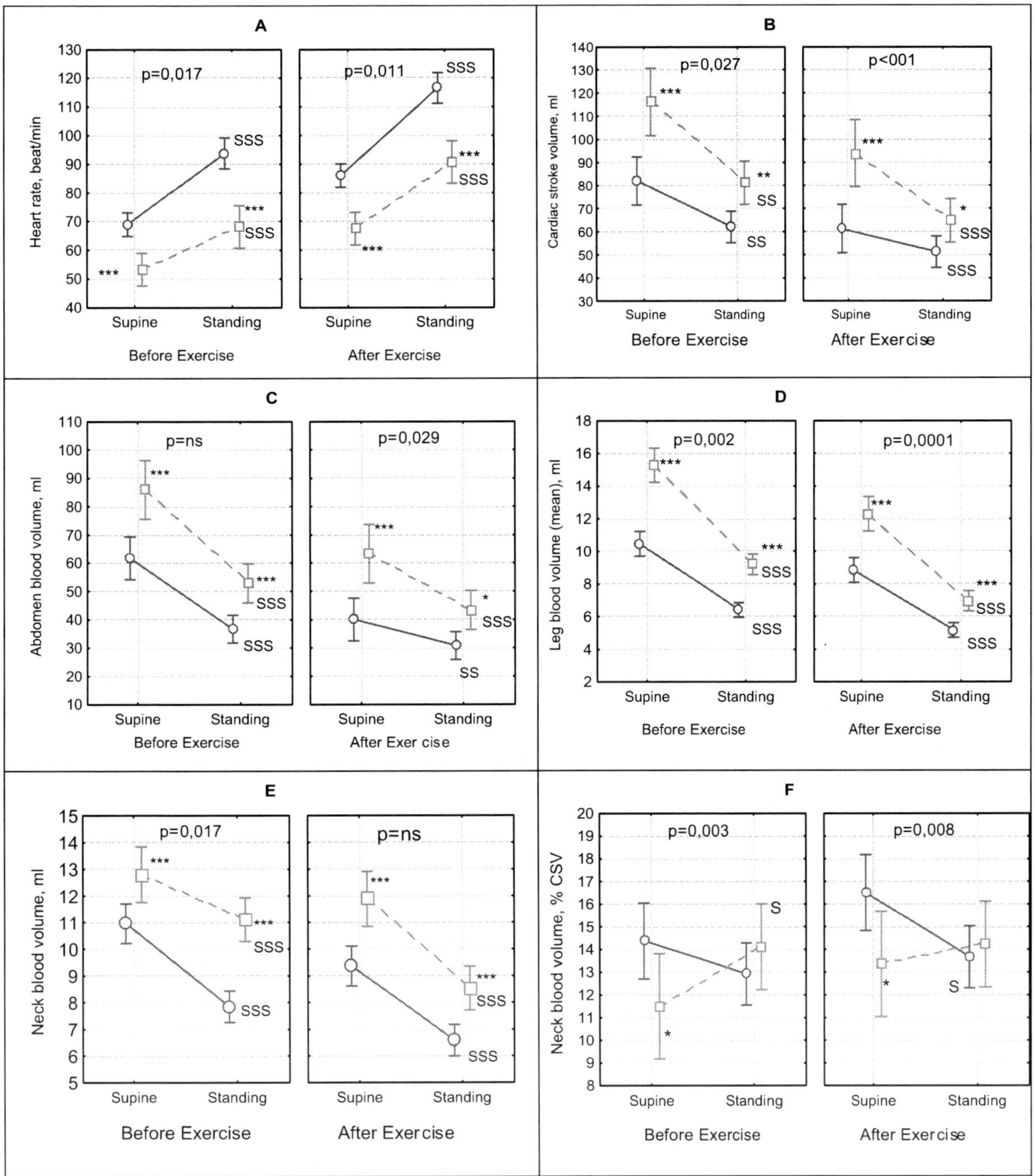

Figure 1. Blood volume reactions to orthostatic stress in groups before and after exercise. The p-values are indicated a difference in reactions to the tilt-testing between controls and athletes. *, **, *** - p<0.05; 0.01; 0.001 between athletes and controls in the corresponding positions. S, SS, SSS - p<0.05; 0.01; 0.001 compared to supine in the corresponding groups according; ns – nonsignificant Solid lines – controls, dotted line – athletes.

3.1. Cardiovascular reactions to orthostatic stress before exercise session

In the supine baseline state athletes had lower HR ($p<0.001$. Figure 1-A), higher values of CSV ($p<0.001$. Figure 1-B), BVabdomen ($p<0.001$. Fig. 1-C), BVleg ($p<0.001$. Figure 1-D), BVarm (4.08 ± 0.67 ml vs 3.06 ± 0.49 ml, $p<0.001$) and BVneck ($p<0.001$. Figure 1-E). Before exercise session HR increment in response to orthostatic stress was significantly lower in athletes (p=0,017. Figure 1-A), but falls of CSV ($p=0.027$. Figure 1-B) and cardiac output (-0.68 ± 0.75 l in athletes and $+0.30\pm1.20$ l, $p=0.015$) were significantly higher than in controls. Athletes had higher reduction of BVleg ($p=0.002$. Figure 1-D) and lower fall of BVneck ($p=0.017$. Figure 1-E). Though the BVneck expressed as a percentage of CSV was significantly lower in supine in athletes ($p<0.05$) but increased in response to tilting in athletes ($p<0.05$) and did not change in controls (figure 1-F). Reactions of mean arterial pressure were not different in the groups (results are not shown).

3.2. Cardiovascular reactions to orthostatic stress after exercise session

Ten minute after exercise session in the supine state HR was higher in both groups compared to before exercise ($p<0.001$), mean arterial pressures and cardiac outputs did not change, and the values of CSV, BVabdomen, BVleg, BVarm and BVneck were all lower (all $p<0.01$) than those before exercise session. As well as before exercise in the supine position athletes had lower HR ($p<0.001$. Figure 1-A) and higher peripheral and cardiac blood volumes ($p<0.001$. Figure 1-B,C,D,E), but cardiac output and mean arterial pressure were the same in the groups. HR rise during the tilting was lower ($p=0.011$. Figure 1-A), but reductions of cardiac stroke volume ($p<0.001$. Figure 1-B), BVabdomen ($p=0.029$. Figure 1-C) and BVleg ($p=0.0001$. Figure 1-D) were higher in athletes than in controls. Cardiac output increased in controls (from 5.16 ± 1.38 l lying to 5.99 ± 1.78 l standing, $p<0.05$), but it did not change in athletes (from 6.0 ± 0.89 l lying to 5.57 ± 0.92 l standing, $p<0.1$). Although BVneck decreased equally in both groups during the orthostatic test (p=ns. Figure 1-E), the percentage of CSV entering into the neck did not change in athletes and decreased in controls (figure 1-F). As a result the difference between these reactions of neck blood volumes was significant ($p=0.008$. Figure 1-F).

4. Conclusion.

In general, our results indicate an increased orthostatic tolerance in athletes, which is contrary to some studies [3]. The study of the blood volume distribution in the peripheral regions by our impedance method indicates an increased ability to maintain cerebral blood flow during tilt testing in athletes in baseline state. This peculiarity of blood flow regulation in athletes was determined despite the increased fall of stroke volumes that may induce cerebral hypoperfusion [3]. These results are consistent with other authors [4], who noted improved mechanisms of cerebral autoregulation in physically active adults. In fatigue state after exercise session maintenance of blood volumes in the neck vessels was deteriorated, but it remained more favourable in athletes compared with controls. Thus we can conclude that endurance trained athletes have a more effective distribution of blood volumes aimed at maintaining cerebral blood flow. This may be due to higher vasoconstriction of leg and abdomen vascular beds.

Reference

[1] Kubicek W G, Kottke J, Ramos M U, Patterson R P, Witsoe D A Labree J W, Remole W, Layman T E, Schoening H, Garamela J T. 1974 *Biomed. Eng.* **9** 410-16
[2] Porter J M, Swain I D, Shakespeare P G. 1985 *Ann. R. Coll. Surg. Engl.* **67** 169-72
[3] Raven P B, Pawelczyk J A 1993 *Med. Sci. Sports Exerc.* **25** 713–21
[4] Formes K, Zhang P, Tierney N, Schaller F, Shi X 2010 *Am. J. Physiol. Heart Circ. Physiol.* **298** H1029–H37

Journal of Physics: Conference Series **434** (2013) 012056 doi:10.1088/1742-6596/434/1/012056

Cervical cancer detection by electrical impedance in a Colombian setting

David A. Miranda[1], Sandra P. Corzo[1], C.A. González Correa[2]

1. CIMBIOS, CMN, Universidad Industrial de Santander, Carrera 27 calle 9 Bucaramanga-Colombia, South América.
2. GRUBIE (Grupo de Bio-Impedancia Eléctrica), Universidad de Caldas, Manizales-Caldas-Colombia-South América.
E - mail: dalemir@uis.edu.co

Abstract. Electrical properties of normal and neoplastic cervical tissues in a heterogeneous group of 56 Colombian women were studied by electrical impedance spectroscopy and a model based on the Generalized Effective-Medium Theory of Induced Polarization (GEMTIP). Differences between the electrical bioimpedance spectra were correlated with cellular and tissue parameters. The analysis performed by the proposed model suggest that the number of different types of cellular layers that form the biological tissue, the intracellular and extracellular conductivity could be used to explain the differences between electrical bioimpedance spectra in normal and neoplastic tissues.

1. Introduction

Cancer is a disease associated to a high morbidity and mortality that causes about 12.5% of all deaths worldwide [1]. Cervical cancer is a public health problem in Colombia and the first cause of death by cancer in women [2]. The Colombian government promotes the Papanicolau test as a screening tool for early detection of cervical cancer. However, the incidence and prevalence of cervical cancer in Colombian women are high because the quality of the diagnostic and monitoring programs is not good enough [3]. Electrical characterization of cervical tissue has been proposed as a tool to improve the sensibility and specificity of cervical cancer screening [4, 5, 6].

Electrical characterization usually is performed by electrical impedance spectroscopy (EIS), which allows the study of electrical properties of tissue at different frequencies. Experimental data have been analysed by empirical or theoretical models as Cole–Cole [7], the three-element RC [8] and physiological models [9,10].The Cole – Cole relaxation model is the most widely used for data analysis of electrical characterization techniques [5, 6, 11, 12], as it is useful to identify tissue changes and has simple mathematics behind it. However, it is not easy to quantitatively associate all Cole–Cole parameters to structural tissue properties [10].

Recently, Michael Zhdanov developed a physical – mathematical model of heterogeneous conductive media based on the Generalize Effective – Medium Theory of Induce Polarization (GEMTIP) to study the heterogeneity, multiphase structure and polarization in rocks [13]. After a rigorous mathematical work, Zhdanov [14] obtained the following equation to describe the effective resistivity of a composite model formed by a homogeneous host medium with resistivity ρ_0 and volume fraction f_0 filled with N types of spherical inclusions of radius a_l (see eq5), volume fraction of inclusion $f_l$and resistivity ρ_l:

$$\rho = \rho_0 \left\{ 1 + 3 \sum_{l=1}^{N} \left[f_l \frac{\rho_0 - \rho_l}{2\rho_0 + \rho_l + 2k_l a_l^{-1}} \right] \right\}^{-1} \tag{1}$$

Here k_l is a surface polarizability factor that usually takes the form of $k_l = \alpha_l (i\omega)^{-C_l}$. Then,

$$\rho = \rho_0 \left\{ 1 + \sum_{l=1}^{N} \mathcal{M}_l \left[1 - \frac{1}{1 + (i\omega\tau_l)^{C_l}} \right] \right\}^{-1} \tag{2}$$

Where, $\mathcal{M}_l = 3f_l \frac{\rho_0 - \rho_l}{\rho_0 + 2\rho_l}$; f_l is the volume fraction of cell of lth type; τ_l, a relaxation time and C_l, a relaxation parameter.

The Cole–Cole equation for resistivity can be derived from GEMTIP data, under the assumption that all spheres have the same ratio and the same resistivity. From this, we hypothesize that the behaviour of biological tissues can be described in an analogue way as Zhdanov did to describe rocks inclusions in an electrolytic media. We assume that cells are spheres with resistivity ρ_l immersed in the extracellular medium of resistivity, ρ_0. In this paper we analyse some experimental data using GEMTIP with above assumptions.

2. Methods and Materials

Electrical impedance spectra (EIS) of human cervical tissues were measured at seven different frequencies between 9.6 KHz and 614 KHz using the MARK-III bioimpedanciometer [4] and a tetrapolar probe. EIS were collected into 56 Colombian women. From EIS experimental spectra one spectrum per classification group was selected: normal (NO), low grade squamous intraepithelial lesion (LSIL), high-grade squamous intraepithelial lesion (HSIL) and carcinoma (CA). Abnormal tissue type was determined histopathologically. Experimental data were fitted to the GEMTIP model by mean of a genetic algorithm to obtain the parameter for each layer, l: Resistivity of extracellular medium, ρ_0 ;M_l; relaxation time, τ_l and relaxation parameter, C_l.

Intracellular resistivity ρ_l of lth layer, was calculated by:

$$\rho_l = \frac{3-\eta_l}{3+2\eta_l}\rho_0 \tag{3}$$

and

$$\eta_l = \frac{\mathcal{M}_l}{f_l} \tag{4}$$

Where, $f_l \leq 1$and $f_0 + \sum f_l = 1$ were selected according to the tissue structure.

The radius of lth cell, was calculated as:

$$a_l = \frac{2\alpha_l \tau_l^{C_l}}{\rho_0 + 2\rho_l} \tag{5}$$

Whereα_l is the surface-polarizability coefficient. The values for surface-polarizability coefficient were selected to obtain cellular radius between $10 - 23\mu m$ considering that typical eukaryotic cell sizes range from ~10 to ~100 μm [15]:

3. Results and Discussion

Figure 1 shows the real and the imaginary part of the electrical impedance spectrum for selected EIS readings from samples classified as NO, LSIL, HSIL and CA. The imaginary part was obtained by fitting from the GEMTIP model. This figure shows clear differences in the spectra due to tissue condition.

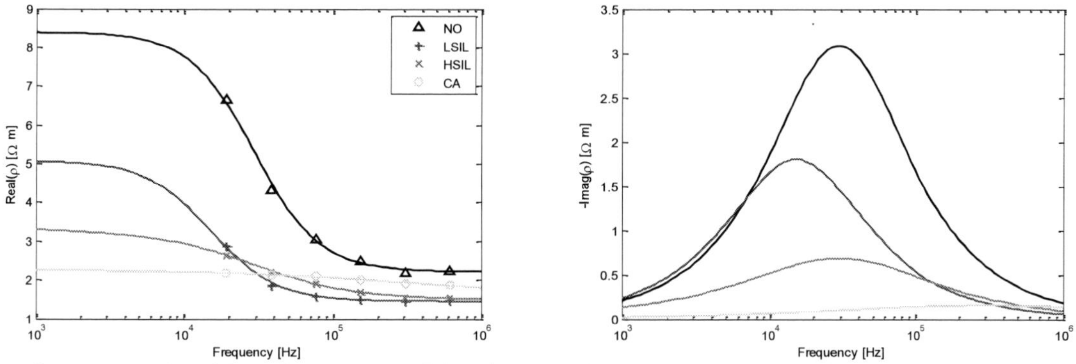

Figure1. Real and imaginary parts of the electrical impedance spectrum for selected EIS readings from samples classified as NO, LSIL, HSIL and CA.

Numerical values of the model parameters are shown in table 1. These values evidence the relations between abnormalities and GEMTIP parameters as, for instance: $\rho_0$tend to decreases as abnormality in tissue increase. The intracellular resistivity in NO tissue data shows a resistivity below to the extracellular resistivity. Otherwise, intracellular resistivities in LSIL exhibit two different situation: two cell types have a similar behavior than NO and one have a intracellular resistivity near to extracellular resistivity value. The behavior of LSIL data suggests an electrical differentiation in the intracellular compartment that, we theorize, could be associated with cancer field effect [16]. This hypothesis could explain the high difficulties to analyze experimental data of LSIL cervical tissue [17] using computational models, where a similar resistivity is assumed to the intracellular compartment of all cells. In HSIL and CA data, the differentiation in the resistivity of cells decreases related to LSIL However, HSIL data show two intracellular resistivities which, we theorize could be associated with the cancer field effect present in an advanced lesion. CA data only show one type of behavior. This could be associated with a homogeneous distribution of cellular sizes and electrical properties in the tissue.

Tissue state	$a_l[\mu m]$	$\tau_l[s]$	c_l	M_l	$\rho_l[\Omega m]$
NO	10.6	1.18E-06	0.999	0.289	0.05
ρ_0 =**8.40** $[\Omega m]$	15.7	1.40E-06	1.000	1.400	0.08
	20.9	1.50E-06	1.000	1.094	0.11
LSIL	10.8	2.63E-06	0.999	1.094	0.03
ρ_0 =**5.08**$[\Omega m]$	15.4	3.29E-06	0.999	1.400	0.01
	22.7	0.671036	0.998	0.001	4.84
HSIL	13.5	2.01E-06	0.834	1.247	0.22
$\rho_0 = 3.49\ [\Omega m]$	24.2	3.46E-05	0.269	0.119	2.33
CA $\rho_0 = 2.28[\Omega m]$	26.0	2.63E-07	0.548	0.430	1.38

Table 1. Numerical values of the model parameters.

GEMTIP model allows us to understand the meaning of the differences between the resistivity in the intra- and the extra-cellular medium as an induced polarization effect due to charge re-distribution [13,14]. This effect could be associated with a potential difference in the cell membrane. Note that the differences in the resistivity of inner and outer cellular compartment change with the neoplastic state of the tissue.

4. Conclusions
Electrical impedance spectra were collected from a heterogeneous Colombian population. In order to understand the relation between physiological parameters of tissue and electrical properties, four experimental data were selected to be analyzed in the framework of the GEMTIP model. Differences between structure and electrical properties of NO, LSIL, HSIL and CA cervical tissue were observed in the numerical values of the model parameters. We theorize that the different numerical values of the intracellular resistivity for each cell type could be associated with the cancer field effect.

5. Acknowledgements
This work was supported by the "Universidad Industrial de Santander" under VIE-5742 Grant. Authors gratefully acknowledge the support of "Liga de Lucha Contra el Cancer, Bucaramanga", "Hospital Universitario de Santander (HUS)", "Hospital de Floridablanca" and COLCIENCIAS.

References

[1] American cancer society 2102, "Cancer facts and figures 2012"*Atlanta: American Cancer Society*
[2] American cancer society "Cancer incedence, mortality and prevalence wordwide" *GLOBOCAN 2008*.
[3] Palacio-Betancourt D. et. al. 2010 "Plan Nacional para el Control del Cáncer en Colombia 2010-2019,República de Colombia, Ministerio de la Protección Social, Institutode Cancerología".
[4] Brown B. H. et. al. 2000 *Lancet***355** 892-95.
[5] Brown BH, Milnes P, Abdul S, Tidy JA 2005*BJOG***112** 802-6.
[6] D A Miranda, S P Corzo and C-A González-Correa 2012*J. PhysConf. Ser.***407** 012010
[7] Cole K and Cole R 1941 *J. chem. Phys.* **9** 341-51).
[8] Ruigang Liu. et. al. 2007*IFMBE Proceedings***14** 3964-7.
[9] D C Walker et al 2000 *Electronics letters***36** No. 19
[10] Ivorra A et all 2005 *Physiol.Meas***26** 1-9
[11] Abdul S et. al. 2005*Gynec. Oncol.* **99** S64.
[12] Grimnes S and Martinsen 2008 *Elsevier***Second edition** Chap 3,4.
[13] Zhdanov M 2008 *Geophys***73** no. 5 F197-211
[14] Zhdanov M 2009*Methods in Geochemistry***43**
[15] T Kotnik, F Bobanovic, D MMiklavcic,1997*Bioelectrochemistry and Bioenergetics***43** 285-291
[16] Slaughter 1953*Cancer***6**, Issue 5, 963–8.
[17] WalkerD*et al.* 2000*Electronics Letters* ***36***(19).

Journal of Physics: Conference Series **434** (2013) 012057 doi:10.1088/1742-6596/434/1/012057

Impedance-based detection of extracellular DNA in wounds

Anna Schröter[1], Angela Rösen-Wolff[2], Gerald Gerlach[1]

[1] Institute of Solid State Electronics, Technische Universität Dresden, 01062 Dresden, Germany

[2] Department of Paediatrics, University Hospital Carl Gustav Carus, 01307 Dresden, Germany

anna.schroeter@tu-dresden.de

Abstract. Wound infection status is a relevant diagnostic parameter to enhance wound treatment towards better healing rate. Impedance evaluation is a powerful tool to measure the inflammatory response like the released DNA of neutrophils. In our research we investigated the dielectric behaviour of neutrophils settled on electrodes in vitro. The cells have been stimulated to react in the same way as in a wound infection. The result is a significant impedance deviation of about 50 % with comparable amount of cells like in an infected wound. Microscopic fluorescence verifications acknowledge these findings.

1. Role of extracellular DNA in wound healing

Evaluation of wound healing status is an ongoing issue in clinical praxis, especially considering the treatment of chronic wounds. Wound infections are one of the most feared complications because of the threat of invading pathogens in the blood, which possibly lead to systemic inflammatory response syndrome and finally to death [1]. An infection of a wound goes by with several immune reactions. Main actors are the leukocytes, e.g. granulocytes and monocytes, which invade the wound. One of the first reactions on pathogens is the release of deoxyribonucleic acid (DNA) packaged in chromatin by neutrophil granulocytes [2]. These neutrophil extracellular traps (NETs) are able to capture some bacteria and fungi to avoid further spreading. NETs are nearly always released, when a granulocyte senses a pathogen. They consist mainly of dispersed chromatin with double-stranded (ds) DNA, histones and other antimicrobial proteins [3]. Due to the massive release of NETs to the exudate and the immediate reaction of the neutrophils, NETs are an excellent marker. A technique to detect extracellular DNA would give the physicians a proper diagnostic value to decide whether an infection is going on in a wound. This again would lead to an improvement of wound treatment.

2. Impedance measurement in wounds

The impedance of wound fluid is affected by the different constituents and their dielectric behaviour. The constituents vary with the wound status and can be divided in three categories, which seem interesting from a dielectric point of view: ionic, molecular and cell components. As the origin of the exudate is blood, in the non-infected stage the exudate composition is close to blood plasma. During an infection, which means mainly an accumulation of harmful pathogens, leukocytes are enriched in the exudate. The first arriving neutrophils immediately form NETs. An impedimetric infection sensor must be sensitive to the dielectric properties of NETs and their components, mainly the DNA with bound proteins. The leukocytes itself and the varying ion strengths during the immune response affect the measurement, too. The dielectric behaviour of dispersed DNA has been studied in various publications beginning from the 1980s [4] up to now, e.g. [5][6]. Most publications concentrate on DNA-detection principles based on hybridization, labelling, specific binding and combinations of these in lab environment. *In vivo* detection is more challenging due to the impure natural form of neutrophil derived DNA with bound proteins. Nevertheless, the basic dielectric behaviour of DNA is also of interest. To understand the dielectric behaviour it is important to know that DNA has a strong negative charge with a positive counter ion shell. The first impedance characterization experiments for DNA were done by Sakomoto et al. [4]. They found a low frequency dispersion (in the range of some Hz) of double-stranded DNA, which was later explained by Molinary et al. as a displacement of the counter ions around the polymer backbone of DNA [7]. A second dispersion between 0.5 to 70 kHz was described by Tomić et al. as a result of counter ion relaxations along a single DNA chain [8]. A third, smaller dispersion can be explained as a hopping of counter ions between DNA chains of one DNA molecule [8][9]. It occurs in the range between 0.1 MHz and 15 MHz.

As NETs do not consist of condensed DNA - ds DNA can be elongated over some hundreds of nanometres - only the interactions directly related to a single DNA stretch can be taken into account. One fibril of NETs contains several DNA chains in an uncommon conformation. Mainly the second dispersion in the medium kHz range is expected to encounter again. Before NET-release, the highly non-conductive cell membrane shields the intracellular space. Especially in the low frequency range dielectric behaviour is dominated by cell membrane effects. This again leads to the α-dispersion in the frequency range from a few mHz up to several kHz in cell suspensions [10]. This permittivity increment is caused mainly by negatively charged cell membranes and is therefore not dominant for eukaryotic cells [11]. The β-dispersion occurs up to several hundred MHz and results from capacitive effects on the cell membranes [12]. Dielectric characteristics from both cells and extracellular DNA are expected in pathologic wound exudate.

3. Sensor concept and parameter extraction

Measuring impedance with impedimetric methods has several advantages. Comparing to other methods, like optical or other electrochemical methods, this method is easy to integrate, has a low power consumption and demands only low-price and biocompatible materials. Direct electric measurement on the body with a low current and the cross-sensitivity to other body tissues also lower the signal quality. Nevertheless, a low current avoids a spreading of the electric field in the surrounding tissue and enables a more local measurement in the wound. Therefore, sensor design is a crucial issue. The integration of the wound sensor in a miniaturized chip enables a connection with a RFID structure. An antenna can easily be integrated in a flat wound cover. Modern wound care products come along with hydrogels and hydrocolloids as absorbing layers [13]. Such materials serve as a passive delivery system. Figure 1a displays how such a sensor can be integrated into the wound cover. To develop an impedimetric wound sensor, a suitable parameter of interest has to be found. Therefore, in a first step, we started with cell cultures of neutrophils in a commercial electrode-well system as shown in Figure 1b. As a model environment we used cultivated human neutrophils and stimulated them to form NETs. Using cultivated cells causes limitations with regard to reliability. However, such cells are easily available and animal tests can be avoided at this stage.

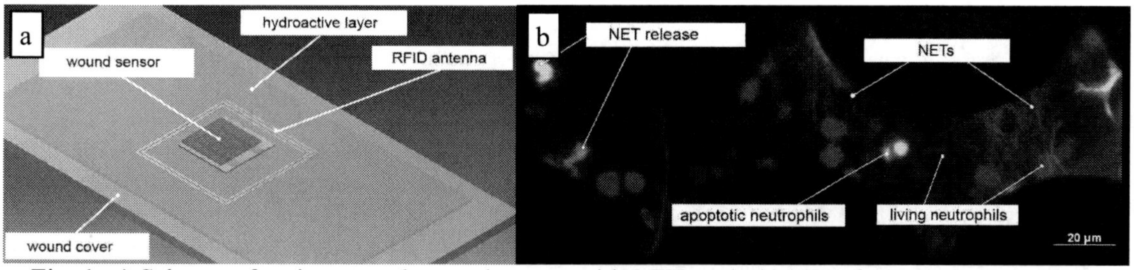

Fig. 1: a) Scheme of an integrated wound sensor with RFID unit, b) NET-formation on electrode structures (Sytox Green 1µM stains extracellular and DAPI 100µM stains also intracellular DNA).

4. Impedance characterization of a wound model

The first step to get impedance spectra from NET-forming neutrophils is the isolation of the cells from human blood. It is done by separating the cell constituents, repeated erythrocyte lysis and centrifugation steps. For the impedance characterization the electrodes (Roche xCelligence®) were coated with poly-L-lysine. For fluorescence verification and for impedance measurements, $5 \cdot 10^4$ and $2 \cdot 10^6$ cells were seeded on top, respectively. We used RPMI medium with donor serum as a cultivation medium. For fluorescence verification incubation took 1 hour, else 4 hours. Cells were then stimulated with phorbol 12-myristate 13-acetate (PMA) medium, which was exchanged. A self-made adapter connects the electrodes with an impedance analyzer (ScioSpec ISX-3). Impedance was measured in the range between 100 and 1 MHZ using an excitation amplitude of 12 mV. For a microscopic verification experiment we stained the DNA with Sytox Green fluorescent dye and took images at constant exposure time of 500 ms with a Zeiss Axio Vision Camera mounted on the microscope. The NETs were then quantified using a self-written threshold and object labelling algorithm in Matlab.

Stimulation of neutrophils with the chemical stimulant PMA leads to an immediate and drastic impedance rise in the range of 30 to 60% of the starting value as shown in Figure 3a. After changing the cell medium in the negative control culture the impedance stays nearly constant with deviations of less than 5%. The time course strongly depends on cell number as it is shown in Figure 3b. For lower cell numbers, a second maximum occurs at the very beginning of stimulation. We interpret this as a cascade effect of NET formation and the diffusion of the stimulant through the cell layer. For $5 \cdot 10^4$ cells, the cell layer roughly consists of 1 to 2 cells. For $2 \cdot 10^6$ cells the neutrophils are in a closed thick layer. Therefore, some neutrophils sense the stimulant later, which results in a continuously rising impedance. For $5 \cdot 10^4$ cells, the maximum deviation after 1 to 3 hours is more consistent with the literature value of beginning NET formation [3].

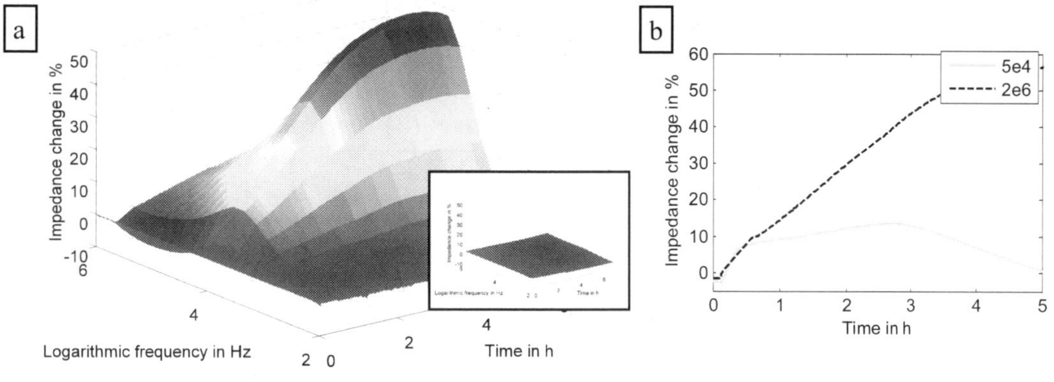

Fig. 3: a) Time course of impedance change over frequency for $2 \cdot 10^6$ stimulated neutrophils (little box: unstimulated control) and b) comparison of impedance changes at 20 kHz for $2 \cdot 10^6$ and $5 \cdot 10^4$ neutrophils.

A comparison of these results with fluorescent microscopic images indicates a dependency between the impedance change and the NET area measured in the same wells. Figure 4 displays this correlation. Strong deviations are near the maximum of fluorescence NET area and impedance change, respectively. Before stimulation (time t = 0) a NET area is already measured. Microscopic observations confirm the presence of NETs before stimulation. We explain this with the harsh preparation steps for granulocytes which result in a pre-stimulation. This explains the existence of NETs in unstimulated samples which cannot be seen in the impedance deviations. Decomposition is observed in both stimulated and unstimulated samples.

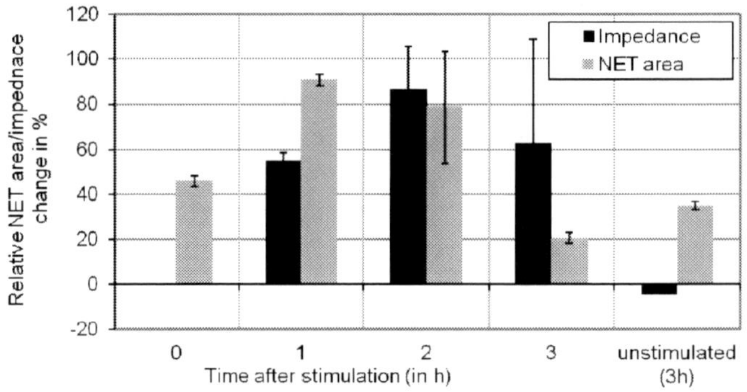

Fig. 4: Impedance change vs. fluorescent area of NETs. Base values for the relations are the maximum values (99.3 Ohms and 403,044 pixels).

5. Conclusions

NET-formation of neutrophil granulocytes resulted in a significant rise in impedance at around 20 kHz. A comparison with fluorescent evaluation indicates a significant relationship. But, this fluorescent verification has to be enhanced by a higher time resolution and an object recognition. One aim will be to understand the NET-specific dielectric mechanisms which lie behind the deviations in impedance behaviour. This can be achieved by detection of the NET formation phases of different in fluorescent images and a correlation with impedance measurements. To reach a higher specificity for the wound sensor we also concentrate our investigations on other measurands, e.g. pH, temperature and elastase.

References

[1] Gardner SE, Frantz RA and Doebbeling BN 2001 *Wound Repair Regen.* **9** (3) 178–86
[2] Martin P and Leibovich SJ 2005 *Trends in Cell Biology* **15** (11) 599–607
[3] Brinkmann V, Reichard U, Goosmann C, Fauler B, Uhlemann Y, Weiss D, Weinrauch Y and Zychlinsky A 2004 *Science* **303** (5663) pp 1532–35
[4] Sakamoto M, Fujikado T, Hayakawa R and Wada Y 1980 *Biophys. Chem.* **11** (3–4) 309–16
[5] Berdat D, Rodriguez ACM, Herrera F, Gijs MAM 2008 *Lab. Chip.* **8** (2) 302–8
[6] Lillis B, Manning M, Hurley E, Berney H, Duane R, Mathewson A and Sheehan MM 2007 *Biosens. and Bioelectron.* **22** (7) 1289–95
[7] Molinari RJ, Cole RH and Gibbs JH 1981 *Biopolymers* **20** (5) 977–90
[8] Tomić S, Babić SD, Vuletić T, Krča S, Ivanković D, Griparić L and Podgornik R 2007 *Phys. Rev.* E **75** (2) 21905
[9] Bone S and Small CA 1995 *BBA - Gene Structure and Expression* **1260** (1) 85–93
[10] Schwan HP 1957 *Adv Biol Med Phys* **5** 147–209
[11] Davey CL and Kell DB 1995 *Bioelectrochemistry of Cells and Tissues* 159–207
[12] Schwan HP 1999 *Annals of the New York Academy of Sciences* **873** (1) 1–12
[13] Lippert H (ed.) 2006 *Wundatlas. Kompendium der komplexen Wundbehandlung* (Stuttgart Thieme)

Reduction of anisotropy influence and contacting effects in in-vitro bioimpedance measurements

M. Guermazi [1-2], O. Kanoun[1], N. Derbel [2]
[1]Chair for Measurement and Sensor Technology, Technische Universität Chemnitz
Reichenhainer Straße 70, 09126 Chemnitz, Germany
[2]Control and Energy Management laboratory (CEM Lab)
National School of Engineers of Sfax (ENIS), PB W 3038 Sfax, Tunisia
E-mail: mahdi.guermazi@etit.tu-chemnitz.de; kanoun@ieee.org

Abstract. Experimental procedure is a decisive part in in-vitro bioimpedance measurement in order to get reproducible measurements. An electrode configuration is proposed to avoid several disadvantages produced by needle electrodes and circular non-penetrating electrode. The proposed electrode geometry reduces the influence of anisotropy and allows simultaneously a good probe contacting. We propose an experimental method to avoid the appearance of bacteria and to reduce water loss in meat during experiment post-mortem. The results show that electrode configuration with the developed experimental method have ensured reproducible measurements during a long period of 14 days post-mortem.

1. Introduction

An important challenge facing consumers is to obtain reliable and fast information on meat quality and freshness providing a higher safety for the consumers. In this scope, we aim to develop an in-vitro bioimpedance measurement method to get information about the state of vacuum packed meat in supermarkets. Sensitive measurements with high reproducibility represent an important objective in the development of the method. For that the electrode configuration is decisive factor, because meat has an anisotropic structure and electrode contact is very important. Different probes are used for in-vitro bioimpedance measurements such as needle electrodes and non-penetrating circular electrodes. Needle electrodes [1-4] assure a very good contact but have a big disadvantage considering reproducibility in anisotropic materials. Biological tissues, particularly meat have an anisotropic structure, so that impedance varies according to whether the current runs parallel or perpendicular to muscle fibres [5]. In 'figure 1' the Nyquist plot of impedance is shown using needle electrodes in a beef muscle in 4 directions in relation to the fibres direction: 0° (parallel), 45°, 90° (perpendicular) and 135°. The figure shows a high non reproducibility. The conductivity of meat increases in the case of measurement in parallel to the muscle fibres. In this case the current is better conducted by the fibres than in other cases where it is obliged to cross them. Non-penetrating circular electrode, used in [6], overcomes the fibers direction dependency. However it has a bad contact to the rough surface of the meat and needs a sufficient force applied to get acceptable results. 'Figure 2' illustrates the Nyquist plot of bioimpedance measurement using circular electrode for 2 different beef muscles (N 'neck' and LD 'longisimus dorsi'). The result changes according to force applied on the electrode. This effect becomes more serious if the meat surface is drying.

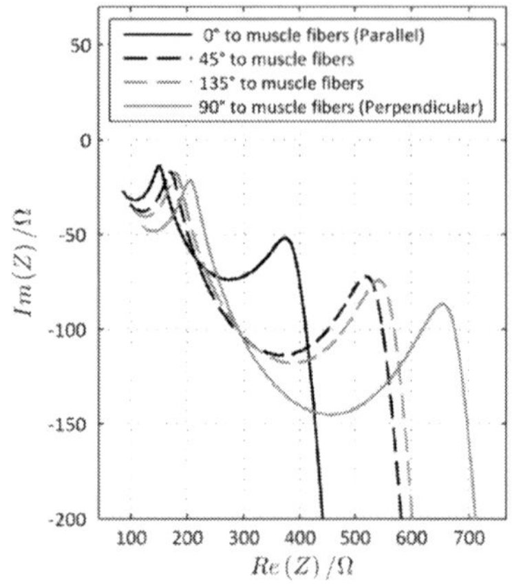

Figure 1. Nyquist plot in 4 positions to beef muscle fibres direction in the impedance plane (*f*: 40Hz - 110MHz)

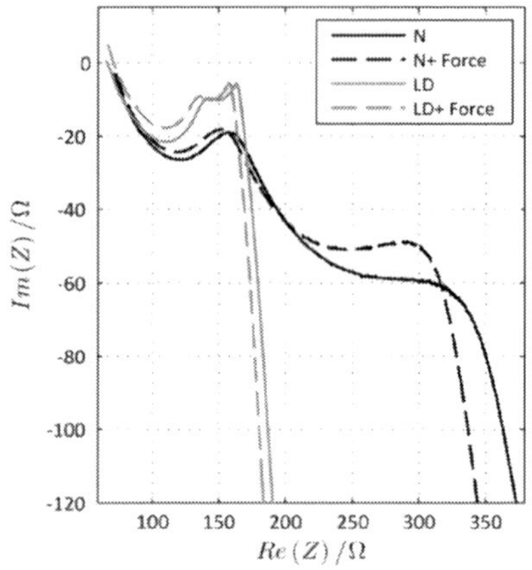

Figure 2. Nyquist plot of beef muscle N and LD with and without force using circular electrode in the impedance plane

2. Electrode configuration and experimental procedure

To reduce the dependence on anisotropy and to assure a good contact between meat and electrodes, we propose the use of a probe consisting of steel needle electrodes coated with gold positioned in a circular geometry. The excitation signal is delivered from the inner electrode to the meat and has a distance of 15 mm to the surrounded electrodes which are grounded. The electrode configuration is chosen according to a research carried out in [7] and supported by simulations using Finite Element Methods that provide information about the distribution of the electric field in the meat sample.

The meat phantom used in simulation is chosen according to the following geometric details; length 80 mm, width 80 mm, height 40 mm and to the physical details according to literature [8], [9] and [10] for the conductivity and the relative permittivity with 0.3 S/m and 100 respectively. In the presence of an electric field, the meat is subjected to a displacement of free charges (conduction current) and a movement of charges (displacement current) depending on the electrical conductivity σ and the dielectric permittivity ε_r respectively. The displacement field correspond to the following equation (1).

$$D = \varepsilon_0 \cdot E + P \qquad (1)$$

Where E is the electric field, ε_0 the vacuum permittivity (also called permittivity of free space), and P is the polarization density. The simulation 'figure 4' show that the electric displacement field D varies in the frequency range [10KHz-10MHz] between 2.87 10^{-7} to 2.1 10^{-16} and propagate in the hole volume of the meat in all possible direction for the meat only and not for the surrounding.

Simulation results 'figure 3' show that 9 electrodes with a diameter of Ø 1 mm with 10 mm depth deliver an optimal solution for a homogeneous field distribution and a good shielding to environment at a reduced number of electrodes. Vacuum capacity serves to define the electrode characteristic according the electrodes geometry following equation (2).

$$C_0 = 8 \cdot \frac{\pi \cdot \varepsilon_0 \cdot l}{\ln\left(\frac{d}{2a} + \sqrt{\frac{d^2}{4a^2} - 1}\right)} \qquad (2)$$

With d the distance between electrodes with the value 15 mm, l represents the length of electrode inserted into the meat with the value 10 mm and a is the radius of electrode with the value 0.5 mm.

Figure 3. Displacement field

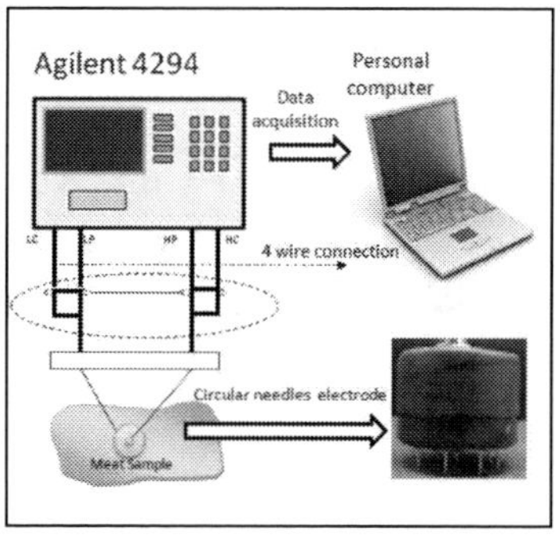

Figure 4. Measurement setup

The measurement setup used in in-vitro bio impedance measurement 'figure 4' consists of a laboratory impedance analyzer (Agilent 4294 A) connected to a personal computer for data acquisition, at a frequency ranging from 40 Hz to 110 MHz. A calibration of electrode effect is done before measurement. The meat is vacuum packed using a vacuum sealer system and was placed in climate chamber in the temperature of 2° C. The choice of temperature is similar to the meat conservation temperature in supermarket that it is around 2° C. Measurements are made simultaneously regarded to a reference time corresponding to the time of slaughter with a very low fat content during two weeks post-mortem. The vacuum meat avoids the appearance of bacteria and reduces the water loss in meat. This allows getting information about the meat for long period of time and maintains the same state of the meat as in supermarkets.

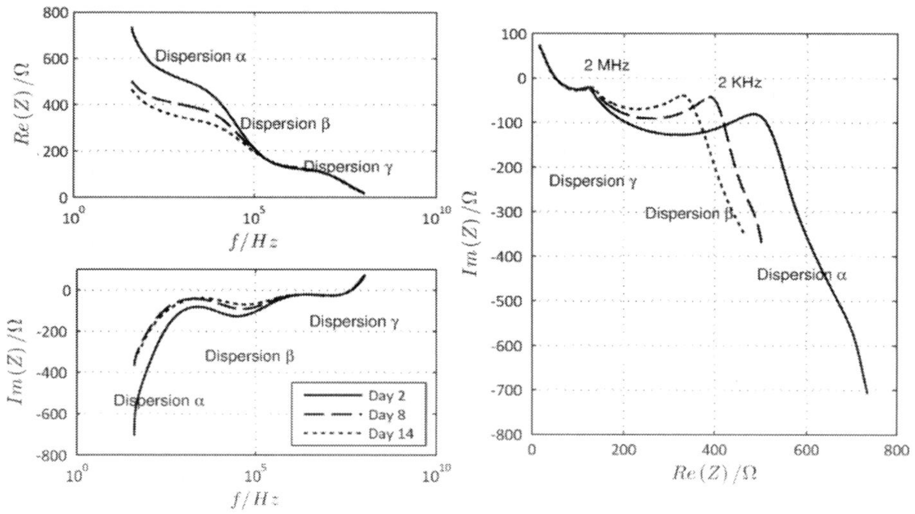

Figure 5. Muscle LD spectra in the frequency range [40Hz-110MHz]

By using the new electrode configuration and the experimental procedure, it is possible to reduce some difficulties especially bacterial contamination, bad contact between meat and electrodes and the anisotropy dependency. In addition the spectra of the meat samples guarantee a sufficient reproducibility even after two weeks. The three areas of dispersion α, β and γ described in [11-12] are clearly observed in the frequency range [40 Hz-110 MHz] (see 'figure 5'). At low frequencies in which dispersion α occurs, the meat cells are not conductive and the current flows outside the cells.

The dispersion α describes also the electrode/electrolyte interface and is therefore very sensitive to annoying effects. In the frequency range of the β dispersion, the current flows intracellular and extracellular. This allows an accurate assessment of the biological tissue. β dispersion includes information about the state of cell membrane integrity and cell aggregations during meat aging. In γ-dispersion, the dispersion observed at high frequency is mainly due to the permanent dipole relaxation of small molecules such as water. This information is not directly representative for characterization and leads to high costs for electronic development.

3. Conclusion

The theoretical and experimental investigations have proved that the proposed electrode configuration ensure a reduction of the inadequate electrode contact and the influence of anisotropy at the same time. The use of electrodes with a circular geometry allows an independency on the anisotropy so long we measure in the same plane perpendicular to the meat fibres. The use of small needles to contact the meat allows a good contact to meat and shows a good reproducibility over time. The developed measurement method allows a reproducible measurement for a long period of 14 days post-mortem.

References

[1] Guermazi M, Kanoun O, Tröltzsch U and Derbel N 2011 Assessment of beef meat aging using impedance spectroscopy *Transactions on systems, signals & devices* vol. **6** No. 4 pp 411–23.

[2] Altmann M, Pliquett U, Suess R and Von Borell E 2004 Prediction of lamb carcass composition by impedance spectroscopy *Journal of animal science* vol **82** pp 816–25.

[3] Lepetit J, Salé P, Favier R and Dalle R 2002 Electrical impedance and tenderisation in bovine meat *Meat science* vol **60** pp 51–62.

[4] Damez J, Clerjon S, Abouelkaram S and Lepetit J 2008 Beef meat electrical impedance spectroscopy and sensing for non-invasive early assessment of meat ageing *Journal of food engineering* vol **85** pp 116–22.

[5] Swatland H-J 1980 Anisotropy and postmortem changes in the electrical resistivity and capacitance of skeletal-muscle *Journal of animal science* vol **50** pp 67–74.

[6] Guermazi M, Kanoun O, Tröltzsch U and Derbel N 2012 Influence of meat aging on modified Fricke models parameter *Lecture Notes on Impedance Spectroscopy (Editor O. Kanoun) CRC-Press Taylor and Francis ISBN 978-0-415-64430-3* vol **3** pp 77–84.

[7] Guermazi M, Tröltzsch U and Kanoun O 2011 Quantitative investigation of electrode geometry for biological tissues measurements *Lecture notes on impedance spectroscopy (Editor O. Kanoun) CRC-Press Taylor and Francis ISBN 978-0-415-69838-2* pp 157–66.

[8] Andretzko J-P 2007 Contribution à l'optimisation et à la modélisation d'un banc de mesure CEM - Application à la caractérisation de l'immunité des stimulateurs Cardiaques *Phd thesis in instrumentation and microelectronic Université Henri Poincaré Nancy-I* 27.

[9] Gabriel C, Gabrieland S and Corthout E 1996 The dielectric properties of biological tissues I. Literature survey *Phys Med Biol* vol **41** pp 2231–49.

[10] Gabriel S, Lau R-W and Gabriel C 1996 The dielectric properties of biological tissues II Measurements in the frequency range 10 Hz to 20 GHz *Phys Med Biol* vol **41** pp 2251–69.

[11] Schwan H 1957 Electrical properties of tissue and cell suspensions *Advances in biological and medical physics* **5** pp 147–209.

[12] Foster K and Schwan H 1989 Dielectric properties of tissues and biological materials: a critical review *Critical reviews in biomedical engineering* vol **17** Issue I pp 25–104.

Multi-Electrode Impedance Method for Detection of Regional Ventilation

Norio Furuya[1], Katsuyuki Sakamoto[2]

[1] Dept. of Electronic Engineering, Tokyo Metropolitan College of Industrial Technology
[2] Dept. of Clinical Engineering, Nihon Institute of Medical Science
8-17-1 Minamisenjyu, Arakawa, Tokyo Japan

E-mail: nfuruya@acp.metro-cit.ac.jp

Abstract. By means of computer simulation and experiment, we investigated the feasibility of simultaneously measuring the transfer impedance changes in the right apex, left apex, right base and left base of the lungs using the multi-electrode impedance method. To obtain the transfer impedance in each region, while suppressing the effects of other regions, changing the amplitude and polarity of the applied current must localize the high sensitivity areas in the interest region. Twelve current and eight voltage electrodes were equidistantly arranged on the anterior and posterior chest walls. The amplitudes and polarities of the currents that were simultaneously applied to the current electrodes, and which provided the appropriate sensitivity distribution, were theoretically obtained. The effects of the localized sensitivity distribution were verified by comparing the simulation results of the investigated method with the results of the conventional four-electrode method. From the results of the computer simulation, we developed a multi-electrode impedance pneumography and applied it to healthy adult volunteers who were both in sitting position and in left decubitus. We found that the measurement results were physiologically reasonable.

1. Introduction

Currently, RI scintigraphy is primarily used to evaluate the regional ventilation in lung. However, it is almost impossible to continuously monitor pulmonary physiological changes for a long duration by this method. Therefore, in this study, we propose a multi-electrode impedance method [1] (MEIM) for lengthy and non-invasive monitoring the pulmonary physiological changes of patients under intensive care. To measure the regional impedance change in the lung, the sensitivity in the interest region must be high and localized. In measuring impedance by MEIM, currents are simultaneously applied to 12 current electrodes arranged on the chest walls, and the regional transfer impedance changes ΔZ are sagittally detected as differential voltage changes between pairs of voltage electrode arranged on the chest wall in the same region. The sensitivity distribution (SD) is theoretically calculated by Geselowitz's (1971) sensitivity theorem [2]. From the results of the computer simulation, we developed multi-electrode impedance pneumography and applied it to health adult volunteers. In this paper, we discuss the feasibility of this method and the superiority of MEIM, both theoretically and experimentally, for the simultaneous measurements of ΔZ in the right apex (*RA*), left apex (*LA*), right base (*RB*) and left base regions (*LB*) of the lungs.

2. Simulation studies

2.1. Theory

In order to measure ΔZ in *RA*, *LA*, *RB*, and *LB* of lung, we equidistantly arranged twelve current electrodes on the right and left upper, middle, and lower sites on the anterior and posterior chest walls as shown in Figure 1. Eight voltage electrodes were equidistantly arranged on the same level as the current electrodes on *RA*, *LA*, *RB*, and *LB* sites as shown in Figure 1. Because the currents simultaneously applied to the current electrodes arranged on the anterior and posterior, the currents flow sagittally as shown by the red arrows in Figure 1.

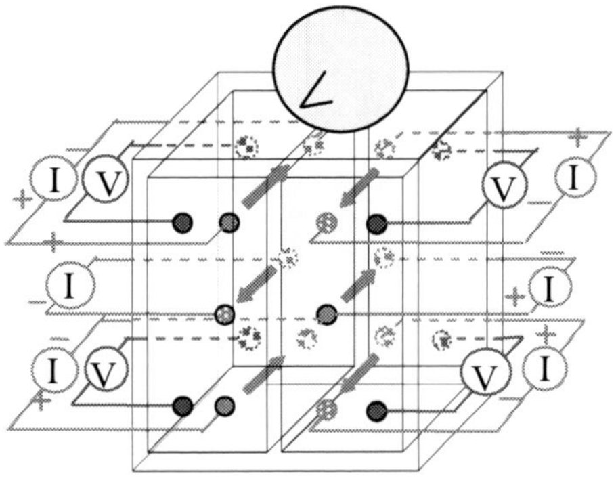

Figure 1. Electrode arrangement
Ⓘ: constant current source
Ⓥ: voltage detector

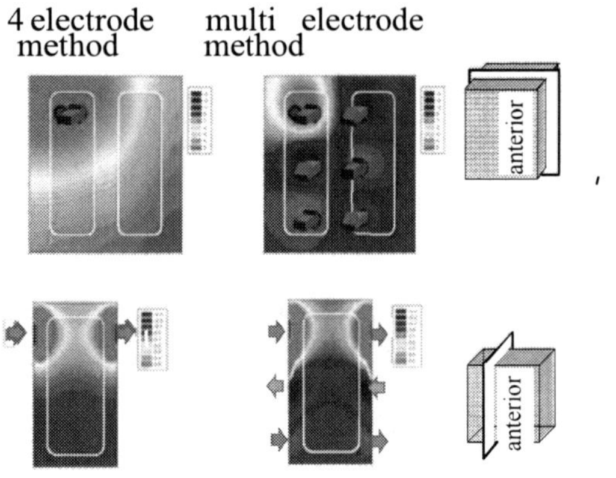

(a) (b) (c)
Figure 2. Sensitivity distribution

The red and blue circles in the figure are the current and voltage electrodes, respectively. The amplitudes of all currents are identical. Because the polarity of the adjacent current electrode pair is opposite to that of the current supplied to the pair at the interest region, as shown in Figure 1, a zero sensitivity region is produced in the boundary between the right and left lungs, and consequently, the high sensitivity area in the region of interest (RI) is localized. Two pair of current electrodes arranged on the right and left middle sites of the chest wall are used to isolate the apex and base. We can simultaneously obtain ΔZ in right apex, left apex, right base and left base of the lung by measuring the voltage difference sagittally as shown in Figure 1. We assume that the spatial conductivity changes of the four regions are $\Delta \boldsymbol{g}_{RA}$, $\Delta \boldsymbol{g}_{LA}$, $\Delta \boldsymbol{g}_{RB}$, and $\Delta \boldsymbol{g}_{LB}$, respectively. The ΔZ at *RA* is theoretically given by the following equation (1). $\boldsymbol{L}_I$ is the electric field that is generated by the unit current supplied to all the current electrode pairs and is referred to as the lead vector. The lead vector $\boldsymbol{L}_V$ is the electric field that is generated by the unit current supplied to the voltage electrode pair in RI. Therefore, the dot product of $\boldsymbol{L}_I$ and $\boldsymbol{L}_V$ gives the sensitivity. $\boldsymbol{L}_{I\text{-}RA}$, $\boldsymbol{L}_{I\text{-}RB}$, $\boldsymbol{L}_{I\text{--}LA}$, and $\boldsymbol{L}_{I\text{-}LB}$ are the lead vectors in *RA*, *RB*, *LA* and *LB*, respectively. $\boldsymbol{L}_{V\text{-}RA\text{-}RA}$, $\boldsymbol{L}_{V\text{-}RA\text{-}RB}$, $\boldsymbol{L}_{V\text{-}RA\text{-}LA}$ and $\boldsymbol{L}_{V\text{-}RA\text{-}LB}$ are the lead vectors at *RA, RB, LA* and *LB*, respectively, produced by the current supplied to the voltage electrode pair at the *RA*. From the superposition theorem, ΔZ in the other regions can be expressed by equations similar to equation (1).

$$\begin{aligned}\Delta Z_{RA} = &\int_{V_{RA}} -\Delta g_{RA} L_{I-RA} \cdot L_{V-RA-RA} dv + \int_{V_{RB}} -\Delta g_{RB} L_{I-RB} \cdot L_{V-RA-RB} dv \\ &+ \int_{V_{LA}} -\Delta g_{LA} L_{I-LA} \cdot L_{V-RA-LA} dv + \int_{V_{LB}} -\Delta g_{LB} L_{I-LB} \cdot L_{V-RA-LB} dv\end{aligned} \quad (1)$$

2.2. Simulation model and results

To obtain SD by MEIM, simulation studies are carried out with the cuboid-model, the conductivity $\boldsymbol{g}$ of which is uniform. The cuboid-model consists of 729 points and 512 cubic elements. SD is calculated by the 3-D finite element method. Figure 2- (b) shows one of the calculated SD for the measurements of ΔZ in RA of the lung. Arranging two pairs on the right and left apex and base can produce the zero sensitivity regions (ZS) in the boundaries between the right and left apex and base of the lungs. Two pair of current electrodes arranged on the right and left middle sites of the chest wall produces ZS between apex and base lungs. For comparison, SD obtained by the conventional four electrode method is shown in Figure 2-(a). The upper and lower diagrams of Figure 2-(b) respectively show SD on the cross section through the center of the anterior and posterior chest walls and on the sagittal plane over the voltage electrode pair arranged on the anterior and posterior chest wall as depicted in Figure 2-(c). Table 1 shows the rate of the influence of $\Delta \mathbf{g}$ in each of the four regions to $\Delta Z_{RA,}$ when the influence of $\Delta \mathbf{g}$ in the RA region is 100%. For comparison, the table shows the corresponding rates to ΔZ_{RA} measured by the conventional four electrode method. The two results indicate that the regional conductivity changes in RI can be effectively detected. From the geometrically symmetry of electrode arrangement, the three other regions have the same sensitivity distribution characteristics as previously described.

Table 1. The per cent rate of the influence of Δg in each of the four regions to ΔZ_{RA}

Region	Four electrode method	Multi-electrode method
Right apex	100%	100%
Right base	12.5%	2.6%
Left apex	17.8%	-9.7%
Left base	5.1%	-1.3%

3. Experiment

3.1. Hardware design

The MEIM system consists of 12 floating-type constant-current sources (six positive and six negative) outputting 37μA at 10kHz and four voltage detectors (differential amplifiers). All the current electrodes are connected to the floating-type constant-current sources while the voltage pickup electrodes are connected to the differential amplifiers, SN ratios of which are greater than 75dB.

3.2. Electrode arrangement

Tidal volume measurements by the MEIM system were conducted on 12 healthy volunteer students between 18 and 22 years old, with their informed consent. For monitoring, the tidal volume was simultaneously measured by the spirogram. The electrode arrangement on the human body shown in Figure 3 is the same as that of the simulation model. Furthermore, as in the simulation, ΔZ was simultaneously obtained for the four regions in the sitting position and left decubitus.

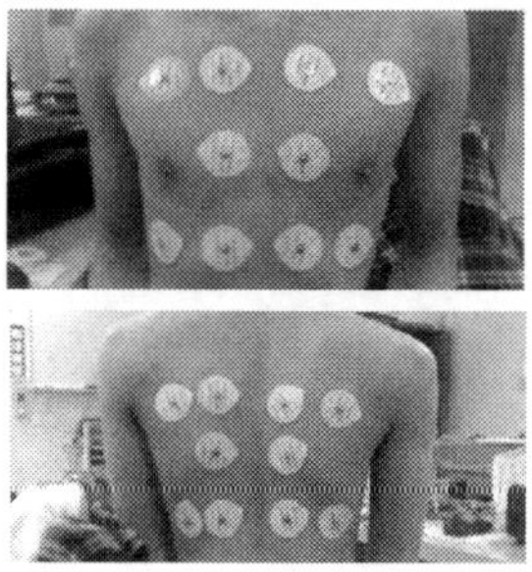

Figure 3. Multi electrode arrangement

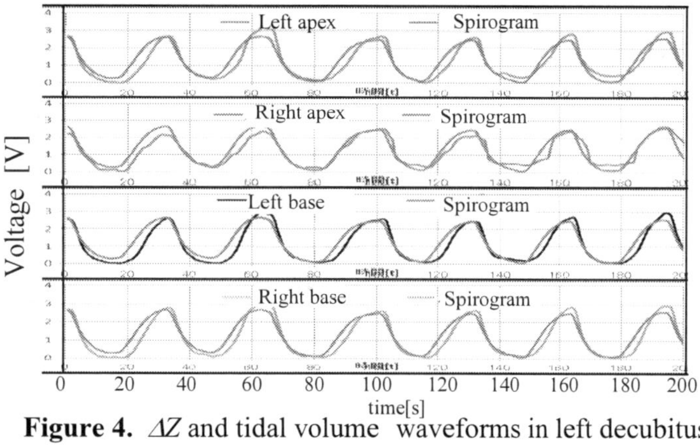

Figure 4. ΔZ and tidal volume waveforms in left decubitus

3.3 Experimental results

Figure 4 and Table 2, 3 and 4 show one of the experimental results. Figure 4 shows the relationship between the tidal volume and the ΔZ waveform in the left decubitus. The waveforms from top to bottom in figure are the simultaneous measurement results at *LA*, *RA*, *LB*, and *RB*. The amplitude of ΔZ is an indicator of the regional lung activity; in fact, it is proportional to the ventilation volume in each region. The ΔZ waveforms in each region correlate well with the tidal volume waveform. Table 2 shows the correlation factors of the tidal volume and ΔZ in each region. The factors are over 0.95% in all four regions. The results indicate that ΔZ waveform reflects the variation in the ventilation volume of each region. We define the ratio of the value of ΔZ in other regions and the value of ΔZ in right base

Table 2. Correlation between tidal volume and ΔZ in each region

	Left apex	Right apex	Left base	Right base
Correlation factor	0.95	0.96	0.97	0.98

as the activity contribution rate to the regional ventilation volume (AC). Table 3 shows AC of each lung in each posture. The amplitudes of ΔZ waveforms at the base are slightly greater than those at the apex in the sitting position. However, there is no difference between AC of the left and right lungs at both the apex and the base. In the left decubitus, AC by the apex of the left lung is significantly greater than that by the apex of the right lung. There is no difference between AC by left and right base.

Table 3. Activity contribution rate of each lung in each posture for tidal volume

Posture	Left apex	Right apex	Left base	Right base
Sitting position	0.93	0.90	1.02	1.00
Left decubitus	1.14	0.92	1.01	1.00

Table 4 shows the ratios of the total value of ΔZ in left apex and base and the total value of ΔZ in right apex and base, to the tidal volume (i.e., AC). In the sitting position, there is no difference between the contributions in the apex and base of the lungs. It is clear that AC of the left apex of the lungs is greater than that of the right apex in the left decubitus.

Table 4. Activity contribution rate of left lung referred to that of right lung in each posture

Posture	Apex	Base
Sitting position	1.00	1.02
Left decubitus	1.24	1.01

4. Discussion

In order to effectively measure the regional transfer impedance, high sensitivity region must be localized by the appropriate SD. From the computer simulation results, it has been found that the proposed MEIM system can be used to obtain reasonably localized high sensitivity area in RI and to carry out simultaneous and independent measurements of regional ΔZ. The results of simultaneous sagittal measurements of ΔZ in both the sitting position and left decubitus were found to be physiologically reasonable. Thus, from the theoretical and experimental study, it is clear that the proposed MEIM can be used to simultaneously and independently detect the physiological changes in a particular region of the lungs

References

[1] Furuya N, Nakayama K 1986 *MBEC* 24, 62-70
[2] Geselowitz D B 1971 *IEEE Trans.* BME-18 38-41

Design and Development of a Portable WiFi enabled BIA device

D Križaj[1], M Baloh[2], R Brajkovič, T Žagar[3]

[1]University of Ljubljana, Faculty of Electrical Engineering, Tržaška cesta 25, 1000 Ljubljana, Slovenia
[2]University of Ljubljana, Faculty of Computer and Information Science, Tržaška cesta 25, 1000 Ljubljana, Slovenia
[3]Geodetic Institute of Slovenia, Jamova cesta 2, 1000 Ljubljana, Slovenia

E-mail: dejan.krizaj@fe.uni-lj.si

Abstract. A bioimpedance device (BIA) for evaluation of sarcopenia – age related muscle mass loss – is designed, developed and evaluated. The requirements were based on lightweight design, flexible and user enabled incorporation of measurement protocols and WiFi protocol for remote device control, full internet integration and fast development and usage of measurement protocols. The current design is based on usage of a microcontroller with integrated AD/DA converters. The prototype system was assembled and the operation and connectivity to different handheld devices and laptop computers was successfully tested. The designed BIA device can be accessed using TCP sockets and once the connection is established the data transfer runs successfully at the specified speed. The accuracy of currently developed prototype is about 5% for the impedance modulus and 5 deg. for the phase for the frequencies below 20 kHz with an unfiltered excitation signal and no additional amplifiers employed.

1. Introduction

One of the first practical uses of the bioimpedance measurement technique (BIA) was evaluation of the content of water in the human body [1], [2]. The basic underlying principle for estimation of water content is based on a fact that water is much better conductor of electric current than the neighboring tissues. An improvement of the technology was achieved by using four electrode measurements with surface electrodes [3]. Usually, for BIA measurements of the whole body mass, two current injecting electrodes are placed on the hand and foot and close to them are placed two voltage electrodes. From measured bioimpedance data several parameters are determined such as index of fat free mass (FFM), total body water (TBW), extracellular and intracellular water content (ECW and ICW) etc. (see for instance [4]).

Our aim was to develop a simple and yet sufficiently accurate device that would be particularly adapted for BIA use for determination of TBW in elderly persons that are hospitalized and more or less immobilized. There is a serious concern in evaluation of loss of muscle mass due to prolonged immobility. The term used for age related loss of muscle mass is sarcopenia [5]. This loss of muscle mass with aging is hypothesized to have negative consequences for health and physical functioning in old age. Several models for evaluation of loss of muscle mass using BIA are known from the literature

[6]. As a result of our investigation of the state of the art technologies, devices on the market and the requirements of the end users (medical doctors) we decided to base our BIA system on the following requirements:

- the device should be low weight and portable, battery powered
- use of low cost technology based on microcontroller with possible integrated AD/DA converters
- the measurement based on digital lock-in technology
- measured data should be transferred by WiFi to handheld devices or laptops
- models for evaluation of sarcopenia using BIA should be modified by the user or even self-declared
- the transfer protocol should enable creation of an internet enabled database

In the following we present current state of design and development and some issues indicating further research and development.

2. Hardware

The presented device design is based on physical separation of the measurement part and the data analysis part. The measurement part (Figure 1) is based on a commercial 32-bit microcontroller with embedded A/D and D/A converter. D/A converter operates at 1 MHz with resolution of 10 bits and is used for sine wave generation. The generated sine wave is low pass filtered and applied to the current electrodes. The resulting voltage drop is picked by an instrumentation amplifier with a high CMRR and the resulting current is led to an operational amplifier operating as a current to voltage converter. Both signals are digitized by microcontroller's A/D converter with resolution of 12 bits and frequency of approximately 90 kHz. Undersampling was used for the frequencies above 45 kHz. The signal generation and the data acquisition are performed simultaneously using timer interrupts for the signal generation and direct memory access for the A/D conversion. Communication between the computer and the measuring device is established through an Ad-Hoc WiFi connection. Once the voltage and current wave forms are acquired they are sent to the computer for analysis and presentation.

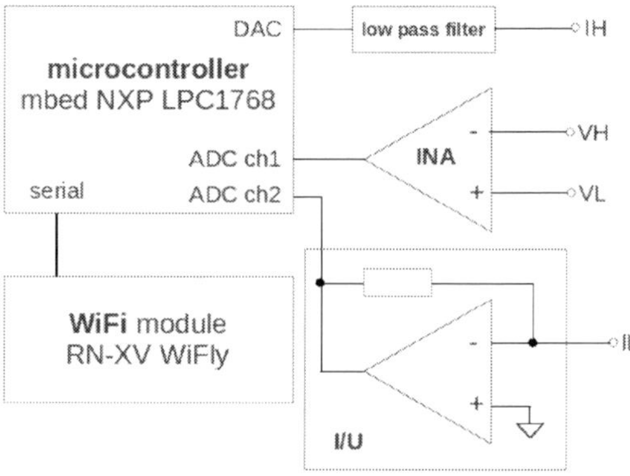

Figure 1. Hardware building blocks.

The bioimpedance measurement is based on a digital lock-in method. We decided not to perform data analysis on the microcontroller but rather on a remote computer that can be either a notebook, a smart phone or a tablet, because these devices are already common equipment in every laboratory or institution and very powerful portable computers can be obtained at low cost. Another advantage is that we are able to modify or improve the algorithms according to users' requirements and wishes without reprogramming the microcontroller.

3. Software

Communication between the device and the computer is performed through a telnet connection. The application listens on an open port and connects to the device, once the device requests a connection. The communication protocol was tailored to the device. It uses a 4 byte instruction word. Each 4 byte instruction is made of two sections: a 2 byte instruction identifier, followed by 2 bytes of data. The required sequence of events for a completed measurement is also defined by the protocol. The device announces the start of a measurement with the MEASURE instruction. After this, the device sends one-time parameter information. These include the parameters the device is using in the current measurement. After all the one-time parameters are sent, a DTST (Data start) instruction is sent by the device, which signals the end of the parameter gathering phase. In the next phase, all received data is

interpreted as measurement data, and no longer as instruction-data combinations. Telnet ensures packets are received reliably and in their proper order, so indexing of the data is not required during transport. Each data value is an unsigned 2 byte integer. The allowed values are from 1 to FFFF(hex). This phase continues until the STOP instruction is received, which is a 2 byte value 0. A query-response style of communication also exists in the opposite direction. The application can query the device for configuration values, without announcing itself. It sends a 4 byte instruction to the device and receives a 4 byte instruction-data combination as a reply.
The software is written in Java programming language so it is easily portable to different operation systems including modern handheld devices.

4. Usage
Usage of the device is similar to usage of other BIA devices with the difference that all the settings and measurements are set remotely on a laptop or a handheld device (Figure 2). The BIA device is only switched on or off, performs the measurements in accordance to the received commands using the communication protocol described in Chapter 3 and reports (sends) the measured data to the remote device for further evaluation. The designed BIA device can be accessed using TCP sockets over a wireless Ad-Hoc connection to the RN-XV WiFly module which is connected to the microcontroller via hardware UART interface and operates at a baud rate of 115200 bps. Tests performed using experimental device showed about one second is needed to establish the connection between a computer or a handheld device and the WiFly module when connecting for the first time. Once the connection is established there are no further delays and the data transfer runs successfully at the specified speed (115200 bps). This approach brings several advantages to the BIA measurements and evaluation such as improved portability, decreased cost, flexibility in software design and processing, direct internet connectivity, etc.

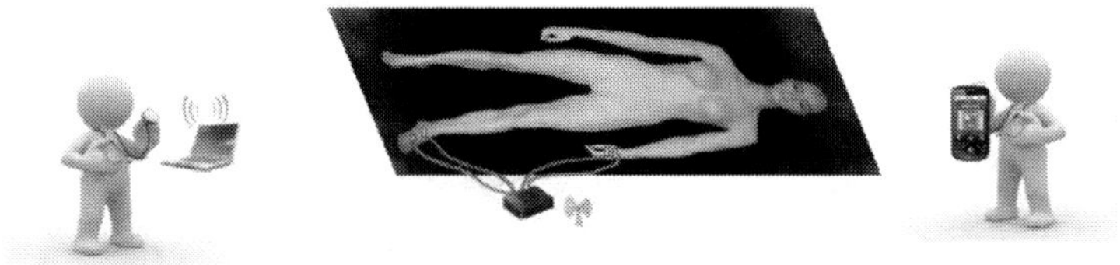

Figure 2. Schematic representation of a developed system for BIA analysis based on WiFi connectivity.

5. Test results
Preliminary measurements using a simple UI method (Figure 3) were performed to estimate the accuracy of the device when using only the microcontrollers embedded ICs. The reference resistor (R) of approximately the same value as the expected impedance under test (IUT) modulus was connected in series between IUT and the ground. The voltage drop on the reference resistor and the IUT (the applied voltage) and the voltage drop only on the reference resistor was measured.

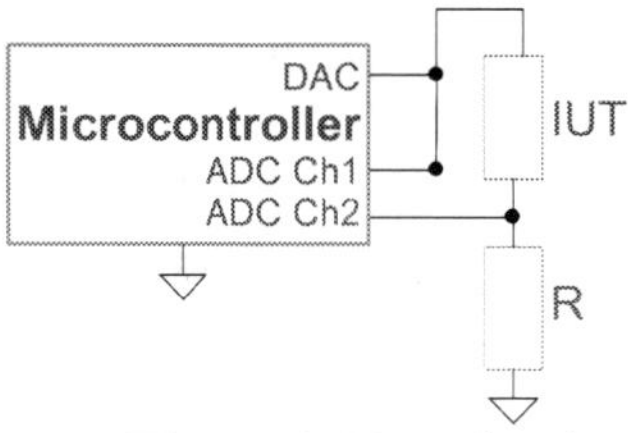

Figure 3. Test circuit

The test was performed on the three different RC circuits (Figure 4) with nominal values of 33 Ω for R1 and 100 nF for C1. The value of R2 was 10 Ω, 100 Ω and 1 kΩ for each test case (I, II and III), respectively.

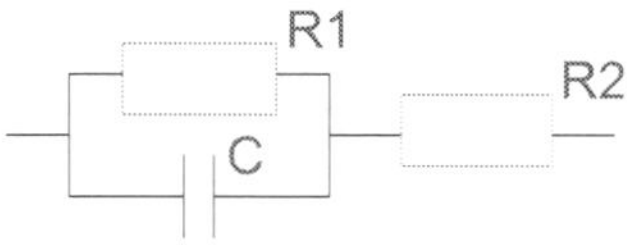

Figure 4. The impedance under test

The measurements were performed at 6 different frequencies in the range from 1 kHz to 40 kHz. The upper frequency limit was set by the sampling frequency of the microcontroller as for higher frequencies undersampling was required but for the test circuit without a low pass filter the results were erroneous and not comparable to the low frequency range. The measured values were compared to the reference values measured by a Quadtech 1920 LCR meter. The results of the test are summarized in Table 1 and Table 2.

Table 1. Mean modulus and phase measurement error (calculated from the mean of 10 repeated measurements)

	Relative modulus error [%]			Absolute phase error [°]		
Frequency	I	II	III	I	II	III
1 kHz	-1.66	0.00	-0.61	-0.0	0.1	-0.2□
2 kHz	1.27	-0.15	0.27	-2.0	0.1	-0.0□
5 kHz	0.28	-0.33	-0.57	1.8	0.7	0.3□
10 kHz	2.43	2.10	0.48	6.0	1.0	0.6□
20 kHz	0.11	0.48	0.80	9.9	2.1	-0.1□
40 kHz	9.36	2.16	-9.08	14.1	3.2	18.7

Table 2. Standard deviation of modulus and phase measurement error (calculated from 10 repeated measurements)

	Relative modulus error [%]			Absolute phase error [°]		
Frequency	I	II	III	I	II	III
1 kHz	5.59	0.78	1.62	3.3	0.5	0.7□
2 kHz	2.29	1.10	1.04	5.8	1.2	0.5□
5 kHz	1.28	1.31	1.33	1.5	0.4	0.6□
10 kHz	2.76	1.55	2.04	3.7	0.9	0.8□
20 kHz	6.95	1.93	2.28	2.3	0.6	0.5□
40 kHz	5.06	4.04	1.03	2.1	1.7	0.8

6. Conclusion

Design, development and evaluation of a bioimpedance measurement device (BIA) to be used in particular for estimation of sarcopenia - loss of muscle mass in elderly population – is presented. The concept is based on use of low cost technology based on a microcontroller with integrated AD/DA converters and WiFi protocol to transfer measured data to a remote computer for data analysis and evaluation. The remote device (can also be a handheld smart phone) is used to set all necessary anthropometric data, select or even design users own evaluation protocol (model) and trigger the measurements. The connection between a laptop or a handheld device and the BIA device using WiFi technology is established in less than one second and after that operates practically without delays (at baud rate of 115200 bps). With current design it is possible to develop a BIA device with accuracy in the frequency range below 20 kHz of approximately 5 % for the impedance modulus and 5° for the impedance phase using only the microcontroller's embedded ICs. The frequency range and the accuracy could be extended with appropriate low pass filtering and undersampling methods. Further improvements in accuracy could be achieved by usage of external fast and more accurate AD/DA converters.

References

[1] Thomasset A 1963 *Lyon Med.* **209** 1325–52
[2] Thomasset A 1962 *Lyon Med.* **207** 107–18
[3] Hoffer EC et al 1969 *J. Appl. Physiol.* **27** 531–4
[4] Kyle U G, et al 2004 *Clin. Nutr.* **23** 1226–43
[5] Visser M 2009 *J. Nutr Health Aging* **13** 713–6
[6] Tengvall M 2009 *Clin. Nutr.* **28** 52–8

Bioelectrical impedance analysis (BIA) for sarcopenic obesity (SO) diagnosis in young female subjects

C H Gonzalez-Correa[1,2], **J C Caicedo-Eraso**[1,2,3], **J S Villada-Gomez**[1,2]
[1]Research Group on Nutrition and Body Composition, University of Caldas, Colombia
[2]Research Group on Electrical Bioimpedance, University of Caldas, Colombia
[3] Department of Systems and Informatics, University of Caldas, Colombia

E-mail: clara.gonzalez@ucaldas.edu.co

Abstract. Sarcopenia is defined as a loss of muscle mass depending of ageing and affecting physical function (definition A). A new definition considers excluding mass reduction criterion (definition B). Obesity is pandemic and occurs at all ages. Sarcopenic obesity (SO) implies both processes. The purpose of this study was to compare the results obtained after applying these two definitions to 66 overweight or obese college women aged 22 ± 2.8 years. Percentage body fat (%BF) and skeletal mass index (SMI) were estimated by BIA, muscle function by handgrip strength test (HGS) and physical performance by Harvard step test (HST). There were 9.1% overweight and 90.9% obese subjects. Twenty nine subjects (43.9%) had decreased HGS and 22 (33.3%) had impaired physical performance. One obese subject (1.5%) met the criteria for sarcopenic obesity by definition A and 9 (13.6%) by definition B. Although a linear regression ($\alpha<0.05$) showed a very weak association between these variables (r^2 =0.094, 0.037 and 0.275 respectively) it was observed a tendency for HGS, HST and SMI deterioration when %BF increases. However, other confounding factors must be investigated. Probably as the population gets more obese, the problematic of SO will be found earlier in life.

1. Introduction

Traditionally, sarcopenia has been defined as a set of signs and symptoms characterized by a progressive loss of all skeletal muscle groups associated to aging that can lead to physical disability and death (definition A) [1]. To make the diagnosis is required to have a reduced skeletal muscle plus low muscle strength or weak physical performance [2]. Nowadays, there is a proposal for changing this definition as there is growing evidence that muscle mass is not entirely related to muscle strength and that this variable is a better marker of functional limitation than muscle mass reduction. Thus, sarcopenia should be diagnosed by functional and physical limitation only (definition B) [3].

Obesity is diagnosed with a BMI greater than 30 kg/m^2 [3] or, more precisely, with a fat mass percentage greater than 37% in women aged 18 to 79 [4]. Obesity is pandemic [3], occurs at all ages and can affect muscle mass function [5, 6]. The global prevalence of overweight is 1 in 4 adults and it will be about 1 in 2 by 2015 [8]. Obesity generates muscle dysfunction due both, fat infiltration in muscle cells [9] and reduction of muscle protein synthesis [5, 10]. Sarcopenic obesity implies both processes: sarcopenia and obesity [7].

The use of Bioelectrical impedance analysis (BIA) to assess skeletal muscle mass, is relatively new but its use is increasing due to the results obtained by this method are comparable to those yielded by dual energy x-ray absorptiometry (DEXA) [11]. BIA results allow to estimate fat free mass (FFM) and from it calculate the skeletal muscle mass index (SMI). The hand dynamometry has been used to evaluate muscle strength of individuals. It has correlated well with malnutrition by both, excess or deficit of energy. The method, easy of access and operation is very useful in clinical and population studies [12]. The estimation of physical performance by the Harvard step test (HST) is appropriate to examine the cardio respiratory reserve [13], one important conditional skill. The test is useful in population studies and easily applied in the laboratory. We hypothesized that, given the high percentage of overweight and obesity in the world, sarcopenic obesity is could start at earlier stages of life. If this is true, preventive measures could be taken to retard or prevent muscle wasting in later. Thus, this study aimed to compare the results obtained after applying these 2 definitions for sarcopenia in a group of young college women having different levels of adiposity.

2. Materials and methods

2.1. Subjects

The methods, classified as minimum risk, were approved by the Bioethics Committee of the University of Caldas. The purpose and procedures of the study were explained to a sample of 66 young college females. Inclusion criteria were: being female, aged 18 to 24 years and with no comorbidities. Exclusion criteria were: being a smoker, have metal implants or pacemakers, be pregnant or be using diuretics. Subjects who had undergone surgery for weight reduction or silicone breasts implants were also excluded.

2.2. Data acquisition

Measurements were performed early in the morning. Relative humidity and environment temperature were controlled with an electric heater (BFH416 by Bionaire™) and a dehumidifier (BMD100 by Bionaire™). Relative humidity and environmental temperature were measured with a thermo-hygrometer (13307 by DeltaTrak®, ±0.1 ºC). Volunteers were asked to comply with standardized requirements before the test and all volunteers were evaluated out of menses.

2.3. Anthropometric measurements

Height (Heightronic-235 by Seca®, ±0.01 cm) and weight (W) (PP2000 by Icob-Detecto®, ±0.1 kg) were measured twice. A third measurement was taken if there was a difference greater than 0.5 cm or 0.1 kg respectively [14].

2.4. BIA measurements

Multi frequency Bioelectrical impedance was measured on the dominant side of body for three times on a nonconductive surface (Hydra 4200 by Xitron Technologies©). FFM was obtained with internal equation of the bioimpedance analyzer and %BF-BIA was calculated with %BF-BIA=((W-FFM)/W)*100%. Estimation of skeletal muscle mass was made by using the data at 50 kHz with the equation developed by Jansen et al, 2000 [15] then; the skeletal muscle index (SMI) was calculated [2]. The classification by %BF was made using cutoffs by Gallagher et al, [4].

2.5. Handgrip strength

The handgrip (HG) was measured 3 times using a hydraulic dynamometer Baseline® (± 1 kg) (Fabrication Enterprises Inc. USA) and following the guidelines of the American Association of Hand Therapists [16].

2.6. Harvard step test

HST was performed by using a metronome Yamaha YM-2000 © to keep pace and three Casio HS3 © stopwatches to measure time [13]. The bench height was 35 cm for women. Results were expressed by an index.

2.7. Statistical methods
Characteristics of the subjects and laboratory conditions were expressed as mean and standard deviation. Subjects with decreased SMI, HGS and HST according %BF were expressed as percentage. A linear regression (α=0.05) was used to evaluate the association between body fat and the 3 parameters.

3. Results
Environmental conditions were kept stable: temperature was 19.5 ºC (±0.5), relative humidity was 72.4 % (±1.1). Table 1. shows subject's characteristics. SMI, HGS, HST and %BF reference values were extracted from previous studies [2, 4, 13, 19]. Table 2 shows the number and percentage of subjects with values of SMI, HGS and HST out of reference values, according to their %BF.

One obese subject (1.5%) met the criteria for sarcopenic obesity by definition A and 9 (13.6%) by definition B. Although a linear regression (α <0.05) showed a very weak association between these variables (r^2 =0.094, 0.037 and 0.275 respectively) it was observed a tendency for HGS, HST and SMI deterioration when %BF increases (Figure 1).

Table 1. Subject characteristics

Variables	Mean	SD
Age (years)	22.0	2.8
Height (cm)	156.7	5.3
Weight (kg)	55.0	8.3
BMI (kg/m^2)	22.4	2.9
%BF-BIA	32.6	4.7
HST	81.5	19.7
%HGS (%)	88.9	15.3
SMI (kg/m^2)	7.9	0.6

Table 2. Subjects with decreased SMI (Skeletal muscle index), HGS (Hand grip strength) and HST (Harvard step test) according %BF-BIA (% Body fat by bioelectrical impedance analysis)

%BF-BIA level	**n** by %BF-BIA level (%)	SMI	HGS	HST
Thin	**0** (0%)	0	0	0
Normal	**0** (0%)	0	0	0
Overweight	**6** (9.1%)	0	1	0
Obese	**60** (90.9%)	1	28	22
Total group	**66** (100%)	1	29	22
% of subject respect n=66		1.5%	43.9%	33.3%

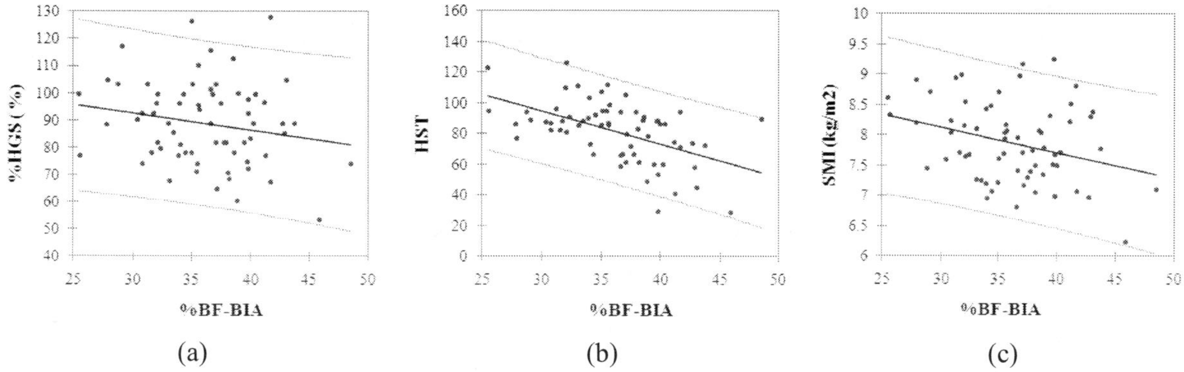

Figure 1. Regression of **(a)** % Hand grip strength (%HGS) (r^2=0.037), **(b)** Harvard step test (HST) (r^2=0.275) and **(c)** Skeletal muscle index (SMI) (r^2=0.094); by % Body fat by bioelectrical impedance analysis (%BF-BIA)

4. Discussion and conclusion
Some authors define sarcopenic obesity as low muscle mass associated to high fat mass [7]. Muscle mass has been evaluated by different methods but cut off points to define these parameters are not unanimous [7]. On the other hand, although historically, sarcopenia has been related to a decline in

muscle strength and functionality in older people, the word sarcopenia comes from the Greek meaning "poverty of flesh" and it does not imply old age. In addition, several researchers have shown hat quantity of muscle mass is not necessarily associated to loss of muscle strength and that more importantly is to define if a subject has less functionality and performance. Thus, detecting changes in muscle composition and quality is the primary concern for nutritionists and geriatricians and a predominant goal in medical care [3]. Thereby, definition of sarcopenic obesity may be controversial [7]. We hypothesized that as obesity produces low-grade inflammation, and hormonal changes affecting the muscle function and metabolism, obese young people could have similar changes and meet the criteria for the diagnose of sarcopenic obesity.

We chose two definitions to evaluate a group of young female college whose body fat percentage was between 27.9 and 37.3. By using the traditional definition of sarcopenia in this group of apparently healthy female college, one subject would be classified as having SO by definition A. By applying the definition B proposed by Stenholm et al [3], 9 young people obtained the same diagnosis. The low association between SMI, HGS, HST have been reported by others and evidence that muscle mass quantity is not entirely related to muscle strength is now clear [3].

It may be that subjects in this study were less physically fit or less active but these are conditions strongly related to obesity. Nevertheless, factors as nutritional conditions, physical activity, occupation, and body size associated to SO must be investigated as suggested by Parr et al , 2013 [5,10, 17]. It was found a total prevalence of 15.2% of SO subjects. Different definitions of SO may be produce polemical results. It is probably that as the population gets more obese, the problematic of SO will be found earlier in life and more research is desirable. This study is a contribution to a topic that is just beginning to be explored.

References

[1] Baumgartner RN, Koehler KM, Gallagher D, Romero L, Heymsfield SB, Ross RR, Garry PJ, And Lindeman RD 1998 *Am J Epidemiol.* **147** 755-63
[2] Cruz-Jentoft AJ, Baeyens JP, Bauer JM, Boirie Y, Cederholm T, Landi F, Martin FC, Michel JP, Rolland Y, Schneider SM, Topinková E and Zamboni M 2010 *Age Ageing.* **39** (4) 412-23
[3] Stenholm S, Harris TB, Rantanen T, Visser M, Kritchevsky SB, and Ferrucci L 2008 *Curr Opin Clin Nutr Metab Care.* **11** 693-700
[4] Gallagher D, Heymsfield SB, Heo M, Jebb SA, Murgatroyd PR and Sakamoto Y 2000 *Am J Clin Nutr.* **72** 694-701
[5] Parr EB, Coffey VG, Hawley JA 2013 *Maturitas.* **74** 109-13
[6] Ford ES and Mokdad AH, 2008. *J Clin Endocrinol Metab.* **93** S1-8.
[7] Prado CM, Wells JC, Smith SR, Stephan BC and Siervo M 2012 *Clin Nutr.* **31** 583-601
[8] WHO, 2011. *Global status report on non-communicable diseases 2010 Annex 4*
[9] Hilton TN, Tuttle LJ, Bohnert KL, Mueller MJ and Sinacore DR 2008 *Phys Ther.* **88** 1336-44
[10] Evans WJ, 2010 *Am J Clin Nutr.* **91** 1123S-27S.
[11] Chien MY, Huang TY and Wu YT 2008 *J Am Geriatr Soc.* **56** 1710-5
[12] Hulens M, Vansant G, Lysens R, Claessens AL, Muls E and Brumagne S 2001 *Int J Obes Relat Metab Disord.* **25** 676-81.
[13] Johnson TJ, Brouha L and Gallagher JR 1943 *Yale J Biol Med.* **15** 781-5
[14] Lohman TG, Roche AF and Martorell R 1988 *Anthropometric standardization reference manual*
[15] Janssen I, Heymsfield SB, Baumgartner RN and Ross R 2000 *J Appl Physiol.* **89** 465-71
[16] Innes E 1999 *Aust Occup Ther J.* **46** (3) 120-140
[17] Cruz-Jentoft AJ, Baeyens JP, Bauer JM, Boirie Y, Cederholm T, Landi F, Martin FC, Michel JP, Rolland Y, Schneider SM, Topinková E and Zamboni M 2010 *Age Ageing.* **39** (4) 412-23

Estimation of limb adiposity by bioimpedance spectroscopy in lymphoedema

LC Ward[1], T Essex[2], R Gaw[2], S Czerniec[3], E Dylke[3], B Abell[3] and S L Kilbreath[3]

[1] School of Chemistry and Molecular Biosciences, The University of Queensland, Brisbane, QLD 4072, Australia
[2] ImpediMed Ltd., 50 Parker Court, Pinkenba, QLD 4008, Australia
[3] Faculty of Health Sciences, University of Sydney, Lidcombe, NSW 2141, Australia

l.ward@uq.edu.au

Abstract. Lymphoedema is a chronic debilitating condition that may occur in approximately 25% of women treated for breast cancer. As the condition progresses, accumulated lymph fluid becomes fibrotic with infiltration of adipose tissue. Bioelectrical impedance spectroscopy is the preferred method for early detection of lymphoedema based on the measurement of impedance of extracellular fluid. The present study assessed whether these impedance measurements could also be used to estimate the adipose tissue content of the arm based on a model previously used to predict whole body composition. Estimates of arm adipose tissue in a cohort of women with lymphoedema were found to be highly correlated ($r > 0.82$) with measurements of adipose tissue obtained using the reference method of dual energy X-ray absorptiometry. Paired *t*-tests confirmed that there was no significant difference between the adipose tissue volumes obtained by the two methods. These results support the view that the method shows promise for the estimation of arm adiposity in lymphoedema.

1. Introduction

Lymphoedema is a chronic debilitating condition occurring in approximately 25% of women post treatment for breast cancer. It results from inadequate functioning of the lymphatic system that leads to accumulation of lymph fluid, an extracellular fluid (ECF) [1]. As the condition worsens, cellular infiltration of the fluid occurs ("stagnation") including development of fibrosis and accumulation of lipid material which may also present as a specific condition known as lipidema [2]. In addition, obesity is a common co-morbidity with lymphoedema. Over the past two decades, bioimpedance spectroscopy (BIS) has found an important clinical role in the early detection of lymphoedema due to its ability to quantify ECF as a change in impedance measured at a low, ideally zero, frequency [3].

The aim of this study was to assess whether the bioimpedance protocol used for assessment of unilateral breast cancer-related lymphoedema could be adapted to estimate concurrently arm adiposity.

2. Experimental

2.1 Participants. Studies were performed in 21 women, 16 of whom had clinically ascribed unilateral lymphoedema (LE) and 5 who had undergone axillary node surgery for cancer but had no diagnosis of lymphoedema (non-LE), and 5 male healthy control participants. Participants were all enrolled in

Published under licence by IOP Publishing Ltd

larger research programs approved by the University of Sydney or the University of Queensland Medical Human Research Ethics Committees.

2.2 Protocol. Bioimpedance measurements were performed on both arms along four consecutive 10-cm segments commencing at the ulnar styloid using an SFB7 (ImpediMed, Brisbane) bioimpedance spectrometer. Current drive electrodes (Ag-AgCl, ImpediMed) were located on the hand at the 3rd metacarpal phalange joint) and on the foot between the 2nd and 3rd metatarsal phalange joints. Voltage sense electrodes were at 10-cm intervals along the dorsal surface of the arm commencing in-line with the ulna styloid. Impedance was measured, in triplicate, at 256 discrete current frequencies in the range 4-1000 kHz. The impedance of the whole arm, wrist to shoulder, was also determined according to the principle of equipotentials. Impedance data were analysed according the Cole model using Bioimp version 4.8.0.0 (ImpediMed Ltd) and the resistance at 10 kHz extracted. Arm circumference was measured at each electrode location to 0.1 cm resolution using a flexible tape. The height of each participant was recorded (to 0.1 mm) using a wall-mounted stadiometer.

Each participant at the time of the impedance measurements underwent a whole-body dual-energy X-ray absorptiometry (DXA) scan. Nine subjects were scanned using an Hologic Explorer instrument; the remainder using a Norland XR36 scanner. The tissue composition (fat mass (FM), bone mineral content (BMC) and lean mass (LM)) of the whole arm and each 10 cm segment was determined for each arm. DXA provides tissue masses; these were converted to volumes using the density factors of Brorson et al. [2].

2.3 Calculation of tissue adipose content from impedance measurements. The method is based on that described by Biggs et al. [4]. Briefly, the model assumes that both skeletal muscle (lean tissue) and fat are conductive while other tissues are non-conductive and that the total resistance of a body segment is the sum of fat and lean resistances according to Kirchoff's law.

$$\frac{1}{R_{total}} = \frac{1}{R_{fat}} + \frac{1}{R_{lean}} \tag{1}$$

Equation (1) can be re-written taking into account the relationship between resistance, segment volume (length, L, x cross-sectional area, A) and resistivity, ρ, as

$$\frac{1}{R_{total}} = \frac{A_{fat}}{L.\rho_{fat}} + \frac{A_{lean}}{L.\rho_{lean}} \tag{2}$$

Since total segmental cross-sectional area equals the sum of the conductive and non-conductive tissue areas and substituting and rearranging equation 2 gives

$$\frac{1}{R_{total}} = \frac{A_{conductive} \cdot \rho_{fat} - A_{fat} \cdot \rho_{fat} + A_{fat} \cdot \rho_{lean}}{\rho_{lean} \cdot \rho_{fat} \cdot L} \tag{3}$$

Further rearrangement provides a quotient (Q) which relates the segmental fat to the sum of fat plus lean which can be calculated from

$$Q = \frac{\rho_{lean} \cdot \rho_{fat} .L}{A_{conductive} \cdot R_{total} \cdot (\rho_{lean} \cdot \rho_{fat})} - \frac{\rho_{fat}}{(\rho_{lean} - \rho_{fat})} \tag{4}$$

Segmental volumes and cross-sectional areas were calculated from the circumferential measurements for each 10 cm segment assuming that the segments can be represented as truncated cones. For the whole arm, the wrist to shoulder length, L, was calculated, based on published anthropometric data, as 0.33 x the participant's height with a circumference equivalent to that of mean segmental circumferences. The cross-sectional of the conductive tissue was calculated as the difference between total cross-sectional area and non-conductive area derived from measurements of bone area by pQCT [5]; resistivity values were those quoted by Biggs et al. [4], i.e. 5000. Ohm.cm for adipose tissue and 120 Ohm.cm for lean tissue. Thus the volume of adipose tissue was calculated from the total segment volume and the adiposity quotient, Q, derived using equation 4.

2.4 Data analysis. Segmental adipose tissue volumes computed from impedance measurements were compared to those measured by DXA by concordance correlation analysis and the significance of differences using a paired *t*-test.

3. Results

The mean bioimpedance inter-arm ratio for the participants with lymphoedema was 1.20, exceeding the established threshold values confirming the presence of lymphoedema in these participants. The mean overall volume, measured by DXA, of the affected arm was 17.5 ± 16.8% larger than that of the unaffected arm for these participants with similar increases in both adipose (18.7 ± 21.2%) and lean (17.7 ± 21.4%) tissues; comparable data (dominant:non-dominant arm) for the 5 female control participants were impedance ratio, 1.02, volume excess,10.9 ± 8.5%; adipose tissue excess, 14.7 ± 12.2% and 3.7 ± 12.6% increase in lean respectively. The comparable values for the 5 control male participants were an impedance ratio of 1.03 and a volume excess of 5.2 ± 9.4%. Excess adipose volume was 22.2 ± 10.9% and excess lean volume 0.5 ± 10.9%.

Pooled data for adipose volume of all segments estimated from impedance was strongly correlated with the DXA measurements, $r_c = 0.82$ ($P < 0.001$), Fig 1.

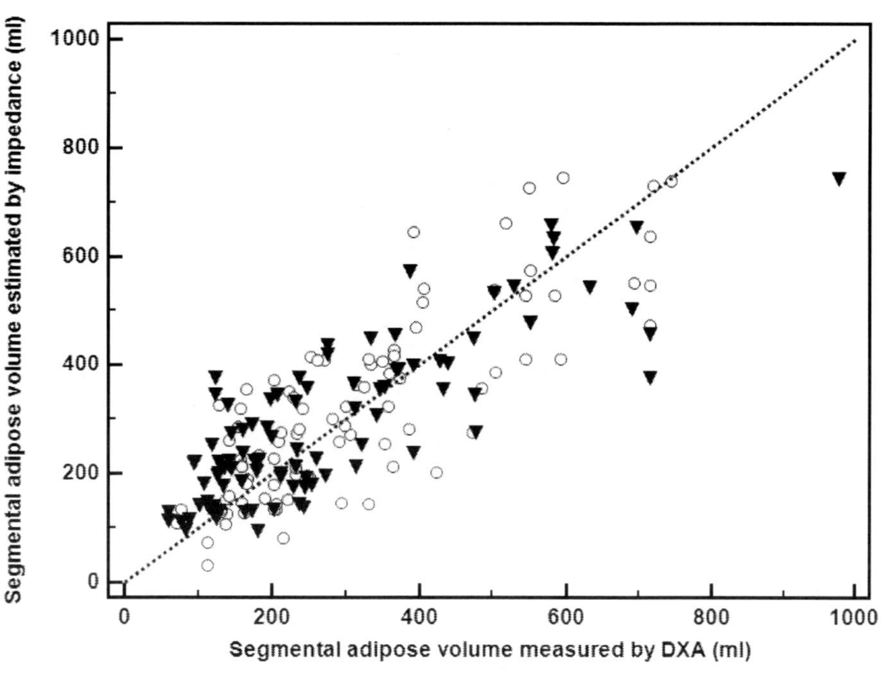

Figure 1. Correlation of segmental adipose tissue volume estimated by impedance with that measured by DXA. Legend: ---, the line of identity; ▼, unaffected or non-dominant arm for controls; O, affected or dominant arm.

For the lymphoedema participants only, segmental adiposity estimated from impedance was equally strongly correlated ($P < 0.001$) with the DXA measurements for both the affected arm and the contralateral control arm, $r = 0.82$ and $r = 0.86$ respectively. Poorer correlations were found for whole arm measurements; $r = 0.67$ and $r = 0.75$ respectively.

There was no significant difference in segmental adipose tissue volume estimated by impedance compared with that measured by DXA, 303 ± 160 and 291 ± 177 ml respectively. Mean of all segmental adipose volumes were larger for the affected limbs of the lymphoedema participants only, again with no significant difference between the impedance and DXA measurements, 345 ± 175 and 341 ± 178 ml respectively. Similarly the mean adipose volume of the whole arm estimated by impedance was not significantly different to that measured by DXA, 1545 ± 465 and 1636 ± 742 ml respectively.

4. Discussion

This study demonstrates the feasibility of estimating arm adipose tissue content from impedance measurements simultaneously with measurement of the excess fluid accumulation characteristic of the presence of lymphoedema. The presence of excess body fat or frank obesity is a common co-morbidity in women with lymphoedema.

Although, the protocol for impedance measurements is the same as that currently used for measurement of lymph accumulation, in order to estimate adipose volume additional anthropometric measurements are required. The cross-sectional areas of the non-conductive and conductive tissues of the arm need to be determined from arm length and circumference measurements. Since the arm is not a simple cylinder, the arm was divided into 10-cm segments approximating truncated cones and the adiposity of these discrete segments estimated. These were found to be not significantly different to those measured using the reference technique of DXA. Although this procedure of segmenting the arms is in common clinical use for calculation of arm volume in lymphoedema assessment, it adds substantially to the number of measurements required and the time taken may be inconvenient to the patient and clinician. Therefore, the possibility that total arm adiposity could be similarly estimated from whole arm measurements only was investigated. Again the results were encouraging in that no significant differences in adiposity values were observed. The precision of limb impedance measurements is well established [3]; further work is required to determine if better anthropometric measurements could improve further accuracy of the method.

In conclusion, the method has the potential to extend the application of impedance assessment of lymphoedema from measuring fluid accumulation to simultaneous measurement of adipose tissue volume; clinical utility, however, remains to be established in larger scale clinical trials.

References

[1] Murdaca G, Cagnati P, Gulli R, Spano F, Puppo F, Campisi C and Boccardo F 2012 Current views on diagnostic approach and treatment of lymphedema *Am. J. Med.* **125** 134-140
[2] Brorson H, Ohlin K, Olsson G and Karlsson MK 2009 Breast cancer-related chronic arm lymphedema is associated with excess adipose and muscle tissue *Lymph. Res. Biol.* **7** 3-10
[3] Ward LC 2006 Bioelectrical impedance analysis: proven utility in lymphedema risk assessment and therapeutic monitoring *Lymph. Res. Biol.* **4** 51-56
[4] Biggs J, Cha K and Horch K 2001 Electrical resistivity of the upper arm and leg yields good estimates of whole body fat *Physiol. Measur.* **22** 365-376
[5] Ward LC, Essex T, Bartlett M, Kilbreath SL and Brookes D 2010 Bioimpedance profiling of the limbs: update. J Physics: Conference Series **224** doi:10.1088/1742-6596/224/1/012105

Percentile curves for body fatness and cut-offs to define malnutrition in Russians

D V Nikolaev[1], S G Rudnev[2], O A Starunova[3], T A Eryukova[1], V A Kolesnikov[1], E G Ponomareva[4], N P Soboleva[5], S A Sterlikov[5]

[1]Scientific Research Centre 'Medas'
[2]Institute of Numerical Mathematics, Russian Academy of Sciences
[3]Moscow State University, Department of Computational Mathematics and Cybernetics
[4]Ministry of Health of the Russian Federation, Department of Medical Prophylaxis, Acute, Primary Care, and Sanatorium-and-Health Resorting Treatment
[5]Central Research Institute for Health Care Organization and Informatization
all – Moscow, Russian Federation

Email: sergey.rudnev@gmail.com

Abstract. Here, we report first results of the large-scale ongoing bioelectrical impedance body composition study in Russians. By the end of 2012, 216 out of 800 Russian Health Centres submitted raw bioimpedance data on 844,221 adults and children aged 5-80 years, representing nearly 0.6% of the Russian population, who were accessed cross-sectionally using the same type of bioimpedance meter, ABC-01 Medas. Estimates of overweight, obesity, and normal weight obesity prevalence in the general population, as well as characteristics of diagnostic sensitivity and specificity of the conventional WHO BMI-based criteria of obesity depending on age are obtained. The smoothed reference centile curves for percentage fat mass are constructed, and localized cut-offs for fatness and thinness are provided that can be used both at the individual and epidemiological levels.

1. Introduction

One of the disadvantages of using body mass index for defining obesity, undernutrition and associated health risks lies in its low diagnostic sensitivity, or high rates of false-negative results [1]. Bioelectrical impedance analysis is a widely used method that enables more direct assessment of body composition [2,3]. In this study, we aimed at the construction of population reference centile curves for bioimpedance percentage body fat mass, as well as at the providing localized cut-offs for classification of fatness levels using large population-based sample of Russian citizens.

2. Subjects and methods

844,221 adults and children aged 5-80 years (see Table 1), who were visitors of the Russian Health Centres in 2010-2012, were accessed cross-sectionally by the same type of bioimpedance analyzer, namely, ABC-01 Medas (SRC 'Medas', Russia). Our sample group represented nearly 0.6% of the Russian population from the all eight Federal Districts of the Russian Federation and 48 out of 83 Federal Subjects in them.

Table 1. Characteristics of the study group (the data as submitted by the end of December, 2012)

	Males	Females	Both sexes
Children and adolescents (5-19 years)	159,293	149,979	309,272
Adults (20-80 years)	123,616	411,333	534,949
Total number (5-80 years)	282,909	561,312	844,221

Standing height and weight were measured using the same type of stadiometer and digital scale with an accuracy 0.5 cm and 0.1 kg, respectively. Body mass index (BMI) was calculated as body mass (BM) divided by standing height (Ht) squared. The whole-body impedance was measured on the right hand side of the body according to a conventional tetrapolar scheme at electrical current frequency 50 kHz. Fat-free mass (FFM) was assessed using appropriate equations for healthy adults and children [4,5]. Fat mass was calculated by subtraction of FFM from the BM, and the percentage fat mass (%FM) was obtained as (FM/BM)×100%.

The smoothed reference centile curves for BMI and %FM were constructed using LMS-method [6] in the software program BIAStatistica [7]. The conventional WHO criteria of malnutrition according to BMI [8,9] were explored to obtain the correspondent localized cut-offs in terms of the %FM. Sensitivity and specificity of the BMI cut-off values for obesity were obtained as a percentage of true-positive and true-negative results, respectively, compared to the bioimpedance %FM criteria taken as a reference.

3. Results and discussion

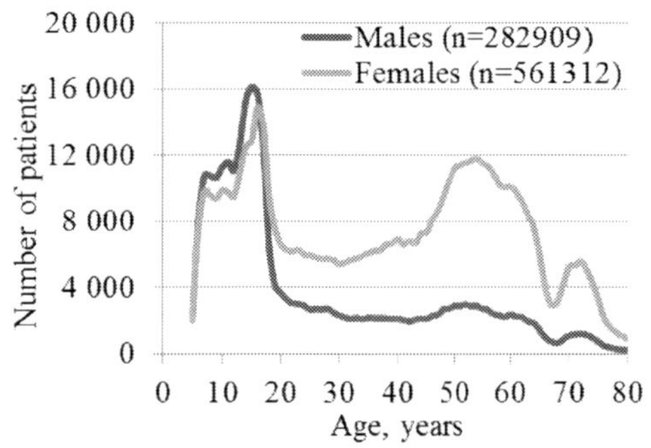

Fig. 1. Distributions of males and females in our sample according to age.

Fig. 1 shows distributions of males and females under study according to age. It can be seen that women visited Health Centers almost twice as much as men ($p<0.01$). An observed first peak of attendance reflects a primary focus of attention on childhood and adolescent years, and the second peak at the interval between 50 and 65 years of age corresponds to the age dependence of cardiovascular morbidity and mortality. The local minimum of attendance at the age interval between 65 and 70 years may reflect significant depopulation of the Russian nation during the WW2.

Fig. 2 shows that the prevalence of overweight according to the conventional WHO criteria increases steadily with age reaching a maximum of 76% and 83% in men and women, respectively, by 60-65 years. This result is in agreement with the high prevalence of hypertension, diabetes, cancer, metabolic syndrome and other chronic non-infectious diseases in Russians in this age group.

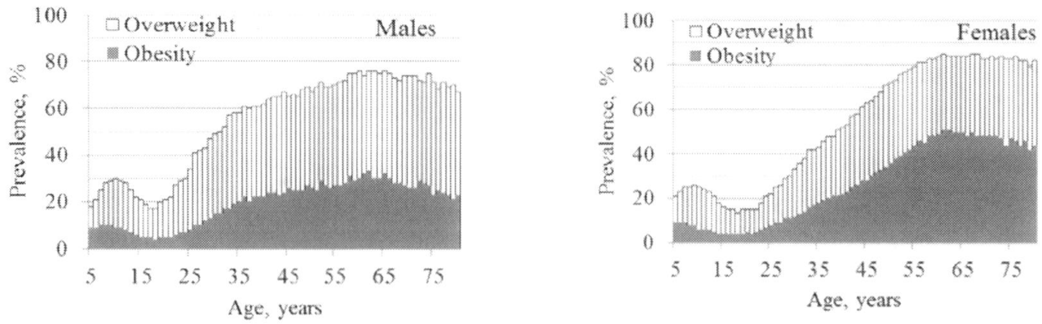

Fig. 2. Estimates of overweight and obesity prevalence in Russians according to age.

Our data show that the diagnostic sensitivity of the conventional WHO criteria for obesity according to BMI differs significantly in adult men and women (55% and 70%, respectively) with the specificity decreasing down to 75-85% with age (see Fig. 3).

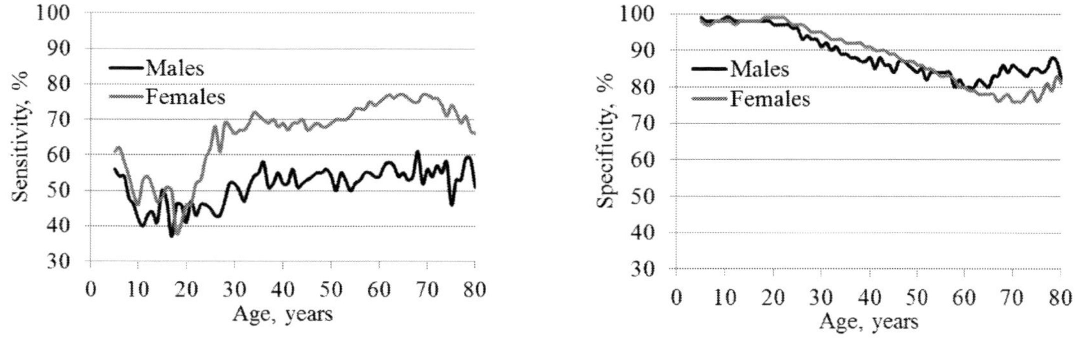

Fig. 3. Sensitivity and specificity of the conventional WHO criteria for defining obesity in Russians.

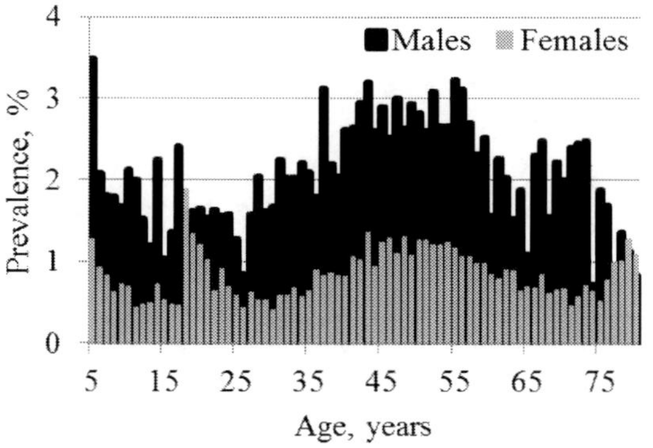

Fig. 4. The prevalence of normal weight obesity in Russians according to age.

As a consequence of low BMI sensitivity, the normal weight obese (NWO) individuals were present in the study group who had high %FM content and normal BMI. Similarly to 'usual' obesity, NWO is associated with an increased risk of metabolic syndrome and cardiovascular morbidity [10,11]. Fig. 4 shows that the prevalence of NWO in Russian population was 1.5-3% in males and 0.5-1.3% in females depending on age (for comparison, see [12]).

The smoothed reference centile curves for %FM, from 10th to 90th, with the median depicted in bold, are presented in Fig. 5.

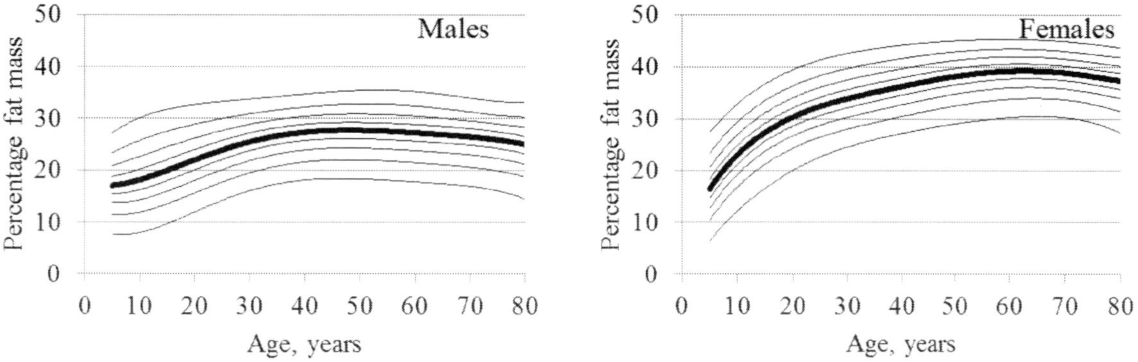

Fig. 5. The smoothed reference centile curves for percentage fat mass in Russians.

The cut-off values of the percentage fat mass for defining underfatness, overfatness and obesity were constructed by the comparison of BMI and %FM age specific distributions in males and females (Table 2). These were defined using the conventional BMI cut-offs 18.5, 25 and 30 kg/m^2, respectively.

Table 2. Bioimpedance %FM cut-off values to define malnutrition in Russians.

Age, years	Females			Males		
	Underfat	Overfat	Obese	Underfat	Overfat	Obese
5	12.4	21.6	26.1	9.1	18.7	22.2
6	11.9	22.4	27.0	9.2	19.5	23.4
7	10.9	23.2	28.2	9.0	19.6	24.3
8	10.3	24.1	29.8	7.8	20.5	25.5
9	12.2	25.4	31.4	8.5	21.5	27.0
10	13.2	26.4	32.7	9.4	22.5	28.4
11	15.1	27.2	33.3	10.1	23.5	30.1
12	15.7	27.6	33.9	10.4	23.7	31.1
13	16.3	29.3	35.2	9.6	22.9	31.2
14	18.0	31.3	36.9	9.3	23.4	31.4
15	19.3	32.9	37.9	9.5	24.0	32.2
16	20.9	33.9	39.0	10.9	24.8	32.3
17	21.0	34.4	39.2	12.2	26.9	33.9
18	22.4	36.5	41.1	12.5	27.9	34.0
19	25.8	38.5	42.7	13.8	28.0	34.0
20-30	24.1	37.7	42.3	13.2	27.6	33.4
30-40	21.2	36.4	41.3	10.9	26.2	32.0
40-50	17.5	34.9	40.3	9.7	25.4	31.2
50-60	16.3	33.9	39.7	9.4	24.6	30.7
60-70	15.2	32.7	38.9	8.6	23.2	29.8
70-80	14.7	31.8	38.3	7.9	22.8	29.5

4. Conclusion

Using our dataset on the whole-body bioimpedance measurements of 844,221 adults and children from the Russian Health Centres, we obtained estimates of obesity and overweight prevalence at the national level, as well as diagnostic %FM thresholds for use in epidemiological and clinical studies. Development and validation of BIA predictive equations based on representative sample of healthy Russian adults and children is needed for more accurate data interpretation and comparisons. Further developments may include localized cut-offs for malnutrition in terms of other bioimpedance body composition screening indexes, such as fat-free mass and skeletal-muscle mass relative to height and, probably more importantly, of phase angle which is directly measurable and not depends on any prediction formula.

References

[1] Baumgartner R N, Heymsfield S B and Roche A F 1995 *Obes. Res.* **3** 73-95
[2] Kyle U G, Genton L, Slosman D O and Pichard C 2001 *Nutrition* **17** 534-541
[3] Chumlea W C, Guo S S, Kuczmarski R J *et al* 2002 *Int. J. Obes.* **26** 1596-1609
[4] Kushner R F and Schoeller D A 1986 *Am. J. Clin. Nutr.* **44** 417-424
[5] Houtkooper L B, Going S B, Lohman T G *et al* 1992 *J. Appl. Physiol.* **72** 366-373
[6] Cole T J and Green P J 1992 *Statistics in Medicine* **11** 1305-1319
[7] Starunova O A 2011 *A collection of papers of young scientists of Lomonosov Moscow State University CMC department* **8** 129-134 (in Russian)
[8] World Health Organization. Growth reference data for 5-19 years. www.who.int/growthref/en/
[9] De Onis M and Lobstein T 2010 *Int. J. Pediatr. Obes.* **5** 458-460
[10] De Lorenzo A, Martinoli R, Vaia F and Di Renzo L 2006 *Nutr. Metab. Cardiovasc. Dis.* **16** 513-523
[11] Romero-Corral A, Somers V K, Sierra-Johnson J *et al* 2010 *Eur. Heart J.* **31** 737-746
[12] Marques-Vidal P, Pecoud A, Hayoz D *et al* 2008 *Eur. J. Nutr.* **47** 251-257

Journal of Physics: Conference Series **434** (2013) 012064 doi:10.1088/1742-6596/434/1/012064

Bio-impedance body composition comparisons between athletes and healthy subjects

A. Clarion[1], E. Ribbe[1], J. Rebeyrol[1], M.-P. Rousseaux-Blanchi[2], C. Dechavanne[2], M.-V. Moreno[1]

[1]Research Department, BioparHom, 30 allée du Lac d'Aiguebelette BP238, Le Bourget du Lac cedex, France.
[2]Centre de médecine du sport de haut niveau d'Albertville, 88bis rue de la République 73200 Albertville, France.

E-mail: antoine.clarion@bioparhom.com

Abstract. Body composition is a useful means for athletes' body composition assessment, relying on reference population data. This study aims at comparing body composition multifrequency impedance data of athletes and healthy adult populations. Differences were found in tissular, hydration and metabolic indices. They were significant, in the expected direction, but quite weak and additional data from reference technologies would set if specific equations are needed. The current ones are nevertheless suitable for reliable follow-up studies.

1. Introduction

Body composition follow-up is useful to athletes to help in physical preparation. Bioimpedance provides a non-invasive, economic and reliable means of measuring body composition indices. Bio-impedance body composition models rely on reference population data. Reliability of body composition indices is thus dependant on how good the measured individual's profile fits to reference population's. Peculiar population such as infants, elderly or athletes may need specific reference database to allow meaningful follow-ups and comparisons.

The aim of this study is to compare body composition data of athletes and reference healthy adult populations and evaluate the pertinence of building a dedicated model for athletes.

2. Materials & Methods

2.1. Subjects

Data was collected among highly trained athletes (89 men, 26 women) and a reference healthy adult population (30 men, 24 women). Athletes practiced different short and long distance sports (e.g. alpine skiing, cross-country skiing / biathlon, ski-mountaineering, running), from local to olympic level. Healthy adults' body mass index (BMI) was lower than 30.

2.2. Measurements

Data was collected using a multifrequency (1-1000 kHz range) bio-impedance measuring device (Z-Métrix®, BioparHom, Le Bourget du Lac, France) validated by a French clinical study (n°2008-A01373-52). Raw complex electrical impedance data is computed from measured voltage caused by

injection of low intensity (77 μA max) electrical currents. It is then turned into body composition indices using a Cole-Cole-derived model [1,2].
Indices are of three categories: tissular, hydration and metabolic. Tissular indices are fat mass (FM/W: fat on body mass ratio), lean mass (LM/W: lean on body mass ratio), and muscular mass (MMus/W: muscular on body mass ratio [3]). Hydration indices are total body water (TBW/W: total body water on body mass ratio [4]), intracellular water (ICW/TBW: intracellular on total water ratio), and extracellular water (ECW/TBW: extracellular on total water ratio). Metabolic indices are metabolic activity index (MAI [5], phase-related), body cell mass (BCM/W: body cell mass on body mass ratio [6]) and its protein content (BCM prot.content [7]), and extracellular mass (ECM/W: extracellular on body mass ratio).
Measurements were taken in upright position and on the right body side.

2.3. Analysis
Comparisons between the two populations were evaluated with unpaired Student tests with a p=.05 significance threshold.

3. Results
Comparisons results are listed in Table 1 and

Table 2. They showed tissular, hydration and metabolic differences: significantly higher MMus/W in athletes than in healthy adults in both men and women, higher LM/W-and lower FM/W in women only; higher BCM/W and BCM prot. content in both genders, lower ECM/W-and higher MAI-in men only; higher TBW/W in women only, lower ICW/TBW-in men only, higher ECW/TBW-in both genders. Electric data showed higher reactance in athletes of both genders, and lower resistance in men only (X/H, R/H: 50 kHz reactance and resistance on height ratios).

Table 1. Comparisons results in women.

	Athletes (mean±sd)	Healthy adults (mean±sd)	Student unpaired test
Age (years)	35.1±12.2	43.5±16.4	t=-2.05, p=.046
Height (cm)	165.0±7.1	163.3±6.5	t=.88, p=.383, NS
Weight (kg)	56.8±8.3	60.7±8.2	t=-1.64, p=0.108, NS
BMI (kg/m²)	20.8±2.0	22.8±3.0	t=-2.71, p=0.009
MAI	6.4±0.5	5.9±1.1	t=1.97, p=0.054, NS
ECM/W (%)	35.1±3.0	34.6±5.4	t=0.46, p=0.649, NS
BCM/W (%)	42.2±3.2	37.3±4.8	t=4.30, p<0.001
BCM prot. content (%)	11.1±0.9	9.7±1.3	t=4.34, p<.001
FM/W (%)	24.1±4.2	29.4±7.3	t=-3.16, p=0.003
LM/W (%)	70.4±6.5	64.7±11.2	t=2.19, p=0.033
MMus/W (%)	35.5±2.9	31.3±4.1	t=4.22, p<0.001
TBW/W (%)	63.0±4.8	58.2±6.5	t=2.95, p=0.05
ECW/TBW (%)	46.0±0.8	44.6±1.6	t=4.06, p<0.001
ICW/TBW (%)	53.3±3.2	54.6±6.1	t=-0.94, p=0.352, NS
X/H (Ω/m)	30.4±3.6	30.2±5.5	t=0.17, p=0.865, NS
R/H (Ω/m)	271.3±27.7	292.4±26.8	t=-2.70, p=0.010

Table 2. Comparisons results in men.

	Athletes (mean±sd)	Healthy adults (mean±sd)	Student unpaired test
Age (years)	37.0±10.5	39.8±14.3	t=-1.15, p=0.253, NS
Height (cm)	178.5±5.8	176.8±6.6	t=1.37, p=0.174, NS
Weight (kg)	76.9±9.7	78.1±10.7	t=-0.54, p=0.592, NS
BMI (kg/m²)	24.1±2.5	25.0±2.9	t=-1.54, p=0.126, NS
MAI	7.5±0.8	6.9±1.2	t=2.94, p=0.004
ECM/W (%)	34.1±2.5	35.5±4.2	t=-2.16, p=0.033
BCM/W (%)	54.3±4.0	51.8±4.2	t=2.88, p=0.005
BCM prot. content (%)	14.4±1.1	13.7±1.1	t=3.04, p=0.003
FM/W (%)	18.4±2.8	19.7±5.9	t=-1.67, p=0.097, NS
LM/W (%)	69.7±6.6	69.9±7.2	t=-0.09, p=0.930, NS
MMus/W (%)	49.3±3.5	47.0±4.0	t=3.04, p=0.003
TBW/W (%)	67.2±4.9	68.2±5.2	t=-0.92, p=0.361, NS
ECW/TBW (%)	45.3±1.2	43.3±1.4	t=7.61, p<0.001
ICW/TBW (%)	54.2±4.2	56.7±4.9	t=-2.66, p=0.009
X/H (Ω/m)	26.4±24.2	24.2±4.3	t=2.58, p=0.011
R/H (Ω/m)	201.6±22.3	201.2±27.8	t=0.09, p=.929, NS

4. Discussion

Most of these differences confirm that athletes are in a better shape than healthy adults. They show a better tissular distribution in athletes. Male athletes differ by more muscular mass; female athletes also have a more favorable fat/fat free balance.

At a metabolic level, athletes' indices suggested a better physical working capacity as evidenced by higher BCM/W [8], and a better cellular membrane exchange capacity as evidenced by higher MAI [9]. Surprisingly, MAI was higher in female athletes than healthy ones but did not reach significance threshold. It is believed that considering parameters such as overtraining, fatigue and performance could help understanding MAI variation factors in this peculiar population [10,11]. The significant (p=.046) difference in the two women groups may also represent a bias as it is known phase angle lowers with age [12].

All groups were fairly hydrated. Intra- extracellular water balance leaned towards extracellular compartment in athletes. Their higher extracellular water is thought to help them coping with effort-related water loss.

Electrical indices such as those used for BIVA analysis (e.g.[13]) are consistent with metabolic and hydration results since resistance tends to be lower and reactance (capacitive dimension of impedance) higher.

Although differences between athletes and healthy adults were found in the expected direction, they were rather weak [14]. Separating different sports could show stronger differences. Alternatively, the model possibly skewed the indices towards median values.

Journal of Physics: Conference Series **434** (2013) 012064 doi:10.1088/1742-6596/434/1/012064

5. Conclusion

Differences were found between athletes and healthy subjects at tissular, hydration and metabolic levels. An athlete-specific data base will allow displaying acute comparisons among specific reference population for individual athletes. It would be interesting to build such database for specific sports or sports type (endurance, sprint). Data from reference body composition technologies should tell if athletes' indices were accurate, thus indicating if new equations are needed. In all cases, repeatability is not affected and follow-up studies are reliable even without it.

References

[1] Cole KS and Cole RH 1941 Dispertion and adsorption in dielectrics. *J. Chem. Rev.* 209-41
[2] Moreno M-V 2007 Étude de la composition corporelle par bioimpédancemétrie sur des adultes et des enfants sains et pathologiques *Doctoral dissertation, Université de Technologie de Compiègne*
[3] Wang Z, Zhu S, Wang J, Pierson RN and Heymsfield SB 2003 Whole-body skeletal muscle mass: development and validation of total-body potassium prediction models. *Am. J. Clin. Nutr.* **77** 76-82
[4] Jaffrin MY, Fenech M, Moreno M-V and Kieffer R 2006 Total body water measurement by a modification of the bioimpedance spectroscopy method. *Med. Biol. Eng. Comput.* **44** 873-82.
[5] Rebeyrol J, Moreno M-V, Ribbe E, Buttafoghi L, Pédron O and Dechavanne C 2010 Bioimpedance data monitoring in physical preparation: A real interest for performance of elite skiers for Winter Olympic Games. *Procedia Engineering* **2** 2881−7
[6] Kotler DP, Burastero S, Wang J and Pierson RN 1996 Prediction of body cell mass, fat-free mass, and total body water with bioelectrical impedance analysis: effects of race, sex, and disease *Am. J. Clin. Nutr.* **64** 489-97S
[7] Wang Z, St-Onge M-P, Lecumberri B, Pi-Sunyer FX, Heshka S, Wang J, et al. 2004 Body cell mass: model development and validation at the cellular level of body composition. *Am. J. Physio.l Endocrinol. Metab.* **286** E123-8
[8] Varlet-Marie E, Brun JF, Blachon C and Orsetti A 1997 Relations entre la composition corporelle mesurée par impédancemétrie et la performance motrice sur ergocycle. Science & Sports **12** 204-6
[9] Baumgartner RN, Chumlea WC and Roche AF 1988 Bioelectric impedance phase angle and body composition. *Am. J. Clin. Nutr.* **48** 16-23
[10] Rebeyrol J, Moreno M-V, Ribbe E and Vannicatte A 2010 Case study: using monitoring of body composition data obtained by bioimpedance, in training of an elite male runner. *Procedia Engineering* **2** 3059-64
[11] Martinelli Y 2012 La bioimpédancemétrie corporelle totale comme outil de monitoring des performances en natation. *Master dissertation, Université de Lausanne*
[12] Barbosa-Silva MCG, Barros AJ, Wang J, Heymsfield SB and Pierson RN 2005 Bioelectrical impedance analysis: population reference values for phase angle by age and sex. *Am. J. Clin. Nutr.* **82** 49-52
[13] Piccoli A, Pillon L and Dumler F 2002 Impedance vector distribution by sex, race, body mass index, and age in the United States: standard reference intervals as bivariate Z scores. *Nutrition* **18** 153-67.
[14] McArdle W, Katch F and Katch V 2001 Chapter 28 - Physique, performance et activité. Physiologie de l'activité physique, Paris: Maloine 533-58.

Bioelectrical impedance analysis (BIA) equations validation against hydrodensitometry in a Colombian population

J C Caicedo-Eraso[1,2], C A Gonzalez-Correa[1], C H Gonzalez-Correa[1]
[1] Research Group on Electrical Bioimpedance, University of Caldas, Colombia
[2] Department of Systems and Informatics, University of Caldas, Colombia

E-mail: julioc.caicedo@ucaldas.edu.co

Abstract: Several studies have shown that the accuracy of BIA results depends of ethnicity, age, gender, hormonal and genetic variations and, so far, there are not specific equations for Colombian population. The purpose was to evaluate reported BIA equations to determine their usefulness in body composition assessment in young females from Colombia using hydrodensitometry as the reference method. A sample of 30 young females was evaluated. Inclusion and exclusion criteria were defined to minimize the variability of BIA. Height, weight, multi-frequency BIA, residual lung volume (RV) and underwater weight (UWW) were measured. Five BIA equations met the inclusion criteria of this study. Three equations overestimated and two equations underestimated body fat (BF). Paired Student t-test and Bland and Altman analysis ($p<0.05$) showed significant differences in four BIA equations. However, all standard error of estimate (SEE) to BF was greater than 2.7 kg. This study showed that the five selected BIA equations are not valid for estimation of body composition in young females from Colombia. It is recommended to develop BIA equations to improve BF fat assessment in our population.

1. Introduction

The high prevalence of overweight and obesity in adults in Colombia [1] makes it necessary to have more appropriate methods for nutritional assessment of people with these diseases. Body composition can be determined by numerous methods, as hydrodensitometry, considered as reference method for in vivo measurements, but a better accuracy involves higher costs and complexity, which would limit its use for large-scale research in the community [2]. BIA is a simple, safe, noninvasive, relatively inexpensive method that is practical and suitable for individual use, large-scale researches and field studies in the community [3]. Variables as gender, age, ethnicity, medical and physiological conditions, affect the predictive validity of BIA equations [4,5]. This has resulted in a variety of BIA equations developed for populations varying in one or more of these variables [3,5]. Before undertaking a new study to develop a specific population equation, a mandatory step is to validate the existing BIA equations since manufacturer [6] or reported BIA equations may be suitable [7,8] or not [9,10] for a different population for which they were developed. The purpose of this study was to evaluate reported BIA equations to determine their usefulness in body composition assessment in young females from Colombia, using hydrodensitometry as the reference method.

2. Materials and methods

2.1. Subjects

The methods, classified as minimum risk by the Colombian Ministry of Health, were approved by the Bioethics Committee of the University of Caldas. Variations in body composition related to gender, age, ethnicity and genetic influence that affect BIA measurements [4,5] were minimized by using an adequate measurement protocol [11] and selecting a homogeneous group of young females [1,5]. A sample of 30 young females was evaluated. The purpose and procedures of the study were explained to the volunteers and the inclusion/exclusion criteria were verified after completing a questionnaire. Afterwards, the volunteers signed an informed written consent. Characteristics of subjects by mean and standard deviation (SD) are shown in table 1. The same subjects and the same measurement data were used in other study to develop a preliminary BIA equation for our population [12].

Table 1. Characteristics of subjects.

Variables (n=30)	Mean	SD
Age (years)	20.4	2.0
Height (cm)	158.1	4.8
Weight (kg)	54.8	8.0
BMI (kg/m^2)	21.9	2.7

2.2. *Data acquisition*

Measurements were performed in one session early in the morning to minimize environmental [13] and biological [4] variations. Relative humidity (RH) (70.5 ± 3.5%) and environmental temperature (20.5 ± 1.0 ºC) were measured with a thermo-hygrometer (13307 by DeltaTrak®, ±0.1 ºC / ±1% RH) and controlled with an electric heater (BFH416 by Bionaire™) and a dehumidifier (BMD100 by Bionaire™). Volunteers were asked to comply with the following requirements before the test: 48 hours without drinking alcohol, 12 hours without vigorous exercise and 12 hours of fasting but keeping normal water hydration. All subjects were asked to evacuate their bladder and colon 30 minutes before the tests and were evaluated out of menses [5,14]. Other possible sources of error like hydration status, subject position, limb abduction, electrode position, wearing jewellery and skin alcohol use, were also controlled in this study [5,14,15]. All volunteers wore underwear and a hospital gown.

2.3. *Anthropometric measurements*

Height (Heightronic-235 by Seca®, ±0.01 cm) and weight (PP2000 by Icob-Detecto®, ±0.1 kg) were measured twice, and a third measurement was taken if it was found a difference greater than 0.5 cm or 0.1 kg respectively [16].

2.4. *BIA measurements*

Bioelectrical impedance was measured (Hydra 4200 by Xitron Technologies©), using the manufacturer's recommended electrodes (292-STE by Impedimed©), on the dominant side of body for three times at the end of an exhalation on a nonconductive surface. Raw R_{50} and Xc_{50} data were stored in a computer. Subjects remained in a supine position by 5 minutes and BIA measurements were made between minutes 6 to 10, with the arms comfortably separated from the body 15 degrees, and the legs comfortably separated about 45 degrees [5,17]. Dorsal hand and anterior foot surfaces were cleaned with alcohol and dried with a paper towel [5]. For reproducible measurements, four landmarks were made for the placement of the electrodes: midline between prominent ends of radius and ulna of wrist, midline of third metacarpal-phalangeal joint on dorsal hand surface, midline between the medial and lateral malleolus of ankle and midline of third metatarsal-phalangeal joint on anterior surface of foot [5,17]. Current was applied at distal electrodes and voltage was measured at proximal electrodes [5,17]. Distance between current and voltage electrodes was always greater than 5 cm [5]. A blanket was used for comfort and homogeneous skin temperature [5,18].

2.5. *Hydrodensitometry measurements*

This technique obtains the body density and implies measuring RV and UWW. RV was measured by the nitrogen washout technique (Quark PFT-2 by COSMED, ±0.01 L) and taking the average of two measurements whose difference was less than 0.1 L [19]. UWW was measured (PROGAN-1500SS by Prometalicos, ±0.001 kg) within a tank filled with water (35.0 ± 2.0 °C) and taking the average of two measurements whose difference was less than 0.1 kg [20]. Subject was in sitting position and wore a form-fitting swimsuit and a ballast weight in the waist [20]. BF was obtained applying the model of two components by Siri equation [21].

2.6. Selected BIA equations.
The criteria for BIA equation selection were: they have been developed for young females using hydrodensitometry as reference technique, a coefficient of determination (r^2) greater than 0.60 and a SEE reported by the authors, lesser than 2.7 kg, considered as acceptable in BF prediction [22].

2.7. Statistical methods
Mean and SD were used to evaluate the characteristics of subjects and laboratory conditions. The difference between BF obtained in our population with each BIA equation against the results obtained with hydrodensitometry was evaluated by SEE, paired Student's t-test ($p<0.05$) and the graphical method of Bland and Altman [23]. Finally, the equation meeting the three criteria was considered valid for the target population of this study.

3. Results
Five BIA equations met the inclusion criteria and the results are shown in table 2. All SEE were greater than 2.7 kg (table 2). Paired Student's t-test (table 2) and Bland and Altman plots (figure 1) showed significance difference; except with Lukaski et al, 1986 equation, which showed no significant differences; however, the SEE was greater than 2.7 kg.

Table 2. Five BIA equations results.

Equation	SEE (kg) Reported by author	SEE (kg) Colombian young female	Significant difference p_t	Significant difference Bland and Altman
Lukaski et al, 1986	2.06	$3.63^{(+)}$	0.56	Not
Segal et al, 1988	2.43	$3.82^{(-)}$	0,0002	Yes
Deuremberg et al, 1991	2.63	$5.66^{(+)}$	< 0,0001	Yes
Lohman, 1992	2.10	$3.49^{(-)}$	0.001	Yes
Xitron Technologies	0.96	$4.13^{(+)}$	< 0,0001	Yes

(+) overestimation and (-) underestimation

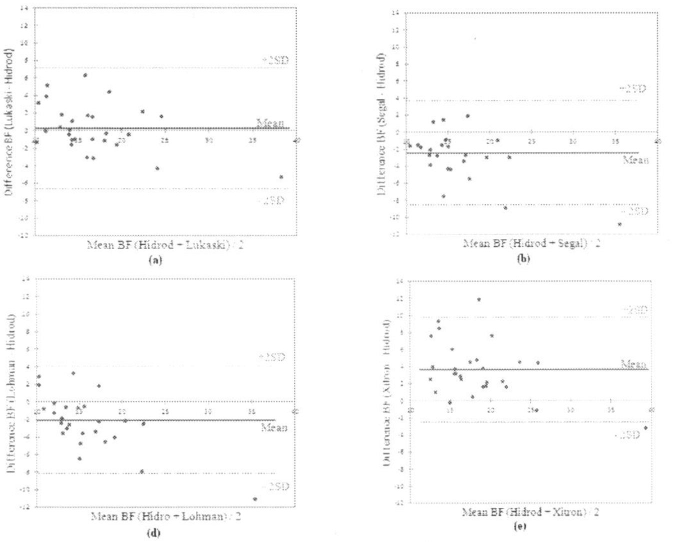

Figure 1. BF obtained by BIA equations against BF obtained by hydrodensitometry (n=30). **(a)** Lukaski et al, 1986, **(b)** Segal et al, 1988, **(c)** Deuremberg et al, 1991 **(d)** Lohman, 1992, **(e)** Xitron Tech.

Journal of Physics: Conference Series **434** (2013) 012065 doi:10.1088/1742-6596/434/1/012065

4. Discussion and conclusion

Practice has shown that a BIA equation developed for a specific population may or may be not useful for others. The five selected BIA equations showed unacceptable SEE. In this study, use of the specific equation for our population continues to be an important issue for BIA results. It is possible that in the future, the new developments of the SBIA allow the specific population equations to be minor an issue [25]. While models and systems of BIA with 4-electrodes to assess body composition remain [26], and before embarking on the development of new equations, validation of existing equations is highly recommended. In conclusion, five chosen BIA equations against hydrodensitometry were not valid to estimate BF in young women in Colombia and the development of specific in young women in Colombia equation is recommended.

References

[1] WHO, 2011. *Global status report on non-communicable diseases 2010* Annex 4
[2] Ellis KJ, 2001. *J Nutr* May **131** (5) pp 1589-1595
[3] Kyle UG, Bosaeus I, De Lorenzo AD, Deurenberg P, Elia M, Gómez JM, Heitmann BL, Kent-Smith L, Melchior JC, Pirlich M, Scharfetter H, Schols AM, Pichard C and Composition of the ESPEN Working Group 2004 *Clin Nutr.* **23** (5) pp 1226-1243
[4] Deurenberg P and Deurenberg-Yap M 2003 *Forum Nut.r* **56** pp 299-301
[5] Kyle UG, Bosaeus I, De Lorenzo AD, Deurenberg P, Elia M, Manuel Gómez J, Lilienthal Heitmann B, Kent-Smith L, Melchior JC, Pirlich M, Scharfetter H, M W J Schols A, Pichard C and ESPEN 2004 *Clin Nutr.* **23** (6) pp 1430-1453
[6] Haroun D, Taylor SJ, Viner RM, Hayward RS, Darch TS, Eaton S, Cole TJ and Wells JC 2010 *Obesity* Jun; **18**(6):1252-9
[7] Mattar L, Godart N, Melchior JC, Falissard B, Kolta S, Ringuenet D, Vindreau C, Nordon C, Blanchet C and Pichard C 2011 *Clin Nutr.* 2011 Dec; **30**(6):746-52
[8] Moon JR 2013 *Eur J Clin Nutr.* Jan; **67** Suppl 1:S54-9
[9] Ramsey R, Isenring E and Daniels L 2012 *J Nutr Health Aging.* Jan; **16**(1):26-30
[10] Nasreddine L, Naja F, Hills AP, Youssef SK, Chahine J and Hwalla N 2012 *Clin Nutr.* Jun; **31**(3):364-71
[11] Gonzalez-Correa CH and Caicedo-Eraso JC 2012 *J. Phys.: Conf. Ser.* **407** 012018
[12] Caicedo-Eraso JC, Gonzalez-Correa CH and Gonzalez-Correa CA 2013 Preliminary BIA equation... *ICEBI Proceedings. J. Phys.: Conf. Ser.* (In press)
[13] Cornish BH, Thomas BJ and Ward LC 1998 *Appl Radiat Isot.* May-Jun **49** (5-6) pp 475-476
[14] Mitchell CO, Rose J, Familoni B, Winters S and Ling F 1993 *Basic Life Sci.* **60** pp 189-191
[15] Deurenberg P, Weststrate JA, Paymans I and van der Kooy K 1988 *Eur J Clin Nutr.* Dec **42** (12) pp 1017-1022.
[16] Lohman TG, Roche AF and Martorell R 1988 *Anthropometric standardization reference manual*
[17] Xitron Technologies 2007 *Hydra ECF/ICF (Model 4200) Operating Manual*
[18] Liang MT, Su HF and Lee NY 2000 *Med Sci Sports Exerc.* Jan **32** (1) pp 221-227
[19] Wilmore JH 1969 *J Appl Physiol.* Jul;**27**(1):96-100
[20] Akers R and Buskirk ER 1969 *J Appl Physiol.* 1969 May; **26**(5):649-52
[21] Siri WE, 1956 *Adv Biol Med Phys.* **4**:239-80
[22] Baumgartner RN 1996. *Human Body Composition* Chap 5 (Human Kinetics Ed)
[23] Bland JM and Altman DG 1986 *Lancet.* Feb **8**, 1 (8476) pp 307-310
[24] Gonzalez-Correa CA, Gonzalez-Correa CH and Ramos-Rodriguez A 2007 *IFMBE Proceedings* **17** pp 803-806
[25] Rush EC, Bristow S, Plank LD and Rowan J 2013 *Eur J Clin Nutr.* Feb; **67**(2):214-7
[26] Bosy-Westphal A, Schautz B, Later W, Kehayias JJ, Gallagher D and Müller MJ 2013 *Eur J Clin Nutr.* Jan **67** Suppl 1:S14-21

Journal of Physics: Conference Series **434** (2013) 012066 doi:10.1088/1742-6596/434/1/012066

Preliminary bioelectrical impedance analysis (BIA) equation for body composition assessment in young females from Colombia

J C Caicedo-Eraso[1,2], C H Gonzalez-Correa[1], C A Gonzalez-Correa[1]
[1] Research Group on Electrical Bioimpedance, University of Caldas, Colombia
[2] Department of Systems and Informatics, University of Caldas, Colombia

E-mail: julioc.caicedo@ucaldas.edu.co

Abstract: A previous study showed that reported BIA equations for body composition are not suitable for Colombian population. The purpose of this study was to develop and validate a preliminary BIA equation for body composition assessment in young females from Colombia, using hydrodensitometry as reference method. A sample of 30 young females was evaluated. Inclusion and exclusion criteria were defined to minimize the variability of BIA. Height, weight, BIA, residual lung volume (RV) and underwater weight (UWW) were measured. A preliminary BIA equation was developed (r^2=0.72, SEE=2.48 kg) by stepwise multiple regression with fat-free mass (FFM) as dependent variable and weight, height and impedance measurements as independent variables. The quality of regression was evaluated and a cross-validation against 50% of sample confirmed that results obtained with the preliminary BIA equation is interchangeable with results obtained with hydrodensitometry (r^2=0.84, SEE=2.62 kg). The preliminary BIA equation can be used for body composition assessment in young females from Colombia until a definitive equation is developed. The next step will be increasing the sample, including a second reference method, as deuterium oxide dilution (D_2O), and using multi-frequency BIA (MF-BIA). It would also be desirable to develop equations for males and other ethnic groups in Colombia.

1. Introduction

Total body water (TBW) by isotope dilution, hydrodensitometry, magnetic resonance imaging (MRI), computerized tomography (CT) and dual-energy X-ray absorptiometry (DXA) provide precise and reliable measurements of body composition and could be considered as reference methods for in vivo body measurements [1]. However, cost, accessibility, complexity and the radiation exposure with DXA, MRI and CT limit the use of these techniques [1,2]. BIA is based on measuring tissue conductivity and the relation between the volume of the conductor and its electrical impedance [3]. BIA is a simple, safe, noninvasive, relatively inexpensive method that is practical and suitable for individual use, large-scale researches and field studies in the community [4,5]. However, practice has shown that a prediction BIA equation developed for one population are valid for that specific population and can accurately predict body composition, but it is not valid when applied to a different ethnic group [6,7]. A previous study showed that available BIA equations in the literature are not suitable for our population [7]. Therefore, the purpose of this study was to develop and validate a preliminary BIA equation for body composition assessment in young females from Colombia.

2. Materials and methods

2.1. Subjects

The methods, classified as minimum risk by the Colombian Ministry of Health, were approved by the Bioethics Committee of the University of Caldas. Variations in body composition related to gender, age, ethnicity and genetic influence which affect BIA measurements [5,6] were minimized by using an adequate protocol [8] and selecting a homogeneous group of young females [5]. A group of 30 young females was evaluated. The purpose and procedures of the study were explained to the volunteers and the inclusion/exclusion criteria were verified after completing a questionnaire. Afterwards, the volunteers signed an informed written consent. The same subjects and the same measurement data were used in other study to show that available BIA equations in the literature are not suitable for our population [7]. Characteristics of subjects by mean and standard deviation (SD) are shown in table 1.

Table 1. Characteristics of subjects.

Variables (n=30)	Mean	SD
Age (years)	20.4	2.0
Height (cm)	158.1	4.8
Weight (kg)	54.8	8.0
BMI (kg/m^2)	21.9	2.7

2.2. Data acquisition

Measurements were performed in one session early in the morning to minimize environmental [9] and biological [6] variations. Relative humidity (RH) (70.5 ± 3.5%) and environmental temperature (20.5 ± 1.0 ºC) were measured with a thermo-hygrometer (13307 by DeltaTrak®, ±0.1 ºC / ±1% RH) and controlled with an electric heater (BFH416 by Bionaire™) and a dehumidifier (BMD100 by Bionaire™). Volunteers were asked to comply with the following requirements before the test: 48 hours without drinking alcohol, 12 hours without vigorous exercise and 12 hours of fasting but keeping normal water hydration. All subjects were asked to evacuate their bladder and colon 30 minutes before the tests and were evaluated out of menses [5,10]. Other possible sources of error like hydration status, subject position, limb abduction, electrode position, wearing jewelry and skin alcohol use, were also controlled in this study [5,10,11]. All volunteers wore underwear and a hospital gown.

2.3. Anthropometric measurements

Height (S) (Heightronic-235 by Seca®, ±0.01 cm) and weight (W) (PP2000 by Icob-Detecto®, ±0.1 kg) were measured twice, and a third measurement was taken if it was found a difference greater than 0.5 cm or 0.1 kg respectively [12].

2.4. BIA measurements

Bioelectrical impedance was measured (Hydra 4200 by Xitron Technologies©), using the manufacturer's recommended electrodes (292-STE by Impedimed©), on the dominant side of body for three times at the end of an exhalation on a nonconductive surface. Raw resistance (R_{50}) and capacitive reactance (Xc_{50}) at 50 kHz data were used. Subjects remained in a supine position by 5 minutes and BIA measurements were made between minutes 6 to 10, with the arms comfortably separated from the body 15 degrees, and the legs comfortably separated about 45 degrees [5,13]. Dorsal hand and anterior foot surfaces were cleaned with alcohol and dried with a paper towel [5]. For reproducible measurements, four landmarks were made for the placement of the electrodes: midline between prominent ends of radius and ulna of wrist, midline of third metacarpal-phalangeal joint on dorsal hand surface, midline between the medial and lateral malleolus of ankle and midline of third metatarsal-phalangeal joint on anterior surface of foot [5,13]. Current was applied at distal electrodes and voltage was measured at proximal electrodes [5,13]. Distance between current and voltage electrodes was always greater than 5 cm [5]. A blanket was used for comfort and homogeneous skin temperature [5,14].

2.5. Hydrodensitometry measurements

This technique obtains the body density and implies measuring RV and UWW. RV was measured by the nitrogen washout technique (Quark PFT-2 by COSMED, ±0.01 L) and taking the average of two measurements whose difference was less than 100 cc [15]. UWW was measured (PROGAN-1500SS by Prometalicos, ±0.001 kg) within a tank filled with water (35.0 ± 2.0 ºC) and taking the average of two measurements whose difference was less than 100 gr. Subject was in sitting position and wore a form-fitting swimsuit and a ballast weight in the waist [16]. Body fat (BF) was obtained applying the model of two components by Siri equation [17].

2.6. Statistical methods

The mean and the standard deviation (SD) were used to evaluate the characteristics of subjects and laboratory conditions. A preliminary BIA equation was developed by multivariate normal distribution, stepwise multiple regression analysis with multiple and partial correlation coefficient with FFM as dependent variable and W, S and impedance measurements R_{50} and Xc_{50} as independent variables. Quality regression was evaluated by coefficient of determination $r^2 \geq 0.60$, SEE<2.7 kg, Fisher analysis of variance F≤1%, Malows index Cp, still valid Cp±1, reliability of 95%, and statistical power PW≥80%. A cross-validation against 50% of the sample was used and a paired Student t-test and the Bland & Altman analysis [18] were used to determine significant differences ($p<0.01$) against hydrodensitometry and only SEE lesser than 2.7 kg was accepted as valid [19].

3. Results

Six predictors were analyzed: S, W, R_{50}, Xc_{50}, S^2 and S^2/R_{50}. A preliminary equation (1) was developed to determine FFM (and BF) for young females from Colombia by stepwise multiple regression analysis. BF was determined by subtracting FFM from W. Equation (1) had $r^2=0.72$ ($r=0.85$), SEE=2.48 kg and F=0.05%; with a PW of 80%, IC of 95% and Cp=5.

$$FFM = 0.240W + 0.022R_{50} + 0.118Xc_{50} + 0.939\frac{S^2}{R_{50}} - 34.09 \quad (1)$$

A cross-validation was performed with 50% randomly selected subjects and SEE of BF was 2.62 kg with $r^2=0.84$ ($r=0.92$, $p<0.0001$) (figure 1). Paired Student's t-test ($p_t=0.57$) and Bland & Altman plot (figure 2) show no significant difference between BF obtained with the preliminary equation (1) against BF obtained with hydrodensitometry.

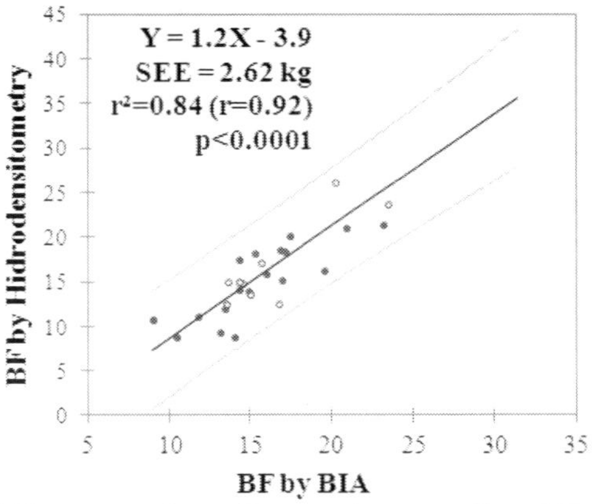

Figure 1. Regression of BF-Hydrodensitometry by BF-BIA (n=15).

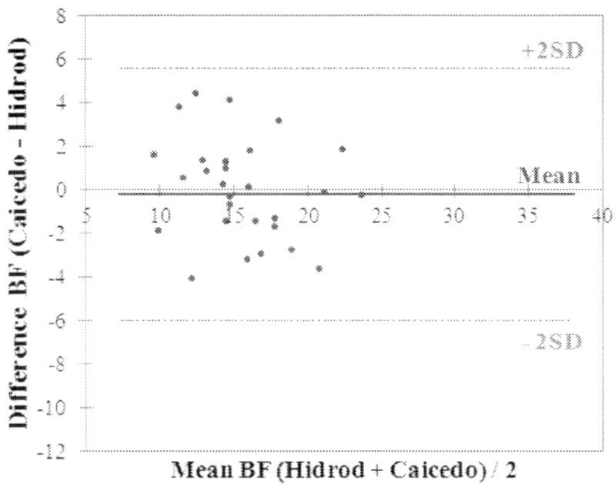

Figure 2. Bland & Altman plot of BF-Hydrodensitometry against BF-BIA (n=15).

4. Discussion and conclusion

BIA has higher precision and accuracy than anthropometric techniques [4, 5] if the proper equation is available [6] and an adequate protocol are used [8]. Additionally, cost and complexity is much smaller than the reference techniques [1,5]. Single frequency BIA (SF-BIA) was used as Moon et al (2013) found good agreement between DXA and SF-BIA (50 kHz) measurements when estimating FFM and BF [20] and Haas et al (2012) determined also that SF-BIA agreed well with D_2O to estimate TBW [21]. Furthermore, in 2013, SF-BIA equations to prediction of FFM still are developed [22]. The preliminary BIA equation met the proposed evaluation criteria and it could be used for estimating FFM and BF in young females from Colombia until a definitive equation is developed. The next step will be to improve r^2, PW and SEE, by increasing the sample, including a second reference method, as D_2O, and using MF-BIA. It would also be desirable to develop equations for males and other ethnic groups in Colombia. Maybe, in the future, the specificity of the population would be of minor importance due to the recent developments of segmental BIA with 8-electrodes [23]. However, the development of specific equations is still needed when using whole-body BIA with 4-electrodes [7].

References

[1] Ellis KJ, 2001 *J Nutr* May **131** (5) pp 1589-1595
[2] Chien MY, Huang TY and Wu YT 2008 *J Am Geriatr Soc.* Sep **56** (9) pp 1710-1715
[3] Houtkooper LB, Lohman TG, Going SB and Howell WH 1996 *Am J Clin Nutr.* Sep **64** (3 Suppl) pp 436-448
[4] Kyle UG, Bosaeus I, De Lorenzo AD, Deurenberg P, Elia M, Gómez JM, Heitmann BL, Kent-Smith L, Melchior JC, Pirlich M, Scharfetter H, Schols AM and Pichard C and Composition of the ESPEN Working Group 2004 *Clin Nutr.* **23** (5) pp 1226-1243
[5] Kyle UG, Bosaeus I, De Lorenzo AD, Deurenberg P, Elia M, Manuel Gómez J, Lilienthal Heitmann B, Kent-Smith L, Melchior JC, Pirlich M, Scharfetter H, M W J Schols A, Pichard C and ESPEN 2004 *Clin Nutr.* **23** (6) pp 1430-1453
[6] Deurenberg P and Deurenberg-Yap M 2003 *Forum Nutr.* **56** pp 299-301
[7] Caicedo-Eraso JC, Gonzalez-Correa CA and Gonzalez-Correa CH 2013 BIA equations validation… *ICEBI Proceedings. J. Phys.: Conf. Ser.* (In press)
[8] Gonzalez-Correa CH and Caicedo-Eraso JC 2012 *J. Phys.: Conf. Ser.* **407** 012018
[9] Cornish BH, Thomas BJ and Ward LC 1998 *Appl Radiat Isot.* May-Jun **49** (5-6) pp 475-476
[10] Mitchell CO, Rose J, Familoni B, Winters S and Ling F 1993. *Basic Life Sci.* **60** pp 189-191
[11] Evans WD, McClagish H and Trudgett C 1998 *Appl Radiat Isot.* May-Jun **49**(5-6):485-7
[12] Lohman TG, Roche AF and Martorell R 1988 *Anthropometric standardization reference manual*
[13] Xitron Technologies 2007 *Hydra ECF/ICF (Model 4200) Operating Manual*
[14] Liang MT, Su HF and Lee NY 2000 *Med Sci Sports Exerc.* Jan **32** (1) pp 221-227
[15] Flesch JD and Dine CJ 2012 *Chest.* Aug **142** (2):506-510
[16] Akers R and Buskirk ER 1969 *J Appl Physiol.* 1969 May; **26**(5):649-52
[17] Siri WE 1956 *Adv Biol Med Phys.* **4**:239-80
[18] Bland JM and Altman DG 1986 *Lancet.* Feb **8**, 1 (8476) pp 307-310
[19] Baumgartner RN 1996. *Human Body Composition.* **Chap 5** (Human Kinetics Ed)
[20] Moon JR, Stout JR, Smith-Ryan AE, Kendall KL, Fukuda DH, Cramer JT and Moon SE 2013 *Eur J Clin Nutr.* Jan; **67** Suppl 1:S40-6
[21] Haas V, Schütz T, Engeli S, Schröder C, Westerterp K and Boschmann M 2012 *Eur J Clin Nutr.* Sep; **66**(9):994-7
[22] Rush EC, Bristow S, Plank LD and Rowan J 2013 *Eur J Clin Nutr.* Feb; **67**(2):214-7
[23] Bosy-Westphal A, Schautz B, Later W, Kehayias JJ, Gallagher D and Müller MJ 2013 *Eur J Clin Nutr.* Jan **67** Suppl 1:S14-21

Journal of Physics: Conference Series **434** (2013) 012067 doi:10.1088/1742-6596/434/1/012067

A comparison between handgrip strength, upper limb fat free mass by segmental bioelectrical impedance analysis (SBIA) and anthropometric measurements in young males

C H Gonzalez-Correa[1,2], **J C Caicedo-Eraso**[1,2,3], **D R Varon-Serna**[1]
[1]Research Group on Nutrition and Body Composition, University of Caldas, Colombia
[2] Research Group on Electrical Bioimpedance, University of Caldas, Colombia
[3]Department of Systems and Informatics, University of Caldas, Colombia

E-mail: clara.gonzalez@ucaldas.edu.co

Abstract: The mechanical function and size of a muscle may be closely linked. Handgrip strength (HGS) is used as a predictor of functional performing. Anthropometric measurements are used to estimate arm muscle area (AMA) and physical muscle mass volume of upper limb (ULMMV). Electrical volume estimation is possible by segmental BIA measurements of fat free mass (SBIA-FFM), mainly muscle-mass. Relationship among these variables is not well established. We aimed to determine if physical and electrical muscle mass estimations relate to each other and to what extent HGS is to be related to its size measured by both methods in normal (n=43) or overweight (n=12) young males. Regression analysis was used to determine association between these variables. Subjects showed a decreased HGS (65.5%), FFM (85.5%) and AMA (74.5%). It was found an acceptable association between SBIA-FFM and AMA (r^2=0.60) but poorer between physical and electrical volume (r^2=0.55). A paired Student t-test and Bland and Altman plot showed that physical and electrical models were not interchangeable (*pt*<0.0001). HGS showed a very weak association with anthropometric (r^2=0.07) and electrical (r^2=0.192) ULMMV showing that muscle mass quantity does not mean muscle strength. Other factors influencing HGS like physical training or nutrition require more research.

1. Introduction

Health professionals as nutritionists, sport scientists and physiotherapists need to assess upper limb muscle mass quantity and functionality to monitor progress during nutritional treatments, physical training and rehabilitation. Accurate methods to perform these evaluations as computed tomography (CT), dual energy x-ray absorptiometry (DEXA) or magnetic resonance imaging (MRI), among others, are cumbersome and expensive. Handgrip strength (HGS) test is widely used as a functional parameter and is a good predictor of mortality from all-causes, cardiovascular disease and nutritional status [1]. The test is portable, quick, inexpensive and easy to measure [2]. HGS results may be influenced by the arm muscle mass volume (AMMV) [3]. Traditionally, anthropometric measurements of arm-muscle-area (AMA) or physical upper limb muscle mass volume (ULMMV) have been used to this purpose but errors behind their assumptions limit their use [4]. An alternative is the estimation of electrical ULMMV by segmental BIA (SBIA). Several studies have been performed to study lower limb estimations of body composition due to the importance of these limbs to independent life. Studies of the upper limb, composition, are scare but equally important in health assessment. Therefore, we wanted to know whether ULMMV estimated by these two methods associate to each other and to what extent HGS is to be related to its muscle size when measured by both methods in a group of normal or overweight young males.

2. Materials and methods

2.1. Subjects and data acquisition

The procedures were approved by the Bioethics Committee of the University of Caldas. The purpose and methods of the study were explained to a sample of 55 young college males. Inclusion criteria were: being male, aged 18 to 30 years and with no comorbidities. Exclusion criteria were: having muscle diseases or arthritis, being a smoker, having metal implants or pacemakers or be using diuretics. All measurements were performed in one session by the same researcher. Relative humidity and environment temperature were kept stable (63.9±1.5% and 22.4±1.0 °C) by using an electric heater (BFH416 by Bionaire™) and a dehumidifier (BMD100 by Bionaire™). These variables were measured with a thermo-hygrometer (13307 by DeltaTrak®, ±0.1 °C). Volunteers were asked to comply with standardized requirements before the test [5].

2.2. Anthropometric measurements

Height (Heightronic-235 by Seca®, ±0.01 cm) and weight (PP2000 by Icob-Detecto®, ±0.1 kg) were measured twice. A third measurement was taken if there was a difference greater than 0.5 cm or 0.1 kg respectively [6]. Subjects were characterized as normal or overweight according to BMI [7]. Mid upper circumference was measured twice at the mid point length between acromion and olecranon. The triceps skin fold was measured in the midline of posterior aspect of the arm, at a point midway between the lateral projection of the acromion process and the inferior margin of the olecranon. AMA was calculated and the reference values were taken from Frisancho et al, 1974 [8]. For SBIA measurements the circumferences at the site of voltage-electrodes were measured. The distance-length between the centers of voltage-detector electrodes were measured accurate to 0.1 cm. Physical volume (Vp) was calculated as by $Vp = [(\pi L)/3].[r_1^2 + r_1 r_2 + r_2^2]$, where L is the distance between the detector electrodes, and $r_{1,2}$ is the radius of the circumferences of the upper limb at the detector electrodes [9].

2.3. SBIA measurements

Bioelectrical impedance measurements (Hydra 4200 by Xitron Technologies©) were made on the dominant side of body for three times on a nonconductive surface and after resting for 5 minutes [10]. Subjects were in a supine position with the limbs comfortably abducted. Two current-injection electrodes (2228 by 3M®) were placed on the dorsal surface proximal to the metacarpal-phalangeal and the lateral projection of acromion process and the center of two-voltage-detector electrodes were placed on the mid-line between the prominent ends of the right radius and ulna of the wrist and in the axillary fold [11]. Raw data of resistance and reactance (800 µA at 50 kHz) were used to calculate impedance (Z_{50}) and estimation of FFM was made by equations for segmental measurements from the BIA analyzer manufacturer. Because reference values for SBIA-FFM were not available for this population, we used the FFM mean of our subjects with normal BMI, HGS and AMA as reference. This value was FFM=2.93 kg. In addition, the values of impedance (Z_{50}) were compared with values from Organ, 1994 [12]. The electrical volume (Ve) was calculated as $Ve = p(L^2/R)$, where p is the resistivity of skeletal mass (118 Ω.cm) [9], L is the distance between the detector electrodes and R is the measured resistance.

2.4. Handgrip strength

HGS was measured on the dominant side 3 times using a hydraulic dynamometer (Baseline®; ±1 kg) and following the guidelines of the American Association of Hand Therapists [13]. Subjects were categorized as normal if they had greater than 85% from the reference values extracted from Mathiowetz [14].

2.5. Statistical methods

Characteristics of the subjects and laboratory conditions were expressed as mean and standard deviation (SD). Subjects with %HGS, FFM, AMA, and Z_{50} out of references values were expressed as percentage. Regression analysis method (α=0.05) was used to determine association of SBIA-FFM with AMA and Vp with Ve; and later between %HGS by AMA, SBIA-FFM, Vp and Ve. A paired Student t-test and a Bland and Altman analysis were used to determine significant differences between Vp and Ve (p<0.05).

3. Results

Table 1 shows subject's characteristics and percentage of individuals having decreased or increased values for different variables according to reference values. The data distribution was normal for subjects with normal (n=43) and overweight (n=43) BMI. Except for %HGS, all variables showed significant differences among the 2 groups. When values of Z_{50} for upper limb were compared with values from Organ, 1994 [12], it was found that subjects with normal BMI had more regional fat mass than subjects categorized as overweight (83.7% vs. 58.3%). Figures 1a and 1b show the regression of SBIA-FFM by AMA (r^2=0.60) and Vp by Ve (r^2=0.55). A paired Student t-test and a Bland and Altman plot (Figure 1c) show that Vp and Ve were not interchangeable (p_t<0.0001). Regression analysis of %HGS by anthropometry showed a very weak association with AMA (r^2=0.219) and with Vp (r^2= 0.070) as well as with electrical variables, SBIA-FFM (r^2=0.204) and Ve (r^2=0.192).

Table 1. Characteristics of subjects (n=55).

Variables	Mean	SD	Normal BMI group (A)	Overweight BMI group (B)	All subjects (A+B)	Significant differences A vs B
Age (years)	19.8	1.9				
Height (cm)	171.1	6.2				
Weight (kg)	66.8	8.3	# (%All)	# (%All)	# (%All)	p_t
BMI (kg/m^2)	22.7	2.3	43 (78.2%)	12 (21.8%)	55 (100%)	< 0,0001
*%HGS (%)	80.9	13.2	30 (54.5%)	6 (10.9%)	36 (65.5%)	0.092
*FFM (kg)	2.6	0.3	39 (70.9%)	8 (14.5%)	47 (85.5%)	0.002
*AMA (mm^2)	5162.2	910.4	35 (63.6%)	6 (10.9%)	41 (74.5%)	0.003
*Physical volume (L)	3.3	0.4	30 (54.5%)	5 (9.1%)	35 (63.6%)	0.005
*Electrical volume (L)	1.2	0.2	31 (56.4%)	5 (9.1%)	36 (65.5%)	0.009
**Z (Ω)	212.9	24.0	36 (65.5%)	7 (12.7%)	43 (78.2%)	0.003

*Increased variable; **Decreased variable; #: number of subjects; %All: % of all subjects

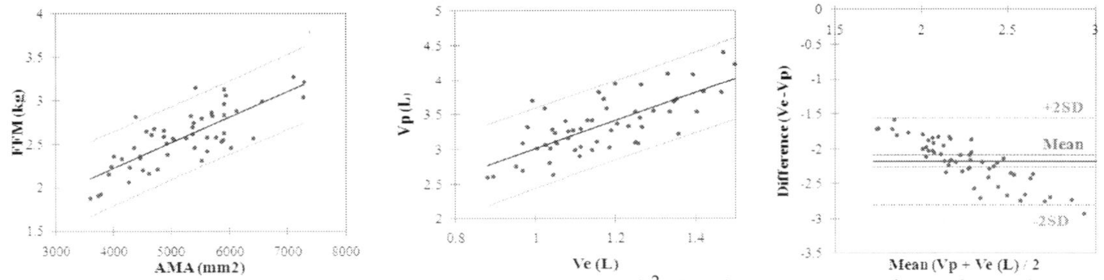

Figure 1. (a) Regression of SBIA-FFM by AMA (r^2=0.60), **(b)** Regression of Vp by Ve (r^2=0.55), **(c)** Bland and Altman plot of Ve against Vp.

4. Discussion and conclusion

The study tested the hypothesis that the greater the upper limb muscle mass volume (ULMMV), measured by two methods, the greater %HGS. We also expected that overweight subjects had affected HGS because of muscle fat infiltration [15]. It was remarkable that subjects with normal BMI had decreased %HGS and AMA, and increased arm Z_{50} in a greater proportion than overweight subjects.

We did not measured total body fat; however, according to the two level body composition model, the results suggest that overweight was due to a larger amount of lean body mass and not because a higher fat mass as it was corroborated by Z_{50}. Maybe, they had more physical training than subjects with normal BMI but this issue was not addressed in this study. This highlights the need for a nutritional assessment beyond the BMI. The results showed an acceptable association between SBIA-FFM and AMA (r^2=0.60, r=0.77) and between anthropometric and electrical volume (r^2=0.55, r=0.74). The former association would suggest a discrete similitude of muscle mass estimations by the two methods as some authors have shown [16] but the methods were not interchangeable. Evaluations of thigh physical and electrical muscle mass volume against DEXA [9] or MRI [17], showed that Vp correlated with fat mass and Ve was a better predictor of muscle mass volume. This would lead us to suggest using SBIA instead of anthropometry for lean tissue assessment. Nonetheless some studies have found that SBIA tends to underestimate ULMMV in men from 5% [18] up to 20% [19] and more research is required to define the accuracy of SBIA. It appears that subjects with normal and overweight BMI had consistent results for %HGS and ULMMV and that a loss of HGS was accompanied by decreased physical and electrical volumes. Nevertheless, HGS as functional parameter showed a very weak association with SBIA-FFM, AMA,Vp and Ve parameters, and these variables did not explain satisfactorily the high percentage of subjects with loss of HGS. Therefore, the results suggest that muscle mass quantity does not mean muscle strength, agreeing with some authors [20]. Other variables, like physical training or nutrition could explain the loss of HGS and prospective studies to exam the association between loss of muscle mass and physical function may be warranted. Many studies have been performed to exam and associate lower limb composition and muscle strength and have shown little association between muscle mass size and function but fewer studies have made for upper limb. This study suggests that results for upper limb are similar to those examining lower limb.

References

[1] Gale CR, Martyn CN, Cooper C, Sayer AA 2007 *Int J Epidemiol* **36** 228-35
[2] Bohannon RW 2008 *J Geriatr Phys Ther.* **31**(1) 3-10
[3] Cruz-Jentoft AJ, Baeyens JP, Bauer JM, Boirie Y, Cederholm T, Landi F, Martin FC, Michel JP, Rolland Y, Schneider SM, Topinková E and Zamboni M 2010 *Age Ageing.* **39** (4) 412-23
[4] Lukaski HC 1996 *Human Body Composition.* Chapter 6. (Human kinetics Ed)
[5] Gonzalez-Correa CH, Caicedo-Eraso JC 2012 *J. Phys. Conf. Ser.* **407** 012018
[6] Lohman TG, Roche AF and Martorell R 1988 *Anthropometric standardization reference manual*
[7] WHO, 2011 *Global status report on non-communicable diseases* 2010 *Annex 4*
[8] Frisancho AR, 1974 *Am J Clin Nutr.* **27**(10) 1052-8
[9] Lukaski HC 2000 Ann N Y Acad Sci. **904** 154-8
[10] Xitron Tech 2007 *Hydra ECF/ICF (Model 4200) Operating Manual*
[11] Chumlea WC, Baumgartner RN and Roche AF 1988 *Am J. Clin Nutr.* **48** (1)7-15
[12] Organ LW, Bradham GB, Gore DT, Lozier SL 1994 *J. Appl Physiol.* **77** (1) 98-12
[13] Innes E 1999 *Aust Occup Ther J.* **46** (3) 120-40
[14] Mathiowetz V, Kashman N, Volland G, Weber K, Dowe M and Rogers S 1985 *Arch Phys Med Rehabil.* **66** (2) 69-74
[15] Delmonico MJ, Harris TB, Visser M, Park SW, Conroy MB, , Boudreau R, Manini TM, Nevitt M, Newman AB and Goodpaster BH 2009 *Am J. Clin Nutr.* **90**(6) 579-85
[16] Casanova M, Rodríguez I, Rico S and Casanova M 2004. *An Pediatr.* **61**(1) 23-31
[17] Stahn A, Terblanche E and Strobel G 2007 *J Appl Physiol.* **103** (4) 1428-35.
[18] Mally K, Trentmann J, Heller M and Dittmar M 2011*Eur J Appl Physiol.* **111** (8)1879-87
[19] Jung-Jui C, Ming-Feng K, Chih-Lin C, Hsueh-Kuan L, Ming-Chang W, Yu-Yawn C and Kuen-Chang Hsieh 2011 *Acad. J.* **6** (24) 5131-7
[20] Visser M, Deeg DJ, Lips P, Harris TB and Bouter LM 2000 *J Am Geriatr Soc* **48** (4)381-6

Coupling IR Thermography and BIA to analyse body reaction after one acupuncture session

M Piquemal[1]
[1]Independent investigador,
Casilla Correo 2899, Assomption, Paraguay

E-mail:piquemal@gmail.com

Abstract. Coupling both thermography and bio-Impedance, some biophysical acupuncture mechanisms are statically studied on a small population of 18 subjects. Results show that a possible way of understanding acupuncture, in an electrical way, should be to consider ionic flux redistribution between vascular and extra cell compartments. This is a two steps mechanism. The first one is starting with needles insertion and the second one is lasting with more intensity after removing them from skin.
Key words: Bio-Impedance, I.R. thermography, acupuncture, cell membrane.

1. Introduction

In previous studies, we have noticed that the phase angle, captured by single-frequency four-electrode bioelectrical impedancemeter, varies as high as 0.2° during one acupuncture session [1]. Coupling thermography and BIA, the current investigation attempts to understand phase angle variation mechanisms based on physiology.

2. Material and Method

2.1.Material

IR thermography is captured by a thermal Inframetrics camera, model 525, 8 bits resolution 2].
Bioimpedancemeter is model BIA 2000 S, from Data-Input (Frankfurt, Germany) [3].
We use Gel electrodes, model Red Dot, (3M), compatible with BIA 2000 S [4, 5].
Image processing is done by Matlab software [6].

2.2.Method

The study group consists of 18 patients (32 ±16 years, 56% female) suffering different kinds of pathologies. Group control (healthy and without needle) includes 10 people (54±5 years, 40% female). After a detailed explanation of the purposes and the protocol of the experiment, all subjects signed an informed consent. Thermal and bioimpedance analysis take place in the morning, at the same hour, patient with empty stomach [7]. The protocol begins with a fifteen minute rest for each patient, lying in supine position, in close metabolism test conditions, in accordance with the recommendations chart provided by the manufacturer. Data acquisition bases are double and done three times. Both techniques surround a single ten minutes acupuncture session based on three needled zones (acupuncture points) according to symptomatology and Chinese rules of acupuncture, but never done in the thermal region of interest (ROI), to avoid direct segmental spinal cord reflex.
In this way, local vasomotricity change will not be related to segmental arc reflex but reflects more integrative brain reflex. ROI is selected, too, not to interfere in acupuncture session.

Four gel electrodes are settled on the posterior face of right hand and on the anterior region of the right foot. Skin electrodes remain stuck, during the protocol. Thermography is used upon the anterior region of the right foot (metatarsus), from skin between the two gel electrodes, before and after a single acupuncture session.Then, a single-frequency four-electrode bioimpedance test is performed (resistance and reactance). The last coupled data acquisition is collected ten minutes after the needles are removed.

ROI is partitioned into three equal surfaces, where the pixels are standardized and averaged to transform qualitative data into quantitative ones. These three partitions are as follows: close to injection electrode, central and close the reception electrode. Phase angle improves by 0.2° each time after acupuncture session

Three distinct periods seem relevant: before (period 1), after a ten minute acupuncture session (period 2) and finally, ten minutes after removing needles (period 3).

3. Results

Simple statistical and correlation analysis are done between thermal data base (reflecting change in local vasomotricity) and bioimpedance analysis results (only reactance and resistance standing for cell membrane) for each time of this three periods protocol. Thermal and BIA changes are explored according to error percentage method [(Initial value- final value)/ Initial value]. To conclude, relationship between thermal and BIA data base is established according to correlation study process.

3.1.Correlation analysis between starting point and immediately after a ten minutes acupuncture session.

3.1.1.Simple Statistical analysis. For ten subjects over the whole population, vasomotricity just shows a 2% change. Speaking about BIA, table1 shows percentage of changes for both, resistance, reactance, phase angle α

Table 1. Percentage of variation of BIA values. Period1 versus Period 2. a) With acupuncture b) Without acupuncture (N=10, 54±5 years; 40% Female)

Statistical Parameters	%Variation of Resistance		% Variation of Reactance		% Variation of α	
Maximum	a) 2.73	b) 1.2	a) 4.69	b) 2.4	a) 2.92	b) 3.6
Minimum	a) -0.52	b) -0.3	a) -2.04	b) – 3.7	a) -2.25	b) -2.6
Average	a) 0.62	b) 0.3	a) 1.60	b) 0.3	a) 0.59	b) -0.1
Standard deviation	a) 0.77	b) 0.005	a)1.66	b) 0.02	a) 1.12	b) 0.02

3.1.2. Correlation between changes in vasomotricity in the three foot zones and BIA values. (Table 2)

Table 2. Coefficient correlation between thermography of the three rectangular zones and BIA values (Resistance and Reactance).

Thermal zones	Resistance	Reactance
Injection zone	- 0.84	- 0.53
Central zone	- 0.60	- 0.44
Voltage zone	- 0.32	- 0.32

During acupuncture session, vasomotricity is better correlated with resistance than reactance, for zone close to injection current electrode.

Following the same procedure, we analyze what is happening after removing the three needles from skin insertion.

Journal of Physics: Conference Series **434** (2013) 012068 doi:10.1088/1742-6596/434/1/012068

3.2. Correlation analysis immediately after acupuncture and after 10 minutes of rest (needles out off skin)

3.2.1. Simple Statistical analysis. For the whole population minus one people, vasomotricity change is homogeneous. From BIA, results are quite different. Table 3 shows percentage of changes for both, resistance and reactance.

Table 3. Percentage of variation of BIA values. Period 2 versus Period 3.

Statistical Parameters	%Variation of Resistance	% Variation of Reactance	% Variation of α
Maximum	5.14	4.69	2.87
Minimum	-0.34	-0.00	-1.35
Average	1.40	2.70	1.09
Standard deviation	1.11	1.37	1.10

Even without needles insertion, we notice that electrical tissue properties still changing

3.2.2 Correlation between changes in vasomotricity of the three ROI versus BIA values. (Table 4)

Table2 . Coefficient correlation between thermography of the three rectangular zones and BIA values (Resistance and Reactance).

Thermal zones	Resistance	Reactance
Injection zone	- 0.21	- 0.20
Central zone	- 0.00	- 0.19
Voltage zone	- 0.06	- 0.34

After removing needles, we cannot observe any relationship between vasomotricity and electrical changes in tissue. We have to consider changes in electrical tissue properties (table 3) to cell membrane ionic flows, as no vascular change is noticed from previous state.

4. Interpretation

Changes in resistance are due to ionic flow in the extra cellular matrix. Reactance relies on spherical capacitor effect of cell membrane so transmembrane ionic flow.

During acupuncture session both mechanisms (blood flow and electrical properties of tissue) are correlated. After removing needles, electrical mechanisms are dissociated from thermal data, that means with blood flow. Both modifications (resistance and reactance) involve ionic flow. The first one is related to vasomotricity as thermal change is noticed with respect to previous vascular state. Second one seems to be more correlated with ionic membrane transfer as previously extracellular matrix has been modified, in an electrical sense, by needle nociceptive actions on central nervous system. Removing needle seems to have left new dynamics in ionic flow balance. No more vasomotricity dominance, physiologically speaking. From now starts the next step: transmembrane ionic equilibrium to reach a new balance regarding electrolytes distribution, at a cellular level. This step is triggered by the local electrical charges concentration. So acupuncture mechanisms seem to be, at least, a two steps physiological process in a schematic way of biophysics thought. First, acupuncture process is modifying extracellular matrix by changing blood flow pattern, thanks to the local needles insertion (central nervous system integration based on nociceptive message with autonomic nervous system response). Then, after removing the stimuli (needles), begins the second auto-regulated phenomenon.

This time, electrical changes occur at cellular level. High concentration ionic charges, present in extra-cellular compartment, modify transmembrane equilibrium, increasing in that way, the capacitive properties of the cells.

Journal of Physics: Conference Series **434** (2013) 012068 doi:10.1088/1742-6596/434/1/012068

5. Conclusion

The study group consists of 18 patients (32 ±16 years, 56% female) suffering different kinds of pathologies. To understand better acupuncture mechanisms on both physics and physiological basis, coupled IR thermal (Inframetrics model 525, Usa) and bioimpedancemetry (BIA,model 200- s, Data-input, Germany) are monitoring changes in the physiological process, after a unique three needling acupuncture session.

The protocol begins with a fifteen minute rest for each patient, lying in supine position, in close metabolism test conditions, in accordance with the recommendations chart provided by the manufacturer. Data acquisition bases are double and done three times. Both techniques surround a single ten minutes acupuncture session based on three needled zones (acupuncture points) according to symptomatology and Chinese rules of acupuncture, but never done in the thermal region of interest (ROI). Thermography is the first of two tests providing quantifiable data. Thermography is used upon the anterior region of the right foot (metatarsus), from skin between the two gel electrodes, before and after a single acupuncture session. Then, a single-frequency four-electrode bioimpedance test is performed (resistance and reactance). The last coupled data acquisition is collected ten minutes after the needles are removed. Phase angle improves by 0.2° each time after acupuncture session.

The results from the correlation between thermal and bioimpedance data show that a possible method of understanding acupuncture, from an electrical point of view, could be to first analyse ionic flux redistribution in tissues from the vascular system and then ionic change on both side of cellular membrane. This is a two-step mechanism, the first starting with the insertion of the needles (vasomotricity versus resistance) and the second (reactance) taking place after the removal of said needles from the skin due to local ionic cell membrane redistribution, as phase angle changes.

References

[1] Piquemal M 2005 Acupuncture et membrane cellulaire au vue de la bioélectricité. *Acupuncture & Moxibustion* **4** 307-13

[2] Inframetrics model 525, Infrared imaging radiometer. Available from URL: http://www.flir.com/uploadedFiles/CVS/Markets/Legacy/Documents/

[3] BIA 2000-s. Available from URL: http://www.data-input.de/site/_data/pdf/ ga_BIA_2000S_2008.pdf

[4] Red dot wet gel monitoring electrodes. Available from URL: http://www.3m.com/product/information/Red-Dot-Wet-Gel-Monitoring-electrode.html

[5] U. Pliquett et al. 2009 Evaluation of electrodes for impedance spectroscopy. *IFMBE Proceedings*; **25** 251-4

[6] Matlab. Available from URL: http://www.mathworks.com/products/ image/

[7] Lukashi HC 1999 Requirements for clinical use of bioelectrical impedance analysis . *Ann NY Acad Sc.* **873** 72-6

Electrical impedance myography for the assessment of children with muscular dystrophy: a preliminary study.

S B Rutkove and B T Darras

Department of Neurology, Beth Israel Deaconess Medical Center, and Department of Neurology, Children's Hospital Boston, Harvard Medical School, Boston, MA, USA

Email: srutkove@bidmc.harvard.edu

Abstract. Electrical impedance myography (EIM) provides a non-invasive approach for quantifying the severity of neuromuscular disease. Here we determine how well EIM data correlates to functional and ultrasound (US) measures of disease in children with Duchenne muscular dystrophy (DMD) and healthy subjects. Thirteen healthy boys, aged 2-12 years and 14 boys with DMD aged 4-12 years underwent both EIM and US measurements of deltoid, biceps, wrist flexors, quadriceps, tibialis anterior, and medial gastrocnemius. EIM measurements were performed with a custom-designed probe using a commercial multifrequency bioimpedance device. US luminosity data were quantified using a gray-scale analysis approach. Children also underwent the 6-minute walk test, timed tests and strength measurements. EIM and US data were combined across muscles. EIM 50 kHz phase was able to discriminate DMD children from healthy subjects with 98% accuracy. In the DMD patients, average EIM phase measurements also correlated well with standard functional measures. For example the 50 kHz phase correlated with the Northstar Ambulatory Assessment test ($R = 0.83$, $p = 0.02$). EIM 50 kHz phase and US correlated as well, with $R = -0.79$ ($p < 0.001$). These results show that EIM provides valuable objective measures Duchenne muscular dystrophy severity.

1. Introduction

Duchenne muscular dystrophy (DMD) is a progressive neuromuscular disease that affects approximately 1 in 3000 male births [1]. Caused by mutations in the dystrophin gene that encodes a membrane structural support protein, DMD begins impacting boys at just 2-4 years of age [2]. Initially subtle weakness affecting the proximal lower extremities progresses until the child has increasing difficulty walking. By 12 years, most boys are wheelchair-bound and have limited upper extremity function; nearly all die from a combination of respiratory and heart failure before age of 30 [1]. Currently, only corticosteroids have been shown to slow the loss of function in this disease [3]. However, since the genetic basis of the disease is now well-understood, a variety of new potential treatments are being tested in clinical trials [4]. One major limitation to this testing is that the outcome measures or biomarkers used to assess drug efficacy are limited. Functional measures, such as the distance a boy can walk in 6 minutes, the 6-minute walk test, is one major measure used, but it is limited by a variety of factors [5]. New, non-invasive, methods for assessing disease status are needed. One potential method is the use of localized impedance measurements of muscle, or electrical impedance myography (EIM) [6]. Studies in other neuromuscular diseases, including amyotrophic

Journal of Physics: Conference Series **434** (2013) 012069 doi:10.1088/1742-6596/434/1/012069

lateral sclerosis [7] and spinal muscular atrophy [8], have shown that EIM is sensitive to the severity of disease and can serve as a useful biomarker of disease change over time. In this study, we preliminarily evaluate the potential use of EIM in the assessment of DMD by comparing EIM data to standard functional measures in a group of DMD boys and a group of healthy boys while also assessing the relationship between EIM and ultrasonography.

2. Methods

A total of 19 boys, aged 2 to 12 years, mean age 7.3 years with genetically confirmed DMD and a total of 20 healthy boys, ages 2 to 12 years, mean age 7.4 years, were recruited for study at Children's Hospital Boston. The study was approved by the hospital's institutional review board; parents provided written informed consent and children verbal assent. All boys underwent functional testing, including the 6MWT, the North Star Ambulatory Assessment, and quantitative strength testing.

2.1. EIM Testing

All subjects had EIM testing of unilateral deltoid, biceps, wrist flexors, quadriceps (rectus femoris), tibialis anterior, and medial gastrocemius. EIM measurements were performed with the Imp SFB7 (Impedimed, Inc, Sydney Australia), configured with a custom-designed hand-held electrode array, based on a design that has been previously described [9] with current flow parallel to the muscles. Three different array sizes were used depending on the size of the child. Although multifrequency data were collected from each muscle, only the 50 kHz data were analyzed here for convenience sake. Additional analysis of the multifrequency data will be performed separately.

2.2. Ultrasound Measurements

Ultrasound was also performed with the Terason t3000 system (Teracorp, Inc, Burlington, MA) using a 10 MHz probe. Measurements were performed on the identical set of muscles on which EIM was performed in the identical locations were the EIM probe was placed. The ultrasound probe was placed transversely on the muscle. Single images were obtained from each muscle. The images were then downloaded into Photoshop ® (Adobe Inc, San Jose, CA) and a gray-scale analysis was performed [10]. In this analysis, a selected area of muscle was chosen from the image and a mean luminosity value measured; in disease, the luminosity increases [10].

2.3. Data Analysis

Data was evaluated both for individual muscles and averaged over the 6 muscle examined. Unpaired t-tests and Spearman correlations were performed; significance was set at $p < 0.05$, two-tailed.

3. Results

3.1. Overall EIM Data

Figure 1A provides a histogram of 6-muscle averaged EIM 50 kHz phase values for 20 healthy boys and 19 DMD boys. Figure 1B provides the same data for quadriceps alone. As can be seen there is a clear differentiation between the groups. A receiver operating characteristic analysis demonstrated EIM's ability to discriminate healthy and sick muscle with an area under the curve, or accuracy, of 98% for the 6-muscle mean EIM values. Figure 2 shows phase correlated with age with DMD and controls, respectively, revealing a gradual increase in healthy children and stability or a slight decrease with age in the DMD boys. For DMD, $R = -0.30$, $p = 0.21$; for control $R = 0.59$, $p = 0.008$.

3.2. Relationship between EIM data and function measures

Figure 3 shows the correlation between the mean EIM values and the North Star Ambulatory Assessment for the mean EIM data for only the DMD boys ($R = 0.83$, $p = 0.02$).

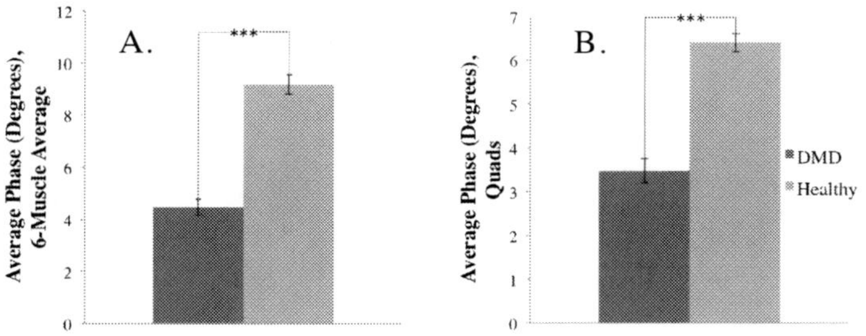

Figure 1. Comparison of EIM 50 kHz phase data for DMD and healthy boys for **A.** Six-muscle average and **B.** Quadriceps alone

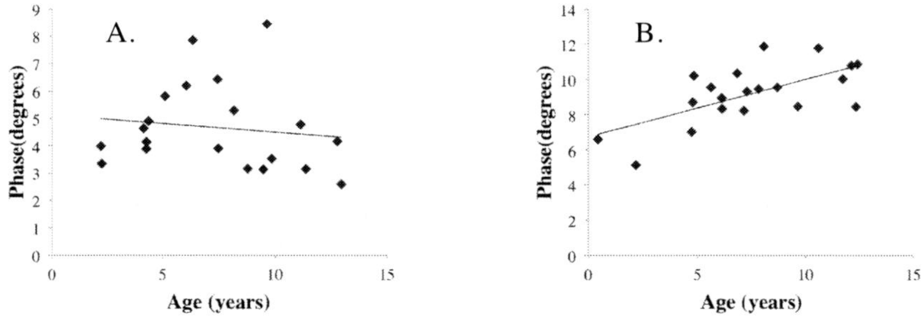

Figure 2. Average 6-muscle EIM phase data versus age. **A.** DMD boys and **B.** Normal subjects. Note the difference in overall trajectories

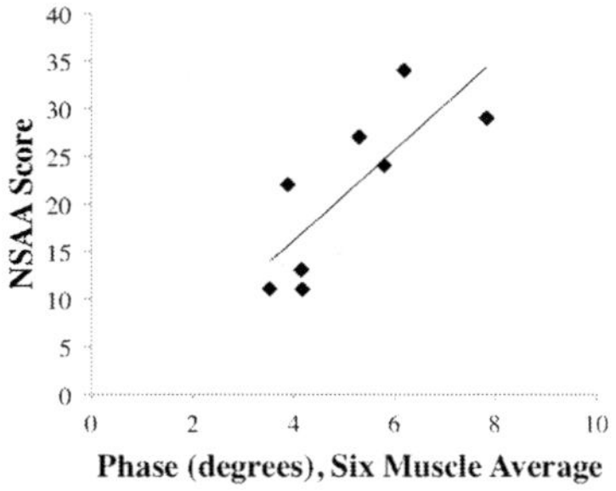

Figure 3. Correlation between North Star Ambulatory Assessment (NSAA) and EIM phase in 8 DMD boys in whom this assessment was obtained.

3.3. Relationship between EIM and ultrasound luminosity

Figure 4A shows the correlation between the 6-muscle mean EIM data and 6-muscle mean luminosity ($R = -0.79$, $p < 0.001$); Figure 4B shows data for quadriceps alone ($R = -0.67$, $p = < 0.001$).

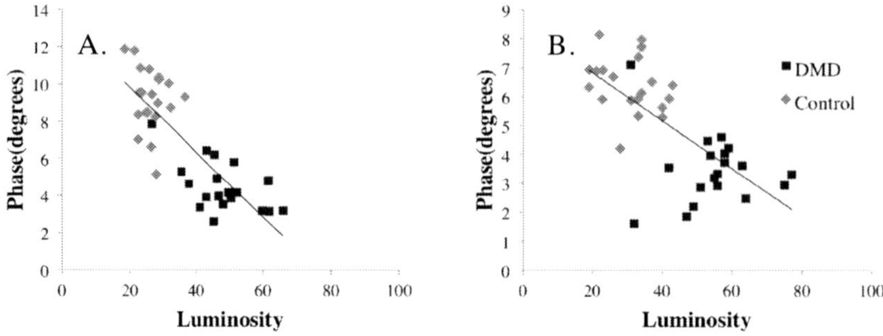

Figure 4. Correlation between EIM phase values and muscle luminosity obtained from ultrasound images for **A.** 6-muscle average and **B.** Quadriceps alone.

4. Discussion

These results show that EIM is sensitive to disease and can readily differentiate normal children from those with DMD. In addition, there is an age dependence to the values in that the data suggest progressively lower values children with DMD but progressively higher values in healthy controls. Moreover, EIM is clinically meaningful in that there is a strong correlation between EIM values and a standard functional assessment tool. Finally, this study confirms that the changes in EIM actually relate directly to tissue pathology. The increasing fibrosis and fatty infiltration, observed on muscle ultrasound, correlate strongly with the impedance data.

There are several limitations to this work. First, we are only looking at a relatively small group of children. Second, the impedance device and handheld array have considerable limitations in that we have to use different size devices for different children. This could be affecting our analysis to some extent and a separate study comparing the characteristics of the data obtained from the 3 different sized arrays will also need to be performed. Third, we have only evaluated one characteristic of the impedance data: the longitudinal 50 kHz phase. We have also acquired multifrequency data in both the longitudinal and transverse directions and thus much data still needs to be analyzed. It is possible that other frequencies may provide more powerful markers of disease status than these 50 kHz values. Similarly the reactance and resistance may also me useful to assess and will be analyzed as well.

In conclusion, EIM appears to be a promising marker of disease in boys with Duchenne muscular dystrophy. Further study utilizing this simple approach to disease assessment are warranted.

Acknowledgments

This work was funded by NIH grant AR060850

References

[1] Bushby K, Finkel R, Birnkrant DJ, Case LE, Clemens PR, et al C 2010 *Lancet Neurol* **9** 77-93.

[2] Prior TW and Bridgeman SJ 2005 *J Mol Diagn* **7** 317-26.

[3] Manzur AY, Kuntzer T, Pike M and Swan A 2008 *Cochrane Database Syst Rev* CD003725.

[4] Nelson SF, Crosbie RH, Miceli MC and Spencer MJ 2009 *Curr Opin Neurol* **22** 532-8.

[5] Bushby K and Connor E 2011 *Clinical investigation* **1** 1217-1235.

[6] Rutkove SB 2009 *Muscle Nerve* **40** 936-946.

[7] Rutkove SB, Caress JB, Cartwright MS, Burns TM, Warder J, David WS, et al 2012 *Amyotrophic Lateral Sclerosis*

[8] Rutkove SB, Gregas MC and Darras BT 2012 *Muscle & nerve* **45** 642-7.

[9] Narayanaswami P, Spieker AJ, Mongiovi P, Keel JC, Muzin SC and Rutkove SB 2012 *Muscle Nerve* **46** 257-63.

[10] Jansen M, van Alfen N, Nijhuis van der Sanden MW, van Dijk JP, Pillen S and de Groot IJ 2012 *Neuromuscular disorders* **22** 306-17.

Journal of Physics: Conference Series **434** (2013) 012070 doi:10.1088/1742-6596/434/1/012070

Motion discrimination of throwing a baseball using forearm electrical impedance

Takao Nakamura[1], Toshimasa Kusuhara[1] and Yoshitake Yamamoto[2]

[1] Graduate School of Health Sciences, Okayama University, Shikata-cho, Okayama 700-8558, Japan

[2] School of Health Care Sciences, Himeji Dokkyo University, Kamiohno, Himeji 670-8524, Japan

E-mail: nakamura@md.okayama-u.ac.jp

Abstract. The extroversion or hyperextension of elbow joint cause disorders of elbow joint in throwing a baseball. A method, which is easy handling and to measure motion objectively, can be useful for evaluation of throwing motion. We investigated a possibility of motion discrimination of throwing a baseball using electrical impedance method. The parameters of frequency characteristics (Cole-Cole arc) of forearm electrical impedance were measured during four types of throwing a baseball. Multiple discriminant analysis was used and the independent variables were change ratios of 11 parameters of forearm electrical impedance. As results of 120 data with four types of throwing motion in three subjects, hitting ratio was very high and 95.8%. We can expect to discriminate throwing a baseball using multiple discriminant analysis of impedance parameters.

1. Introduction

We have proposed biodynamics analysis using equivalent series resistance of bioelectrical impedance at 50 kHz [1, 2]. We have also developed a measuring system of frequency characteristics of Bioelectrical impedance with high time resolution [3] and showed that a motion discrimination (hitting ratio 95.0%) in the following six motions: palmar flexion, dorsi flexion, ulnar deviation and radial deviation of wrist joint, and pronation and supination of forearm using four parameters of frequency characteristics of forearm electrical impedance [4].

As an application for sports, we are interested in throwing a baseball. The extrovesion or hyperextension of elbow joint cause disorders of elbow joint in throwing motion Improvement of throwing motion, which causes less injury, can prevent the disorders [5]. Though a video camera system is needed in order that a quantitative analysis of throwing motion, it is evaluated visually or subjectively using video picture because the video camera system is complicated to set up and high-priced. A method, which is easy handling and to measure motion objectively, could be useful for evaluation of throwing motion. We investigate a possibility of discrimination of throwing a baseball included easily-hurt motion using parameters of frequency characteristics (Cole-Cole arc) of forearm electrical impedance using high time resolution measurement.

2. Materials and methods

2.1. Characteristics of bioelectrical impedance

The frequency locus of bioelectrical impedance Z is well known to be circular arc on a complex plane which consists of equivalent series resistance R_s (real part) and equivalent series reactance X_s (imaginary part). The frequency locus is called Cole-Cole circular arc [6] and is expressed by the following equation that is Cole-Cole circular arc's law

$$Z = R_s - jX_s = Z_\infty + \frac{Z_0 - Z_\infty}{1 + (jf/f_m)^\beta} \quad (1)$$

where f is frequency, Z_0, Z_∞ are both resistances based on ion conductivity. $Z_0 = \lim_{\omega \to 0} Z$, $Z_\infty = \lim_{\omega \to \infty} Z$. f_m is frequency which gives maximum reactance of X_s. The central relaxation time τ_m, which shows dielectric relaxation phenomena, equals $1/(2\pi f_m)$. β represents degree of distribution of relaxation time. The extracellular fluid resistance R_e and the intracellular fluid resistance R_i are expressed by $R_e = Z_0$ and $R_i = Z_0 Z_\infty/(Z_0 - Z_\infty)$, respectively.

2.2. Measuring system and throwing method

The measuring system of frequency characteristics of bioelectrical impedance with high time resolution was used [3, 4]. The 10 frequency points were the following 4, 8, 10, 16, 32, 40, 64, 80, 100 times harmonics as fundamental wave (0.9765625 kHz). The parameters of impedance were calculated by an optimized calculation method (KH method) [7]. In order to confirm relations between these impedance parameters and throwing motions, images of 125 fps of throwing motion were taken a high-speed camera (VFC–1000SC, FOR-A). Figure 1 shows locations of electroeds. The measured part of forearm were a part between potential electrodes(P+, P−).

The forearm electrical impedances of three subjects, who were Sub. A, B and C, were measured during four types of throwing a baseball (Th1, Th2, Th3 and Th4). These four types of throwing were the following different motions after ball release: • Th1 excessive pronation; • Th2 hyperextension of elbow joint; • Th3 excessive spination; and • Th4 natural pronation as shown in Fig. 2. From Th1 to Th3 are easily-hurt motion and Th4 is normal motion.

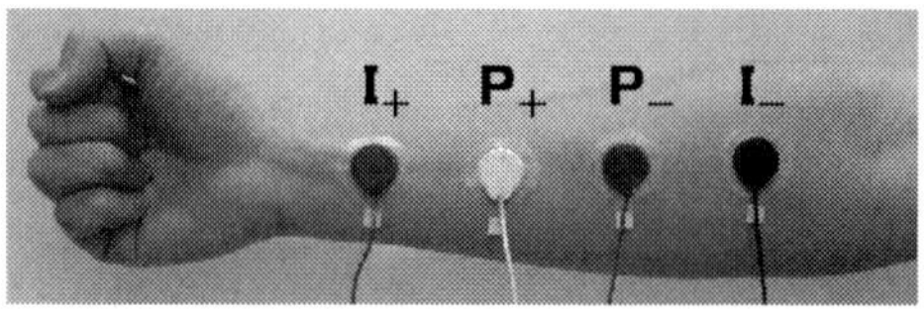

Figure 1. Locations of electrodes. Current electrode (I+), potential electrode (P+), potential electrode (P−) and current electrode (I−) are shown in order from the left.

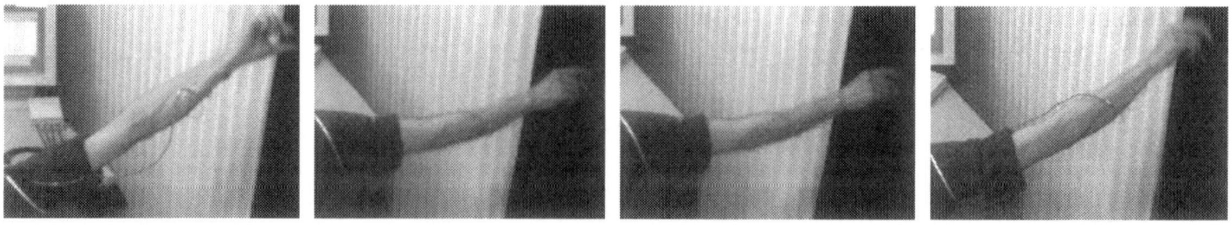

(a) Th1: excessive pronation (b) Th2: hyperextension of elbow joint (c) Th3: excessive spination (d) Th4: natural pronation(normal)

Figure 2. Four types of throwing a baseball. Each picture shows just after release a baseball.

2.3. Multiple discrimination analysis

The change ratio of impedance parameters, — ΔR_e, ΔR_i, ΔZ_∞, Δf_m, $\Delta\beta$ —, were calculated [4]. In time-series parameters, we defined three time point as follows: • t_1 when wrist joint were fully dorsiflexed; • t_2 when elbow joint were fully extended and fingers were also flexed; and • t_3 when maximum motion of forearem after ball release. We also defined T which is time difference from t_1 to t_3. The 11 parameters, — which consist of five parameters in t_2 ($\Delta R_e(t_2)$, $\Delta R_i(t_2)$, $\Delta Z_\infty(t_2)$, $\Delta f_m(t_2)$, $\Delta\beta(t_2)$), those in t_3 ($\Delta R_e(t_3)$, $\Delta R_i(t_3)$, $\Delta Z_\infty(t_3)$, $\Delta f_m(t_3)$, $\Delta\beta(t_3)$), and T —, were analyzed for multiple disrimination. 10 throwings in each motion were used for calculation for discriminate coefficient, other 10 throwings in each motion were analyzed for discrimination.

3. Results and Disscussions

Fig. 3 shows sample of change ratios of impedance parameters, time points and images of throwing motions. There were ball release between t_1 and t_2. There were two common points of each throwing as follows: • After Δf_mbecame local maximum value around $t = 2.5$ s in Fig. 3, it became local minimum value; • After Δf_m became local maximum value, ΔR_e, ΔR_i and ΔZ_∞ became local maxmum values. There were also different among each local maxmum values of R_e, R_i, Z_∞ after Δf_m became local maximum values. The common points and difference points of throwing a baseball were common among each subject. Therefore a discrimination may be effective using time points t_1–t_3 defined as the mentioned above and change ratio of parameters in that time point.

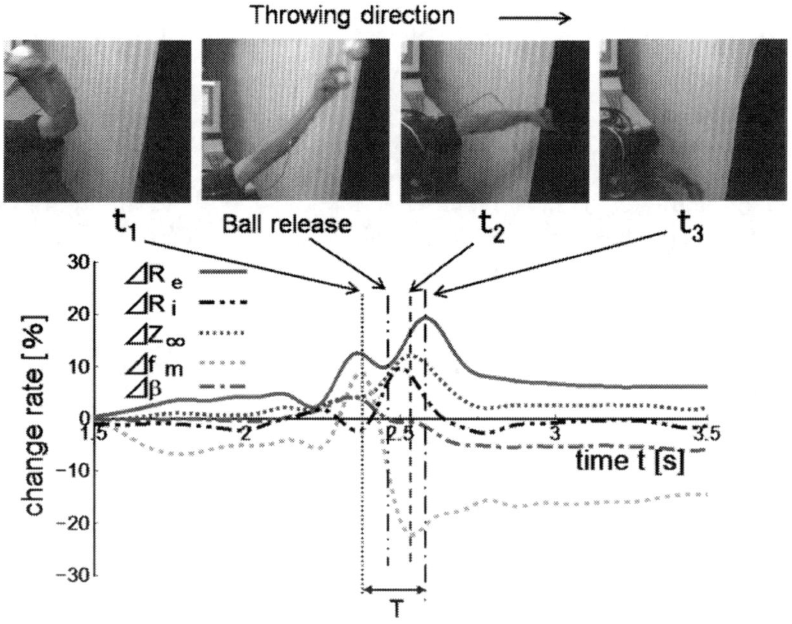

Figure 3. Sample of change ratios of impedance parameters during throwing a baseball.

Table 1 shows standardized discriminate coefficient and contribution ratio of each vector in Sub. A. The absolute values of ΔR_e, ΔR_i and ΔZ_∞ were larger than the other parameters in every vector. These three parameters greatly contributed to discrimination of throwing a baseball. There were same tendency in the othere subjects. The cumulative contribution ratio in 1st and 2nd vectors were 91.7% and quite high.

Table 1. Standardized discriminate coefficient and contribution ratio of each vector in Sub. A.

	1st vec.	2nd vec.	3rd vec.
$\Delta R_e(t_2)$	−14.375	4.573	−5.963
$\Delta R_i(t_2)$	−10.828	4.413	−4.814
$\Delta Z_\infty(t_2)$	17.127	−6.016	7.795
$\Delta f_m(t_2)$	0.447	−1.131	2.157
$\Delta\beta(t_2)$	−0.282	−0.481	−0.423
$\Delta R_e(t_3)$	7.447	0.363	10.356
$\Delta R_i(t_3)$	6.494	−0.526	9.635
$\Delta Z_\infty(t_3)$	−8.668	0.606	−15.047
$\Delta f_m(t_3)$	1.003	1.144	−1.960
$\Delta\beta(t_3)$	0.140	−0.139	1.029
T	0.106	0.800	0.335
contribution ratio[%]	66.3	25.4	8.3

Table 2. Result of discrimination analysis for 120 throwings in three subjects.

		Discriminant motion				total
		Th1	Th2	Th3	Th4	
True motion	Th1	25	0	0	5	30
	Th2	0	30	0	0	30
	Th3	0	0	30	0	30
	Th4	0	0	0	30	30
total		25	30	30	35	120

Th1: excessive pronation
Th2: hyperextension of elbow joint
Th3: excessive spination
Th4: natural pronation(normal)
Hitting ratio is 95.8%

Table 2 shows result of discrimination analysis for 120 throwings in three subjects. We confirmed that the temporal change ratios had individual characteristics and there were different temporal change ratios among throwing a baseball. The individual hitting ratios were 95.0% for Sub. A, 92.5% for Sub. B and 100% for Sub. C. Five Th1, –excessive pronation–, were mistaken for Th4, –natural pronation–. This was caused by the same movement (pronation) in Th1 and Th4.

4. Conclusion

In order to discriminate throwing a baseball, we measured impedance parameters using the measuring system of frequency characteristics of bioelectrical impedance with high time resolution during four types of throwing a baseball. We confirmed that the temporal change ratios had individual characteristics and there were different temporal change ratios among throwing a baseball. The hitting ratio of discrimination analysis for 120 throwings in three subjects were very high and 95.8%. We can expect to discriminate throwing a baseball using multiple discriminant analysis of impedance parameters despite needs of pre-measurement to calculate equations for discrimination.

References

[1] Nakamura T, Yamamoto Y, Yamamoto T, Tsuj H 1992 Fundamental characteristics of human limb electrical impedance for biodynamic analysis, *Med. Biol. Eng. Comput.*, **30** 465–473
[2] Kusuhara T, Nakamura T, Shirakawa Y, Mori K, Naomoto Y and Yamamoto Y 2004 Impedance Pharyngography to Assess Swallowing Function, *J. Int. Med. Res.*, **32**, 606–616
[3] Nakamura T, Kusuhara T and Yamamoto Y 2005 Measuring System of Frequency Characteristics of Bioelectrical Impedance with High Time Resolution for Motion Analysis, *Proc. SICE Annual Conf., (Okayama)* 1668–1673
[4] Nakamura T, Kusuhara T and Yamamoto Y 2007 Motion Discrimination Using Parameters of Bioelectrical Impedance, *Proc., 13th Int. Conf. Elect. Bioimp.*, 671–674
[5] Kawashima T 2004 *IBSN 978-4-8376-7021-6 (in Japanese)* (Tokyo: Makino Shuppan) 14–23
[6] Cole K S 1940 Permeability and impermeabilty of cell membranes for ions, *Cold Spring Harbour Symposia Quant. Biol.*, *8*, 110–122
[7] Kuroda H 2002 *Engineering calculation program using Visual Basic (in Japanese) ISBN 4-7898-3693-2* (Tokyo: CQ Publishing), 87–88

Journal of Physics: Conference Series **434** (2013) 012071 doi:10.1088/1742-6596/434/1/012071

Facts and artefacts regarding correlation between skin electrical impedance spectroscopy (EIS) and blood glucose

Stig Ollmar*, Ingrid Nicander*, Peter Åberg*, Jan Bolinder**

*Karolinska Institutet, CLINTEC, ANA 10, SE-14186 Huddinge, Sweden
**Karolinska University Hospital, Dept. of Medicine, SE-14186 Huddinge, Sweden

E-mail: stig.ollmar@ki.se

Abstract. Earlier observations on possible co-variation between skin EIS and blood glucose prompted us to map and include other factors at play in the predictive model. Skin pH would be one such factor. A cohort of 20 diabetics was investigated, taking around 30 measurements spread over each of two different days 2-21 days apart. Each measurement comprises skin EIT in the frequency range 1kHz to 2.5MHz, skin pH, and immediately evaluated blood samples. There is a co-variation for some, but not all, test persons. The relationship gets stronger on the group level by adding pH-information, but is still poor or non-existent for some test persons. Non-invasive EIS measurements on skin is influenced by skin hydration, blood glucose, skin pH, body location, season, environmental factors, and variables not yet understood. Since impedance related parameters are used to estimate skin hydration, users of such devices should be aware that skin pH may influence as much as the water content of the stratum corneum.

1. Background

An easy-to-use and reliable device to non-invasively monitor blood glucose (BG) would be the holy grail of diabetes management. In 1999, Elden was awarded a patent [1] involving dielectric properties of skin based on simultaneous measurements with a device intended to estimate skin water content and a common BG meter. The idea has been further investigated in clinical studies in the US, and in a Master thesis by Birgersson and Neiderud 2004 [2]. It was found that there is correlation between information in electrical impedance spectra and BG, but not for every person. Defining more rigorous procedures in the preparation of the skin before measurements improved correlation somewhat, but still correlation was weak or non-existing for many persons. Pondering possible confounding factors revealed that skin pH could be a strong factor modulating the impedance – while pH inside the human body varies very little, skin pH naturally varies between 3.5 and 7, and may rise to about 10 after washing with a hard soap, then slowly returning to baseline during several hours. Skin pH also has diurnal fluctuations [3].

The rational for including skin pH in the model is that many organic acids, in particular fatty acids, but also DNA-residues, would migrate from the living epidermis into the stratum corneum during its continuous rejuvenation, and possibly this acid pool could be represented as pH-value. Other charge carriers such as sodium, potassium, and chlorine ions would also be part of the impedance space, but reasonably under control by a standardized inundation procedure before EIS measurement. In the frequency range used, the impedance spectrum would be dominated by the stratum corneum layer, as has been shown by Birgersson et al. [4]. Since the glucose molecule is uncharged, the measured alterations of electric or dielectric properties elicited in tissue by variation in BG must be attributed to secondary effects, and a coupling mechanism might be sought in composition of interstitial liquid in the skin or in cell membranes, although the stratum corneum consists of dead keratinized cells. Clinically interesting BG-levels (up to about 25 mmol/L) are not sufficient to impede the mobility of sodium ions in solution enough to explain the observed alterations in the impedance spectra. Subtle structural changes in the stratum corneum may be at play [5]. In short: The coupling mechanism between skin EIS and BG is not understood, but the phenomenon exists. A co-variation between pH and impedance is necessary according to the laws of electrochemistry.

2. Materials and methods

The study was approved by the Regional Ethics Committee, and 20 diabetics type 1, 10 men, 10 women, ages 25-70 were recruited. The test persons spent two full days in our lab. High glycemic food was used to increase BG at coffee breaks and lunch, and apart from the somewhat unnatural food intake, the test persons managed their insulin as usual. Skin EIS was measured with a SciBase II instrument (SciBase AB, Stockholm, Sweden). Skin pH was measured with a Skin-pHMeter PH 905 (Courage & Khazaka GmbH, Cologne, Germany). Venous and capillary blood was sampled every 15 minutes (longer intervals at coffee breaks and lunch) by qualified nurses, simultaneously with EIS and pH measurements. All measurements were done by the same operator (IN). Both Venous and capillary BG was measured with HemoCue Glucose 201 (HemoCue AB, Ängelholm, Sweden). One PLS model [6] using all data with or without pH was calculated for the first day, and also individual models. As an index of goodness, the number of predicted points falling into Clark zones A or B was used [7-10].

3. Results

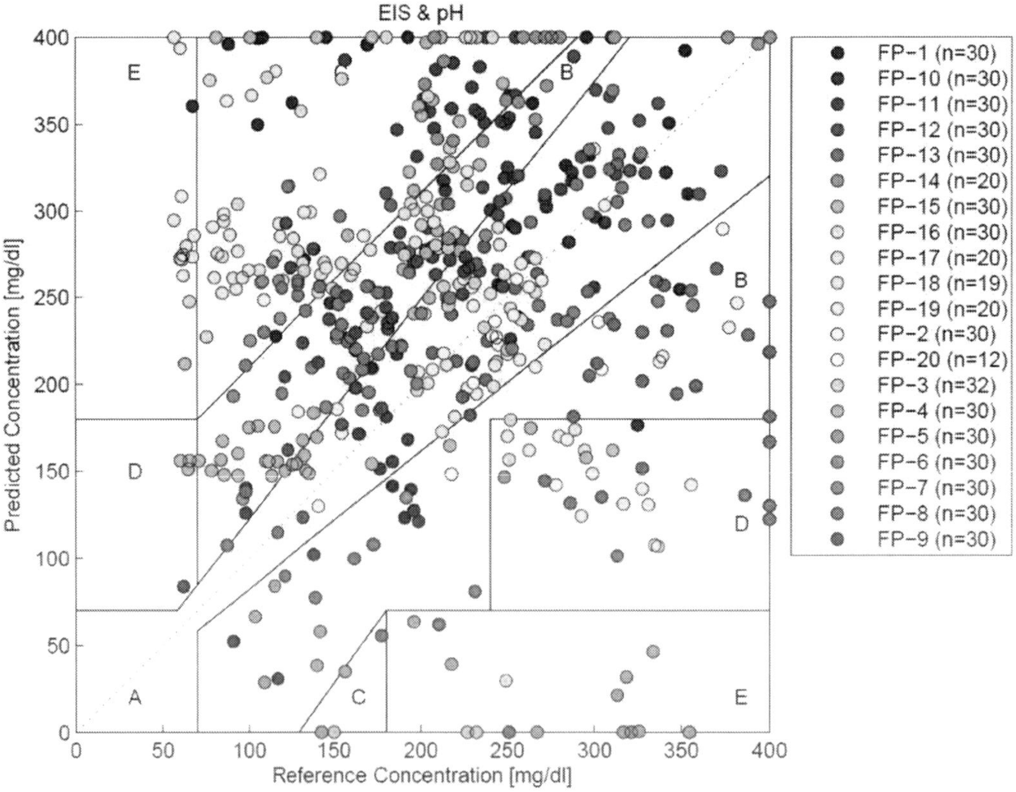

Figure 1: Clarke grid for predicted BG using both skin EIS and skin pH. Note scales: 18mg/dl = 1mmol/L BG. The graphics template places BG in the range 400-500 mg/dl at 400 mg/dl.

In this context, zone A is considered good and an error in zone B is not good but neither dangerous nor does it require any intervention such as administration of insulin or sugar. Clarke grid was chosen because it is commonly used, although the so called BD grid is somewhat more realistic in defining the error zones. For the outcome of this study it does not matter.

		Clarke A+B: EIS only		**Clarke A+B: EIS and pH**	
Test person	n tot	n in A+B	% in A+B	n in A+B	% in A+B
All pooled	543	326	60.0	348	64.1

An ocular inspection of the above plot (Figure 1) is not encouraging, there is no obvious correlation in the pooled data, and for a couple of test persons the correlation appears to be negative. The number of data points falling in Clarke zones A or B increases 4 percent units after including pH in the model based on all 20 test persons. However, as we have seen good correlation in some test persons in earlier studies, a closer look on an individual level using the number of data points falling in zones A or B is presented/was performed, with or without pH data in the model.

		Clarke A+B: EIS only		**Clarke A+B: EIS and pH**	
Test person	n tot	n in A+B	% in A+B	n in A+B	% in A+B
FP-1	30	21	70.0	21	70.0
FP-2	30	25	83.3	25	83.3
FP-3	32	20	62.5	20	62.5
FP-4	30	24	80.0	24	80.0
FP-5	30	27	90.0	27	90.0
FP-6	30	27	90.0	27	90.0
FP-7	30	28	93.3	28	93.3
FP-8	30	26	86.7	26	86.7
FP-9	30	29	96.7	29	96.7
FP-10	30	30	100.0	30	100.0
FP-11	30	28	93.3	28	93.3
FP-12	30	29	96.7	29	96.7
FP-13	30	29	96.7	29	96.7
FP-14	20	13	65.0	13	65.0
FP-15	30	20	66.7	20	66.7
FP-16	30	7	23.3	6	20.0
FP-17	20	3	15.0	3	15.0
FP-18	19	8	42.1	8	42.1
FP-19	20	10	50.0	10	50.0
FP-20	12	9	75.0	9	75.0

Using individual models for calibration, the average number of points in zones A or B increased to 76% while the influence of skin pH now became negligible.

4. Discussion

Predicted values of BG falling into Clarke error zones C-E do have clinical consequences and can be fatal, e.g. a predicted value of, say, 25 mmol/L (450 mg/dl) while the true value is 1.5 mmol/L (27 mg/dl) would suggest administration of an insulin dose that may be highly dangerous for the patient. Any reliable device should give values only within error zone A, although a few points falling into error zone B may be acceptable. There are a number of skin research instruments intended to estimate skin hydration (water content of the stratum corneum) based on EIS or subsets of such data (detecting conductivity or capacity only, at one or more frequencies), and using a variety of probes/electrode systems [11,12]. Some observed variation in skin hydration studies might be explained by the pH contribution, in particular on intact healthy skin which can be quite acid.

5. Conclusions

There is a co-variation between skin EIS and blood glucose. The co-variation is not strong enough to merit a device based on this phenomenon only. Skin surface pH is one confounding factor, but adding this parameter to the model is not enough to merit a device based on measurement of both EIS and pH. However, impedance related parameters are used to estimate skin hydration, and investigators using such devices should be aware that alterations in skin pH may influence impedance parameters as much as the water content of the stratum corneum, depending on actual baseline of pH changes.

References

[1] Elden HR. United States Patent 1999, No. 5,890,489
[2] Birgersson U and Neiderud F 2004 Bioelectrical parameters related to glucose level. M.Sc. thesis. (*Royal Institute of Technology (KTH), Stockholm*)
[3] Yosipovitch G, Xiong GL, Haus E, Sackett-Lundeen L, Ashkenazi I and Maibach HI 1998 Time-dependent variations of the skin barrier function in humans: Transepidermal water loss, stratum corneum hydration, skin surface pH, and skin temperature. *J. Invest. Dermatol.* **110** 20–4
[4] Birgersson U, Birgersson E and Ollmar S 2012 Estimating electrical properties and the thickness of skin with electrical impedance spectroscopy: Mathematical analysis and measurements. *J. Electrical Bioimpedance* **3** 51–60
[5] Iwai I, Han HM, den Hollander L, Svensson S, Öfverstedt LG, Anwar J, Brewer J, Bloksgaard M, Laloeuf A, Nosek D, Masich S, Bagatolli LA, Skoglund U and Norlén L 2012 The human skin barrier is organized as stacked bilayers of fully extended ceramides with cholesterol molecules associated with the ceramide sphingoid moiety. *J. Invest. Dermatol.* **132** 2215–25
[6] Eriksson L, Johansson E, Kettaneh-Wold N and Wold S 1999 Introduction to Multi- and Megavariate Data Analysis using Projection Methods (PCA & PLS). (*Umetrics, Umeå*)
[7] Clarke WL, Cox D, Gonder-Frederick LA, Carter W and Pohl SL 1987 Evaluating clinical accuracy of systems for self-monitoring of blood glucose. *Diabetes Care* **10** 622–8
[8] Cox DJ, Richards FE, Gonder-Frederick LA, Julian DM, Carter WR and Clarke WL 1989 Clarification of error-grid analysis *Diabetes Care* **12** 235–6
[9] Gough DA and Botvinick EL 1997 Reservations on the use of error grid analysis for the validation of blood glucose assays *Diabetes Care* **20** 1034–6
[10] Parkes JL, Slatin SL, Pardo S and Ginsberg BH 2000 A new consensus error grid to evaluate the clinical significance of inaccuracies in the measurement of blood glucose *Diabetes Care* **23** 1143–8
[11] Fluhr JW, Gloor M, Lazzerini S, Kleesz P, Grieshaber R and Berardesca E 1999 Comparative study of five instruments measuring stratum corneum hydration (Corneometer CM820 and CM 825, Skicon 200, Nova DPM 9003, Dermalab) Part I. In vitro *Skin Res. Technol.* **5** 161–70
[12] Fluhr JW, Gloor M, Lazzerini S, Kleesz P, Grieshaber R and Berardesca E 1999 Comparative study of five instruments measuring stratum corneum hydration (Corneometer CM820 and CM 825, Skicon 200, Nova DPM 9003, Dermalab) Part II. In vivo *Skin Res. Technol.* **5** 171–8

In vivo assessment of corneal barrier function through non-invasive impedance measurements using a flexible probe.

A. Guimera[1*], X. Illa[1], E. Traver[3], S. Marchan[3], C.Herrero[3], C.Lagunas[3] MJ. Maldonado[2], A. Ivorra[4], R. Villa[1]

[1] Institut de Microelectrònica de Barcelona IMB-CNM (CSIC), Bellaterra 08193, Barcelona, Spain. CIBER-BBN, Networking Center on Bioengineering, Biomaterials and Nanomedicine, Zaragoza, Spain.
[2] IOBA-Eye Institute, University of Valladolid. Valladolid, Spain.
[3] Laboratorios SALVAT S.A., Esplugues de Llobregat, 08950. Barcelona, Spain.
[4] Department of Information and Communication Technologies, Universitat Pompeu Fabra, Carrer Tánger 122-140, 08018 Barcelona, Spain

*E-mail: anton.guimera@csic.es

Abstract. The cornea is a transparent structure composed of three layers: the epithelium, the stroma and the endothelium. To maintain its ransparency the stroma remains in a constant state of dehydration. Consequently, any ion flow disorder through the covering layers can compromise the barrier function and, therefore the corneal homeostasis. Since ionic permeability has a fundamental impact on the passive electrical properties of living tissues, in this work it is proposed and demonstrated a diagnosis method based on tetrapolar impedance measurements performed by electrodes placed on the corneal surface. The contribution of each cornea layer to the total measured impedance has been analysed over a frequency range. Following the obtained guidelines, a flexible probe with integrated electrodes has been developed and manufactured using SU-8 photoresin. The feasibility of the proposed method has been evaluated in vivo by monitoring corneal epithelium wound healing. Obtained impedance measurements have been compared with measurements of permeability to sodium fluorescein from different excised corneas. Successful results demonstrate the feasibility of this novel flexible sensor and its capability to quantify corneal permeability *in vivo* in a non-invasive way.

1. Introduction

The corneal epithelium is the outer surface of the cornea and serves as a barrier to prevent chemical, biological, and physical insults from the environment from entering into the eye. This barrier function is critical for corneal homeostasis, including maintenance of corneal transparency [1]. Because of this, quantitative measurements of the epithelium permeability are of special clinical interest. In animal models, the most commonly used technique is based on the measurement of the corneal permeability to carboxy fluorescein. However, this technique requires the excision of the corneas from euthanized animals. As an alternative, approaches based on measurement of passive electrical properties for *in-vivo* monitoring of corneal tissues were envisioned in the past [2]. In fact, such strategy was employed for assessing the corneal barrier function *in-vitro* [3–5]. Recently, Trans-layer Electrical Resistance (TER) measurements, which have consistently been used in *in vitro* studies, have been modified in order to be implemented in living animals [6,7]. Nevertheless, invasiveness of these last approaches makes them inconvenient for continuous monitoring of corneas and absolutely banned for clinical use.

To overcome the invasiveness of the above referred techniques some numerical studies have been carried out [8,9]. Here, a novel development based on the use of a flexible substrate integrating the methodology presented in a previous work from the authors is disclosed [10]. With this new flexible probe, no pressure to ensure the electrical contact of the electrodes with the corneal surface is needed. Hence, the maximum separation between the electrodes is not limited as it was when using a rigid substrate, which increases its sensitivity. Moreover, the non-invasiveness of the presented method allows monitoring the corneal barrier function evolution in the same individual. This fact, apart from reducing the quantity of animals required to perform corneal barrier function studies, lays the first stone for its further clinical use.

2. Impedance sensor

2.1. Sensor design

The measurement method is based on tetrapolar impedance measurements in order to minimize the parasitic effects of the electrode-electrolyte interface impedances. In a tetrapolar system, the electrical properties of each sub-volume of the tissue have a different contribution to the measured impedance [11]. Consequently, the correct choice of each electrode size and the position relationship between them is important for focusing the measurement on the region of interest. In parallel, in previous studies, it was concluded that the most important parameter of the sensor geometry is the sensor width due to its relation with the spatial resolution of the measurement [9,12]. Furthermore, the contribution of each corneal layer to total measured impedance depends on the frequency range, being within the low frequency range (Freq. < 50 kHz) where the corneal epithelium has the largest contribution. However, at this frequency range the contribution of the tear film is also important, constituting a limitation in the resolution of the proposed method since little variations in the tear film thickness could produce variations on the performed measurements[10]. For that reason, among the 4 different electrode configurations available in the fabricated sensing device, the one with the biggest distance between the pair of electrodes was chosen for this study (geometry scheme shown in figure 1).

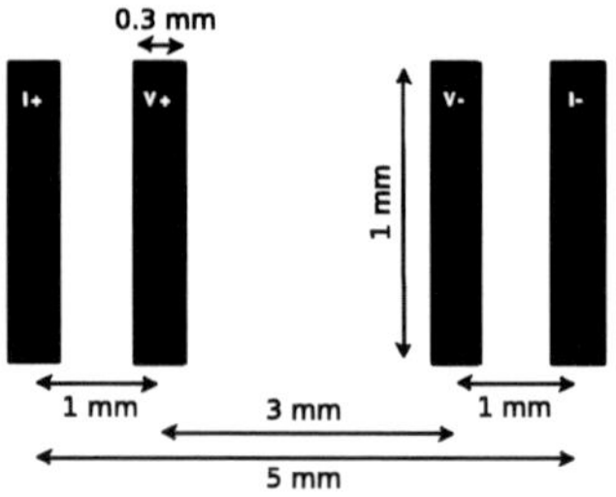

Figure 1. Schematic representation of the used geometrical relationship between electrodes.

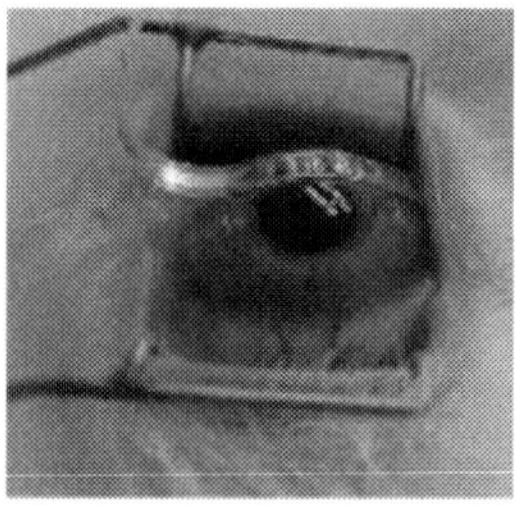

Figure 2. Image of the sensing probe applied to rabbit's eye during the *in vivo* experiments.

2.2. Sensor fabrication

The sensor fabrication is based on the construction of SU-8 free standing structures over a SiO_2 sacrificial layer grown on a 4 inch Si wafer. The fabrication process consists of three standard photolithographic steps. The first stage defines the device structure on a 25 μm of SU-8 layer. Secondly, electrodes and metal connections were patterned on a previously evaporated 20/200 nm Ti/Au layer. Finally, a thin layer of 1 μm of SU-8 was patterned to insulate the metal tracks and define the electrodes and connections. The structures are finally obtained by etching the SiO_2 sacrificial layer in a HF-based chemical bath. Moreover, electrodes have been electrochemically coated with black platinum to minimize the parasitic effects produced by the electrode-electrolyte impedance [10].

3. Experimental procedures

3.1. *Corneal epithelial wound*
To study the wound healing of the corneal epithelium, a total of 30 New Zealand white rabbits weighing between 2.5 and 3 kg were used. The rabbits were deeply anaesthetized with an intramuscular injection (50 mg/kg of ketamine plus 7 mg/kg of Xilacine). Then, both eyes were kept open by a blepharostat and a circular wound was performed in the central corneal epithelium by applying a 6 mm diameter paper disc soaked with 10 μl of n-heptanol for 30 seconds. After removing the paper disk, the corneas were rinsed with 10 ml of sterile saline solution. It has been reported that this protocol generates an epithelial wound with little or even no damage to the underlying stroma [13,14].

3.2. *Epithelial permeability to sodium fluorescein measurements*
Corneal epithelial permeability to fluorescein was evaluated as an indicator of corneal epithelial integrity. After sodium fluorescein instillation (50 μl, 1 mg/ml) to the ocular surface, animals were euthanized with intravenous pentobarbital at 200 mg/kg and corneas were excised. Homogenization of each cornea was performed in 300 μl of PBS and samples were centrifuged at 14000 rpm and 4°C during 10 minutes. Fluorescence in supernatants and in a fluorescein standard was measured using a Victor II plate reader (Perkin Elmer) by duplicate.

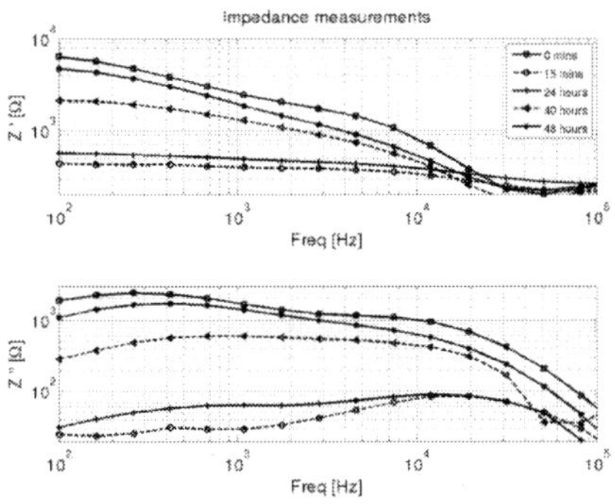

Figure 3. Representation of the impedance measurements for the following-up of the corneal epithelial wound healing process. Data is presented as the mean value of each group (n=6).

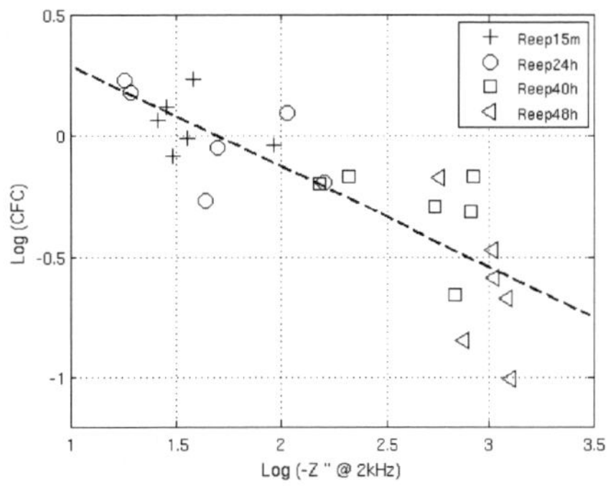

Figure 4. Representation of the Corneal Flourescein Content (CFC) in function of the impedance imaginary part at 2 kHz. Dash line shows a linear regression of the logarithmic values of both magnitudes.

3.3. *Impedance measurements*
A custom made tetrapolar impedance analysis system was used to perform impedance spectroscopy measurements as it was described in a previous work [10] To evaluate the wound healing, both impedance and permeability to sodium fluorescein measurements were carried out before and after (at 15 min, 24 h, 40 h and 48 h) damage. Accordingly, each group of measurements is composed by 6 eyes.

4. Results and discussion
Fig. 3 shows the impedance measurements performed to monitor the wound healing process of the corneal epithelium. These results are presented by plotting the mean of all the measurements for each group. As expected, the measured impedance increases with the time after wounding can be observed. It is interesting to note that the last measurement do not recover the basal value, which denotes that the healing process has not finished yet. The imaginary part of the impedance measured at 2 kHz has been proposed as an indicator of the corneal epithelium permeability. This choice is based on previously

simulation results [9], where is shown that, within this frequency range, there is an optimum compromise between the corneal epithelium and the tear film contributions. Thus, the measurement is minimally affected by the tear film. Moreover, the tear film should not contribute to the imaginary part as it is mainly composed by a saline solution. It can be observed that its value after wounding is almost null, which denotes that it is mainly affected by the corneal epithelium.

Table 1 show the data from both the fluorescein measurements and the proposed indicator mentioned above, which were performed in the same individuals and at the same time. As far as both measurements correspond to the same phenomena, a similar behaviour is expected. To show this relation both logarithmic values are depicted in Fig. 4, where a trend can be perceived although the variability of the data in both measurements methods. In particular, the lack of sensitivity of fluorescein quantification in corneas at 40 and 48 hours post wounding, which had very low fluorescein content. Moreover and to better show this trend, a linear regression of the logarithmic values of both magnitudes is represented (dash line). These results indicates that the non-invasive tetrapolar impedance measurements described in this work can be used for *in-vivo* experiments, paving the way for further clinical use of this technique.

Table 1. Summary of the experimental data. Data presented as mean ± Std from 6 eyes per group.

Time		Corneal fluorescein (CFC) [mg]		Impedance [kΩ] at 2kHz	
		Mean ± Std	Δ (%)	Mean ± Std	Δ (%)
Pre-wounding		0,11 ± 0,05	0	1,22 ± 0,23	0 %
Post-wounding	15 minutes	1,12 ± 0,33	918 %	0,04 ± 0,03	-96,7 %
	24 hours	1,00 ± 0,47	809 %	0,07 ± 0,06	-94,3 %
	40 hours	0,50 ± 0,17	354 %	0,55 ± 0,30	-54,9 %
	48 hours	0,24 ± 0,21	118 %	0,99 ± 0,27	-18,9 %

References

[1] Kinoshita S, Adachi W, Sotozono C, Nishida K, Yokoi N, Quantock A J and Okubo K 2001 *Prog. Retin. Eye Rse.* **20** 639–7
[2] Biermann H, Boden K and Reim M 1991 *Fortschr. Ophthalmol.* **88** 17–20
[3] Jurgens I, Rosell J and Riu P J 1996 *Physiol. Meas.* **17** A187–A195
[4] Ma L, Kuang K, Smith R W, Rittenband D, Iserovich P, Diecke F P J and Fischbarg J 2007 *Exp. Eye Res.* **84** 790–8
[5] Watanabe M, Mokudai Y, Ueno H, Ando M and Irimajiri A 1993 *Nippon Ganka Gakkai Zasshi* **97** 569–74
[6] Uematsu M, Kumagami T, Kusano M, Yamada K, Mishima K, Fujimura K, Sasaki H and Kitaoka T 2007 *Ophthalmic. Res.* **39** 308–14
[7] Chen W, Hu J, Zhang Z, Chen L, Xie H, Dong N, Chen Y and Liu Z 2012 *PLoS ONE* **7** e40893
[8] Mandel Y, Laufer S and Rubinsky B 2012 *Med. Eng. Phys.* **34** 195–201
[9] Guimera A, Ivorra A, Gabriel G and Villa R 2010 *Med. Eng. Phys.* **32** 1107–15
[10] Guimera A, Gabriel G, Plata-Cordero M, Montero L, Maldonado M J and Villa R 2012 *Biosens. Bioelectron.* **31** 55–61
[11] Grimnes S and Martinsen Ø G 2007 *J. Phys. D: Appl. Phys.* **40** 9–14
[12] Yang F and Patterson R P 2008 *Ann. Biomed. Eng.* **36** 762–8
[13] Talarico E F 2010 *Mol. Vis.* **16** 2259–72
[14] Burgalassi S, Nicosia N, Monti D, Falcone G, Boldrini E, Fabiani O, Lenzi C, Pirone A and Chetoni P 2011 *Curr. Eye Res.* **36** 21–8

Journal of Physics: Conference Series **434** (2013) 012073 doi:10.1088/1742-6596/434/1/012073

A GREIT-type linear reconstruction algorithm for EIT using eigenimages

Christoph Hoog Antink, Robert Pikkemaat, Steffen Leonhardt

Chair for Medical Information Technology, Helmholtz-Institut, RWTH Aachen, Pauwelsstr. 20, D-52074 Aachen, Germany

E-mail: hoog.antink@hia.rwth-aachen.de

Abstract. Reconstruction in electrical impedance tomography (EIT) is a nonlinear, ill-posed inverse problem. Based on point-shaped training and evaluation data, the "Graz consensus Reconstruction algorithm for EIT" (GREIT) constitutes a universal, homogenous method. While this is a very reasonable approach to the general problem, we ask the question if an optimized reconstruction method for a specific application of EIT, i.e. thoracic imaging, can be found. Instead of point-shaped training data we propose to use spatially extended training data consisting of eigenimages. To evaluate the quality of reconstruction of the proposed approach, figures of merit (FOMs) derived from the ones used in GREIT are developed. For the application of thoracic imaging, lung-shapes were segmented from a publicly available CT-database (www.dir-lab.com) and used to calculate the novel FOMs. With those, the general feasibility of using eigenimages is demonstrated and compared to the standard approach. In addition, it is shown that by using different sets of training data, the creation of an individually optimized linear method of reconstruction is possible.

1. Introduction

Electrical Impedance Tomography (EIT) seeks to reconstruct the electrical impedance distribution inside a body from measurements on its surface. It has proven to be a valuable tool in various areas, including medical science [1, 2]. One great advantage of EIT is its high temporal resolution with up to 50 frames per second, a drawback of this technology is it relatively poor spatial resolution. This is due to the fact that the problem of reconstruction, i.e. the calculation of the underlying impedance distribution from the measured voltage data, is ill-posed. As such, it is very sensitive to noise and can only be solved incorporating a-priori knowledge. Several iterative [3, 4] and non-iterative reconstruction algorithms exist. While the former show advantageous behavior in terms of reconstruction non-smooth impedance distributions, they are expensive from a computational point of view. Algorithms of the latter type have the advantage of being computationally inexpensive and can be calculated on-line, supporting EITs high frame rate. The first algorithm of this type is the "Sheffield Backprojection" algorithm. Though its basic assumptions are borrowed from Computed Tomography (CT) and hold only very limited in EIT, it showed reasonable results and was widely used in the clinical and experimental EIT community. More rigorous approaches are based on the regularized inversion of the forward-problem, i.e. the calculation of resulting voltage data from the known impedance distribution. The "Newton's One Step Error Reconstructor (NOSER)" algorithm [5] inverts a well-posed modification of the Jacobian; other regularization methods demand the solution to be smooth

Journal of Physics: Conference Series **434** (2013) 012073 doi:10.1088/1742-6596/434/1/012073

(Laplace) or small (Tikhonov). An implementation of the different algorithms can be found in the EIDORS-package [6].

A recent development in the category of non-iterative reconstruction algorithms is the "Graz consensus Reconstruction algorithm for EIT" (GREIT) [7]. Here, instead of *directly* inverting the forward problem, a set of impedance distributions and the associated voltage data are used to learn a linear reconstruction matrix, i.e. *indirectly* inverting the forward problem. Special attention needs to be given to the handling of noise by incorporating it as virtual measurements. Several metrics are implemented to assess the quality of the GREIT-algorithm and showed its superior performance and homogeneity compared to other linear reconstruction algorithms.

The training data used in GREIT consists of point-shaped changes in the homogenous background impedance. If we assume the impedance-space to be discretized, we can consider these changes in impedance as a set of unit vectors. As such, they form an orthonormal set that spans the impedance-space. This renders the training data neutral, as they are equally distributed through the whole impedance-space and give no advantage to a specific location or shape. On the other hand, this choice of training data is an arbitrary one with respect to real world medical applications, where the observed impedance distribution will likely be spatially extended and not point-shaped. Due to the ill-posed non-linear nature of the problem, it is reasonable to assume that the choice of an application specific set of training data will result in an application specific improvement of the quality of reconstruction. To assess this, the following steps were undertaken in this work: *Section 2.1* Definition of a realistic test data set & FOMs, *Section 2.2* Development of a modular framework, *Section 2.3* Finding of realistic training data.

2. Materials & Methods

2.1. Realistic Test Data Set & FOMs

As a potential area of application, pulmonary EIT is chosen. To assess the quality of the reconstruction algorithm objectively, a novel shape-based evaluation scheme is created. On the one hand, a point-based evaluation method as used in GREIT is likely to favor a reconstruction algorithm that was trained using point-shaped training data. At the same time, as argued above, the occurrence of point-shaped impedance distributions is not a realistic scenario in pulmonary EIT applications. Thus, the algorithms real-world performance may not be evaluated objectively. To counter this, an evaluation set of lung-shapes was created from a publicly available set of human thorax CT scans (*www.dir-lab.com*, [8]). The database consists of 4DCT scans of 10 patients, from which a total of 5518 lung shapes were segmented automatically and rescaled to 32 by 32 pixels using a custom algorithm. Figure 1 shows a selection of 60 shapes, sorted by

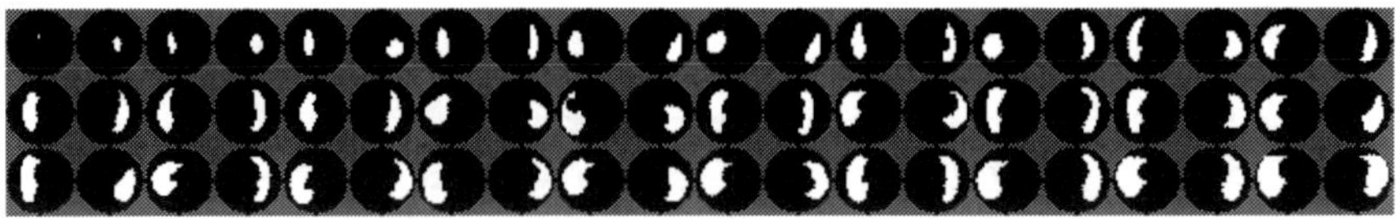

Figure 1. Examples of automatically segmented lung shapes from the 4DCT-database available at *www.dir-lab.com*, sorted by area.

size. The extreme changes in size origin from the different position of the transversal slices and the differently sized patients.

Since the evaluation data is not of point-shaped nature, the FOMs defined in [7] need to be adapted:

Journal of Physics: Conference Series **434** (2013) 012073 doi:10.1088/1742-6596/434/1/012073

Table 1. Novel FOMs, their correspondent in [7] and description.

Novel FOM	Similar to	Description
Widening	Resolution	Ratio of area of reconstructed to test shape (mean)
Overshoot	Ringing	Ration of neg. to pos. signal in reconstructed image (mean)
Distortion	Shape Deformation	Absolute difference of reconstructed shape and original shape over area of original shape (mean)
Dislocation	Position Error	Euclidian distance of reconstructed shape to true position of test data (mean, relative to diameter of background)
Area/Impedance Ratio	Amplitude Response	Quotient of size of test shape and sum of impedances in reconstructed image (standard deviation (std) over mean)

2.2. Modular Framework
The algorithm is implemented using MATLAB and the EIDORS framework. The training data is defined on the 32 x 32 pixel impedance-image-space. From this, an interpolation to a 2D FEM-mesh is done, which in turn allows the calculation of the resulting voltage vectors. These sets of impedance-images are spatially filtered and with the voltage vectors used to calculate the reconstruction matrix. A noise figure optimization as described in [7] is implemented, for all calculations a noise figure of 0.5 was chosen.

2.3. Training Data
GREIT-PT In order to compare the approach to the GREIT-algorithm, a set of point-shaped training data was created. Here, the conductivity of one of the 1024 pixels is increased against the background. Since only changes inside a circular region are considered, the resulting dataset consists of 748 impedance-images.

GREIT-SVD-GT To create the non-point-shaped training data, the singular value decomposition (SVD) is used on the 5518 lung-shape test data. This yields to a set of eigenimages, which also form an orthonormal set. Since the number of linear independent voltage measurements is limited to 104 per frame when using a configuration of 16 electrodes, it is sufficient to use the first 104 eigenimages.

GREIT-SVD-DR As a second source of spatially extended training data, the same approach was used on a set of 4800 frames from a pulmonary EIT sequence. These frames were acquired in an animal trial using the *Dräger EIT Evaluation Kit 2* and the proprietary method of reconstruction. Thus, they represent a set of training data completely independent of the test data.

An overview of the respective first 20 eigenimages can be found in Figure 2.

Figure 2. First 20 eigenimages of the test data set (top row, *GREIT-SVD-GT*) and the pulmonary EIT sequence (bottom row, *GREIT-SVD-DR*).

Table 2. Comparison of *GREIT-PT*, *GREIT-SVD-GT* and *GREIT-SVD-DR* using the novel FOMs (less is better). Values printed boldly denote the optimum.

FOM	GREIT-PT	GREIT-SVD-GT	GREIT-SVD-DR
Widening [area ratio]	1.8589	**1.4929**	1.7961
Overshoot [impedance ratio]	0.0038	**0.0015**	0.0063
Distortion [area ratio]	1.8595	**1.4958**	1.7970
Dislocation [distance ratio]	0.0257	**0.0198**	0.0246
Area/Impedance Ratio [std/mean]	0.1219	0.1350	**0.1108**

3. Results

In Table 2, a comparison of the three different GREIT-type algorithms can be found. For most FOMs, using eigenimages calculated from the test data yield the best results. In addition, using spatially extended data that is completely independent of the test data also leads to an improvement in some FOMs. It is interesting to observe that especially the *Area/Impedance Ratio* seems to benefit from using the *GREIT-SVD-DR* training data set.

4. Conclusion & Future Work

Firstly, the general feasibility of using spatially extended, orthonormal training data can be deduced. In addition, it can be stated that a-priori information about the morphology of the impedance distribution to be reconstructed can be incorporated into a GREIT-type algorithm by the use of eigenimages. Interestingly, the *Area/Impedance Ratio* shows the best results when neither point-shaped nor the decomposed evaluation data but a completely independent training data set is used. One explanation could be found when examining Figure 2: The eigenimages used for *GREIT-SVD-GT* show sharp edges as one finds in point-shaped training data. The ones used for *GREIT-SVD-DR*, however, show a much smoother behavior. Thus, smoothness of the training data might be an important factor when the *Area/Impedance Ratio* is to be optimized.

While the proposed FOMs suggest an increased quality of reconstruction for a specific application, the practical value of this approach has yet to be proven. This could be achieved by retrospectively re-evaluating animal trials using the proposed method of reconstruction. In addition, an extension of the approach to a 2.5D-FEM-model as well as a full integration into the EIDORS/GREIT framework seems reasonable.

References

[1] Meier T, Luepschen H, Karsten J, Leibecke T, Grossherr M, Gehring H and Leonhardt S 2008 *Intensive Care Medicine* **34** 543–50
[2] Adler A, Amato M B, Arnold J H, Bayford R, Bodenstein M, Böhm S H, Brown B H, Frerichs I, Stenqvist O, Weiler N and Wolf G K 2012 *Physiological Measurement* **33** 679–94
[3] Borsic A, Graham B, Adler A and Lionheart W 2007 *MIMS EPrint* **92**
[4] Rahmati P, Soleimani M, Pulletz S, Frerichs I and Adler A 2012 *Physiological Measurement* **33** 739–50
[5] Cheney M, Isaacson D, Newell J, Simske S and Goble J 1990 *International Journal of Imaging Systems and Technology* **2** 66–75
[6] Polydorides N and Lionheart W 2002 *Measurement Science and Technology* **13** 1871
[7] Adler A, Arnold J H, Bayford R, Borsic A, Brown B, Dixon P, Faes T J C, Frerichs I, Gagnon H, Gärber Y, Grychtol B, Hahn G, Lionheart W R B, Malik A, Patterson R P, Stocks J, Tizzard A, Weiler N and Wolf G K 2009 *Physiological Measurement* **30** S35–55
[8] Castillo E, Castillo R, Martinez J, Shenoy M and Guerrero T 2010 *Physics in Medicine and Biology* **55** 305

Journal of Physics: Conference Series **434** (2013) 012074 doi:10.1088/1742-6596/434/1/012074

High spatial and temporal resolution 4D FEM simulation of the thoracic bioimpedance using MRI scans

Mark Ulbrich[1], Bastian Marleaux[1], Jens Mühlsteff[2], Felix Schoth[3], Ralf Koos[3], Daniel Teichmann[1], Steffen Leonhardt[1]

[1] RWTH Aachen University, Philips Chair for Medical Information Technology, Pauwelsstrasse 20, 52074 Aachen, Germany

[2] Philips Research Eindhoven, High Tech Campus 34, 5656 AE Eindhoven, The Netherlands

[3] University Hospital Aachen, Pauwelsstrasse 30, 52074 Aachen, Germany

E-mail: ulbrich@hia.rwth-aachen.de

Abstract. In this work, a finite element model was created using MRI scans of the main author to analyze sources of the dynamic thoracic bioimpedance. This model can be used to identify limitations of impedance cardiography (ICG) in practice. Heart beat (8.3 ms temporal resolution) and aortic wave propagation (2.6 ms temporal resolution) were implemented. The static volume contains all major organs of the thorax in high spatial resolution. Simulations were successfully conducted and a high correlation (r = 0.9) between the simulated aortic ICG signal and a measured signal of the same subject was obtained.

1. Introduction

ICG is a simple, cheap and non-invasive method to acquire hemodynamic parameters. Due to several limitations of this technology, it is not used in clinical practice today. The finite element model developed within this work shall provide remedy by analyzing multiple sources of the dynamic thoracic impedance and their influence on the signal. That way, the impedance signal of patients with certain pathologies can be simulated to identify problems occurring during measurements.

Others used MRI scans or volumetric representations of the thorax to investigate the thoracic impedance with a limited temporal resolution and no volumetric changes [1, 2, 3, 4]. This work combines high spatial and temporal resolution to create a simulation model which is a novelty. Another advantage of this work is that the basis for the simulation model and the subject measured for the reference signal are the same person.

2. Methods

To build a geometric representation of the thorax of the main author, three different MRI scans were made: a static scan, a dynamic whole-heart scan and an angiographic scan. During the latter, a contrast agent was injected to visualize major blood vessels. The MRI scans were performed by using a Philips Gyroscan NT device with a magnetic field strength of 1.5 T. All three scans were used to build a dynamic representation of the thorax.

Major organs (12 different tissues) were identified from the static and angiographic scans using manual and semi-automatic segmentation. The volumetric geometry changes of both ventricles were extracted manually from the whole-heart scans, since they were ECG gated, resulting in 10 volumes per heartbeat. Unfortunately, diagnostic whole-heart scans do not contain information about atrial changes. Hence, static volumes were used to represent atria during heartbeat.

For the aortic wave propagation, the end-diastolic volume of the aorta obtained by the angiographic scan was semi-automatically segmented by Region Competition Snake Segmentation (RCSS) using the open source tool itk-SNAP [5] and subdivided into 13 parts. The volume of these parts was altered according to the measured pressure pulse propagating over the aorta. Here, 30 volumes per heartbeat were implemented.

Since the angiographic scans were not ECG gated, the dynamics of the aorta were implemented based on data derived from a project in which the arterial system was rebuilt by using silicone [6]. Within this work, pressure and flow data were acquired at various points on the aorta. Thus, the temporal development of pressure and flow for every point on the aorta was calculated and interpolated using MATLAB® (MathWorks, Inc. MA, USA). Using these results and the end-diastolic volume of the aorta, the radial change of the aortic segments was computed (figure 1).

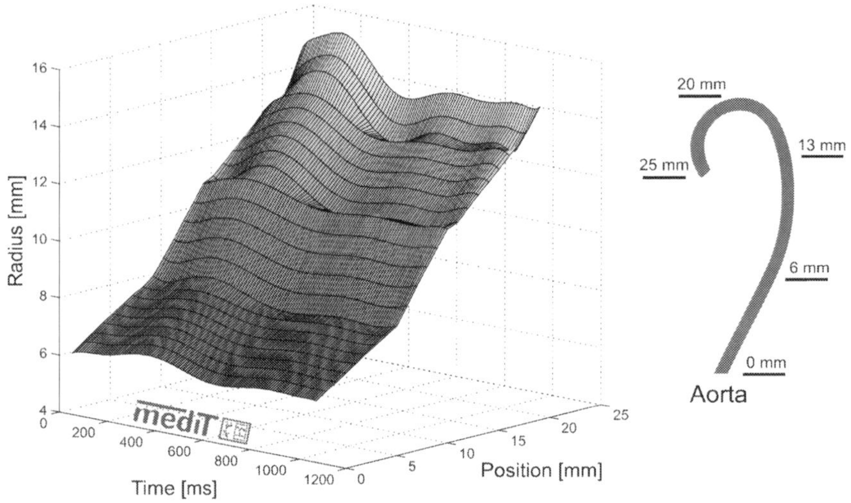

Figure 1. Temporal development of the aortic radius along the aorta

The radius change in this figure was temporally normalized to facilitate the comparison of the curves, but the temporal delay of the propagating pulse was included into the model. All volumes were converted in stereolitography (STL) format and their surfaces were smoothed before being imported into the simulation software. Conductivity and permittivity values for 100 kHz obtained by Gabriel [7] were assigned to all tissues.

Standard electrode positions at neck and abdomen were used to apply a fixed voltage. The resulting complex current flowing through the thorax was calculated by integrating conduction and displacement current density over an area:

$$\underline{I}_{cond} = \int_A (\sigma \cdot \underline{\vec{E}})\, d\vec{A} = \int_A \underline{\vec{J}}\, d\vec{A} = \int_A (Re\{\underline{\vec{J}}\} + jIm\{\underline{\vec{J}}\})\, d\vec{A} \tag{1}$$

$$\underline{I}_{disp} = \int_A (\epsilon \cdot \frac{\partial \underline{\vec{E}}}{\partial t})\, d\vec{A} = \int_A (\frac{\partial \underline{\vec{D}}}{\partial t})\, d\vec{A} = \int_A (j\underbrace{\omega Re\{\underline{\vec{D}}\}}_{Im\{\underline{I}_{disp}\}} - \underbrace{\omega Im\{\underline{\vec{D}}\}}_{Re\{\underline{I}_{disp}\}})\, d\vec{A} \tag{2}$$

$$\underline{I}_{total} = \underbrace{Re\{\underline{I}_{cond}\} + Re\{\underline{I}_{disp}\}}_{Re\{\underline{I}_{total}\}} + \underbrace{Im\{\underline{I}_{cond}\} + Im\{\underline{I}_{disp}\}}_{Im\{\underline{I}_{total}\}} \tag{3}$$

That way, the dynamic complex impedance was calculated. For every calculation volume, a new simulation was conducted. An electroquasistatic solver was used since the wavelength of alternating currents with a frequency of 100 kHz is much bigger than the calculation volume. For discretization, a hexahedral mesh was used with a maximum mesh density of 50 parts subdividing the simulation volume. A higher density could not be used due to the limitation of the main memory (48 GB RAM). The CST EM Studio® (Computer Simulation Technology, Darmstadt, Germany) was used to perform the simulations.

3. Results

Figure 2 shows the simulation model with voltage sources and areas for current density integration. Fat and muscle tissue are semi-transparent to show the inner organs. The extracted volumes for the dynamic volume changes in combination with static blood vessels are shown in figure 3.

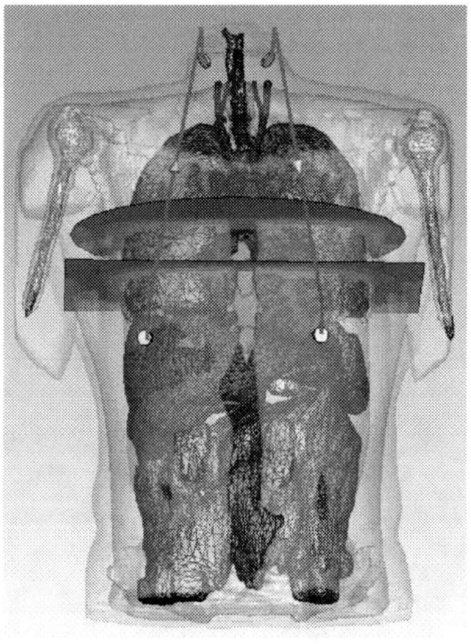

Figure 2. Simulation model.

Figure 3. Blood vessels.

Impedance curves for both dynamic sources were assessed. Because discretization noise for a mesh density of 50 parts was unavoidable, a smoothing spline was calculated for further signal analysis.

Since it is known from previous simulations, that the aortic diameter change defines the characteristic morphology of the ICG signal [8], and since no other dynamic sources besides the mentioned ones are implemented so far in the model, the derivative of the aortic impedance change was compared to a measured ICG signal of the same subject using the Niccomo device (medis, Ilmenau, Germany) showing a high cross correlation factor of $r = 0.9$ (see figure 4).

Despite the discretization noise, the signal is good enough to extract characteristic points, such as closure (X-point) and opening (B-point) of the aortic valve, to calculate left ventricular ejection time (LVET). Table 1 shows the extracted times of characteristic points and LVET.

Although simulated and measured ICG curve show a rather large temporal offset, LVET is the same for both curves.

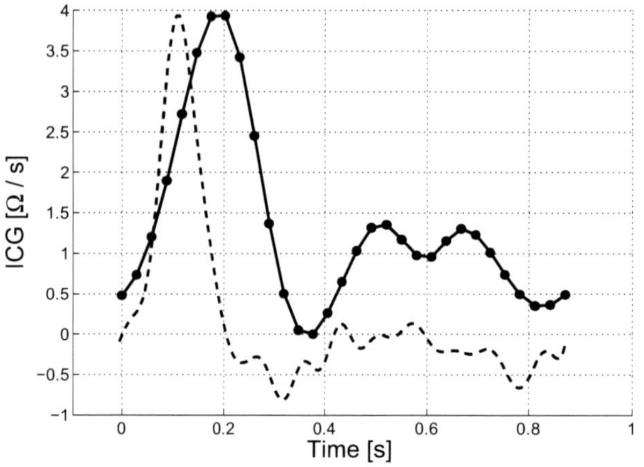

Figure 4. Simulated ● and measured - - - -ICG.

Table 1. Extracted times from ICG curve.

	B-point	C-point	X-point	O-point	LVET
Simulated	93 ms	213 ms	373 ms	506 ms	280 ms
Measured	35 ms	105 ms	315 ms	430 ms	280 ms

4. Conclusion

First, a finite element model of a human thorax with high temporal and spatial resolution was created containing important organs and major blood vessels. Besides these static volumes, heartbeat and aortic wave propagation were implemented to simulate the thoracic complex dynamic impedance change. Second, the impedance change for both dynamics was simulated showing a reasonable development. Third, the morphology of the ICG signal shows high correlation with the morphology of measured signals. Hence, a model was created that provides the possibility to identify changes of the ICG signal due to pathologies in the future.

Improvements of the model comprise the integration of dynamic atria by using additional MRI scans and the implementation of other important dynamics, such as erythrocyte orientation and lung perfusion, into the simulations. Since volumes of large venous vessels were extracted, the impact of volumetric changes of the vena cava on the impedance cardiogram can be analyzed.

References

[1] Patterson R 1985 *Medical & Biological Engineering & Computing* **23** 411–417
[2] Sakamoto K, Muto K, Kanai H and Iizuka M 1979 *Medical & Biological Engineering & Computing* **17** 697–709
[3] Kim D, Baker L, Pearce J and Kim W K 1988 *IEEE Transactions on Biomedical Engineering* **35** 993–1000
[4] Wang L and Patterson R 1995 *IEEE Transactions on Biomedical Engineering* **42** 141–148
[5] Yushkevich P A, Piven J, Cody Hazlett H, Gimpel Smith R, Ho S, Gee J C and Gerig G 2006 *Neuroimage* **31** 1116–1128
[6] Matthys K S, Alastruey J, Peir J, Khir A W, Segers P, Verdonck P R, Parker K H and Sherwin S J 2007 *Journal of Biomechanics* **40** 3476–3486
[7] Gabriel C, Gabriel S and Corthout E 1996 *Physics in Medicine and Biology* **41** 2231–2249
[8] Ulbrich M, Mühlsteff J, Walter M and Leonhardt S 2011 *Computing in Cardiology* vol 38 pp 149–152

Studies of an Adaptive Kaczmarz Method for Electrical Impedance Imaging

Taoran Li[1], David Isaacson[2], Jonathan C. Newell[3], and Gary J. Saulnier[1]

[1] Department of Electrical Engineering, Rensselaer Polytechnic Institute, Troy, NY, USA
[2] Department of Mathematical Sciences, Rensselaer Polytechnic Institute, Troy, NY, USA
[3] Department of Biomedical Engineering, Rensselaer Polytechnic Institute, Troy, NY, USA

E-mail: lit5@rpi.edu

Abstract. We present an adaptive Kaczmarz method for solving the inverse problem in electrical impedance tomography and determining the conductivity distribution inside an object from electrical measurements made on the surface. To best characterize an unknown conductivity distribution and avoid inverting the Jacobian-related term $J^T J$ which could be expensive in terms of memory storage in large scale problems, we propose to solve the inverse problem by adaptively updating both the optimal current pattern with improved distinguishability and the conductivity estimate at each iteration. With a novel subset scheme, the memory-efficient reconstruction algorithm which appropriately combines the optimal current pattern generation and the Kaczmarz method can produce accurate and stable solutions adaptively compared to traditional Kaczmarz and Gauss-Newton type methods. Several reconstruction image metrics are used to quantitatively evaluate the performance of the simulation results.

1. Introduction

Different reconstruction methods proposed to solve the electrical impedance tomography (EIT) inversion problem can be categorized as direct methods and iterative methods. Direct methods are considered to be fast but less accurate. In contrast, iterative methods tend to search for more accurate solutions by iteration. However, these methods, such as the Gauss Newton method, can be expensive in terms of memory storage in large scale problems [3]. The choice of current patterns will also have a great impact on determining the system's distinguishability. A system with limited measurement precision but high distinguishability is more likely to detect small conductivity changes in an object.

We propose the adaptive Kaczmarz method, which solves the inverse problem by applying the optimal current patterns for distinguishing the actual conductivity from the conductivity estimate between each iteration of the block Kaczmarz method. With a novel ordered subset scheme, this reconstruction method appropriately combines the optimal current pattern generation with the block Kaczmarz method and produces accurate solutions memory-efficiently.

2. Method

In this paper the inverse problem of EIT is a least-squares problem with the objective functional:

$$E(\rho) = \left\| V - U(\rho) \right\|_2^2 \quad (1)$$

where ρ is the piecewise constant conductivity distribution, $U(\rho)$ are the voltages predicted by a forward model, and V are the measured voltages on the electrodes.

2.1. Ordered Subset Scheme

The block Kaczmarz method requires that the complete set of both current patterns and measurements should be appropriately divided into subsets. One subset scheme, known as the sequential subset scheme, partitions the complete set in the order that the current patterns are applied and sequentially pick the block at each iteration. Another subset scheme, known as the randomized subset scheme randomly picks the block from a pre-partitioned complete set at each iteration. We propose a novel subset scheme so that we can partition and order the subsets of current patterns and measurements by their ability to characterize the conductivity distribution. Specifically, we partition the current patterns $I = \{I_{sub}^1, I_{sub}^2, \ldots, I_{sub}^n\}$ and measurements $V = \{V_{sub}^1, V_{sub}^2, \ldots, V_{sub}^n\}$ so that the pair (I_{sub}^1, V_{sub}^1) is the best subset to distinguish the actual conductivity from the conductivity estimate; the pair (I_{sub}^2, V_{sub}^2) is the second best subset and so on. In this way, the best "blockwise" conductivity estimate from the current subset will be obtained and passed down as the starting point to the next subset. To obtain the ordered subset of the current patterns, the complete set of the optimal current patterns, which best distinguishes the actual conductivity distribution from a conductivity estimate [2], should be generated and partitioned.

2.2. Adaptive Kaczmarz

An adaptive Kaczmarz method with ordered subsets is proposed based on the block Kaczmarz method [4] and optimal current pattern generation. It begins with an initial guess for the current patterns $I^{(0)}$ and an initial guess for the conductivity distribution $\rho^{(0)}$ and proceeds to generate the optimal current patterns $I^{(1)}$ to best distinguish the actual conductivity from the guess $\rho^{(0)}$. Then $I^{(1)}$ with the ordered subset scheme is used to solve for the conductivity distribution with one iteration of the block Kaczmarz method to obtain $\rho^{(1)}$. The more accurate conductivity estimate $\rho^{(1)}$ is used for the next iteration to find the optimal current patterns $I^{(2)}$ to best distinguish the actual conductivity distribution from the more accurate estimate $\rho^{(1)}$, which are then used to solve for $\rho^{(2)}$. Therefore, a new conductivity estimate and a set of optimal current patterns are obtained adaptively at each iteration. The iteration continues until a reasonable conductivity estimate is obtained.

Let T be an orthonormal current matrix, $\rho^{(0)}$ be an initial conductivity guess, m be the number of iterations. The algorithm is as follows:

1) Initialize $\hat{T} = T, \rho = \rho^{(0)}$;
2) **for** $k = 1{:}m$ **do**
   a. **repeat**
      1. Apply $\hat{T}$ and measure voltages V;
      2. Apply $\hat{T}$ and compute the simulated voltages $U(\rho)$;
      3. Form the voltage difference matrix $D = V - U(\rho)$ and $P = D^*D$;
      4. Eigenvalue decomposition of P and form the eigenvector matrix E so that the corresponding eigenvalues are sorted in descending order;
      5. $\hat{T} = \hat{T}E$;

      until *the current pattern converges;*
   b. Obtained the ordered subsets $I = \{I_{sub}^1, I_{sub}^2, \ldots, I_{sub}^n\}$, $V = \{V_{sub}^1, V_{sub}^2, \ldots, V_{sub}^n\}$, and the row submatrix $J = \{J_1, J_2, \ldots, J_n\}$, $\rho_0 = \rho$;
   c. **for** $i = 1{:}n$ **do**

      $$\rho_i = \rho_{i-1} + a_i J_i^T \left(J_i J_i^T + \lambda_i I\right)^{-1} [V_i - U_i(\rho_{i-1})];$$

      end
   d. $\rho = \rho_n$;

   end

The adaptive Kaczmarz method updates adaptively both the optimal current pattern and a more accurate conductivity estimate at each iteration. It is empirically observed that an accurate and stable solution can be obtained within 5 iterations.

2.3 Model Setup

A normalized 2D circular tank with unit radius and 32 equally-spaced square electrodes is modelled. Two circular targets with radius 0.15 are placed inside the tank. The conductor is placed at (0.4,-0.35) with conductivity 2 S/m, and the insulator is placed (-0.3, 0.2) with the conductivity 0 S/m. Trigonometric current patterns are used with the maximum amplitude of 1 mA and the complete electrode model is assumed. The inverse mesh (2992 elements) is designed to be coarser that the forward mesh (70394 elements) to avoid an *inverse crime*. The non-adaptive Kaczmarz method, the Gauss Newton method, the adaptive Kaczmarz method, and the adaptive Gauss Newton method are compared for their reconstructed images. The adaptive Gauss Newton method combines the Gauss Newton and the optimal current pattern generation in a similar way as the adaptive Kaczmarz method does.

3. Results

3.1 Comparison of different subset schemes

Difference images are reconstructed using the Kaczmarz method with the sequential subset scheme, the randomized subset scheme, and the proposed ordered subset scheme. It is observed in Figure 1 that reconstructed images made with the randomized scheme and the sequential scheme have target deformation and artifacts on different levels, while the image made with ordered subsets more accurately reflects the actual target's location, size, and shape.

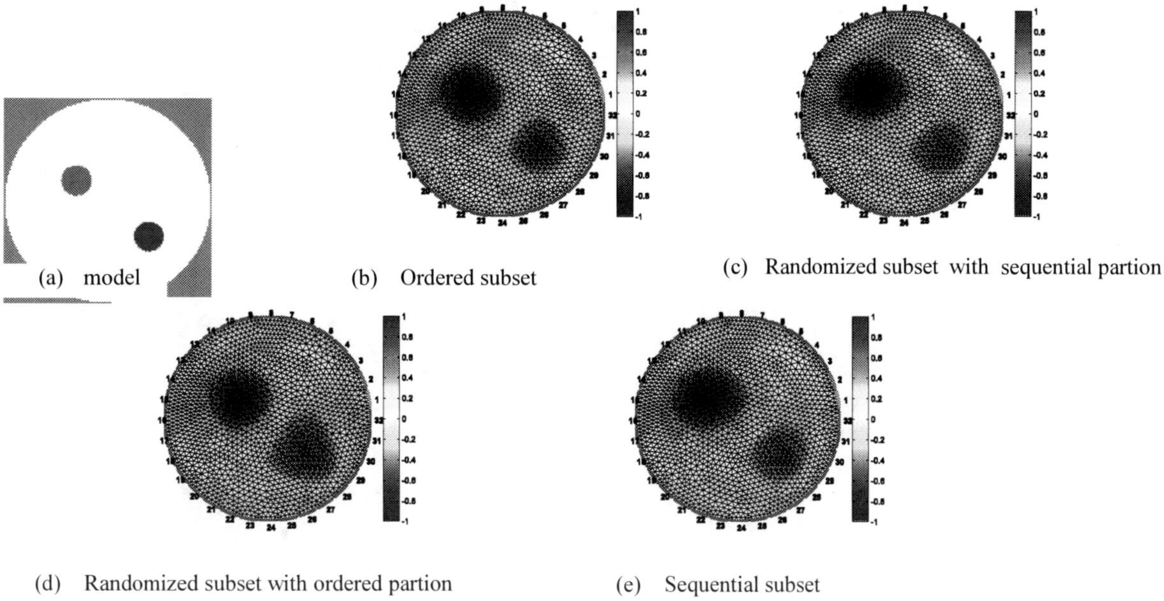

(a) model (b) Ordered subset (c) Randomized subset with sequential partion

(d) Randomized subset with ordered partion (e) Sequential subset

Figure 1. Comparison of reconstructed conductivity distribution using different subset schemes

3.2 Comparison of different reconstruction methods

In Figure 2, the improvement of the adaptive Kaczmarz and the adaptive Gauss Newton over their non-adaptive version can be easily observed. The images are further compared using several consensus figures of merit described in [1]. The comparison of different methods is shown in Table 1. From the error table, it is observed that the adaptive Kaczmarz has the smallest position error and blurring error, and the closest peak value to the true peak value 2. In terms of memory requiremet, the Gauss Newton method with 10 iterations requires 231.4 Megabytes, while the block Kaczmarz method with 10 iterations require only 31.2 Megabytes.

4. Conclusion

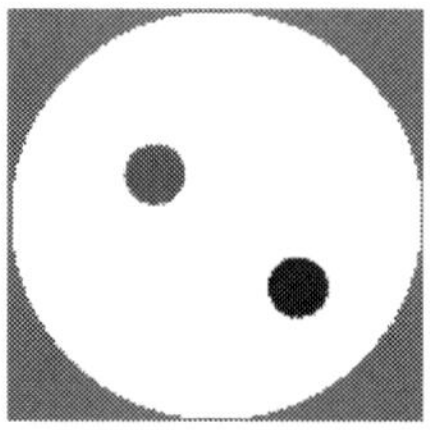

(a) model

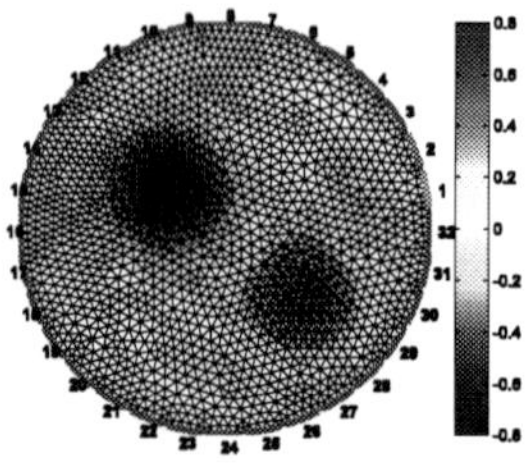

(b) Kaczmarz

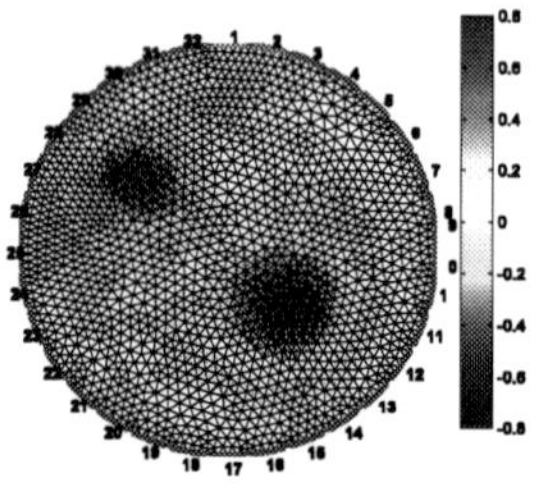

(c) Gauss Newton

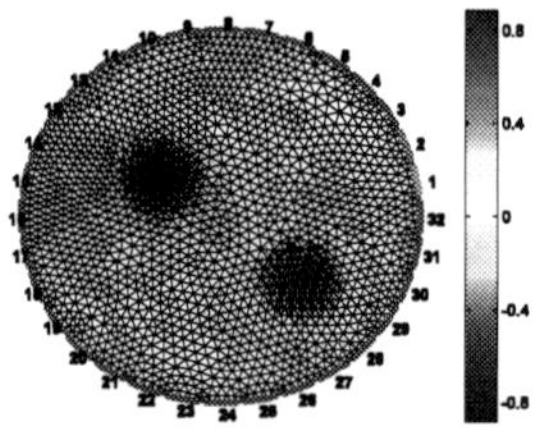

(d) Adaptive Kaczmarz

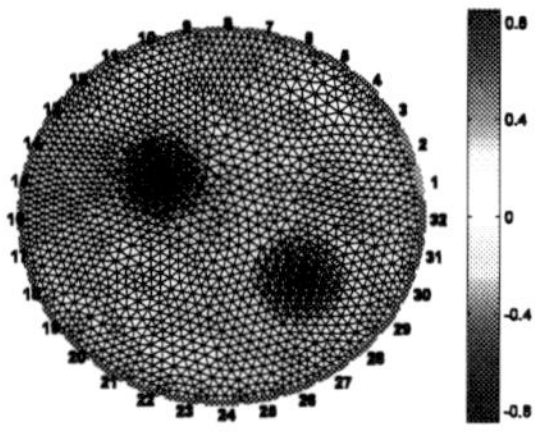

(e) Adaptive Gauss Newton

Figure 2. Comparison of reconstructed conductivity distribution using different reconstruction methods

Table 1. AR stands for Amplitude Response, PE stands for Position Error, BE stands for Blurring Error, and PV stands for Peak Value

	AR	PE	BE	PV
Gauss Newton	0.0222	0.0202	3.0261	0.4499
Kaczmarz	0.0199	-0.0166	2.6690	0.7336
Adaptive Gauss Netwon	0.0233	-0.0166	1.8037	0.7974
Adaptive Kaczmarz	0.0233	0.0032	1.5262	0.9351

We introduce the adaptive Kaczmarz method to solve the inversion in electrical impedance tomography. With a novel subset scheme, the cost-efficient method appropriately combines the block Kaczmarz method and optimal current pattern generation to produce accurate solutions. Unlike the Gauss Newton method, the adaptive Kaczmarz method avoids the expense of inverting large scale matrices and is quite memory-efficient. The simulation results show that the adaptive Kaczmarz is able to provide accurate 2D reconstruction images within a few iterations with less memory requirement.

Future efforts will involve further improvement in the efficiency of the algorithm and the generalization to the 3D case.

Acknowledgements
This work is supported by a NIH grant under the grant number 1R01HL 109854-01.

References

[1] Adler A *et al* 2009 GREIT: a unified approach to 2D linear EIT reconstruction of lung images *Physiol. Meas.* **30** S35-55, 169, 791
[2] Isaacson D and Cheney M 1996 *Process for producing optimal current patterns for electrical impedance tomography* US Patent 5588429
[3] Horesh L, Schweiger M, Arridge S R and Holder D S 2005 Large scale non-linear 3D reconstruction algorithms for electrical impedance tomography of the human brain *Conf. on Applied Inverse Problem*
[4] Gordon R, Bender R and Herman G T 1970 Algebraic reconstruction techniques (ART) for three-dimensional electron microscopy and X-ray photography *J. Theoret. Biol.* **29** 471-481

Monotony based inclusion detection in EIT for realistic electrode models

Bastian Harrach and Marcel Ullrich

Institut für Mathematik, Julius-Maximilians-Universität, 97074 Würzburg, Germany

E-mail: bastian.harrach@uni-wuerzburg.de and marcel.ullrich@uni-wuerzburg.de

Abstract. In electrical impedance tomography (EIT), there is a monotony relation between the conductivity of an imaging subject and the EIT measurements. Such monotony relations can be used for inclusion detection methods. However, up to now, monotony-based methods were only known to reconstruct an upper or lower bound of inclusions.

In a recent preprint, the authors showed that, for continuous boundary data, monotony based methods are in fact capable of reconstructing the exact shape of inclusions. In this work, we discuss how to extend our results to the practically relevant complete electrode model setting and study the possibility of rigorous resolution guarantees.

1. Introduction

The aim of electrical impedance tomography is to image the conductivity inside a conducting subject by boundary current-voltage measurements. In this work we focus on detecting the shape of conducting anomalies (aka inclusions) within a practically relevant setting. To obtain current-voltage measurements we consider the following setting, cf. figure 1.

We are given a conducting object with a set of separated electrodes attached to its boundary. A prescribed current pattern is applied to the electrodes and yields measurement data of the resulting potential on each electrode (up to a real constant representing the ground level). The boundary of the object is isolated, currents can only flow trough the electrodes. The electrodes are highly conducting, such that the potentials can be assumed to be constant on each electrode. Furthermore we assume that the contact impedance of a thin contact layer between each electrode and the object is known.

Mathematically, for this setting, the current-voltage measurements can be modeled by the complete electrode model (CEM) as follows. Let $\Omega \in \mathbb{R}^n$ be a bounded domain with smooth boundaries, $\sigma \in L^\infty_+(\Omega)$ (where $L^\infty_+(\Omega)$ denotes the subspace of $L^\infty(\Omega)$-functions with positive essential infima), the electrodes are identified with their corresponding boundary parts $e_1, e_2, \dots, e_L \subseteq \partial\Omega$ and the contact impedances are real values $z_1, z_2, \dots, z_L \in \mathbb{R}$. Then, for a prescribed current pattern $I = (I_l)_{l=1}^L \in \mathbb{R}^L_\diamond$ there exists a unique pair $(u, U) \in H^1_\diamond(\Omega) \times \mathbb{R}^L_\diamond$ (where the subscript $\diamond$ denotes the subspace of $H^1(\Omega)$-functions resp. $\mathbb{R}^L$-elements with vanishing integral mean resp. sum of components), such that

$$\nabla \cdot \sigma \nabla u = 0 \quad \text{in } \Omega, \tag{1}$$

$$u|_{\partial\Omega} + z_l\sigma\partial_\nu u|_{\partial\Omega} = U_l \quad \text{on } e_l, \quad 1 \le l \le L, \tag{2}$$

$$\sigma\partial_\nu u|_{\partial\Omega} = 0 \quad \text{on } \partial\Omega \setminus \bigcup_{l=1}^{L} e_l, \tag{3}$$

$$\int_{e_l} \sigma\partial_\nu u|_{\partial\Omega} \mathrm{d}S = I_l, \tag{4}$$

where $\sigma\partial_\nu u|_{\partial\Omega}$ denotes the Neumann boundary values and $u|_{\partial\Omega} \in \mathrm{L}^2_\diamond(\partial\Omega)$ is the trace of u. By this formulation of the CEM, we can identify the current-voltage measurements for an object with conductivity σ, with the linear symmetric mapping $R(\sigma) : \mathbb{R}^L_\diamond \to \mathbb{R}^L_\diamond, I \mapsto U$, cf. e.g. [1].

2. Monotony based methods

2.1. The monotony relation

Theorem 1. *Let $\sigma_1, \sigma_2 \in L^\infty_+(\Omega)$ and (v, V) be the solution of the CEM for the conductivity σ_2. Then*

$$\int_\Omega (\sigma_1 - \sigma_2)|\nabla v|^2 \mathrm{d}x \ge \langle I, (R(\sigma_2) - R(\sigma_1))\, I\rangle \ge \int_\Omega \frac{\sigma_2}{\sigma_1}(\sigma_1 - \sigma_2)|\nabla v|^2 \mathrm{d}x, \tag{5}$$

where $\langle\cdot,\cdot\rangle : \mathbb{R}^L \times \mathbb{R}^L \to \mathbb{R}^L$ denotes the euclidean inner product.

Note that there is an analogue monotony results for the continuous boundary model that seem to go back to Ikehata, Kang, Seo and Sheen [2, 3].

Proof. It is known, that $(u, U) \in H^1_\diamond(\Omega) \times \mathbb{R}^L_\diamond$ is the solution of the CEM for the conductivity σ and the prescribed current pattern $I \in \mathbb{R}^L_\diamond$ if and only if the following variational formulation

$$B_\sigma((u, U), (v, V)) := \int_\Omega \sigma\nabla u \cdot \nabla v \mathrm{d}x + \sum_{l=1}^{L} \frac{1}{z_l} \int_{e_l} (u|_{\partial\Omega} - U_l)(v|_{\partial\Omega} - V_l)\mathrm{d}S = \sum_{l=1}^{L} I_l V_l = \langle I, V\rangle$$

holds for all $(v, V) \in H^1_\diamond(\Omega) \times \mathbb{R}^L_\diamond$, cf. e.g. [1].

With this equivalent formulation of the CEM-System we obtain the first inequality. Let (u, U) resp. (v, V) be the solution of the CEM for the conductivity σ_1 resp. σ_2.

$$\begin{aligned}
0 &\le \int_\Omega \sigma_1|\nabla(u - v)|^2 \mathrm{d}x + \sum_{l=1}^{L} \frac{1}{z_l} \int_{e_i} \left((u|_{\partial\Omega} - U_i) - (v|_{\partial\Omega} - V_i)\right)^2 \mathrm{d}S \\
&= B_{\sigma_1}((u, U), (u, U)) + B_{\sigma_2}((v, V), (v, V)) - 2B_{\sigma_1}((u, U), (v, V)) + \int_\Omega (\sigma_1 - \sigma_2)|\nabla v|^2 \mathrm{d}x \\
&= \langle I, (R(\sigma_1) - R(\sigma_2))I\rangle + \int_\Omega (\sigma_1 - \sigma_2)|\nabla v|^2 \mathrm{d}x.
\end{aligned}$$

Thus, we obtain the first inequality.

By interchanging (u, U) and (v, V) as well as σ_1 and σ_2, the equation above yields:

$$\begin{aligned}
&\langle I, (R(\sigma_2) - R(\sigma_1))I\rangle \\
&= \int_\Omega (\sigma_1 - \sigma_2)|\nabla u|^2 \mathrm{d}x + \int_\Omega \sigma_2|\nabla(v - u)|^2 \mathrm{d}x + \sum_{l=1}^{L} \frac{1}{z_l} \int_{e_i} \left((v|_{\partial\Omega} - V_i) - (u|_{\partial\Omega} - U_i)\right)^2 \mathrm{d}S \\
&= \int_\Omega \sigma_1 \left|\nabla u - \frac{\sigma_2}{\sigma_1}\nabla v\right|^2 + \frac{\sigma_2}{\sigma_1}(\sigma_1 - \sigma_2)|\nabla v|^2 \mathrm{d}x + \sum_{l=1}^{L} \frac{1}{z_l} \int_{e_i} \left((v|_{\partial\Omega} - V_i) - (u|_{\partial\Omega} - U_i)\right)^2 \mathrm{d}S.
\end{aligned}$$

Since σ_1 and z_l are positive the second inequality follows. □

A simple consequence is

$$\sigma_1 \leq \sigma_2 \quad \Longrightarrow \quad R(\sigma_1) \geq R(\sigma_2), \tag{6}$$

where the first inequality is meant in the pointwise (a.e.) on Ω sense, and the second one is to be understood in the sense that $R(\sigma_1) - R(\sigma_2)$ has only non-negative eigenvalues.

2.2. A simple monotony based shape reconstruction method

We start by presenting the main idea on a simple case. Let the conductivity σ of the conducting object Ω be given by $\sigma = 1 + \chi_D \in L^\infty_+(\Omega)$, where $D \subseteq \Omega$ is an open set. Then (6) yields that

$$D_{\mathrm{R}} := \bigcup_{\epsilon>0,\ x\in\Omega,\ R(1+\chi_{B_\epsilon(x)})\geq R(\sigma)} B_\epsilon(x)$$

is an upper bound of the inclusion D.

Such monotony based reconstruction methods were numerically studied by Tamburrino and Rubinacci [4]. In the recent preprint [5] (see also [6]), the authors showed that, for continuous boundary data, monotony methods are actually capable of reconstructing the exact shape D (if D has connected complement) and that they allow a fast linearized but still exact implementation.

3. Resolution guaranties for monotony based methods

In the realistic CEM-model with finitely many electrodes, we can no longer expect the monotony based method to reconstruct the exact shape of an inclusion. However, we will now present a new approach that allows us to guarantee certain detectability properties (or resolution guarantees) for the CEM-model, even with imprecisely known backgrounds and measurement noise.

For a resolution given by a partition $(\omega_i)_{i=1}^N$, $\dot{\bigcup}_{i=1}^n \omega_i = \Omega$ ($w_i \neq \emptyset$), we will show how to test whether a set of electrodes $e_1, e_2, \ldots, e_L$ is suitable to ensure some reconstruction guaranties or not.

Similar as in section 2.2, let us consider the case where the conductivity σ is given by $\sigma = \sigma_0 + \chi_D \kappa \in L^\infty_+(\Omega)$ with $\sigma_0, \kappa \in L^\infty(\Omega)$, $\kappa \geq c$ (a.e.) on a set $D \subseteq \Omega$ and $|\sigma_0 - 1| \leq \epsilon$ for some small constants $c > 2\epsilon > 0$.

3.1. A basic resolution guarantee

For an inclusion detection method that yields a reconstruction D_R to D, we say that it fulfills a resolution guarantee with respect to a partition $(\omega_i)_{i=1}^N$ if the following holds.

If a piece ω_i of the resolution partition is completely overlapped by an inclusion D, the method will mark this piece as a part of the reconstruction D_R. On the other hand, if there is no inclusion, no piece of the resolution partition will be marked. Mathematically speaking, this means that

(i) $\omega_i \subseteq D$ implies $\omega_i \subseteq D_R$ for $i \in \{1, 2, \ldots, N\}$, and
(ii) $D = \emptyset$ implies $D_R = \emptyset$.

3.2. Verification of resolution guarantee

Theorem 2. *The inclusion detection method*

$$D_R := \bigcup_{i\in\{1,2,\ldots,N\},\ R(1-\epsilon+c\chi_{\omega_i})\geq R(\sigma)} \omega_i \tag{7}$$

fulfills the resolution guarantee of 3.1 if and only if

$$R(1 - \epsilon + c\chi_{\omega_i}) \not\geq R(1 + \epsilon), \quad \textit{for all } i \in \{1, 2, \ldots, N\}. \tag{8}$$

Proof. By the monotony relation (6), assertion (i) is always satisfied, since $\omega_i \subseteq D$ implies $1 - \epsilon + c\chi_{\omega_i} \leq \sigma$ for all $i \in \{1, 2, \ldots, N\}$. Hence, only assertion (ii) needs to be verified.

First, we assume that $D = \emptyset$. Thus, $\sigma = \sigma_0 \leq 1 + \epsilon$ and by the monotony relation (6) we obtain that $R(\sigma) \geq R(1 + \epsilon)$. Hence, $R(1 - \epsilon + c\chi_{\omega_i}) \not\geq R(1 + \epsilon)$ for all $i \in \{1, 2, \ldots, N\}$ implies $R(1 - \epsilon + c\chi_{\omega_i}) \not\geq R(\sigma)$ for all $i \in \{1, 2, \ldots, N\}$.

For the converse, we assume that $R(1 - \epsilon + c\chi_{\omega_i}) \geq R(1 + \epsilon)$ for at least one $i \in \{1, 2, \ldots, N\}$ and the conductivity be given by $\sigma = 1 + \epsilon$. For this case $D_R \supseteq \omega_i \neq \emptyset$. □

Theorem 2 offers a criterion to check, if the resolution and the electrode setting fit together, such that the reconstruction method, given by (7), fulfills the resolution guarantee or not.

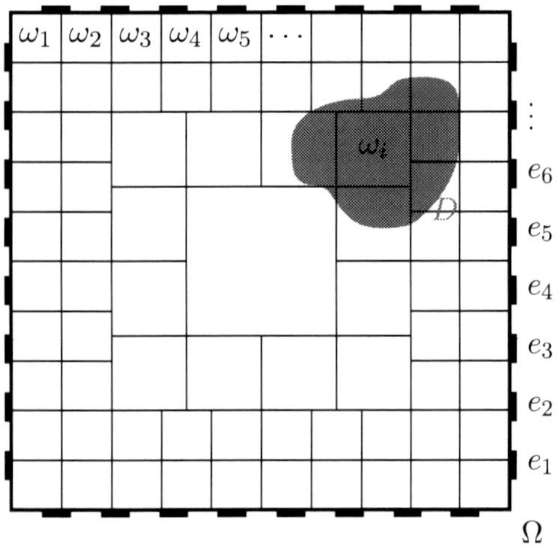

Figure 1. CEM-setting with a potentially resolution $(\omega_i)_{i=1}^N$ and sample inclusion D.

3.3. Presence of measurement errors

We consider noisy measurement data $R^\delta(\sigma)$ with $\|R(\sigma) - R^\delta(\sigma)\| = \delta$. If the reconstruction method with exact data fulfills the resolution guarantee, the test criteria (8) also yield a bound $\Delta > 0$ for the measurement error δ, such that the resolution guarantee is still valid for measurement with noise up to Δ. This bound depend on the smallest eigenvalue of all the matrices $R(1 - \epsilon + c\chi_{\omega_i}) - R(1 + \epsilon)$ from the test criterion.

References

[1] Somersalo E, Cheney M and Isaacson D 1992 *SIAM J. Appl. Math.* **52** 1023–1040
[2] Ikehata M 1998 *J. Inverse Ill-Posed Probl.* **6** 127–140
[3] Kang H, Seo J K and Sheen D 1997 *SIAM J. Math. Anal.* **28** 1389–1405
[4] Tamburrino A and Rubinacci G 2002 *Inverse Problems* **18** 1809–1829
[5] Harrach B and Ullrich M submitted for publication, online available at `http://www.mathematik.uni-wuerzburg.de/~harrach/publikationen.html`
[6] Harrach B and Ullrich M 2010 *AIP Conf. Proc.* **1281** 1975–1978

XV Int. Conf. on Electrical Bio-Impedance & XIV Conf. on Electrical Impedance Tomography IOP Publishing
Journal of Physics: Conference Series **434** (2013) 012077 doi:10.1088/1742-6596/434/1/012077

A Novel Method for Monitoring Data Quality in Electrical Impedance Tomography

Andy Adler[1], Bartłomiej Grychtol[2], Pascal Gaggero[3], Jörn Justiz[3], Volker Koch[3] and Yasin Mamatjan[1]

[1] Systems and Computer Engineering, Carleton University, Ottawa, Canada
[2] German Cancer Research Center, Heidelberg, Germany
[3] Bern University of Applied Sciences, Biel/Bienne, Switzerland

E-mail: `mamatjan,adler@sce.carleton.ca`

Abstract. Electrical impedance tomography (EIT) has the promise to help improve care for patients undergoing ventilation therapy by providing real-time bed-side information on the distribution of ventilation in their lungs. To realise this potential, it is important for an EIT system to provide a reliable and meaningful signal at all times, or alert clinicians when this is not possible. Because the reconstructed images in EIT are sensitive to system instabilities (including electrode connection problems) and artifacts caused by e.g. movement or sweat, there is a need for EIT systems to continuously monitor, recognize and, if possible, correct for such errors. Motivated by this requirement, our paper describes a novel approach to quantitatively measure EIT data quality suitable for online and offline applications. We used a publicly available data set of ventilation data from two pediatric patients with lung disease to evaluate the data quality on clinical data. Results suggest that the developed data quality could be a useful tool for real-time assessment of the quality of EIT data and, hence, to indicate the reliability of any derived physiological information.

1. Introduction

Electrical impedance tomography (EIT) has emerged as a promising technology for continuous monitoring of lung patients. In this application, measurement errors are inevitable due to dynamics of the human body (movement and sweat), electrode detachment and noise or drift in electronics [1] decreasing the clinical value of the acquired data. Although various approaches [2,3] have been proposed to date to detect EIT data errors, no systematic approach to quantifying EIT data quality has been offered. Particularly in an Intensive Care Unit (ICU), where it is vital that an EIT system provides a reliable and meaningful signal at all times, or alert clinicians when this is not possible, there is a strong need for a data quality measure. Motivated by this need, we define the requirements of a data quality metric for EIT, develop a metric that meets them, and an efficient way to calculate it.

2. Methodology

2.1. Overview

Our approach to measure and quantify data quality is based on the jackknife (leave-one-out) concept [4], which is an efficient cross-validation technique suitable for real-time applications. Briefly, we use an FEM model of the investigated domain to predict the value of an individual voltage measurement from all the other measurements (the jackknife estimate). The comparison of the prediction with the actual measurement yields an estimate of the measurement error. The final data quality metric q is a scalar quantity aggregated from all the individual measurement errors, following a suitable transformation.

2.2. Requirements

A useful data quality metric for EIT should satisfy the following requirements:

(i) q should be bound between 0 and 1 ($0 \leq q \leq 1$), with 1 indicating perfect quality.
(ii) adding "good" data to the measurement vector should *increase* q.
(iii) conversely, adding more data of low quality should *decrease* q.
(iv) q should be low if data is wrong (e.g. electrodes were placed incorrectly).

2.3. Formulation of data quality

We consider the case of dynamic (functional) EIT, where images of relative conductivity changes $\mathbf{m} = (\boldsymbol{\sigma} - \boldsymbol{\sigma}_r)/\boldsymbol{\sigma}_r$ are reconstructed from normalized voltage difference data $\mathbf{d} = (\mathbf{v} - \mathbf{v}_r)/\mathbf{v}_r$, where $\mathbf{v}$ and $\mathbf{v}_r$ are the subsequent and reference measurements, respectively, while, $\boldsymbol{\sigma}$ and $\boldsymbol{\sigma}_r$ represent the corresponding conductivity distributions inside the medium.

We define the error to signal ratio* $\boldsymbol{\varepsilon} \in \mathbb{R}^M$ as the estimated measurement error over signal

$$\varepsilon_i = \frac{d_i - \hat{d}_i}{S} = \frac{1}{S}\left(\boldsymbol{\theta}_{\mathrm{i}}^T\left(\mathbf{d} - \hat{\mathbf{d}}_{\mathrm{i}}\right)\right), \tag{1}$$

for $i = 1 \ldots M$, where $\mathbf{d} \in \mathbb{R}^M$ is the vector of measured data, and $\hat{\mathbf{d}}_{\mathrm{i}}$ is a jackknife estimate of $\mathbf{d}$ based on all the data $\mathbf{d}$ but its i-th element, $\boldsymbol{\theta}_{\mathrm{i}}$ is a vector ($M \times 1$) of zeros with a single element $[\boldsymbol{\theta}_{\mathrm{i}}]_i = 1$ and S represents the signal strength. Because the latter is not known *a priori*, we define S as the mean of simulated normalized difference data generated by small (20% medium radius) spherical contrast (2:1 contrast-to-background conductivity ratio) at the center of the medium.

To obtain the jackknife estimates, we devise a reconstruction matrix $\mathbf{R}_{\mathrm{i}}$ based on the one-step Gauss-Newton schema, such that the i-th measurement channel is ignored. This is accomplished by defining a data noise covariance matrix as $\boldsymbol{\Sigma}_{\mathrm{n}} = (\mathbf{I} + \mu\boldsymbol{\Theta}_{\mathrm{i}})$, i.e. such that noise on channel i is assumed to be much greater (by an additive factor $\mu \gg 0$) than on other measurements:

$$\mathbf{R}_{\mathrm{i}} = \boldsymbol{\Sigma}_{\mathrm{m}}\mathbf{J}^T\left(\mathbf{J}\boldsymbol{\Sigma}_{\mathrm{m}}\mathbf{J}^T + \lambda^2\left(\mathbf{I} + \mu\boldsymbol{\Theta}_{\mathrm{i}}\right)\right)^{-1} \tag{2}$$

where $\boldsymbol{\Theta}_{\mathrm{i}}$ is a matrix ($M \times M$) of zeros with $[\Theta_{\mathrm{i}}]_{ii} = 1$.

Hence, we rewrite 1 as

$$\varepsilon_{\mathrm{i}} = \frac{1}{S}\left(\boldsymbol{\theta}_{\mathrm{i}}^T\left(\mathbf{d} - \mathbf{J}\mathbf{R}_{\mathrm{i}}\mathbf{d}\right)\right) = \frac{1}{S}\left(\boldsymbol{\theta}_{\mathrm{i}}^T\left(\mathbf{I} - \mathbf{J}\mathbf{R}_{\mathrm{i}}\right)\mathbf{d}\right) \tag{3}$$

To speed up the calculations, we define a "quality" matrix $\mathbf{Q}$ such that each row i of $\mathbf{Q}$ is given by $\mathbf{Q}_i = \boldsymbol{\theta}_{\mathrm{i}}^T(\mathbf{I} - \mathbf{J}\mathbf{R}_{\mathrm{i}})$. Thus

$$\boldsymbol{\varepsilon} = \frac{1}{S}\left(\mathbf{Q}\mathbf{d}\right). \tag{4}$$

We then express the quality of an individual measurement i as

$$q_i = \left(\frac{1}{2}\right)^{|\varepsilon_i|} = \begin{cases} 1 & \text{if } |\varepsilon_i| \text{ is } 0 \\ \frac{1}{2} & \text{if } |\varepsilon_i| \text{ is } 1 \\ 0 & \text{if } |\varepsilon_i| \text{ is } \infty \end{cases} \tag{5}$$

Journal of Physics: Conference Series **434** (2013) 012077 doi:10.1088/1742-6596/434/1/012077

where 1 indicates good quality, 0 — bad quality, and $\frac{1}{2}$ indicates that $|\varepsilon_i| = 1$. Finally, the proposed metric q is the arithmetic mean of the individual q_i

$$q = \frac{1}{M} \sum_{i=1}^{M} q_i \,, \tag{6}$$

where M is the length of the data vector $\mathbf{d}$. Since $\mathbf{Q}$ can be precalculated, as it does not depend on the data, the calculation of q requires one matrix multiplication per data set (eq. 4), which is comparable to the computational requirement of linear image reconstruction.

2.4. Evaluation Data Set

The developed quality metric q was evaluated using a publicly available data set from two mechanically ventilated patients with acute lung injury or acute respiratory distress syndrome (ALI/ARDS) undergoing a recruitment manoeuver to estimate the extent of regional lung overdistention and atelectasis [5]. The experimental protocol consisted of 1) a baseline ventilation stage, 2) a lung recruitment stage (increased airway pressure) under pressure-controlled (PC) mode, and 3) a PEEP (positive end-expiratory pressure) titration stage with airway pressure sequentially decreased to the lowest possible setting. The Goe-MT II EIT device (CareFusion, Hoechberg, Germany) was used for acquiring measurement data at a single frequency using 16 electrodes, and adjacent stimulation and measurement patterns.

3. Results

Fig. 1 shows q values and EIT images reconstructed with three different algorithms, for the two patients (labeled 1 and 7) during each step of the experimental protocol described above, where R1-R4 stands for the lung recruitment (sequentially increased airway pressure) and T1-T4 for the PEEP titration (sequentially decreased airway pressure).

Patient 1 had consistent high q values, which is reflected in the stable and largely artefact-free reconstructed images for all protocol steps with few image artifacts. These images reflect the expected shifts of ventilation distribution according to the protocol. In contrast, for patient 7 physiologically unrealistic images were obtained, particularly at stages R3 and T1 for the GREIT algorithm. This is reflected by the much lower data quality q of during these two stages. Further analysis of the measurement data showed that the standard deviation of difference data at these stages has 3 times higher noise compared to other protocol steps (not presented).

4. Discussion

We propose a formulation of a data quality metric q for EIT suitable for use in real-time applications. Our intention is to provide an indicator of the trustworthiness of EIT reconstructions that would inform their analysis and thus help preventing incorrect diagnosis. We set out with a list of requirements for a useful quality metric, which we discuss in turn below:

(i) q is guaranteed to be bounded between 0 and 1, as desired, by the formulation of the individual q_i in eq. 5. Preliminary results (not presented) suggest, however, that the threshold for q below which measurements should be rejected is system dependent.

(ii) Because q is the arithmetic mean of the quality q_i of individual measurements, adding additional measurements with high q_i will increase q.

(iii) Conversely, adding measurements with below-average q_i will reduce the value of q.

(iv) Incorrect placement of electrodes or any other substantial difference between the domain and method of measurement and the model on which the Jackknife estimates are calculated will result in the latter being a poor match for the recorded data, causing q to be low.

Journal of Physics: Conference Series **434** (2013) 012077 doi:10.1088/1742-6596/434/1/012077

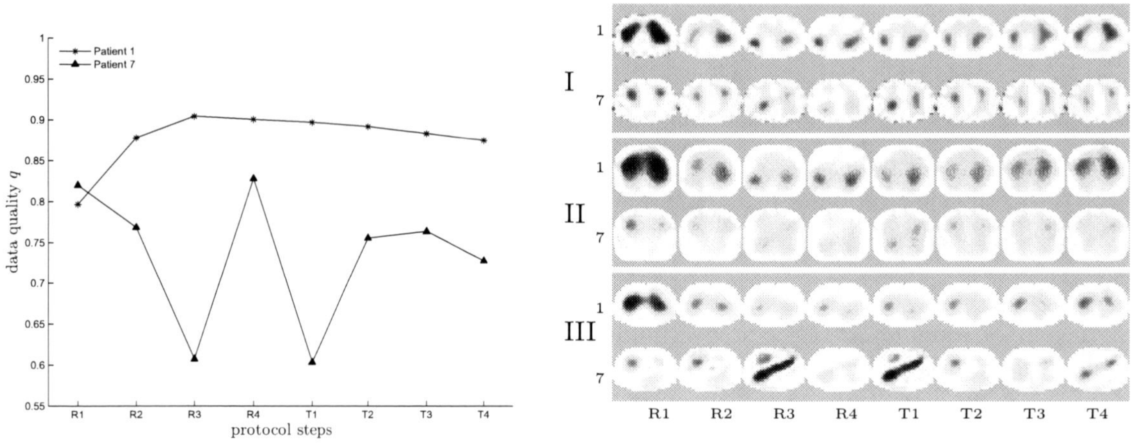

Figure 1: Quality metric evaluation. *Left:* data quality metric for all protocol steps with both patients. *Right:* the corresponding reconstructed images obtained with I: one-step Gauss-Newton solver with Laplace image prior and a 3D forward model, II: one-step Gauss-Newton solver with Gaussian high pass filter image prior and movement compensation [6] and a 2D forward model, and III: GREIT algorithm built on a forward model including lung tissue contrast based on a CT from an adult male [7].

Our example of real clinical data illustrates the case for using a data quality metric. As shown in Fig. 1, images reconstructed from poor quality data may at times appear sufficiently plausible as to be analysed. An objective quality metric independent from the reconstruction algorithm and the appearance of the image could provide clinicians with an indication of the trustworthiness of the images being analysed, thus helping to avoid false diagnosis. Since the proposed quality metric is computationally efficient, it is well suitable for real-time and long-term monitoring of patients.

The proposed metric q could also be useful to evaluate different EIT system performance, to monitor their long term stability, and as a test standard to measure hardware improvements.

5. References

[1] A. Adler *et al* 2012 Whither lung EIT: where are we, where do we want to go, and what do we need to get there? *Physiol. Meas.* **33** 679–694

[2] Y. Asfaw and A. Adler 2005 Automatic detection of detached and erroneous electrodes in electrical impedance tomography, *Physiol. Meas.* **26** 175–183

[3] A. Hartinger *et al* 2009 Real-time management of faulty electrodes in electrical impedance tomography *IEEE Trans. Biomed. Eng.* **56** 369–377

[4] B. Efron 1982 The jackknife, the bootstrap and other resampling plans *Society for Industrial and Applied Mathematics* **38** 49–59

[5] C. Gómez-Laberge, J. H. Arnold, and G. Wolf 2012 A unified approach for EIT imaging of regional overdistension and atelectasis in acute lung injury *IEEE Trans. Med. Imag.* **31** 834–842

[6] C. Gómez-Laberge and A. Adler 2008 Direct EIT jacobian calculations for conductivity change and electrode movement *Physiol. Meas.* **29** 89–99

[7] B. Grychtol, *et al* 2012 Impact of model shape mismatch on reconstruction quality in electrical impedance tomography *IEEE Trans. Med. Imag.* **31** 1754–1760

Experimental/clinical evaluation of EIT image reconstruction with ℓ_1 data and image norms

Yasin Mamatjan[1], **Andrea Borsic**[2], **Doga Gürsoy**[3], **and Andy Adler**[1]

[1] Systems and Computer Engineering, Carleton University, Ottawa, Canada
[2] Thayer School of Engineering, Dartmouth College, USA
[3] Department of Physics, University of Houston, USA

E-mail: mamatjan,adler@sce.carleton.ca

Abstract. Electrical impedance tomography (EIT) image reconstruction is ill-posed, and the spatial resolution of reconstructed images is low due to the diffuse propagation of current and limited number of independent measurements. Generally, image reconstruction is formulated using a regularized scheme in which ℓ_2 norms are preferred for both the data misfit and image prior terms due to computational convenience which result in smooth solutions. However, recent work on a Primal Dual-Interior Point Method (PDIPM) framework showed its effectiveness in dealing with the minimization problem. ℓ_1 norms on data and regularization terms in EIT image reconstruction address both problems of reconstruction with sharp edges and dealing with measurement errors. We aim for a clinical and experimental evaluation of the PDIPM method by selecting scenarios (human lung and dog breathing) with known electrode errors, which require a rigorous regularization and cause the failure of reconstructions with ℓ_2 norm. Results demonstrate the applicability of PDIPM algorithms, especially ℓ_1 data and regularization norms for clinical applications of EIT showing that ℓ_1 solution is not only more robust to measurement errors in clinical setting, but also provides high contrast resolution on organ boundaries.

1. Introduction

Electrical impedance tomography (EIT) shows promise to help patient ventilation in an intensive care unit (ICU) by monitoring the distribution of inspired air in mechanically ventilated patients. However, EIT image reconstruction is an ill-posed inverse problem. Due to limited independent measurements, the spatial resolution of reconstruct images is low. The Jacobian (sensitivity) matrix is ill-conditioned, which implies that measurement noise or errors due to electrode detachment and patient movement cause solutions to be unstable with large image artifacts. In order to stabilize the problem, regularization techniques are commonly used by introducing forms of filtering to reduce these artifacts. Regularization introduces a priori information regarding smooth conductivity changes which limits reconstructing sharp images. However, sharp edges are physiologically realistic and are desired to image better inter-organ boundaries and step changes in the tissue conductivity.

Total Variation (TV) regularization (the use of 1-norm on the regularization term) was demonstrated on simulation and clinical data [1], which allowed the reconstruction of sharp image transitions. The Primal Dual-Interior Point Method (PDIPM) was proposed by [2], where ℓ_1 and ℓ_2 norms were applied to data and regularization terms using simulated data only. The simulated results of the paper suggested that ℓ_1 norm (Least-absolute values) on both data and regularization terms improved image resolution and provided robustness to noise and data outliers (errors).

In this paper, we evaluated the PDIPM algorithms [2] on the experimental and clinical data by applying ℓ_1 and ℓ_2 norms for both the data-fidelity and regularization terms. Subsequently, we analyze and compare the algorithms' management of measurement errors in the data.

2. Methodology

2.1. Formulation of PDIPM

In EIT, the measurements (data) are the surface potentials on the boundary of the object. Let $f(\mathbf{x}) = \mathbf{y}$ represent the nonlinear relationship between the data (voltage data) and model parameters (electrical properties), $\mathbf{x} : \mathbf{x} \in R^N$ and $\mathbf{y} : \mathbf{y} \in R^M$, respectively. In difference EIT, $\mathbf{y}$ is the normalized difference data ($\mathbf{y} = (\mathbf{v} - \mathbf{v}_r)/\mathbf{v}_r$) between the current measured voltages $\mathbf{v}$ and the reference measurements $\mathbf{v}_r$. $\mathbf{x}$ is the image of relative conductivity changes between the background (or reference) conductivity $\boldsymbol{\sigma}_r$ and the corresponding conductivity $\boldsymbol{\sigma}$ for the measurement $\mathbf{v}$.

Generally, the number of the unknown model parameters are much more than the available number of data, *i.e.*, $M < N$, which results in an underdetermined set of equations. The corresponding inverse problem of recovering $\mathbf{x}$ from $\mathbf{y}$ based on the mapping f can be stated in the form of a minimization problem as follows:

$$\arg\min_{\mathbf{x}} \left\{ \mathcal{F}(\mathbf{x}) := ||f(\mathbf{x}) - \mathbf{y}||_p^p + \lambda ||\mathbf{L}(\mathbf{x} - \mathbf{x}_0)||_n^n \right\}, \tag{1}$$

where the $||f(\mathbf{x}) - \mathbf{y}||_p^p$ is the data fidelity term and $||\mathbf{L}(\mathbf{x} - \mathbf{x}_0)||_n^n$ is the regularization term, and $\mathbf{x}_0$ is a prior conductivity distribution (in our case the initial estimate was set to zero); $\mathbf{L} : \mathbf{L} \in R^{M \times N}$ is usually referred to as the regularization matrix and constructed based on the a priori information about the model parameters. The trade-off between these two terms is determined by the regularization parameter λ. p and n respectively specify the types of norms for the data fidelity and regularization terms. The characteristics of the solutions are dependent on the choice of the norms:

- Using ℓ_2-norm for both data and regularization terms is widely used and the most convenient way regarding computational time and convergence stability. The resultant solutions are commonly referred to as the least-squares solution with Tikhonov regularization.
- Using a ℓ_1-norm instead of ℓ_2-norm in the regularization term allows high frequency components in the solutions to be preserved. For instance, when the regularization matrix is constructed as a gradient operator, the variations in the solution are penalized. Such a scheme is referred to as the total variation (TV) regularization with an appropriate choice of $\mathbf{L}$. The regularization method preserves the edge information in the restored images, while filtering out noise.
- The ℓ_1-norm of the data fidelity is much more resilient to outliers than the ℓ_2-norm since it does not square the measurement misfit. Thus, it is usually preferred to suppress errors originating from outliers.
- For $\ell_1\ell_1$ case, the solutions are resilient to data outliers because of the ℓ_1-norm of the data fidelity term and the ℓ_1-norm of the regularization term permits reconstruction of sharp edges. However, it is computationally challenging due to the non-differentiability of the absolute value function at a number of points.

2.2. Evaluation Data Set

We considered difference imaging to reflect the physiological changes during the breathing for a certain time duration. We used EIT data acquired during mechanically ventilation of humans and dogs. These data are known to contain a certain level of electrode errors in the measurement data.

Human data: The human breathing data were based on [3], where pediatric patients with ALI/ARDS were recruited for an open lung treatment protocol. The experimental protocol was consisted of a baseline ventilation stage, a lung recruitment stage with sequentially increased airway pressure, and a PEEP (Positive end-expiratory pressure) titration stage with sequentially decreased airway pressure. A tidal volume of 6 mL/kg of body weight was used as baseline ventilation using volume-controlled mode. During the PEEP titration stage, the PEEP level was decreased sequentially to the lowest possible setting. The EIT system used for the measurement was the Goe-MT II EIT device (CareFusion, Hoechberg, Germany) operated in single frequency, where 16 electrodes were used, and adjacent stimulation and measurement patterns were applied.

Dog data: 16-electrode EIT system was used to take measurements of conductivity changes due to lung ventilation and lung fluid instillation in mongrel dogs [4]. ECG was used to record cardiac activity and synchronize EIT measurements 100 ms after the QRS peak of the ECG. Measurements were taken before and after 100 ml of fluids (5 % bovine albumin and Evans blue dye) were injected to a lobe of the right lung at the presence of 700 ml tidal volume. The same reference data was used before fluid instillation.

3. Results

Fig. 1 shows the time difference EIT image of a human thorax. A series of λ values were tested for the first measurement data (R1) taken from a ventilated patient for the protocol steps (R-R4 and T1-T4). The λ value that produces good reconstructed image was heuristically selected and fixed for the rest of the data sets. Different ℓ-norms have different fixed λ values.

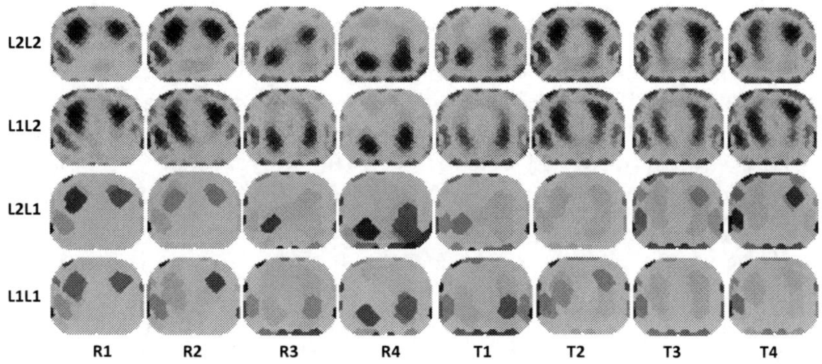

Figure 1: Time difference images of a human thorax for the PDIPM algorithms with four different ℓ-norms on the data and regularization terms. R1–R4 are the protocol steps of lung recruitment stage and T1–T4 are the protocol steps of the PEEP titration stage.

Fig. 1 showed that the ℓ_2-norm on the regularization term produced smooth reconstruction while the ℓ_1-norm on the regularization term produced high contrast conductivity distributions (sharper images). Interestingly, at the protocol steps R3 and T1, $\ell_2\ell_2$ and $\ell_2\ell_1$ produced a clinically unrealistic (failed) reconstruction, while $\ell_1\ell_2$ and $\ell_1\ell_1$ showed expected images. Further analysis showed 3 times higher noise levels at the two steps compared to the rest of the protocol steps.

Fig. 2 illustrates the use of PDIPM algorithms for imaging the conductivity changes in dog's lung immediately after fluid injection (row 1) and 60 minutes after the fluid injection (row 2) with the presence of certain level of electrode errors.

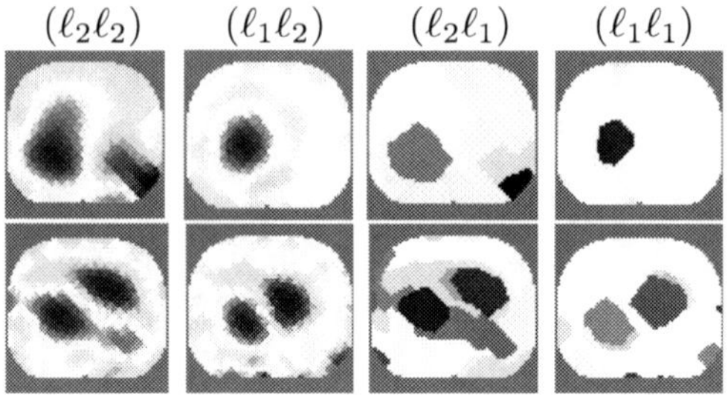

Figure 2: Reconstructed cross-section of conductivity changes in dog's lung right after injection (row 1) and 60 minutes after the injection (row 2). Difference images were reconstructed using PDIPM algorithms, where the presence of electrode errors in the measured data a affected all reconstructions, except those using the l1 norm on the data term (robust reconstruction).

As shown in Fig. 2, $\ell_2\ell_2$ and $\ell_2\ell_1$ showed large image artifacts as they suffered from noise from measurement system. $\ell_1\ell_2$ and $\ell_1\ell_1$ reconstructed cleaner and meaningful images, where particularly $\ell_1\ell_1$ produced sharp images with little image artifacts.

4. Discussion

In this study, we evaluated PDIPM algorithms for the EIT real-measurement on clinical patient and experimental dog data (with electrode errors) using ℓ_1/ℓ_2 data and image norms. The results (Fig. 1 and 2) showed that $\ell_1\ell_2$ and $\ell_1\ell_1$ are less affected by data outliers because of 1-norm for the data term, while $\ell_2\ell_1$ and $\ell_1\ell_1$ produces sharper reconstructed images because of 1-norm on the regularization term. $\ell_1\ell_1$ provided both sharp conductivity image and robustness to data outliers. However, ℓ_1 solution is not efficient compared to Gauss-Newton based solution, since ℓ_1 solution involves with the minimization of a non-differentiable function, thus requires more computation time and more iterations to reach the convergence. Most traditional image reconstruction is based on ℓ_2 norm, which produces smooth solutions without clear edges and is sensitive to measurement noise (data outliers). Nevertheless, $\ell_2\ell_2$ takes least computation time as it is the one-step linearized image reconstruction, which is also widely used in EIT imaging.

PDIPM algorithms with 1-norm ($\ell_1\ell_1$) present significant advantages in medical applications of EIT, since high contrast sharper images are more desired in clinical scenario. In reality, noise from the acquisition system, and patient movement or electrode errors are unavoidable, while measurement noise has been modeled as Gaussian. This greatly affects reconstructed image quality (especially ℓ_2 data norm solution). This paper demonstrated the advantages of ℓ_1 norm solutions in clinical applications of EIT, where ℓ_1 solution is not only more robust to unavoidable measurement errors in clinical setting, but it also could provide high contrast resolution.

5. References

[1] Borsic A, Graham B, Adler A, and Lionheart W 2010 In vivo impedance imaging with total variation regularization *IEEE Trans. Med. Imag.* **29** 44–54

[2] Borsic A and Adler A 2012 A primal dual - interior point framework for using the L1-norm or the L2-norm on the data and regularization terms of inverse problems *Inverse Problems* **28** 27

[3] Gómez-Laberge C, H. Arnold J, and Wolf G 2012 A unified approach for EIT imaging of regional overdistension and atelectasis in acute lung injury *IEEE Trans. Med. Imag.* **31** 834–842

[4] Adler A, Amyot R, Guardo R, Bates J, and Berthiaume Y 1997 Monitoring changes in lung air and liquid volumes with electrical impedance tomography *J. Appl. Physiol.* **83** 1762–1767

Journal of Physics: Conference Series **434** (2013) 012079 doi:10.1088/1742-6596/434/1/012079

Preliminary Results on Different Impedance Contrast Agents for Pulmonary Perfusion Imaging with Electrical Impedance Tomography

D T Nguyen[1], R Kosobrodov[1], M A Barry[2], W Chik[2], J Pouliopoulos[2], T I Oh[3], A Thiagalingam[2] and A McEwan[1]

[1] School of Electrical and Information Engineering, The University of Sydney, New South Wales, Australia

[2] Department of Cardiology, Westmead Hospital, New South Wales, Australia

[3] Impedance Imaging Research Center, Kyung Hee University, South Korea

E-mail: `d.nguyen@sydney.edu.au`

Abstract.

Recent studies in animal models suggest that the use of small volume boluses of NaCl as an impedance contrast agent can significantly improve pulmonary perfusion imaging by Electrical Impedance Tomography (EIT). However, these studies used highly concentrated NaCl solution (20%) which may have adverse effects on the patients. In a pilot experiment, we address this problem by comparing a number of different Impedance Contrast Boluses (ICBs). Conductivity changes in the lungs of a sheep after the injection of four different ICBs were compared, including three NaCl-based ICBs and one glucose-based ICB. The following procedure was followed for each ICB. Firstly, ventilation was turned off to provide an apneic window of approximately 40 s to image the conductivity changes due to the ICB. Each ICB was then injected through a pig-tail catheter directly into the right atrium. EIT images were acquired throughout the apnea to capture the conductivity change. For each ICB, the experiment was repeated three times. The three NaCl-based ICB exhibited similar behaviour in which following the injection of each of these ICBs, the conductivity of each lung predictably increased. The effect of the ICB of 5% glucose solution was inconclusive. A small decrease in conductivity in the left lung was observed in two out of three cases and none was discernible in the right lung.

1. Introduction

Pulmonary perfusion is as important as lung ventilation as both elements are necessary to sufficiently describe the regional behaviour of lung tissues for applications such as the adjustment of ventilator settings [1]. However, the significantly smaller amplitude of the perfusion impedance change signal compared to the respiratory impedance change signal renders perfusion mapping very difficult.

Recently, Borges et al [2] proposed a novel method to determine regional blood flow based on impedance change. They used a bolus of hypertonic sodium chloride (20% NaCl) as an impedance contrast agent. An algorithm was then performed to follow the first–pass kinetics of this impedance contrast bolus to quantitatively measure the relative regional blood flow in the lungs and heart. The regional blood flow measured by EIT using this novel method, in both healthy and injured lung cases in six piglets, is highly correlated with that measured by SPECT.

The use of highly concentrated NaCl solution (20%), however, may have adverse effects on the patients, especially those who are critically ill with pre–existing cardiac or renal conditions.We address this problem by experimentally comparing a number of different Impedance Contrast Boluses (ICBs). The purpose was to find an alternative to the highly concentrated 20% NaCl solution for pulmonary perfusion/blood-flow imaging.

2. Method

2.1. Experimental procedure

The experiment was performed on a male Merino–cross, weighing 64 kg, using the KHU Mark 2.5 Impedance Imaging system [3] with 16 electrodes placed equidistant around the thorax of the animal in a transverse plane. The plane of electrodes was located at approximately the 6th intercostal space of the animal. The sheep was appropriately anaesthetised and mechanically ventilated. The animal was selected to lack any cardiac or respiratory problem. This experiment has been approved by the Westmead Hospital Animal Ethics Committee.

EIT data were acquired using the adjacent electrodes current injections (1 mA p-p, 50 kHz) and voltage measurements protocol at 33 fps[1].

The experimented ICBs, in order of injection, included:

- ICB1: 60 ml of warm (37 °C) 3% NaCl solution,
- ICB2: 60 ml of cold (21 °C) 5% glucose solution,
- ICB3: 10 ml of warm (37,° C) 20% NaCl solution,
- ICB4: 60 ml of cold (21 °C) 3% NaCl solution,

The following procedure was followed for each ICB. Firstly, ventilation was turned off to provide an apneic window of approximately 40s to image the conductivity changes due to the ICB without the interference from the ventilation signal. Each ICB was then injected through a pig-tail catheter directly into the right atrium at no more than 10 seconds after the start of the apnea. The injection of each ICB took less than 5 seconds to complete. EIT images were acquired throughout the apnea to capture the conductivity change due to each ICB.

The experiment was repeated three times for each ICB, with 5 to 10 minutes between two consecutive injections. In order to prevent any residual effect, 60 ml of 5% glucose was injected into the ICB injection catheter before an ICB of a different type was injected, followed by 5 ml of the solution the successive ICB was based on. These injections were firstly, to erase the presence of the preceding ICB in the injection catheter and secondly, to dilute out any residual effect of the preceding ICB in the lungs.

2.2. Reconstruction algorithm and processing of reconstructed images

Standard time difference EIT was employed. The reference voltage measurement set was the computed mean of the first 1000 frames in each case. Notably, these were the frames in which the animal was mechanically ventilated. The GREIT reconstruction algorithm [4] was employed to reconstruct images based on an extruded sheep thorax FEM.

Two regions of interest (ROIs) were visually identified from the ventilation EIT images to represent the left and right lungs. The conductivity change within each ROI throughout the experiment was extracted and averaged to reflect the mean change. A Butterworth low-pass filter with the cut-off frequency of 0.3 Hz was used to reduce high frequency component of the signal, which included cardiac signal and noise.

Relative conductivity change for each ROI in each case was calculated as:

[1] frames per second

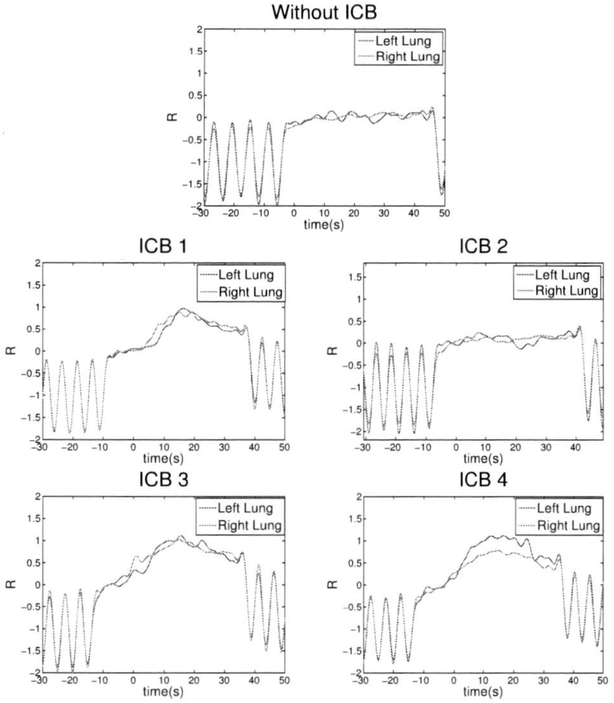

Figure 1. Relative conductivity change over time after the ventilator was turned off followed by injection of ICB1, ICB2, ICB3, ICB4 and without any ICB injection (Without ICB). Time $t = 0$ corresponds to the beginning of injection.

	Left Lung R	Right Lung R
ICB1	1.08 ± 0.10	0.87 ± 0.09
ICB3	1.23 ± 0.11	0.91 ± 0.11
ICB4	1.04 ± 0.12	0.74 ± 0.12

Table 1. Mean and standard deviation of the maximum relative conductivity change (R) induced by the injection of three NaCl-based ICBs (ICB1, 3 and 4). The positive R values indicated an increase in conductivity.

$$\mathbf{R}_{ICBx} = \frac{\sigma_{ICBx} - \bar{\sigma}_{Apnea}}{\bar{\sigma}_{Apnea}} \tag{1}$$

where $\bar{\sigma}_{Apnea}$ was the mean of the conductivities of each ROI from the beginning of apnea to the time at which ICB injection started or the first 10 s of apnea where there is no ICB injection.

3. Results

Figure 1 shows the relative conductivity change in the left and right lungs ROIs throughout the experiment with and without ICB injections. Observably, during apnea and without ICB injection, the conductivity of the two lungs were similar to each other and remained relatively flat throughout. Relative conductivity change after the introduction of contrast shows dependence on the type of ICB. The responses of the two lungs under the influence of the same ICB were similar but not identical, both in magnitude and phase.

4. Discussion and Conclusion

We report the conductivity changes due to four different ICBs in an ovine model. Our preliminary result shows conductivity contrast caused by injection of 10 ml of 20% NaCl (ICB 3) in the ovine model was much higher (up to 123% in the left lung) than the 10% previously reported in [2] in piglet models. We postulate that the discrepancy is due to the intrinsic differences in sizes of the blood volumes and tissue perfusion of each model.

The ICBs 1, 3 and 4 displayed similar dilution curves (based on the conductivity change), which were in agreement with the general shape that was previously reported [2; 5; 6].

The larger bolus size evidently compensated for the lower concentration of NaCl used in ICB 1 and 4 (3%) as the conductivity increases caused by these two ICBs were comparable to that caused by the ICB 1 (10 ml of 20% NaCl) (see Table 1). Furthermore, a warm (37 °C) contrast bolus was marginally more effective than a cold (21 °C) one as indicated by the slightly higher increase in conductivity caused by ICB1 compared with ICB2 (see Table 1). The closely matched results between ICB1 and ICB4 also indicated that the residual effect of previous ICBs, which we had encountered in the past experiments, were mitigated.

The glucose-based ICB effect on conductivity was unexpected based on previous reports that glucose should cause a decrease in conductivity [7]. In our in vivo experiment, the change caused by this ICB was inconclusive. In two out of three experiments, a small relative conductivity decrease was observed in the left lung ($R = -0.14$ and $R = -0.6$). However, no such decrease were discernible in the right lung in all three cases. The small decrease observed were within experimental error (standard deviation for the NaCl-based ICBs were around 10%, see Table 1). Furthermore, even without any ICB injection, the conductivity of the two lungs fluctuate typically between -15% to $+15\%$ (see Figure 1), which was due to the movement of internal organs such as stomach, diaphragm and even the heart.

The results presently reported are in a preliminary stage, limited to one animal. We plan to repeat the experiments in more subjects to investigate the effect of each ICB in more detail.

In conclusion, the preliminary *in vivo* results demonstrate that 3% NaCl can produce comparable conductivity contrast to 20% NaCl for pulmonary perfusion blood flow indication. This is advantageous because 3% NaCl solution is readily available and routinely used in the ICU while highly concentrated 20% NaCl is rarely, if at all, used in clinical settings.

References

[1] Frerichs I, Pulletz S, Elke G, Reifferscheid F, Schadler D, Scholz J and Weiler N 2009 *Respiration; international review of thoracic diseases* **77** 282–91

[2] Borges J B, Suarez-Sipmann F, Bohm S H, Tusman G, Melo A, Maripuu E, Sanstrom M, Park M, Costa E L V, Hedenstierna G and Amato M 2011 *J Appl Physiol* **112** 225–236

[3] Oh T I, Wi H, Kim D Y, Yoo J P and Woo E J 2011 *Physiological measurement* **32** 835–849

[4] Adler A, Arnold J H, Borsic A, Brown B H, Dixon P, Faes T J, Frerichs I, Gagnon H, Garber Y, Grychtol B, Hahn G, Lionheart W R B, Malik A, Patterson R P, Stocks J, Tizzard A, Weiler N and Wolf G K 2009 *Physiological measurement* **30** S35–S55

[5] Brown B H, Leathard A D, Sinton A, McArdle F J, Smith R W M and Barber D C 1992 *Clin. Phys. Physio. Meas* **13** 175–179

[6] Frerichs I, Hinz J, Herrmann P, Weisser G, Hahn G, Quintel M and Hellige G 2002 *IEEE transactions on medical imaging* **21** 646–52

[7] Hellige N C, Meyer B, Rodt T, Vogel-Claussen J, Hahn G and Hellige G 2012 *Biomed Tech* **57 (Suppl. 1)** 525–528

Journal of Physics: Conference Series **434** (2013) 012080 doi:10.1088/1742-6596/434/1/012080

Effect of electrode arrangements on bladder volume estimation by electrical impedance tomography

Thomas Schlebusch, Steffen Leonhardt

Philips Chair for Medical Information Technology
Helmholtz-Institut, RWTH Aachen University
Pauwelsstr. 20, D-52074 Aachen

E-mail: schlebusch@hia.rwth-aachen.de

Abstract. The estimation of bladder volume by electrical impedance tomography is a promising technology to provide continuous volume estimates to patients who suffer from impaired bladder volume sensory. To find an optimized electrode arrangement for electrical impedance tomography of the human bladder, seven different electrode arrangements have been analysed using a finite element model. The arrangements under consideration can be assigned to three general groups: ring arrangements, linear arrangements and matrix arrangements. For each arrangement in these three groups, the condition of the Jacobian matrix and the global impedance for several bladder volumes has been calculated and compared using the EIDORS toolkit.

1. Introduction

Knowledge of current urine volume is of great interest to patients with paraplegia or those suffering from overactive bladder to determine the right moment for bladder emptying. One promising way to estimate current urine volume is by electrical impedance tomography (EIT). First EIT measurements using a circular electrode arrangement around the pelvis have shown a linear correlation of global impedance and bladder volume [1]. Still in question is, if a circular electrode array is the best arrangement to reach high sensitivity of the EIT measurement system. In FEM simulations of a four lead bioimpedance system, it could be shown that vertical electrode arrangements have increased sensitivity to bladder volume changes as compared with horizontal arrangements [2]. In contrast to lung EIT imaging, bladder impedance measurements have to deal with both impedance change and a shift of organ boundaries with increasing bladder volume: the bladder itself is a hollow muscle, located on the pelvic floor in the lesser pelvis. With increasing volume, the bladder stretches and extends upwards along the abdominal wall, displacing surrounding tissue. Although even a classical EIT ring array is able to measure impedance changes outside the electrode plane, an optimized arrangement will be much more suitable for quantification of bladder extension.

To find suitable electrode positions for bladder EIT, existing arrangements from literature have been reviewed. The most popular arrangement is the classical ring array, placing all electrodes around the thorax on the same level. This arrangement has been used by Leonhardt et al. for their bladder volume measurements [1]. This basic arrangement can be extended to the

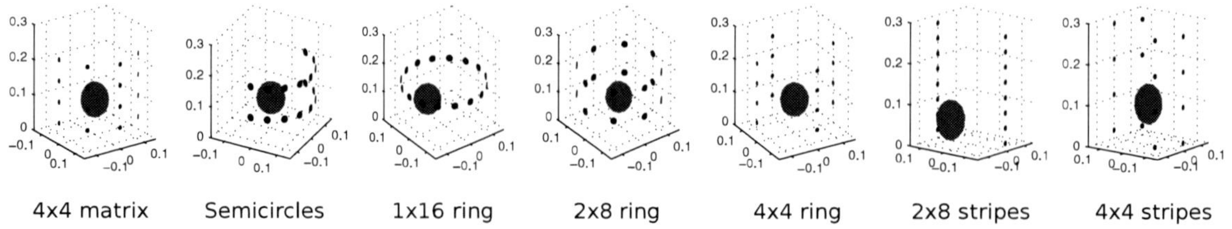

Figure 1: Electrode arrangements

third dimension by applying multiple rings in a fixed distance along the body axis. Graham and Adler showed good performance of this arrangement for lung EIT [3]. A similar arrangement using only ventral semicircles has been presented by Sadleir et al. [4]. Using ventral semicircles results in increased sensitivity in the ventral region in proximity to the electrodes, while dorsal regions show only small sensitivity. For bladder EIT, reduced sensitivity in the dorsal region might be advantageous to reduce influences of the rectum on calculated bladder volume. An electrode arrangement in vertical stripes has been presented by Murphy et al. [5]. Their system shows good resolution for conductivity gradients along the vertical axis. Configurations using some kind of matrix arrangement in the ventral half space have been presented by Hua et al. for spectroscopic measurements [6] and Erol et al. and He et al. for tomographic imaging [7, 8].

2. Materials & Methods

All calculations in this work have been conducted in Matlab using the EIDORS framework [9]. For a first evaluation of the electrode arrangements, a simple cylindrical tank model of 30 cm height and 15 cm radius has been used. Netgen has been used for meshing, producing a fine mesh for forward calculations of about 14,000 tetrahedral mesh cells. The background conductivity has been set to $\sigma_{bgnd} = 1\,S/m$. A sphere of variable radius and conductivity $\sigma_{sph} = 2\,S/m$ has been used to model the bladder.

Two different locations have been used for the bladder sphere: in the first model, the bladder has been located in the centre of the tank. A more physiological arrangement has been used for the second model by locating the sphere at the bottom of the cylinder directly to the wall. With increasing volume, the sphere extends towards centre and top of the cylinder.

It has been shown in literature that image quality is highly dependent on the number of electrodes used [10]. Therefore, 16 electrodes have been used for all arrangements taken into consideration to focus on the influence of different arrangements only. A total of seven different arrangements have been evaluated that can be grouped into the categories of ring arrangements, matrix arrangements and vertical stripe arrangements. The arrangements described in the following paragraphs are shown in Figure 1.

In the group of ring arrangements, a classical 16 electrode ring at position $z = 0.15$ in the middle of the cylinder, two rings of eight electrodes at positions $z = 0.1$ and $z = 0.2$, as well as four rings of four electrodes at positions $z \in \{0.06, 0.12, 0.18, 0.24\}$ have been used. Adjacent excitation has been used to stay compatible with the Goe MF II device.

In the group of matrix arrangements, a model with two ventral semicircles of eight electrodes comparable to the arrangement of Sadleir [4] and a matrix of 4×4 electrodes comparable to the arrangement of Hua [6] have been used. Adjacent excitation in horizontal direction has been used as well.

The last two arrangements are inspired by Murphy et al. [5] with one model having two vertical stripes of eight electrodes on opposite sides of the cylinder wall, and a model using two stripes of four electrodes on each side. Excitation between adjacent electrodes on the same

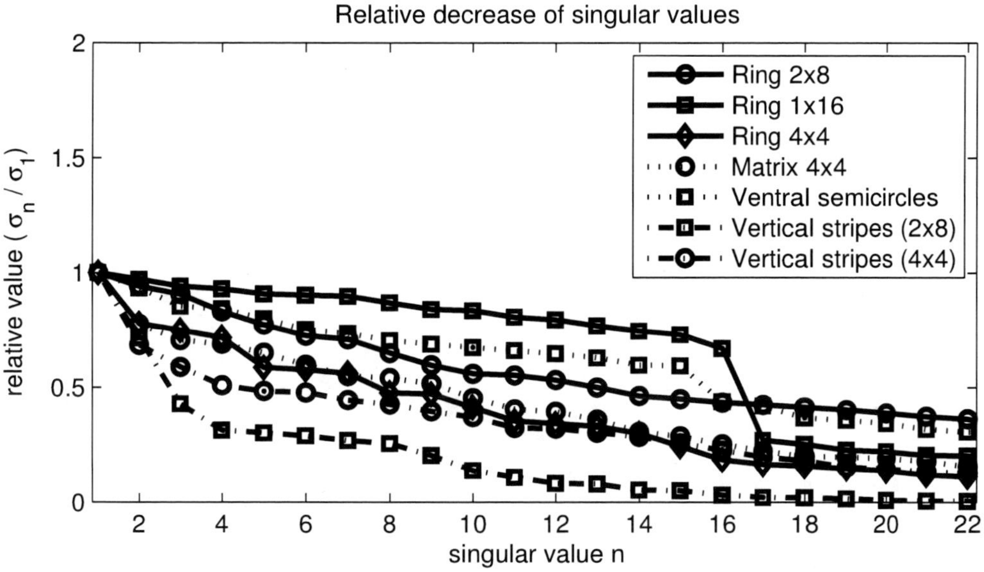

Figure 2: Relative decrease of singular values

z-level has been used which results in a polar through measurement for the 2×8 stripes model and an alternating change of adjacent and polar for the 4×4 stipes model.

In general, the GREIT algorithm available in EIDORS has been used to find a reconstruction matrix for all models. Unfortunately, we did not succeed to compute a GREIT reconstruction matrix for the vertical stripes models, so basic Gauß-Newton reconstruction has been used for these two models.

The performance of each arrangement has been evaluated by comparing the normalized singular values of the Jacobian matrix and by comparing global impedances for varying bladder sphere volumes. The Jacobian matrix J describes how an impedance change of a single mesh element influences the boundary voltage measurement. Analysing the singular values provides a measure for the condition of J [10], which is linked to reconstructed image quality. An electrode arrangement is considered to be good, if the singular values are decreasing slowly. To compare the performance of bladder volume measurements, the global impedance has been calculated for 14 sphere radii from 1 cm to 14 cm, referenced to a baseline measurement of an empty tank.

3. Results

The singular values in Figure 2 are normed to the first singular value σ_1 for better comparison, as suggested by Tang et al. [10]. The slowest decay of singular values show the single 16-electrode ring, the ventral semicircles and two rings of eight electrodes. The best condition number, which is the ratio of greatest to smallest non-zero singular value, is provided by the ventral semicircles arrangement.

The sensitivity to bladder volume changes, as depicted in Figure 3, is highest for the two three dimensional ring arrangements "Ring 2×8" and "Ring 4×4". Ventral semicircles and the 4×4 matrix model show also a strong bladder-related impedance change. The global impedance results of the two vertical stripes alignments can not be compared directly, since they are based on a different reconstruction method.

All arrangements except the single 16 electrode ring and the vertical stripes arrangements are very robust with respect to bladder position, which means that there is only little difference in global impedance for both bladder positions as described in section 2. Especially for bladder

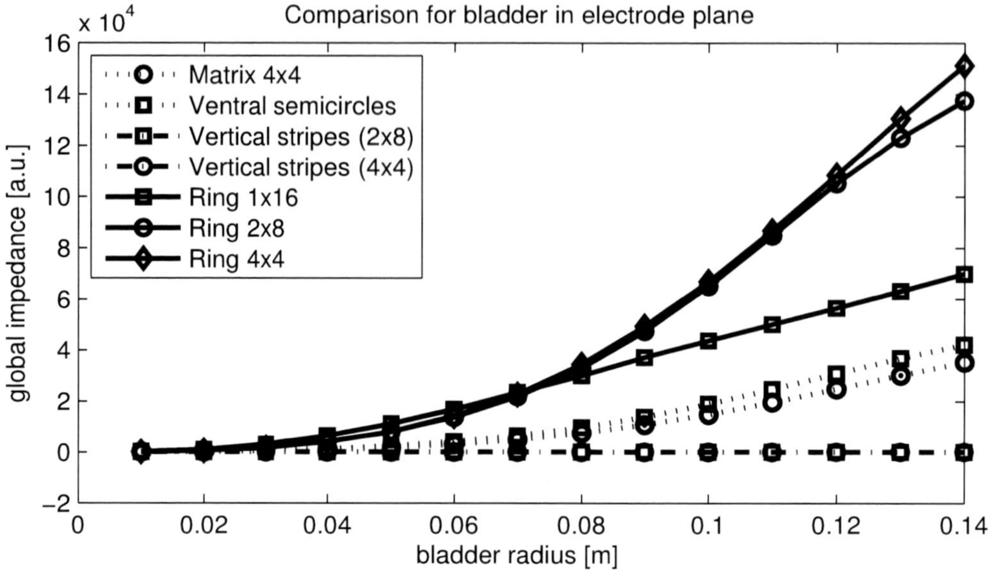

Figure 3: Comparison of global impedance

radii of a few centimetres, the single ring arrangement shows high error. Using two rings of eight electrodes results in a much more homogeneous sensitivity, which produces more stable results. The results for "Ring 4 × 4", matrix and semicircles are comparable.

4. Conclusion

Seven different electrode arrangements of 16 electrodes for 3D EIT measurement of the human bladder have been evaluated using a simplified FEM-model using EIDORS and NetGen. The arrangements have been rated by analysing the decay rate of singular values and by comparison of the correlation of global impedance and volume for simulated EIT measurements. It could be shown that a single 16 electrode ring, as usually applied for lung EIT measurements, may not be optimal for reliable bladder tomography. Instead, multiple ring and ventral semicircle arrangements seem to yield better electrode topologies for this application.

References

[1] Leonhardt S, Cordes A, Plewa H, Pikkemaat R, Soljanik I, Moehring K, Gerner H and Rupp R 2011 *Biomedizinische Technik/Biomedical Engineering* **56** 301–307

[2] Schlebusch T, Nienke S, , Leonhaeser D, Grosse J and Leonhardt S 2012 Optimal electrode positions to determine bladder volume by bioimpedance spectroscopy *International Workshop on Impedance Spectroscopy (IWIS) Chemnitz*

[3] Graham B and Adler A 2007 *Physiological Measurement* **28** S29

[4] Sadleir R, Zhang S, Tucker A and Oh S 2008 *Physiological measurement* **29** 913

[5] Murphy S, Stanley S, Rhodes D and York T 2006 *Measurement Science and Technology* **17** 3053

[6] Hua P, Woo E, Webster J and Tompkins W 1988 Bladder fullness detection using multiple electrodes *Engineering in Medicine and Biology Society, 1988. Proceedings of the Annual International Conference of the IEEE* (IEEE) pp 290–291

[7] Erol R, Smallwood R, Brown B, Cherian P and Bardhan K 1999 *Physiological measurement* **16** A143

[8] He W, Ran P, Xu Z, Li B and Li S n 2012 *Computational and Mathematical Methods in Medicine* **2012**

[9] Adler A and Lionheart W 2006 *Physiological measurement* **27** S25

[10] Tang M, Wang W, Wheeler J, McCormick M and Dong X 2002 *Physiological measurement* **23** 129

FPGA Based High Speed Data Acquisition System for Electrical Impedance Tomography

S Khan[1], A Borsic[2], Preston Manwaring[3], Alexander Hartov[4], Ryan Halter[5]

[1,2,3,4,5] Thayer School of Engineering, Dartmouth College, Hanover, NH, USA

[1] Corresponding author, email: shadab.khan@dartmouth.edu.

Research supported by NIH/NCI grant number 1R01CA14302001A1 and NIH/NIBIB grant number 1R21EB011568-01.

Abstract. Electrical Impedance Tomography (EIT) systems are used to image tissue bio-impedance. EIT provides a number of features making it attractive for use as a medical imaging device including the ability to image fast physiological processes (>60 Hz), to meet a range of clinical imaging needs through varying electrode geometries and configurations, to impart only non-ionizing radiation to a patient, and to map the significant electrical property contrasts present between numerous benign and pathological tissues. To leverage these potential advantages for medical imaging, we developed a modular 32 channel data acquisition (DAQ) system using National Instruments' PXI chassis, along with FPGA, ADC, Signal Generator and Timing and Synchronization modules. To achieve high frame rates, signal demodulation and spectral characteristics of higher order harmonics were computed using dedicated FFT-hardware built into the FPGA module. By offloading the computing onto FPGA, we were able to achieve a reduction in throughput required between the FPGA and PC by a factor of 32:1. A custom designed analog front end (AFE) was used to interface electrodes with our system. Our system is wideband, and capable of acquiring data for input signal frequencies ranging from 100 Hz to 12 MHz. The modular design of both the hardware and software will allow this system to be flexibly configured for the particular clinical application.

1. Introduction

Electrical Impedance Tomography (EIT) is an imaging modality that can be potentially used to distinguish between benign and malignant tissues, through measurement of tissue's bio-impedance. Traditional EIT instruments apply AC current and measure induced potentials on a number of electrodes in contact with a specific tissue or body region of interest. Previous research on EIT has led to development of several instruments to record and image bio-impedance [1], [2], [3], [4], [5].

The motivation of this work was development of a system that can be quickly prototyped, and upgraded. Instrumenting state-of-the-art EIT systems requires the latest embedded controllers and robust software architecture, along with any external circuitry that has been built using the best performing components. This is challenging because with continuous progress in fabrication of better electronics, there always are components in the system that can be upgraded. To overcome these issues, we chose National Instruments' (NI) LabVIEW platform along with NI hardware to design the EIT system control hardware and software. The user control interface, FPGA controls software and calibration code was developed using LabVIEW. Our system comprises a computer as a controller (also called Host) connected via a PCI bus to the NI hardware. For analog signal conditioning of the input signal, we developed an Analog Front End (AFE) board that interfaces the hardware with the electrodes. The inherent compatibility of the LabVIEW programming environment with the NI hardware and the hot-swap-upgrade possible with NI modules, allowed us to develop a system in which future upgrade can be rapidly made. Suitability of the PXI platform for EIT has been previously reported as well [6].

Another significant goal of this project was to develop a system that has high frame rate of acquisition, so that variation of parameters during fast physiological processes, such as heart cycle, can be observed. However, imaging systems with high frame rates require high throughput rates for data transfer from ADC to host, if all of the data processing is performed on the PC. To work around this bottleneck, we implemented signal demodulation and spectral characterization on the FPGA itself.

2. Hardware Design

The DAQ system comprises of the following devices and components:

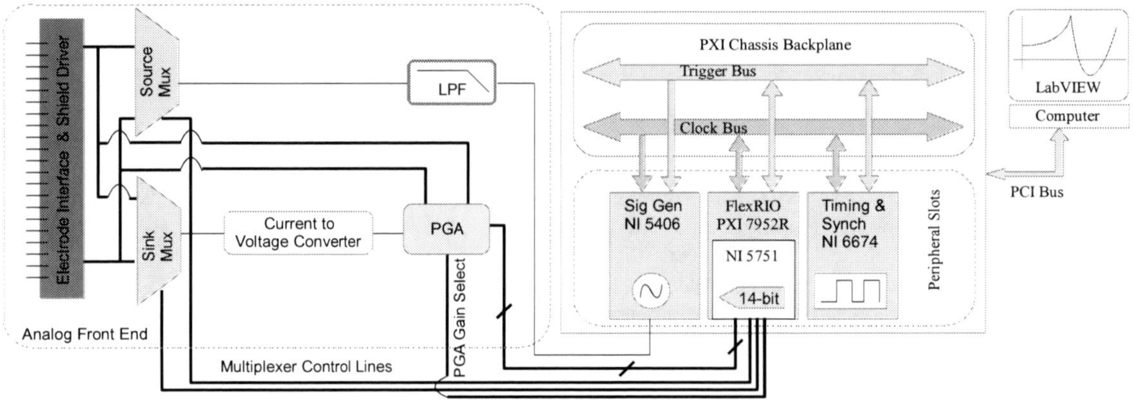

Figure 1. Block diagram of the EIT system hardware.

- System Controller: We have developed our control software using LabVIEW 2012 on a PC.
- FPGA (NI FlexRIO 7952R) and ADC modules (NI 5751): ADC module samples the AC signal at the electrodes and digitizes it. FPGA communicates with ADCs via Serial Peripheral Interface (SPI) protocol and reads the ADC registers at sample clock frequency.
- Signal Generator (SigGen): This module generates the AC signals. SigGen module also generates a trigger pulse synchronized with the start of the waveform cycle. This pulse is read by the FPGA to synchronize start of acquisition with the starting phase of the waveform cycle.
- Timing and Synchronization (TSync) Module (NI 6674t): This module generates a DDS clock running at a user-selectable frequency. TSync module also provides advanced timing and triggering functionality for inter-module synchronization.
- Relay control and Power monitor board (NI X-Series Card 6341): This module controls the relays on the AFE board and monitors the power from DC supply to ensure that the supply doesn't exceed the operating voltage specification.
- Analog Front End (AFE) : We designed an AFE board with filters, multiplexers for signal switching, Programmable Gain Arrays (PGA), shield drivers, relays for power control etc. on it for appropriate analog conditioning of the signal. The AFE board interfaces with the ADC module, SigGen module and TSync module. The bio-impedance probe is interfaced with the AFE board via front end connectors (shown in blue, on the left end of Figure 1.)

3. Software Design

The control software on the computer performs a number of functions including: a) FPGA and ADC module initialization, b) Writing acquisition parameter values in FPGA registers, c) Writing the source-electrode/sink-electrode pattern file in the FPGA memory, d) Triggering start of acquisition, e) Reading data from the FPGA, f) Data calibration and g) Data logging. The FPGA program (realized in hardware, Figure 2) has an initialization circuitry and four blocks performing different functions. These are the Acquisition block, the FFT block, the Spectral Characterization block and the Data Transfer block. Functioning of each block is briefly discussed next.

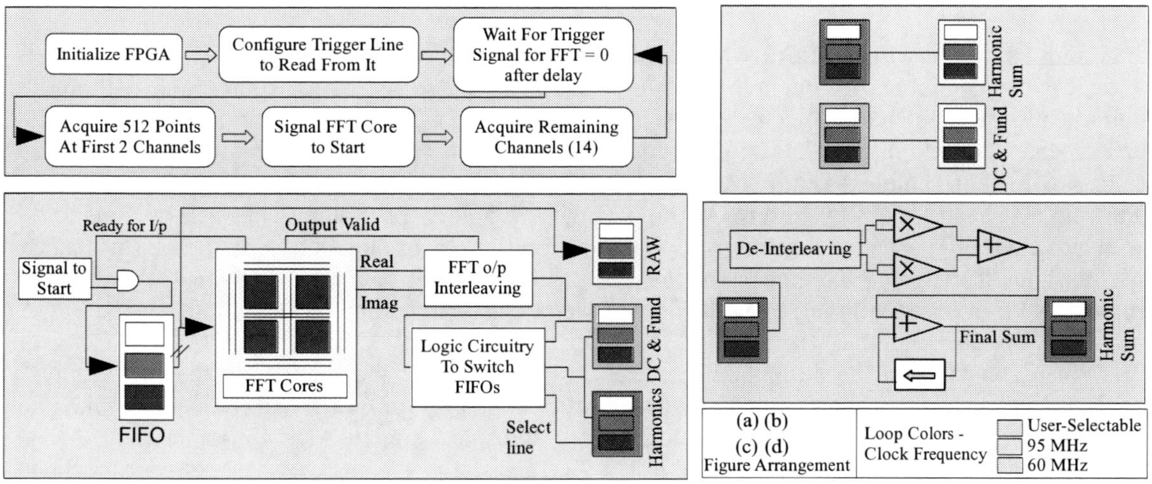

Figure 2. FPGA design block diagram

3.1. Acquisition Block: This block reads the ADC registers and writes the values to a First-In-First-Out (FIFO) buffer. It can be clocked using a clock frequency within 30-50 MHz range. The acquisition functions in semi-parallel mode; FPGA acquires samples from 4 channels simultaneously, and then moves on to the next four channels to acquire samples from all other channels.

3.2. FFT Block: This block computes 512-point FFT on the input samples, **x**, collected from 8 channels and returns **X**, the real and imaginary component of the FFT output. Since the Acquisition Block and FFT Block work asynchronously, FFT cores use handshaking signals to read from FIFOs used for sharing data between blocks.

3.3. Spectral Characterization Block: We define the Total Harmonic Distortion (THD) of the signal by (1). In this equation, the numerator is the sum of squared magnitude of higher order harmonics, and the denominator is the magnitude squared of the input signal's dominant harmonic.

$$THD = \frac{\sum_{i=2}^{\frac{N}{2}-1} ||Xi||_2^2 - ||dominant\ harmonic||_2^2}{||dominant\ harmonic||_2^2} \quad (1)$$

We compute the numerator on the FPGA itself and use it to analyze the transient abnormalities or signal quality deterioration. Division is a computationally expensive operation to perform on the FPGA thus the THD is computed on the PC. Calculation of the numerator on FPGA allows us to keep the throughput low, while having a metric to ensure that the signal quality is satisfactory

3.4. Data Transfer Block: DT block queues the transfer of elements from the intermediate FIFOs to the Direct Memory Access (DMA) FIFOs from which PC control software can read the data. DMA FIFOs are implemented using block memory and thus allow simultaneous read and write access and offer deterministic performance across clock domains for loss-less data sharing.

3.5. Averaging Block: Averaging block (not shown in the block diagram) performs signal averaging on the acquired samples before they are read by the FFT block. The averaging block can average at a user-selectable factor up to 64 waveform cycles.

To reduce fabric usage and operation time on the FPGA, design optimizations were performed. The data-sharing blocks were programmed with a pipelined architecture to shorten the longest combinatorial path of gates and were fine-tuned to clock at highest possible clock frequency (without violating timing constraints and determinism), ranging from 50 MHz to 95 MHz.

4. System Characteristics & Future Work

In the current configuration, our system uses 8 channels and a set of 28 source/sink patterns to alternate injecting and sinking electrodes; we are able to record a full spectrum of data (with FFT and spectral characterization, for 47 discrete frequencies spaced logarithmically from 100 Hz to 11.25 MHz, $\approx$ 8 Million samples) within 14 seconds. At a fixed signal frequency of 100 KHz, to acquire data from 12 channels for 100 patterns with no signal averaging, the system takes nearly 13ms to do the acquisition, FFT, spectral characterization and calibration. With 28 patterns per frame (or image), this translates to roughly 274 frames (or images) per second, with averaging factor 4; this number drops to 188 frames/sec. Further, by hard coding spectral characterization of the samples on the FPGA itself, we were able to reduce the required throughput between the PC and the FPGA by a factor of 32:1, when compared to the throughput required to transfer all of the samples to the PC. Figure 3 shows the voltage measurement SNR vs. Number of Average plot for 6 different frequencies with and without the AFE connected. For the ADC test, a $2V_{pp}$ sinusoidal signal (ADC full-scale) at 100 KHz was applied at the ADC input. For the AFE test, a 2KΩ resistor was connected between a channel and current to voltage converter's virtual ground. The net gain on the source and sink side was 0.28 and 1.11 respectively (current sense resistor is 499Ω). With input signal set to $2V_{pp}$ sine wave at 100 KHz, V_{pp} at the ADC input was 0.56 V. It can be observed that: 1) Increasing the number of averages improved the SNR (by reducing the Gaussian noise power); 2) With the AFE connected, SNR dropped by about ~4 dBs. In future, we plan to comprehensively evaluate the system performance and to identify and reduce the number of noise sources in it.

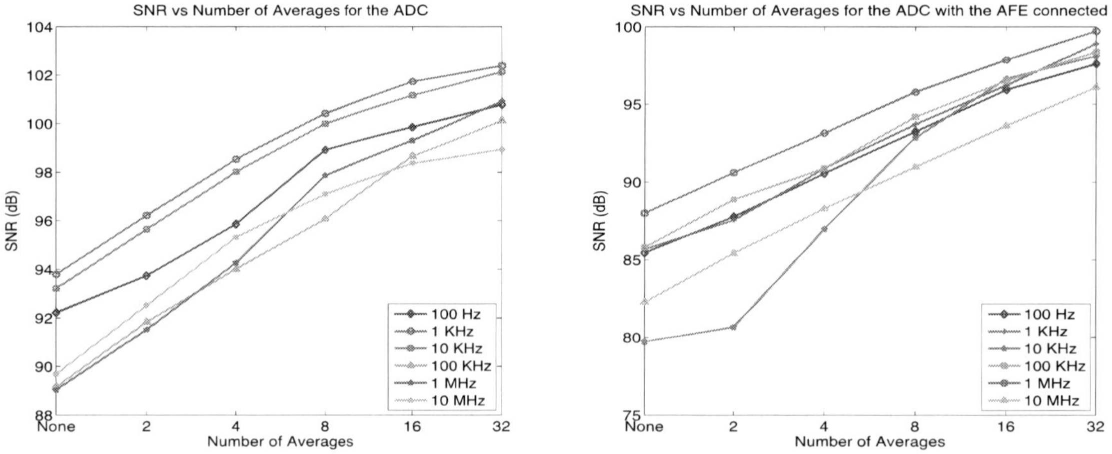

Figure 3. SNR versus number of averages for discrete frequency levels from 6 decades.

References

[1] McEwan A. *et al.* 2006 Design and calibration of a compact multi-frequency EIT system for acute stroke imaging, *Physiol. Meas.*, **27** 199-210

[2] Oh T. I. *et al.* 2008 Validation of a multi-frequency electrical impedance tomography (mfEIT) system KHU Mark1: Impedance spectroscopy and time-difference imaging, *Physiol. Meas.*, **29** 295-307

[3] Halter R. *et al.*, 2008 A broadband high frequency electrical impedance tomography system for breast imaging, *IEEE Trans. On Biomed. Eng.*, **59** 379-390.

[4] Oh T. I. *et al.*, 2011 A fully parallel multi-frequency EIT system with flexible electrode configuration: KHU Mark2, *Physiol. Meas.*, **32** 835-849

[5] Saulnier G. J., 2005 EIT Instrumentation in Electrical Impedance Tomography: Methods, History and Applications, *IOP Publishing*, 67-104.

[6] Kourunen J. *et al.*, 2009 Suitability of a PXI platform for an electrical impedance tomography system, *Meas. Sci. and Technol.*, **20** 015503

Journal of Physics: Conference Series **434** (2013) 012082 doi:10.1088/1742-6596/434/1/012082

Mutual information as a measure of reconstruction quality in 3D dynamic lung EIT

M G Crabb[1], J L Davidson[2], R Little[3], P Wright[2], J H Naish[3], G J M Parker[3], R Kikinis[4], H McCann[2] and W R B Lionheart[1]

[1] School of Mathematics, University of Manchester, UK
[2] School of Electrical and Electronic Engineering, University of Manchester, UK
[3] Imaging Sciences, Biomedical Imaging Institute, University of Manchester, UK
[4] Surgical Planning Laboratory, Brigham and Women's Hospital, Harvard Medical School, Boston, USA

E-mail: `Michael.Crabb@postgrad.manchester.ac.uk`

Abstract. We report on a pilot study with healthy subjects who had an MR scan in addition to EIT data acquired with the Manchester fEITER system. The MR images are used to inform the external shape of a 3D EIT reconstruction model of the thorax, and small changes in the boundary that occur during respiration are addressed by incorporating the sensitivity with respect to boundary shape into a robust reconstruction algorithm. A quantitative comparison of the image quality for different EIT reconstructions is achieved through calculation of their mutual information with a segmented MR image. A shape corrected reconstruction algorithm reduces boundary artefacts relative to a standard reconstruction, and has a greater mutual information of approximately 10% with the segmented MR image.

1. Introduction

The majority of EIT reconstruction algorithms do not exhibit an acceptable level of reproducibility across different patients, or for the same patient at different data acquisition times. The reason being a combination of poor modelling approximations as well as an insufficient assessment of EIT reconstruction quality. An inaccurately known boundary shape is known to result in major artefacts in EIT image reconstruction [1, 2], and so an MR image of the subject is used to inform the boundary shape of a 3D tetrahedral reconstruction model of the thorax. Reconstruction artefacts associated with small changes in the boundary shape that occur during respiration are addressed by calculating a derivative of the EIT forward model w.r.t. the boundary shape. An assessment of the spatial resolution and reconstruction quality of various EIT reconstructions is carried out via image co-registration and calculation of the mutual information with a segmented MR image.

2. Methods

2.1. Data acquisition and model generation

Two planes of 16 electrodes were placed equidistantly around the chest at the 4th and 6th intercostal spaces and EIT measurements were acquired with the fEITER instrument [3] with sinusoidal current injection of 0.5 mA amplitude, 10 kHz frequency operating at 100 fps. A nearest-neighbour current injection and voltage measurement protocol was deployed with 20 independent current patterns and EIT data was recorded whilst the subject carried out normal tidal breathing. MRI fiducial markers were placed on the subject at all EIT electrode locations immediately after the last EIT data collection, before carrying out an MRI scan on a 1.5 T Philips Achieva. The cross sectional shape of the subject was obtained from a selected transverse MR image slice at the fifth intercostal spacing, and a 3D FE forward model generated by extruding along the caudal-distal axis using EIDORS 3.6 [4]. The forward model consisted of typically 140000 tetrahedral elements and 30000 vertices, whilst the coarse reconstruction model consisted of 6000 tetrahedral elements and 1400 vertices, both modelled to a depth of 10 cm.

2.2. Shape corrected reconstruction algorithm

Denote the domain $\Omega \subset \mathbb{R}^3$, with L electrodes and positions $v \in \mathbb{R}^{3L}$, the measured voltages $\mathbf{V} \in \mathbb{R}^m$ and the conductivity $\sigma \in \mathbb{R}^N$. Define a forward operator F_I via $\mathbf{V} = Z_I F(\sigma, v) I$, where Z_I is an operator mapping the voltages for each current pattern, $I \in \mathbb{R}^L$, to the measurement protocol. This is solved numerically using a finite element method (FEM) with piecewise constant and piecewise linear representations of the conductivity and potential respectively.

In the following we assume the boundary is sufficiently well approximated from the MR image and small corrections to the electrode positions are accounted for due to breathing. The forward problem can be linearised using a Taylor series expansion as

$$\mathbf{V}_{\sigma+\delta\sigma,\mathbf{v}+\delta\mathbf{v}} = \mathbf{V}_{\sigma,v} + J^c_{\mathbf{v},\sigma}(\delta\sigma) + J^m_{\mathbf{v},\sigma}(\delta\mathbf{v}) + O(||(\delta\sigma, \delta v)||^2), \tag{1}$$

where $||(x,y)|| := ||x||_\infty + ||y||_\infty$ and $J^m : \mathbb{R}^{3L} \mapsto \mathbb{R}^m$ and $J_c : \mathbb{R}^N \mapsto \mathbb{R}^m$ are the movement and conductivity Jacobians respectively (see [5, 6] for details).

The estimation of conductivity and electrode positions $\mathbf{x} := (\sigma, v)$ from the data is approached from a semi-Bayesian viewpoint. The voltages are assumed to be related by $\mathbf{V} = Z_I F(\sigma, v) I + n$, where $n \in \mathbb{R}^m$ represents the measurement channel noise. We assume mean zero Gaussian noise with covariance matrix $\Gamma_e \in \mathbb{R}^{m\times m}$ and a Gaussian smoothness prior on the conductivity and electrode positions, with covariance matrix $\Gamma_{\sigma,v} \in \mathbb{R}^{3L+N\times 3L+N}$ and mean $\mathbf{x}_r := (\sigma_r, v_r)$. We assume further that measurement noise is independent and identically distributed (i.i.d) and that conductivity and electrode position changes are independent from one another and are separately i.i.d. These assumptions imply that $\Gamma_e^{-1} = I \in \mathbb{R}^{m\times m}$ and $\Gamma_{\sigma,v}^{-1}$ is a diagonal matrix with entries $\Gamma_{\sigma,v}^{-1}(i,i) = \alpha^2$ if $i \le N$ and $\Gamma_{\sigma,v}^{-1}(i,i) = \beta^2$ if $N < i \le N + 3L$. α and β represent the ratios of expected changes of conductivity and electrode position to the measurement noise standard deviation. The maximum *a-posterior* (MAP) estimate of $\mathbf{x}^i$, at the ith temporal data frame, $\mathbf{V}_i$, is equivalent to the following minimisation problem

$$\mathbf{x}_i = \arg\min_{\mathbf{y}} \Big\{ (Z_I F(\mathbf{y}) I - \mathbf{V}_i)' \Gamma_e^{-1} (Z_I F(\mathbf{y}) I - \mathbf{V}_i) + (\mathbf{y} - \mathbf{x}_r)' \Gamma_{\sigma,v}^{-1} (\mathbf{y} - \mathbf{x}_r) \Big\}.$$

An update $\mathbf{x}_i^{(1)}$ is computed from an initial guess, chosen to be the prior, $\mathbf{x}_r$, computed using a one step Gauss-Newton method. Denoting $J = [J_c|J_m] \in \mathbb{R}^{m\times(N+3L)}$ and neglecting second order partial derivatives of the forward problem w.r.t. σ and v, the linearised conductivity difference, $\delta\mathbf{x}_{ij}$, between $\mathbf{V}_i$ and $\mathbf{V}_j$ is then

$$\delta\mathbf{x}_{ij} := \mathbf{x}_i^{(1)} - \mathbf{x}_j^{(1)} = (J^T J + \Gamma_{\sigma,v}^{-1})^{-1} J^T (\mathbf{V}_i - \mathbf{V}_j). \tag{2}$$

2.3. Co-registration and mutual information

The reconstructed EIT image is mapped from the tetrahedral mesh into a voxel based image with 2 mm cubic voxels in the same space as the MRI data. The EIT images were thresholded so approximately 5% of extreme points were discarded from the visualisation and the dataset imported into the 3D Slicer open source visualisation software [7].

Mutual information has been used in previous multi modal imaging studies [8]. We denote two images A and B, each with N cubic voxels, and each voxel having a positive grayscale value. The probability distribution of A, $p_A(a)$, is defined as the number of voxels in image A that have pixel value a, normalised by N. The joint probability distribution of A and B, denoted $p_{A,B}(a,b)$, is then calculated as the number of times out of N that pixel in A contains a and the same pixel in B contains b normalised by N.

The Shannon entropy, in the imaging context, is the ability for an image to convey information and is measured in *bits*. The information content of a single event is proportional to the log of the inverse of the probability of an event. The information content of a single event is weighted by the probability that the event occurs and summed over all events to give the total information content, or entropy. The total entropy of A, and the joint entropy of A and B, are thus

$$H_A = -\sum_a p_A(a)\log(p_A(a)), \qquad H_{A,B} = -\sum_a \sum_b p_{A,B}(a,b)\log(p_{A,B}(a,b)). \qquad (3)$$

It can be shown that $0 \leq H_A \leq \log(N)$, where $H_A = 0$ when the image conveys no information i.e. it is featureless or homogeneous, and $H_A = \log(N)$ for white noise. The mutual information of A and B, $I_{A,B}$, is defined as the total entropy of A and B minus the joint entropy, $I_{A,B} = H_A + H_B - H_{A,B}$. It is also true that $0 \leq I_{A,B} \leq H_A$, where $I_{A,B} = 0$ when A and B have no features in common and $I_{A,B} = H_A = H_B$ when A and B are the same.

To calculate the mutual information, the MR image is first thresholded to create a binary MR image representing just lungs and chest. The voxel based EIT image is then scaled linearly to lie within the same range as the binary MR image. The images have a finite number of voxels, and so the probabilities above are interpreted as histograms with 256 equispaced bins, and if a bin happens to be empty, then $0 \times \log(0)$ is interpreted as 0. The MR and EIT images are both sampled at their voxel centres to estimate H_A, H_B and $H_{A,B}$ via (3), and hence $I_{A,B}$. This calculation is thus measuring how well a linearised EIT reconstruction is able to find the lungs as a whole. We perform the mutual information calculation over a range of regularisation parameters, (α,β), for both a standard ($\beta = 0$) and shape corrected reconstruction algorithm. In do doing so we effectively perform a numerical parameter study to maximise the mutual information between MRI and EIT as a function of α and β.

3. Results

Figure 1 illustrates an image co-registration example of MRI and EIT for the subject in the sitting position at three transverse planes for both a standard and movement reconstruction. The images are at max inhalation (relative to max exhalation) and blue indicates negative conductivity changes. The EIT reconstruction typically underestimates the physiological size of the cross-sectional areas of the lung. Figure 2 illustrates the mutual information with segmented MRI as a function of the parameter α, for a movement ($\beta = 4\times10^{-2}$) and standard reconstruction ($\beta = 0$). The maximum possible mutual information (the lung segmented MR image with itself) was 1.51 bits. There is a peak of mutual information of 1.11 bits at $\alpha \approx 3\times10^{-2}$, for the standard reconstruction and 1.21 bits at $\alpha \approx 10^{-2}$, for the shape corrected reconstruction, corresponding to a 10% increase in mutual information with shape correction. In general, the shape corrected reconstruction had larger mutual information then the standard reconstruction when taken over a range of α and β.

4. Conclusions and future work

The shape corrected reconstruction algorithm deployed results in a visual reduction of boundary artefacts relative to a standard reconstruction, and an increase in mutual information when compared with a segmented MR image. If EIT were to be routinely used as a bedside monitoring

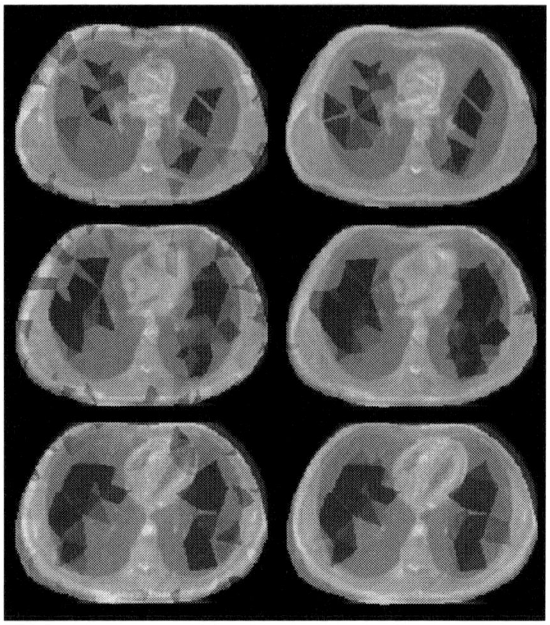

Figure 1. MR and EIT image co-registration. Left to right: Standard and movement reconstruction. Top to bottom: Transverse planes from superior to inferior.

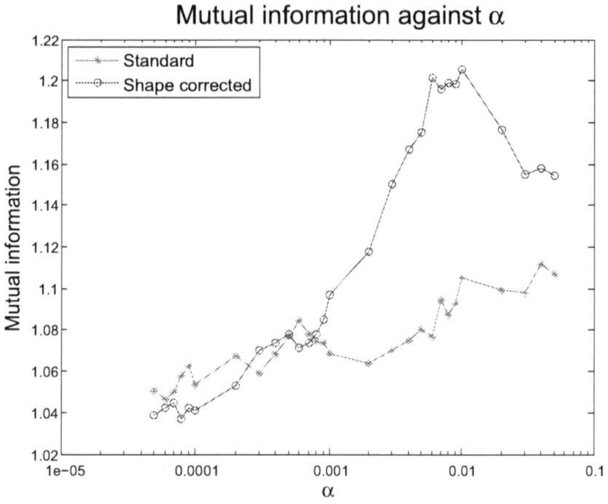

Figure 2. Mutual information parameter study as a function of α for a shape corrected reconstruction ($\beta = 4 \times 10^{-2}$) and a standard reconstruction ($\beta = 0$).

tool, optical data capture techniques could be used to inform the boundary shape in real time. The strength of MRI to inform the shape, however, is that we can also measure reconstruction quality through co-registration and mutual information as proposed here, and begin to assess the quality of EIT reconstructions over a range of available algorithms.

We would like to continue with this work by developing fully 3D models, that have variation in the cross section along the caudal-distal axis, and also incorporating more prior information such as the spine, liver and heart into a reconstruction model. An assessment of the best prior smoothness constraints on the conductivity, to offset the ill-posedness of EIT reconstruction, is another topic of importance and we believe significant improvements to reconstruction could be made from using both generalised Tikhonov regularisation and non-smooth penalisation norms.

References

[1] Adler A, Guardo R and Berthiaume Y 1996 *IEEE Trans. Biomed. Eng.* **43** 414-20

[2] Lionheart W R B 1998 *Inverse Problems* **14** 139-47

[3] McCann H, Ahsan T, Davidson J L, Robinson R L, Wright P and Pomfrett C J D 2011 *Proc. 33rd Annual Int. Conf. IEEE EMBS*, 7029-32

[4] Adler A and Lionheart W R B 2006 *Physiol. Meas.* **27** S25-S42

[5] Soleimani M, Gómez-Laberge C and Adler A 2006 *Physiol. Meas.* **27** S103-S113

[6] Crabb M G, Davidson J L, Little R, Wright P, Naish J H, Parker G J M, McCann H and Lionheart W R B 2012 100 *years of electrical imaging* 83-86

[7] Pieper S, Halle M and Kikinis R 2004 *Proc. 1st IEEE Int. Symp. on Biomedical Imaging: From Nano to Macro* **1** 632-35 http://www.slicer.org

[8] Pluim J P W, Antoine Maintz J B and Viergever M A 2003 *IEEE Trans. Medical Imaging* **22** 986-1004

A Level Set based Regularization Framework for EIT Image Reconstruction

Peyman Rahmati[1], Andy Adler[1]
[1]Dept. of Systems and Computer Eng. Carleton University, Ottawa, ON., Canada
E-mail: {prahmati, adler}@sce.carleton.ca

Abstract. Electrical Impedance Tomography (EIT) reconstructs the conductivity distribution within a medium from electrical stimulation and measurements at the medium surface. Level set based reconstruction method (LSRM) has gained attention during the last decade as an effective solution to address the need of reconstructing structures with limited amount of available data. The classical LSRM is based on the quadratic formulations (L2 norms); however, the L2 norms are not robust to outliers and spatial noise. The L1 norm is a more solid alternative to produce high robustness against outliers and noise. The L1 norm is minimized by Primal dual-interior point method (PDIPM). In this paper, we derive a novel level set (LS) based regularization framework for using the L1 norm independently on the data and the regularization term of an inverse problem. The proposed LS based regularization method, called LS based PDIPM (LS-PDIPM), applies the PDIPM to minimize the L1 norms. We use the LS-PDIPM to reconstruct 2D images from EIT simulated data. The proposed LS-PDIPM with the L1 norms provides sharper and less noisy images, when comparing with the L2 norm based regularization method.

Keywords: Level set, Inverse problem, Primal-Dual, Electrical Impedance Tomography

1. Introduction

Electrical Impedance Tomography (EIT) is an imaging modality to reconstruct the conductivity distribution within a medium by solving an inverse problem. EIT image reconstruction is an ill-posed problem. To provide robustness to measurement noise, regularization techniques are applied. The level set based regularization method (LSRM) is desirable because of its topological based representation of unknown structures using level zero (a front) of a higher dimension function (level set function) (Dorn & Lesselier 2009). The LSRM reconstructs the unknown topological information of structures using the evolution of a level set function which minimizes a predefined cost functional. We (Rahmati et al. 2012) showed the first clinical results of applying the LSRM to reconstruct EIT lung images. Borsic and Adler (2012) discuss the minimization of the L1 norm based regularization problem. They minimize the L1 norms using primal-dual interior point method (PDIPM) and show the L1 norms provide higher robustness against outliers and noise when compared with L2 norm based regularization method, such as Gauss-Newton (GN) method. In this paper, we derive a framework to solve a L1 norm based inverse problem using the LSRM, hereinafter called LS-PDIPM. There are two main advantages to applying the LSRM with the L1 norms: 1) level set techniques preserve the edges (Rahmati et al. 2012), 2) the L1 norms on the data mismatch term and the regularization term of an inverse problem provide robust estimations against outliers and spatial noise, respectively (Borsic & Adler 2012).

Journal of Physics: Conference Series **434** (2013) 012083 doi:10.1088/1742-6596/434/1/012083

2. Generalized PDIPM

A general primal problem can be written as follows

$$(P) = \text{argmin} \left\{ \sum_{i=1}^{D_1} |f_{d_i}(m)| + \sum_{j=1}^{D_2} |f_{p_j}(m)| + \|g_d(m)\|^2 + \|g_p(m)\|^2 \right\} \tag{1}$$

where $f_d(m)$ is a L1 norm based data mismatch term, $f_p(m)$ is a L1 norm based regularization term, $g_d(m)$ is a L2 norm based data mismatch term, and $g_p(m)$ is a L2 norm based regularization term. A primal minimization problem can be formed through any combination of the error terms defined in (1). The non-differentiability of L1 norm is resolved through using a centering condition which is the replacement of the L1 norm by a quadratic norm using an auxiliary variable, called β. The Primal-Dual (PD) framework can be formulated as

$$C_d(m) = f_{d_i}(m) - x_{d_i}\sqrt{f_{d_i}(m)^2 + \beta} = 0, \ \forall i \tag{2}$$

$$C_p(m) = f_{p_j}(m) - x_{p_j}\sqrt{f_{p_j}(m)^2 + \beta} = 0, \ \forall j \tag{3}$$

$$|x_{d_i}| \leq 1, |x_{p_j}| \leq 1 \tag{4}$$

$$F_c(m) = \frac{\partial}{\partial m}(f_d(m))x_d + \frac{\partial}{\partial m}(f_p(m))x_p + \frac{\partial}{\partial m}(\|g_d(m)\|^2) + \frac{\partial}{\partial m}(\|g_p(m)\|^2) = 0 \tag{5}$$

The derived general Primal-Dual (PD) framework above is solved iteratively using an iterative method, such as Newton method as follows

$$\begin{bmatrix} \frac{\partial}{\delta m}F_c(m) & \frac{\partial}{\delta x_d}F_c(m) & \frac{\partial}{\delta x_p}F_c(m) \\ \frac{\partial}{\delta m}C_d(m) & \frac{\partial}{\delta x_d}C_d(m) & \frac{\partial}{\delta x_p}C_d(m) \\ \frac{\partial}{\delta m}C_p(m) & \frac{\partial}{\delta x_d}C_p(m) & \frac{\partial}{\delta x_p}C_p(m) \end{bmatrix} \begin{bmatrix} \delta m \\ \delta x_d \\ \delta x_p \end{bmatrix} = - \begin{bmatrix} F_c(m) \\ f_d(m) - (\sqrt{f_d(m)^2 + \beta})x_d \\ f_p(m) - (\sqrt{f_p(m)^2 + \beta})x_p \end{bmatrix} \tag{6}$$

where two auxiliary variables x_d and x_p are in the range [-1, 1] depending on the absolute value of $f_d(m)$ and $f_p(m)$; respectively. The primal variables (m) are updated with a line search procedure, as $m^{(k+1)} = m^{(k)} + \lambda_m \delta m^{(k)}$, where k is the iteration number, δm is the update value with a descend direction to the optimal point, and λ is the step length (Nocedal & Wright 1999). In a similar manner, the dual variables (x_d and x_p) are also updated using *scaling rule* as follows $x^{(k+1)} = x^{(k)} + \min(1, \varphi^*)\,\delta x^{(k)}$(Andersen et al. 2000), where φ^* is a scalar value such that $\varphi^* = \sup\left\{\varphi : \left|x_i^{(k)} + \varphi\ \delta x_i^{(k)}\right| \leq 1, \quad i = 1, \ldots, n\right\}$.

3. Level Set based Primal Dual - Interior Point Framework

In this section, we formulate the solution of a system defined as (1) using the proposed LS-PDIPM. The solution of the L1L2 problem is derived using the LS-PDIPM. The solutions for the L2L1 and L1L1 problems are analogous.

3.1. LS-PDIPM for the primal problem with L1 norm based data mismatch term and L2 norm based regularization term

The level set function (Ψ) is a signed distance function which is zero at the optimal solution and nonzero otherwise. The minimum distance from the optimal solution is achieved at zero level set function. The evolution of the level set function according to the minimization of a functional objective function (primal problem), which can be a standard least square error function, results in the optimal solution of an inverse problem. A mapping function (Φ) is used to project the level set function onto finite element mesh (FEM). The level set evolution function is as follows

$\Psi_{k+1} = \Psi_k + \lambda(\Delta\Psi)$, where Ψ_{k+1} is the updated level set function, Ψ_k is the current level set function, $\Delta\Psi$ is the update, Φ is the mapping function, λ is the step size. The primal formulation (P) for the L1L2 problem is a special state of the general primal problem in (1) when $|f_p(m)| = \|g_d(m)\|^2 = 0$ and can be written as

$$(P) = argmin_{\Phi(\Psi)} \left[\sum_i W_i |h_i(\Phi(\Psi)) - d_i| + \alpha \|L(\Phi(\Psi) - \Phi(\Psi^0))\|^2 \right] \quad (7)$$

where $|f_d(m)$ is replaced by $f_d(\Phi(\Psi)) = W|h(\Phi(\Psi)) - d|$, $|g_p(m)$ is replaced by $g_p(\Phi(\Psi)) = \alpha\|L(\Phi(\Psi) - \Phi(\Psi^0))\|^2$, W is a weighting diagonal matrix, W_i is the i-th diagonal element, $h_i(\Phi(\Psi))$ is the i-th forward measurement, d_i is the i-th measured data, L is the regularization matrix, $\Phi(\Psi)$ is the model parameter distribution or the primal variables, $\Phi(\Psi^0)$ is a reference model parameter distribution According to the chain rule, the LS Jacobian matrix (J_{LS}) can be written as $J_{LS} = \frac{\partial d}{\partial \Psi} = (\frac{\partial h}{\partial \Phi(\Psi)})(\frac{\partial \Phi(\Psi)}{\partial \Psi}) = (J_{GN})(M)$. To make the algorithm computationally efficient, we restrict the Jacobian computation within a narrow band containing the data (non-zeros). To construct the narrow band, we define the level set function, or the signed distance function, to be negative inside its boundary and positive outside. Matrix M is non-zero within the narrow band and zero otherwise, which is the notion of Dirac delta function. In every iteration of the level set function (Ψ), we calculate the Jacobian matrix (J_{LS}) for the narrow band. We define a dual variable x_i in the range $[-1, 1]$, depending on the absolute value of $W_i(d - h(\Phi(\Psi)))$. The dual problem can be written as

$$(D) = argmin_{\Phi(\Psi)} \left[max_x \quad x^T W(h(\Phi(\Psi)) - d) + \alpha\|L(\Phi(\Psi) - \Phi(\Psi^0))\|^2 \right], with |x| \leq 1 \quad (8)$$

The smoothed version of the LS based PD framework can be obtained through applying the *centering condition* which is the replacement of $|W_i(h(\Phi(\Psi)) - d_i)|$ by $\sqrt{(W_i(h(\Phi(\Psi)) - d_i)^2 + \beta}$, with $\beta > 0$. Replacing $|f_p(m)| = 0|$ and $\|g_d(m) = 0\|$ in (2-5), the smoothed LS based PD framework is achieved as

$$|x_i| \leq 1, \quad (9)$$

$$F_c(\Phi(\Psi)) = J_{LS}^T(\Phi(\Psi))Wx + 2\alpha M^T L^T L(\Phi(\Psi) - \Phi(\Psi^0)) = 0, \quad (10)$$

$$C_d(\Phi(\Psi)) = (h_i(\Phi(\Psi)) - d_i) - x_i\sqrt{(h_i(\Phi(\Psi)) - d_i)^2 + \beta} = 0, \quad \beta > 0, \quad (11)$$

We define $f = h(\Phi(\Psi)) - d$, $F = diag(f)$, $X = diag(x)$, $\kappa = \sqrt{f^2 + \beta}$, $E = diag(\kappa)$. Applying the general solution in (6), the following newton system to be iteratively solved to calculate the updates for the primal variables (Ψ) and the dual variable (x) can be written

$$\begin{bmatrix} 2\alpha M^T L^T L M & J_{LS}^T W \\ (I - XE^{-1}F)J_{LS} & -E \end{bmatrix} \begin{bmatrix} \delta\Psi \\ \delta x \end{bmatrix} = - \begin{bmatrix} J_{LS}^T W x + 2\alpha M^T L^T L(\Phi(\Psi) - \Phi(\Psi^0)) \\ f - Ex \end{bmatrix} \quad (12)$$

the dependency of J_{LS} on Ψ is dropped in the derived LS based PD framework. The derived set of equations in (12) are iteratively solved for the primal variables ($\delta\Psi$) and the dual variables (δx) using an iterative method such as Newton method. A traditional line search procedure (Nocedal & Wright 1999) can be applied to find an appropriate step length λ_Ψ resulting in the update $\Psi^{(k+1)} = \Psi^{(k)} + \lambda_\Psi \delta\Psi^{(k)}$, where k is the iteration number. A *scaling rule* is applied to compute the updates for the dual variables (x) (Andersen et al. 2000).

4. Simulated Data

We used EIT with 16 electrodes on one electrode plane and a circular FEM. The adjacent current stimulation was considered for the evaluation of our simulation. The applied 2D phantom contains two sharp inclusions with the same conductivity of 0.9 S/m. The background conductivity value is 1 S/m. The inverse problem used the mesh density of 576 elements, which was different than the mesh density of the forward problem (1024 elements).

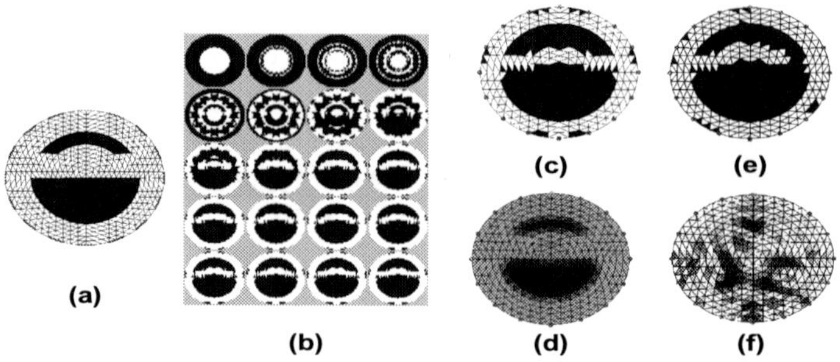

Figure 1. Image reconstruction of EIT simulated data with and without synthetic additive noise. (a) The applied 2D phantom. (b) The iterations of the derived LS-PDIPM with no added noise. (c) The final reconstructed image of the LS-PDIPM with no added noise. (d) Gauss-Newton method with no added noise. (e) The final reconstructed image of the LS-PDIPM with added 60dB Gaussian noise. (f) The Gauss-Newton method with additive 60dB Gaussian noise.

5. Results

In this section, we apply the derived LS-PDIPM to solve a primal problem with L1 norm on both the data mismatch and the regularization term (L1L1 problem). The reason of choosing the L1L1 problem is because the L1 norms has been shown to offer the highest robustness to spatial noise and outliers. The inverse solution is calculated using the proposed LS-PDIPM with $\beta = 1 \times 10^{-12}$. The stopping term to terminate the iterations depends on the value of the primal dual gap computed in every iteration. 20 iterations for the LS-PDIPM were sufficient to reach to convergence. Figure 1(a) shows the applied 2D phantom with two inclusions with conductivity of 0.9 S/m. The reconstructed images for every iteration of the LS-PDIPM are shown in figure 1(b). The final reconstructed image using the LS-PDIPM is demonstrated in figure 1(c). Figure 1(d) is the reconstructed image of GN method over the same simulated data achieved from the 2D phantom. To account for the possible systematic and random errors occurring in EIT data acquisition process, a 60 dB Gaussian noise was added to the EIT simulated data. Figure 1(e) and figure 1(f) present the final reconstructed image of the LS-PDIPM and the image of GN method in the presence of additive noise, respectively. GN method fails due to the usage of the L2 norms on the regularization term of the inverse problem. The LS-PDIPM is highly robust to the additive noise and still provides sharp image. Unlike the GN method, the LS-PDIPM preserves the edges and is not sensitive to the spatial noise.

6. Conclusion

We derive a level set based regularization method which allows any possible combination of norms (L1 norm or L2 norm) on the data term and the regularization term of an inverse problem. The LS-PDIPM contains the benefit of a shape based reconstruction algorithm in producing sharp edges as well as using the L1 norm in the primal problem, increasing the robustness to outliers and noise. we applied the proposed LS-PDIPM to reconstruct EIT 2D images and showed the LS-PDIPM is highly robust to the additive 60dB Gaussian noise. An extended version of the LS-PDIPM to elaborate the derivations of the LS-PDIPM is currently under way.

7. References

Andersen K D, Christiansen E, Conn A & Overton M L 2000 *SIAM J. on Scientific Computing* **22**, 243–262.
Borsic A & Adler A 2012 *Journal of Inverse Problems* **28**.
Dorn O & Lesselier D 2009 *Inverse Problems* **25**, 125.
Nocedal J & Wright S J 1999 *Numerical Optimization* Springer.
Rahmati P, Soleimani M, Pulletz S, Frerichs I & Adler A 2012 *Journal on Physiological Measurement* **33**, 739.

MonitoRing - Magnetic induction measurement at your fingertip

D Teichmann, J Foussier, D Löschcke, S Leonhardt and M Walter

Philips Chair for Medical Information Technology, RWTH Aachen University

E-mail: teichmann@hia.rwth-aachen.de

Abstract. The device presented in this paper is a sensor for monitoring pulse by measuring the bioimpedance of the thumb in an unobtrusive way. The sensor is based on magnetic induction measurement, a non-contact technique for measuring impedance changes of objects [1]. The sensor head of the presented system has the form of a ring and is worn on the finger. The developed technique renders the possibility of easy and unnoticed pulse recording during every day activities without the need for, e.g. electrodes, a pulse belt around the chest, or a pulse photoplethysmographic finger or ear clip.

1. Introduction

Cardiac pulse is an important vital sign for the bedside monitoring of patients as well as for homemonitoring devices. It is also a frequently used parameter in professional and consumer sport monitoring applications. Due to its easy access, the finger has become an established location for pulse plethysmographic recordings. This work presents a novel device for impedance plethysmography of the finger without the need for electrical or optical contact.

Commonly used photoplethysmographic (PPG) sensors applied as fingerclips are sometimes perceived as obtrusive and bothersome. In addition, they strongly suffer from motion artefacts, since LED and photosensor are only loosely coupled to the skin, and are therefore, merely useful in quiet scenarios. Using electrical impedance plethysmographs (EIP), ring electrodes are placed on the finger at fixed locations [2]. The necessary parallel alignement of these electrodes results in a difficult electrode application and is also a factor causing motion artefacts, since the geometrical alignement could change during motion.

The proposed system called MonitoRing is based on a single sensor head that can be worn as a ring around the finger. It is more convenient to wear and less obtrusive in comparison to commonly used finger plethysmographs. It is also less susceptible to motion artefacts, because it can be worn in a more stable contact to the skin (compared to PPG) and is based on only one measurement location instead of two. Additionally, the device does not need conductive or optical contact and could therefore also be worn above a textile layer (e.g. hand glove, sleeve). The sensoric part of the device is based on few electronic components, hence the device can easily be realized as a mobile design.

Journal of Physics: Conference Series **434** (2013) 012084 doi:10.1088/1742-6596/434/1/012084

2. Methods

2.1. Measurement System

The physical principle of the device is based on eddy current induction (also called magnetic induction monitoring [1]). It is illustrated in Fig. 1(a).

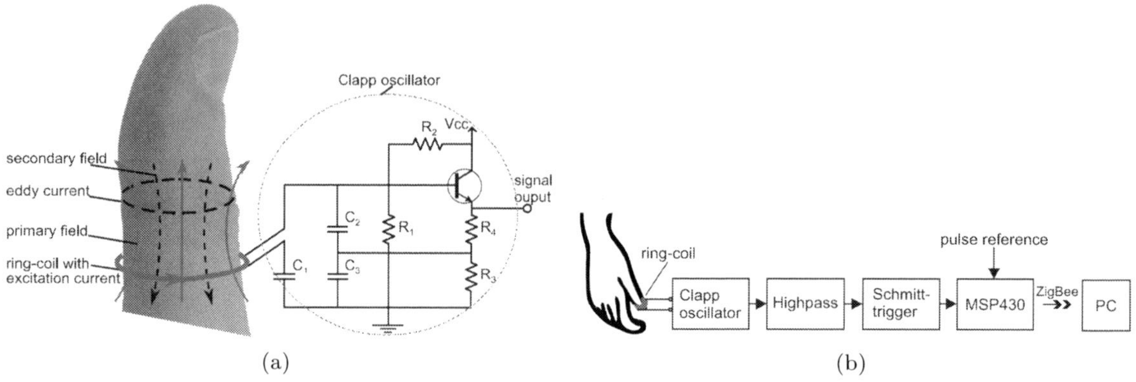

Figure 1. Principle (a) and system overview (b) of the MonitoRing device

A coil formed as a ring around the thumb sends out an alternating magnetic field, which induces eddy currents into the thumb. The ring-coil is a frequency determining part of a Clapp oscillator, which is driving the coil's current [3]. Hence, the sending frequency of the field emitted by the coil equals the oscillatory frequency. The eddy currents within the thumb reinduce a secondary alternating magnetic field affecting the primary one and changing the impedance of the coil. When the impedance distribution within the thumb (and thus the strength and orientation of the eddy currents) changes due to blood pulsation, the impedance of the coil changes and varies the oscillatory frequency of the Clapp oscillator. In this way, the oscillatory frequency is a simply measured signal, which correlates with the impedance distribution changes of the thumb caused by blood pulsation.

A block diagram of the system can be seen in Fig. 1(b). The output signal of the Clapp oscillator is high pass filtered and schmitt triggered before its frequency is measured by the use of a microcontroller (MSP430F5457A, Texas Instruments).The system is battery driven using two 3.3 V lithium batteries and comprises a wireless ZigBee connection to a PC, where the data is stored and vizualized. The sampling frequency of the system is 95 Hz. The MonitoRing device is integrated into a small portable housing (64x42x18 mm), which can be attached to the arm or wrist.

2.2. Measurement Setup

The MonitoRing device was evaluated by measuring the pulse of a subject and comparing it to a simultaneously recorded photoplethysmographic (PPG) pulse reference (ChipOx, Corscience GmbH & Co. KG).

The ring-coil for measuring pulse at the thumb was tested at two different positions on the left thumb (see Fig. 2(a)). It consists of n=12 windings of high frequency litz (HFL) wire, that are wrapped around the finger. The resulting base sending frequency of the device was $f_b = 3.8$ MHz.

Furthermore, the application of a ring-coil as an arm bracelet was examined with a coil positionend around the forearm. For this purpose, three different sensor locations (see Fig. 2(b)) were tested with the coil realized by four windings of HFL wire ($n = 4$, $f_b = 5$ MHz) similiar to the ring-coil used for thumb measurement.

Journal of Physics: Conference Series **434** (2013) 012084 doi:10.1088/1742-6596/434/1/012084

In order to investigate the device's ability to monitor the cardiac pulse without mechanical contact between coil and skin, a pancake coil (n =3, f_b = 6 MHz) was printed on a printed circuit board (PCB) and the forearm was put through a hole in its middle without contact to the board (see Fig. 2(c)).

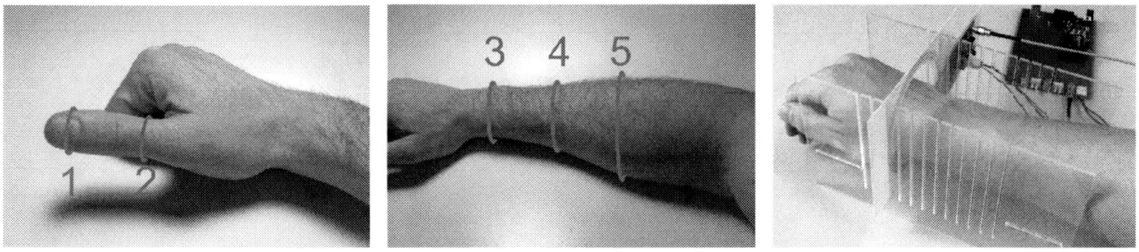

(a) (b) (c)

Figure 2. Measurement positions of the ring coil on the thumb (a) and forearm (b). (c) shows the measurement setup for non-contact recording at the forearm.

3. Results

When the sensor coil is attached to the thumb in form of a ring, the device shows good correlation with the signal recorded by the reference PPG system (Pearson correlation coefficient $\rho_{pos1} = -0.9452$). The result recorded at position 1 is presented in Fig. 3(a) and its corresponding Fourier Transform is given in Fig. 3(b). The average peak-to-peak change ($avPP$) of the oscillator within a cardiac cycle is 258.75 Hz and the signal-to-noise ratio (SNR) of the system is about 56.12 dB. When measuring at position 2 the pulse signal is also recordable but with a lower signal quality ($avPP = 226.71$ Hz, $SNR = 17.52$ dB, $\rho_{pos2} = -0.4157$).

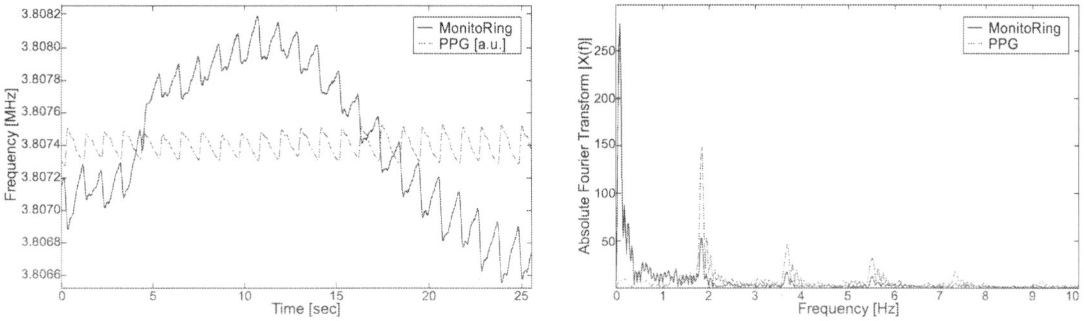

(a) time domain (b) frequency domain

Figure 3. Result of a measurement with the ring-coil worn on the thumb (position 1). solid line: MonitoRing device; dashed line: plethysm. pulse reference (PPG).

Applying the MonitoRing onto the arm by HFL wire as described in section 2.2, the pulse signal could only be recorded on measurement position 4 ($avPP = 76.93$ Hz, $SNR = 13.93$ dB). This result is shown in Fig. 4(a) and 4(b). Using the non-contact measurement setup with a pancake coil printed on a PCB for arm measurement as shown in Fig. 2(c), it was not possible to recognize any signal content related to cardiac pulse.

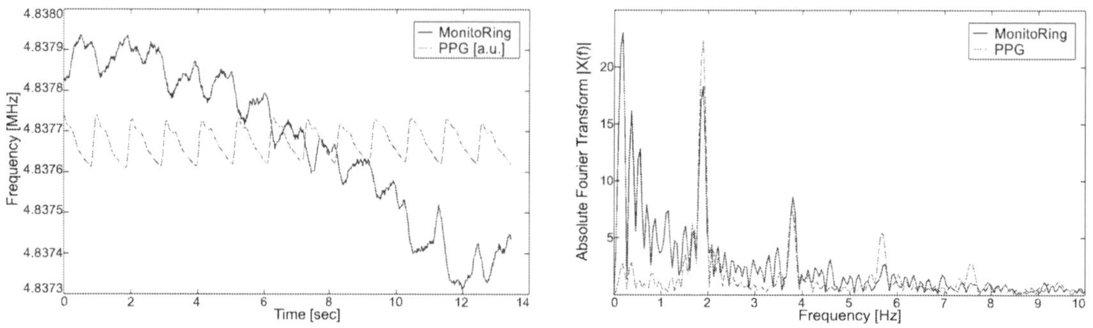

(a) time domain (b) frequency domain

Figure 4. Result of a measurement with the ring-coil worn on the forearm (position 2). solid line: MonitoRing device; dashed line: plethysm. pulse reference (PPG).

4. Discussion

The presented results indicate the system's potential to monitor pulse at the fingertip. However, in order to ensure a comfortable wearibility of the system, the electronic circuitry should be further miniaturized. Furthermore, for a robust and reliabe signal more effort has to be spent into the investigation of the system's sensitivity to motion artefacts. Especially, signals recorded at the measurement positions on the forearm show low quality, probably due to motion artefacts produced by contraction of the lower arm muscles.

When using a coil around the arm without contact between coil and skin, it was not possible to derive any pulse signal. This poses the question if the pulse signal derived using the other measurement setups (where the coil was directly worn on the skin) was actually evoked by impedance changes of the tissue or by changes of the coil diameter due to blood pulsation. This issue should be investigated in future. Nevertheless, in related studies the general sensitivity to impedance changes of objects was proven for similiar magnetic induction sensors [3, 4].

5. Conclusion

Using magnetic induction measurement with the sensor coil worn as a ring around the fingertip, it is possible to monitor pulse with a *SNR* of 56.12 dB. When the ring-coil is located at a measurement position nearer to the base of the thumb, pulse measurement is still possible, but with a lower signal quality. Pulse measurement with the sensor coil worn as an arm bracelet, was only possible with the coil located in the middle of the forearm.

In conclusion, it can be stated that the described device represents a convenient method for plethysmographic pulse monitoring at the finger in an unobtrusive way.

Acknowledgments

This work was funded by the EU project 'HeartCycle'.

References

[1] Tarjan P P and McFee R 1968 *IEEE Trans. Biomed. Eng.* **15** 266–78
[2] Montgomery L D, Moody D L and Williams B A 1977 *IEEE Trans. Biomed. Eng.* **24** 385–6
[3] Wilson D L and Geselowitz D B 1981 *IEEE Trans. Biomed. Eng.* **28** 702–10
[4] Teichmann D, Foussier J and Leonhardt S 2010 *IEEE Biomed. Circuits and Systems Conf. (Paphos)* pp 37–40

A magnetic induction measurement system for adult vital sign monitoring: evaluation of capacitive and inductive effects

H.Mahdavi and J.Rosell-Ferrer

Group of biomedical and electronic instrumentation, department of electronic engineering,Universitat Politècnica de Catalunya (UPC), Barcelona, Spain

E-mail: hadiseh.mahdavi@upc.edu

Abstract. Magnetic induction (MI) measurement technique could provide an unobtrusive contactless method for continuous monitoring of vital signs such as breathing and cardiac activity in bed. In this paper, we present a magnetic induction system to evaluate the feasibility of monitoring heart and lung activity and the preliminary measurement results. The excitation and detection coils are designed to be implemented in a single printed circuit board, allowing the use of the system in a bed with coils under the mattress. The electronic system is based on a 16 bit arbitrary waveform generator (PXI-5422, National instrument) operating at a sample rate of 200 MS/s for the excitation signal and the detected amplified signal is then, sampled at 100 MS/s by a 14 bit digitizer (PXI-5122, National Instruments). The preliminary results at 10 MHz show the ability of the system to detect solutions with different conductivities. However the capacitive effect is in the same order of magnitude as the inductive effect due to eddy currents. Safety of the system has been evaluated to be in accordance with the standards of human exposure to the magnetic fields.

1. Introduction

Unobtrusive methods for monitoring vital signs are of high importance in many medical issues. Home health care applications, monitoring neonates or burn victims and patients with sleep problems are some of those areas in which unobtrusive techniques are of particular interest. Magnetic induction (MI) could provide a method for continuous monitoring of vital signs such as breathing and cardiac activity in bed. The theory on which this method is based is that, when a conductive object is placed in a time varying magnetic field (B), eddy currents are induced in the body. These currents produce a secondary magnetic field that can be detected by a properly designed receiver system. The amplitude of the eddy current is proportional to the magnetic flux density and the conductivity of the material. Based on this theory, monitoring the changes of conductivity in the thorax produced by the inflation and deflation of the lungs and the blood circulation in the trunk, enable us to monitor these vital signs. Since the conductivities of biological tissues are very low, the secondary signals to be measured are very weak which makes the measurement process difficult. For a sample of material between an excitation coil and a sensing coil [1],

$$\Delta B/_{B} \propto \omega(\omega \varepsilon_0 \varepsilon_r - j\sigma) \tag{1}$$

where σ is the conductivity of the sample, ε_r is its relative permittivity, ε_0 is the permittivity of free space, ω is the angular frequency of the excitation and $j = \sqrt{-1}$. For biological tissues, the conductivity (imaginary) component will normally be dominant [2]. From (1) it is seen that the eddy currents produce a component of the secondary signal, which is proportional to σ and ω. The total detected field $(\Delta B + B)$ lags the primary field by an angle φ [3]. The objective of

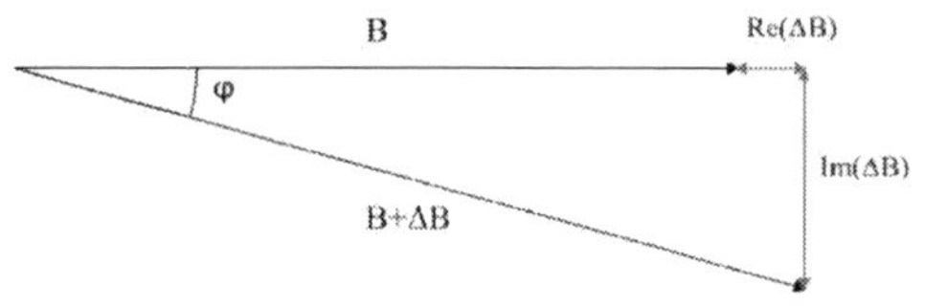

Figure 1. Phasor diagram of the primary B and secondary (ΔB) magnetic fields detected.

the presented work is to evaluate the capacitive and inductive effects in a practical MI system to be used under the mattress for monitoring vital signs. In addition, since the system is based on the exposure of an electromagnetic field, evaluation of the specific absorption rate (SAR) is of interest to be in accordance with international standards.

2. Materials and Method

2.1. SAR Simulation

As stated previously, safety is an important issue to be studied in the systems based on the exposure of the electromagnetic field. In accordance with ICNIRP recommendations [4], the maximum field strength is limited by the specific absorption ratio (SAR) which shall not exceed 2 W/Kg for head and torso. SAR simulation has been done using COMSOL Multiphysics by a simplified anatomical 3D model of a human trunk (figure 3).

2.2. Tank Simulation

As a reference for experiments, the experimental setup has been simulated using the same software. The model consists of a cylinder with a radius of 8 cm and a height of 40 cm (approximate volume of 8 liters). By using the parametric sweep, solutions with different conductivity have been simulated and the magnetic field was obtained for each of the solutions.

2.3. MI system

The system is based on a 16 bit arbitrary waveform generator (PXI-5422, National instrument) operating at a sample rate of 200 MS/s for the excitation signal. The signal detected at the detector coil was passed to a two stage low noise amplifier (AD8432) with a gain of 64. The amplified signal is then, sampled at 100 MS/s by a 14 bit digitizer (PXI-5122, National Instruments). The driving frequency for the measurements is fixed to 10 MHz. As a result of the very low conductivity of biological tissues, the secondary magnetic signals to be measured are very weak and since the secondary coil will receive the superposition of the primary and secondary fields, the dynamic range of the receiver channels would be very wide. Finding an efficient coil configuration is an important issue in the system design process in order to suppress the large primary magnetic field as much as possible. Up to now, several flux cancellation techniques have been proposed from which we plan to use a planar gradiometer to cancel the primary magnetic field [5]. In addition, it is known from the state of the art that capacitive coupling effects exist between the transmitting and receiving coils [6] and also between the object and the sensors [7]. Several methods were reported to minimize this effect. In our experiments the coils are two 8-turn coils with a diameter of 5 cm that could be placed in a common printed circuit board (PCB). The generator (PXI-5422) has been configured to produce a sine wave with a desired frequency and amplitude as excitation signal with an injected current of 22 mA. The signal at the detection coil is then passed through the AD8432 to be amplified to the input level of the digitizer. The amplifier is an AD8432 which is a dual channel, low power amplifier with an input voltage noise of 0.85 $nV\big/\sqrt{Hz}$.

2.4. Measurement protocol

Six saline solutions, in plastic tanks of 8 liters with similar dimensions to the simulated model were made up with conductivities of 0.0005, 0.2, 1, 2, 5, 16 $[Sm^{-1}]$. The experiment steps are as follows; place the tank at a distance of 5 cm from the system, step back, wait for 3 seconds, step forward and touch the tanks for 2 seconds, step back, wait for 3 seconds more and finally take it out. For each experiment, 40 measurements have been done in 22 seconds.

3. Results

The estimation of the absorption rate at a range of frequencies at different distances of excitation coil from the body, shows that even considering the worst case for exposure to the magnetic field (a frequency of 10 MHz and minimum distance), the maximum SAR is orders of magnitude lower than the safety standards' limits (figure 2). That is, for an excitation current of 1 A in a one turn coil at a distance of 5 cm from the chest at a frequency of 10 MHz, we are more than one order of magnitude under the safety limits imposed by ICNIRP standard (Maximum calculated SAR=0.0056 W/Kg) [8]. Figure 4 shows the changes of the detected phase for all solutions following the measurement protocol. Figure 5 shows the increment of the phase angle with respect to the changes in conductivity of the solution, in experiments and simulations. Since the phase changes produced by distilled water have to be zero, this value was subtracted from all the measurements.

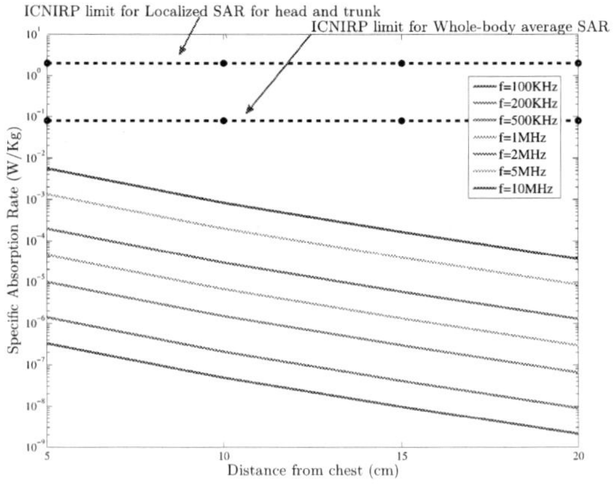

Figure 2. SAR vs. distance from chest

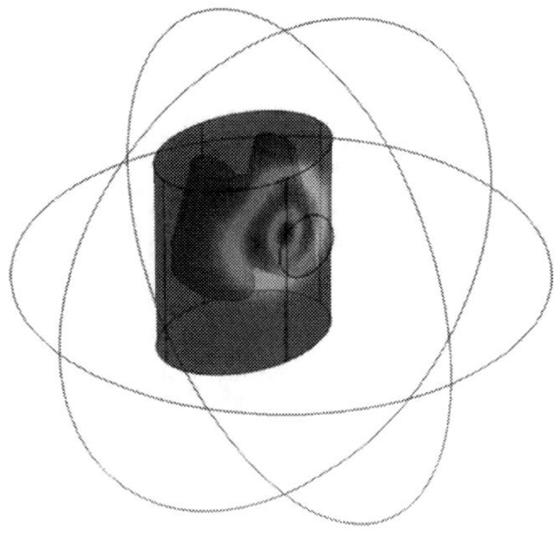

Figure 3. SAR simulation at 10 MHz

4. Discussion and Conclusions

Equation (1) predicts that the imaginary component of the secondary magnetic field (ΔB) must be proportional to σ. The equation is valid when the skin depth of the electromagnetic field is much larger than the thickness of the sample. As the conductivity increased, the skin depth became comparable to the diameter of the tank, i.e. for a conductivity of 16 $[Sm^{-1}]$ the skin depth is 3 cm in comparison with 16 cm of tank diameter. The simulation shows that the skin depth affects the measurement results as expected and it can be seen in figure 5 that, there is a change in the slope for greater values of σ.

Moreover, figure 5 shows that we have an overestimation of the phase angle. When the operator

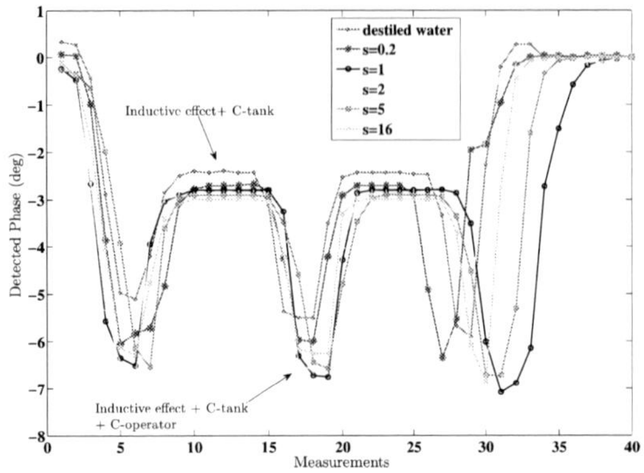

Figure 4. Phase angle of the detected signal

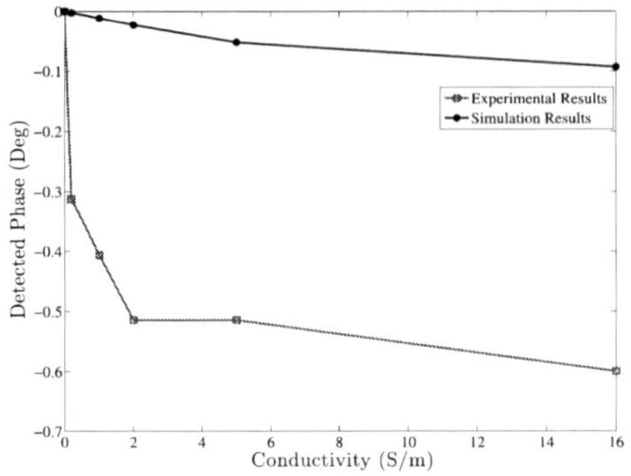

Figure 5.
Phase angle increment due to changes in σ

is touching the tank, a phase change -from free space situation- of about 6° can be seen in the graphs (figure 4) while without touching, the maximum phase change is about 3°. Although the phase was adjusted to the empty space case, these phase changes could still not be explained only by magnetic induction effects, and the capacitive coupling must play an important role that is in accordance with [3] and [6]. There are also measurement errors due to misplacement of the tank over the sensor. Experimentally, we obtained that a misplacement of 1 cm, in the same axis as cylinder's axis, produces a change of about 0.29° and a misplacement of 1 cm, in the axis perpendicular to the cylinder's axis, produces an approximate phase shift of 0.047°. These phase changes could also be explained by capacitive coupling. As capacitive effect is mainly related to the distance between the object and the coils, and magnetic induction is more sensitive to the changes in conductivity, we want to evaluate the possibility of using both magnetic and capacitive effects to monitor vital signs. The advantage of monitoring physiological signals is that they are rhythmic and the artifacts and unwanted static signals, produced by environmental factors, could be suppressed by signal processing techniques. Future studies will be focused on the design of the coils to optimize the balance between capacitive and magnetic effects. Since it is necessary to use a gradiometer as a field cancellation method to suppress the large primary magnetic field and increase the sensitivity of the magnetic induction signal, we will study different topologies of gradiometers and coils suitable for being implemented on a PCB.

Acknowledgments
This work has been funded in part by the Help4Mood project (Contract Number: ICT- 248765) from the European Union's Seventh Framework Program.

References
[1] Griffiths H, Stewart W R and Gough W 1999 *Annals Of The New York Academy Of Sciences* **873** 335–345
[2] Griffiths H 2001 *Measurement Science and Technology* **12** 1126–1131 ISSN 09673334
[3] Barai a, Watson S, Griffiths H and Patz R 2012 *Measurement Science and Technology* **23** 085501 ISSN 0957-0233
[4] International Commission on Non-Ionizing Radiation 1998 *Health Physics* **74** 494–522
[5] Scharfetter H, Lackner H K and Rosell J 2001 *Physiological Measurement* **22** 131–146
[6] Griffiths H, Gough W, Watson S and Williams R J 2007 *Physiological measurement* **28** S301–11 ISSN 0967-3334
[7] Steffen M, Heimann K, Bernstein N and Leonhardt S 2008 *Physiological measurement* **29** S291–306 ISSN 0967-3334
[8] Mahdavi H and Rosell Ferrer J 2012 *Barcelona forum on Ph.D. research in communication and information technologies* (Universitat Politècnica de Catalunya- Barcelona Tech) pp 17–18

Journal of Physics: Conference Series **434** (2013) 012086 doi:10.1088/1742-6596/434/1/012086

Magnetic resonance electrical impedance tomography for determining electric field distribution during electroporation

Matej Kranjc[1], Franci Bajd[2], Igor Serša[2], Damijan Miklavčič[1]

[1]University of Ljubljana, Faculty of electrical engineering, Trzaska 25, Ljubljana SI-1000, Slovenia

[2]Institut "Jožef Štefan", Jamova cesta 39, Ljubljana SI-1000, Slovenia

damijan.miklavcic@fe.uni-lj.si

Abstract. Electroporation is a phenomenon caused by externally applied electric field to cells that results in an increase of cell membrane permeability to various molecules. Accurate coverage of the tissue with a sufficiently large electric field presents one of the most important conditions for successful membrane permeabilization. Applications based on electroporation would greatly benefit with a method for monitoring the electric field, especially if it could be done *in situ*. As the membrane electroporation is a consequence of an induced transmembrane potential, which is directly proportional to the local electric field, we have been investigating current density imaging and magnetic resonance electrical impedance tomography techniques to determine the electric field distribution during electroporation. In this paper, we present comparison of current density and electric field distribution in an agar phantom and in a liver tissue exposed to electroporation pulses. As expected, a region of increased electrical conductivity was observed in the liver tissue exposed to sufficiently high electric field but not in agar phantom.

1. Introduction

Electroporation is a phenomenon in which cells exposed to externally applied electric field of an adequate strength and duration increase their membrane permeability to various molecules, which are otherwise deprived of transmembrane transport mechanism [1–3]. Clinical applications, such as electrochemotherapy and non-thermal irreversible electroporation, both potent procedures used in solid tumor treatment, are being successfully introduced into clinical practice [4,5]. Therapies success depend on adequate electric field distribution in the tissue, which is currently determined by electrode positioning and by applied electric pulse amplitude, both obtained by treatment planning [6–8]. Even though treatment planning has already proved to have a great potential, its applicability is currently limited by conductivity data of tissues, especially within the tumor where heterogeneous and nonlinear conductivity was observed [9]. Poorly determined tissue conductivities and inaccurate electrode positions can result in an insufficient electric field coverage of the tumor and thus in a suboptimal treatment outcome [10].

As the membrane electroporation is a consequence of an induced transmembrane potential, which is directly proportional to the local electric field, we propose current density imaging (CDI) and magnetic resonance electrical impedance tomography (MREIT) techniques to determine the electric field distribution during electroporation [11]. CDI is a magnetic resonance imaging method for acquiring current density distribution inside conductive samples by measuring magnetic field changes caused by applied current [12,13]. Tissue conductivity can be obtained by MREIT, a technique used for reconstruction of electrical conductivity inside a tissue by means of current density [14,15]. As the measurement of current density and electrical conductivity is performed during electric pulse delivery, determined electric field distribution includes all changes, which occur in tissue due to electroporation. In this paper we present comparison of current density and electric field distribution in an agar phantom and in a liver tissue exposed to electroporation pulses. We examined differences in the obtained current density and electric field distribution of non-biological tissue (agar) and biological tissue (chicken liver).

Electroporation applications such as electrochemotherapy and non-thermal irreversible electroporation would greatly benefit from conductivity imaging method such as MREIT. Reconstruction of electrical conductivity during electroporation pulse delivery together with measuring current density distribution using MREIT and CDI, respectively, would enable much needed *in situ* determination of an electric field distribution thus increasing and assuring the effectiveness of electroporation based treatments.

2. Materials and methods

All experiments were done in an acrylic glass container as shown in figure 1. The agar phantom was made of agar powder (Kemika, Croatia), 0.9 % NaCl saline solution (B. Braun, Germany) and distilled deionized water (B. Braun, Germany).

Evaluation of *ex vivo* monitoring of electric field distribution was done on a fresh chicken liver tissue. Temperature of the tissue was maintained at 4° C before the beginning of experiment when they were allowed to adjust to the room temperature. Tissues were sectioned in flat and cylindrical shaped samples with a diameter of 20 mm and placed in an acrylic glass container. Four needle electrodes measuring 1 mm in radius made of platinum-iridium alloy were inserted through the cover of the acrylic glass container into the phantom as shown in figure 1.

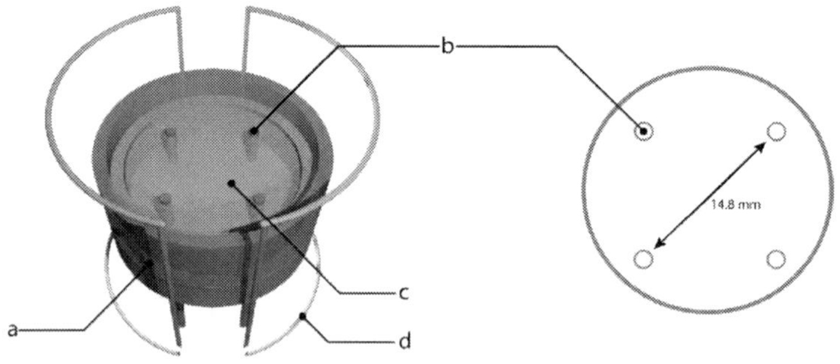

Figure 1: Container (a) used in the study was made of the acrylic glass with four holes for electrodes (b). Either agar phantom or liver tissue (c) was placed inside. The container was inserted in the 25 mm RF probe (d).

Electric pulse generator used for delivering electroporation pulses into the agar phantom and liver tissue was customized Cliniporator Vitae (IGEA, Carpi, Italy). The electric pulses with an amplitude of 1500 V were delivered between the diagonal electrodes of the imaging sample in sequences of four 100 μs long pulses separated by 100 μs intervals. Current density distribution and electrical conductivity were obtained in all experiments by the two-shot RARE current density magnetic resonance imaging sequence [16] and MREIT J-substitution algorithm [15], respectively. When both, current density distribution and electrical conductivity were obtained, Ohm's law was used for calculating electric field inside the imaging sample as described in [11].

3. Results and discussion

As shown in figure 2, current density and electric field distribution within agar and liver during application of electroporation pulses were obtained successfully.

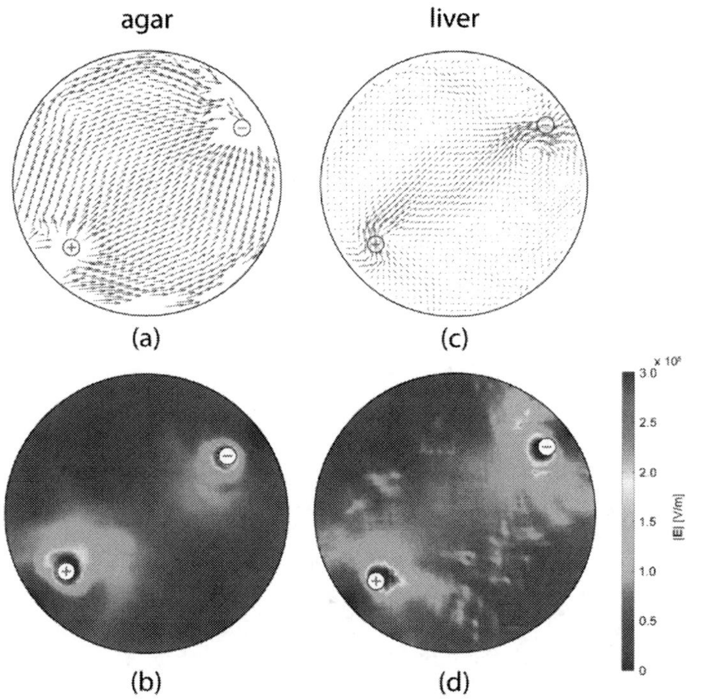

Figure 2: Electric current density distribution presented as a vector field (a and c) and electric field distribution (b and d) in agar phantom and liver tissue obtained by CDI and MREIT, respectively. Agar phantom and liver tissue were exposed to four 100 μs long electric pulses with an amplitude of 1500 V. Pulses were delivered between two needle electrodes (marked with + and −).

When we examined and compared results obtained in agar phantom and liver tissue we observed considerable differences, especially in the current density distributions (figure 2a and 2c). Current density distribution was uniformly distributed throughout the agar as dielectric properties of agar remained unchanged in spite of high electric field applied. On the contrary in the liver tissue, current density distribution was the highest between the electrodes where electrical conductivity was increased due to tissue changes associated to local tissue electroporation. Differences between electric field distributions (figure 2b and 2d) can be seen in pronounced alteration of field's intensity in the case of the liver tissue (figure 2d). We stipulate this is a consequence of a dynamic process during which electric field is changing due to increasing tissue conductivity. Similar alterations can also be observed in sequential numerical modelling of electroporation [17].

As there is a great need for *in situ* determination of electric field distribution during electroporation pulse delivery, magnetic resonance electrical impedance tomography together with current density imaging could be of significant help in monitoring of electroporation based applications. The proposed method has been demonstrated to be feasible for determination of electric field during electroporation [11,18] and therefore a next step towards *in vivo* experiments can be taken.

References

[1] Neumann E, Schaeferridder M, Wang Y and Hofschneider P 1982 Gene-Transfer into Mouse Lyoma Cells by Electroporation in High Electric-Fields *EMBO J.* **1** 841–5

[2] Mir L M and Orlowski S 1999 Mechanisms of electrochemotherapy *Adv. Drug Deliv. Rev* **35** 107–18

[3] Kotnik T, Pucihar G and Miklavčič D 2010 Induced Transmembrane Voltage and Its Correlation with Electroporation-Mediated Molecular Transport *J. Membrane Biol.* **236** 3–13

[4] Serša G and Miklavčič D 2008 Electrochemotherapy of Tumours *JoVE*

[5] Charpentier K P 2012 Irreversible electroporation for the ablation of liver tumors: are we there yet? *Arch. Surg.* **147** 1053–61

[6] Miklavčič D, Čorović S, Pucihar G and Pavšelj N 2006 Importance of tumour coverage by sufficiently high local electric field for effective electrochemotherapy *Eur. J. Cancer Suppl.* **4** 45–51

[7] Miklavčič D, Snoj M, Županič A, Kos B, Čemažar M, Kropivnik M, Bračko M, Pečnik T, Gadžijev E and Serša G 2010 Towards treatment planning and treatment of deep-seated solid tumors by electrochemotherapy *Biomed. Eng. Online* **9** 10

[8] Pavliha D, Kos B, Županič A, Marčan M, Serša G and Miklavčič D 2012 Patient-specific treatment planning of electrochemotherapy: Procedure design and possible pitfalls *Bioelectrochemistry* **87** 265–73

[9] Muftuler L, Hamamura M, Birgul O and Nalcioglu O 2006 In vivo MRI electrical impedance tomography (MREIT) of tumors *Technol. Cancer Res. T.* **5** 381–7

[10] Kos B, Županič A, Kotnik T, Snoj M, Serša G and Miklavčič D 2010 Robustness of Treatment Planning for Electrochemotherapy of Deep-Seated Tumors *J. Membrane Biol.* **236** 147–53

[11] Kranjc M, Bajd F, Sersa I and Miklavcic D 2011 Magnetic Resonance Electrical Impedance Tomography for Monitoring Electric Field Distribution During Tissue Electroporation *IEEE T. Med. Imaging* **30** 1771–8

[12] Joy M, Scott G and Henkelman M 1989 In vivo Detection of Applied Electric Currents by Magnetic-Resonance Imaging *Magn. Reson. Imag.* **7** 89–94

[13] Serša I, Jarh O and Demšar F 1994 Magnetic Resonance Microscopy of Electric Currents *J. Magn. Reson.* **111** 93–9

[14] Woo E J, Lee S Y and Mun C W 1994 Impedance tomography using internal current density distribution measured by nuclear magnetic resonance *SPIE* **2299** 377 – 385

[15] Khang H S, Lee B I, Oh S H, Woo E J, Lee S Y, Cho M H, Kwon O I, Yoon J R and Seo J K 2002 J-substitution algorithm in magnetic resonance electrical impedance tomography (MREIT): phantom experiments for static resistivity images *IEEE T. Med. Imaging* **21** 695–702

[16] Serša I 2008 Auxiliary phase encoding in multi spin-echo sequences: application to rapid current density imaging *J. Magn. Reson.* **190** 86–94

[17] Šel D, Cukjati D, Batiuskaite D, Slivnik T, Mir L M and Miklavčič D 2005 Sequential finite element model of tissue electropermeabilization *IEEE T. Biomed. Eng.* **52** 816–27

[18] Kranjc M, Bajd F, Sersa I, Woo E J and Miklavcic D 2012 Ex Vivo and In Silico Feasibility Study of Monitoring Electric Field Distribution in Tissue during Electroporation Based Treatments ed B Rubinsky *PLoS ONE* **7** e45737

Journal of Physics: Conference Series **434** (2013) 012087 doi:10.1088/1742-6596/434/1/012087

Revealing membrane potential by advanced impedance spectroscopy: theoretical and experimental aspects

M Gheorghiu, D Bratu, A Olaru, C Polonschii, E Gheorghiu
International Centre of Biodynamics, Intr. Portocalelor 1B, Bucharest, Romania

E-mail: egheorghiu@biodyn.ro

Abstract. In spite of recent advancement of novel optical and electrical techniques, availability of non-invasive, label-free methods to assess membrane potential of living cells is still an open issue. The theory linking membrane potential to the low frequency α dispersion exhibited by suspensions of spherical shelled particles (presenting a net charge distribution on the inner side of the shell) has been pioneered in our previous studies with emphasis on the permittivity spectra. We now report on both theoretical and experimental aspects showing that whereas α dispersion is related to a rather large variation exhibited by the permittivity spectrum the decrement presented by impedance magnitude spectrum is either extremely small, or occurs (for large cells) at very low frequencies (~mHz) explaining the lack of experimental bioimpedance data on the matter. Based on the microscopic model we indicate that an appropriate design of the experiment may enable access to membrane potential as well as to other relevant parameters when investigating living cells and charged lipid vesicles. We discuss the effect on the low frequency of permittivity and impedance spectra of: I. Parameters pertaining to cell membrane i.e. (i) membrane potential, (ii) size of the cells/vesicles, (iii) conductivity; II. Conductivity of the outer medium. A novel measuring set-up has recently been developed within the International Centre of Biodynamics allowing for sensitive low frequency (~10mHz) four point (bio)impedance assays. Its capability to test theoretical predictions is reported as well. The far reaching implications of this study applicability for life sciences (non-invasive access to the dynamics of relevant cell parameters) as well as for biosensing applications, e.g. assess the cytotoxicity of a wide range of stimuli, will be outlined.

1. Introduction

The dispersions exhibited by impedance spectra of biosystems over a wide frequency range (α, β, γ), as introduced by Schwan [1], enable thorough characterization of the electrical properties of biomaterials. Since mechanisms behind dispersions are often unclear or unknown, the dispersion grouping has been simply sorted according to the frequency range [2] with alpha dispersion revealed by bio-impedance spectra in the range below 1 Hz and up to 100 kHz.

Even though Gerhard Schwarz seminal paper on α dispersion of colloidal particles (related to displacement of counter-ions) was published five decades ago [3], assessment of living cell properties pertaining to α dispersion is still an open issue. Remarkably, such dispersion could basically be explained by different mechanisms described by microscopic models of cell systems, as the ones pioneered by us based either on shape effects (e.g. exhibited by clusters of interconnected cells [4, 5, 6]), or on displacement of counter-ions [7, 8]. Notably, studies relying on the same approaches (and deriving similar results) continue to be published [9, 10].

We now pinpoint the limitations/constraints and suggest solutions enabling measurement of α dispersion on suspensions of living cells, or liposomes consisting in charged lipids and derivation of the related membrane potential. Membrane potential is related to the charge asymmetry across membrane involving the charge distribution on the inner side of the membrane and the counter-ions on the external side which ensure cell/particle electrical neutrality. The effect of the charges on the inner side was implicitly considered, but not the effect of the fixed external ones. We present the expression of the complex permittivity of a suspension of spherical cells and the related effect on the permittivity and impedance spectra of: I. parameters pertaining to cell membrane i.e. (i) membrane potential, (ii) size of the cells/vesicles, (iii) conductivity of the membrane; II. conductivity of the outer medium (diffusion effect was discussed in [11]). The last section discusses ways to check the theory and the role of parameters pertaining both to cells or vesicles under investigation, and to suspending medium fostering measurable variations of low frequency, α dispersion exhibited by impedance spectra.

2. The model and numerical simulations

Aiming to describe the dielectric behavior of a suspension of spherical living cell, we analyze the effect of a uniform AC field of angular frequency ω (ω=2 π ν) and amplitude E_0 on a suspension of shelled spherical particles (of radius R_1) with a distribution of (negative) charge, $e_0\, n_{01}$, on the inner side of the shell (membrane), of thickness d [7,8]. External medium acts as a charge reservoir enabling counter-ions to migrate to the external side of the membrane to balance the charge from the inner side. Diffusion effects are assumed both in the inner and external/ suspending medium, but not in the membrane. Considering σ_k as conductivity, and ε_k as permittivity within medium k (k=1, 2, 3) the corresponding complex permittivity and conductivity is $\epsilon_k^* = \varepsilon_k + \sigma_k (i\omega)^{-1}; \sigma_k^* = i\omega \cdot \epsilon_k^*$. Solving the set of equations for potential one obtains the potential in the outer medium, [7]:

$$\varphi(r,\theta) = -E_0 \cdot r\cos\theta + \frac{E_0 \cdot H_n}{r^2}\cos\theta \tag{1}$$

Where θ and r are the spherical coordinates in a system whose origin is placed at the center of the particle and the polar axis has the direction of the applied field, H_n comprises the entire information on the electrical and geometric parameters of the cell prototype; $3\times H_n$ equals particle polarizability. Following the Maxwell-Wagner *equivalence* approach the complex permittivity of suspension is [7,8]:

$$\varepsilon_{sus}^* = \varepsilon_3^* \left(1 + p \cdot \frac{3 \cdot \dfrac{H_n}{(R_1+d)^3}}{1 - p \cdot \dfrac{H_n}{(R_1+d)^3}} \right); \; H_n = \frac{\dfrac{g}{h} \cdot \left(\sigma_2 + i\,\omega \cdot \varepsilon_2 \varepsilon_0 (1-Q) + i\,\omega \cdot \varepsilon_3 \varepsilon_0 \cdot Q \cdot \dfrac{\sigma_2}{\sigma_3} \right) + i\,\omega \cdot e_0 \cdot NN \cdot (R_1+d) \cdot (1-Q) - \sigma_3^*}{\dfrac{g}{h} \cdot \left(\sigma_2 + i\,\omega \cdot \varepsilon_2 \varepsilon_0 (1+2Q) - 2\,i\,\omega \cdot \varepsilon_3 \varepsilon_0 \cdot Q \cdot \dfrac{\sigma_2}{\sigma_3} \right) + i\,\omega \cdot e_0 \cdot NN \cdot (R_1+d) \cdot (1+2Q) + 2\sigma_3^*} \cdot (R_1+d)^3; \tag{2}$$

Where p denotes volume concentration of the suspended particles, and:

$$NN = \frac{e_0 \cdot n_{01}}{k_B T} \frac{R_1^2}{(R_1+d)^2} \left(1 + i\omega \frac{(R_1+d)^2}{2 \cdot D_3} \right)^{-1}; \; \delta = \left(\frac{R_1}{R_1+d} \right)^3; \; G_k = \left(\frac{\sigma_k^*}{D_k \varepsilon_k \varepsilon_0} \right)^{-2}; \; k = 1,3; \tag{3}$$

$$g = (1+2\delta)\sigma_1\sigma_1^* + 2(1-\delta)\left(i\omega \cdot \sigma_1 \varepsilon_2 \varepsilon_0 (1+S) + \sigma_2 (\sigma_1 - i\omega \cdot \varepsilon_1 \varepsilon_0 S)\right); \; h = (1-\delta)\sigma_1\sigma_1^* + (2+\delta)\left(i\omega \cdot \sigma_1 \varepsilon_2 \varepsilon_0 (1+S) + \sigma_2 (\sigma_1 - i\omega \cdot \varepsilon_1 \varepsilon_0 S)\right);$$

$$S \approx \frac{\sigma_1 \cdot (G_1 R_1 - 1)}{i\omega \cdot \varepsilon_1 \varepsilon_0 \cdot \left((G_1 R_1)^2 - 2(G_1 R_1 - 1) \right)}; \; Q = \frac{\sigma_3}{i\omega \cdot \varepsilon_3 \varepsilon_0} \cdot \frac{1 + G_3 \cdot (R_1+d)}{G_3^2 \cdot (R_1+d)^2 + 2 \cdot \left(1 + G_3 \cdot (R_1+d)\right)}$$

T represents the absolute temperature, and k_B the Boltzmann constant.

2.1. Effect of cell membrane parameters on the permittivity and impedance spectra

The surface charge distribution n_{01} is related both to membrane potential and to the complex permittivity ε_{sus}^* of the suspension via H_n in equation (2) through the term NN, equation (3), [11].

We consider a suspension of shelled particles with $R_1 \sim 2$ μm ($\sigma_1 \sim 0.2$ S m^{-1}, $\sigma_3 \sim 0.37$ S m^{-1}, $\varepsilon_2 \sim 12$, $\sigma_2 = 10^{-6}$ S m^{-1}, $d \sim 10^{-2}$μm, $D_1 \sim 2\times10^{-10}$ m^2 s^{-1}, $D_3 \sim 2\times10^{-9}$ m^2 s^{-1}, p=0.02) with membrane potential: -150 mV, -75 mV and 0V respectively (figures 1 A, B). Impedance spectra are derived based on ε_{sus}^* using:

Z^*_{sus} = GF $(i\omega \cdot \varepsilon^*_{sus})^{-1}$ where GF represents the Geometric Factor of the measurement chamber. Figure 1 A shows the permittivity spectra and the dependency of α dispersion on membrane potential. However, the spectra of impedance magnitude relative to the value at 1 kHz (the level prior to β dispersion) reveal (figure 1 B) very small decrements ($\Delta Zr \leq 5\ 10^{-3}$ %) related to α dispersion. The same challenge is met for phase variation within α dispersion, $\Delta\theta \leq 2\times10^{-3}$ degree, raising hard experimental constraints.

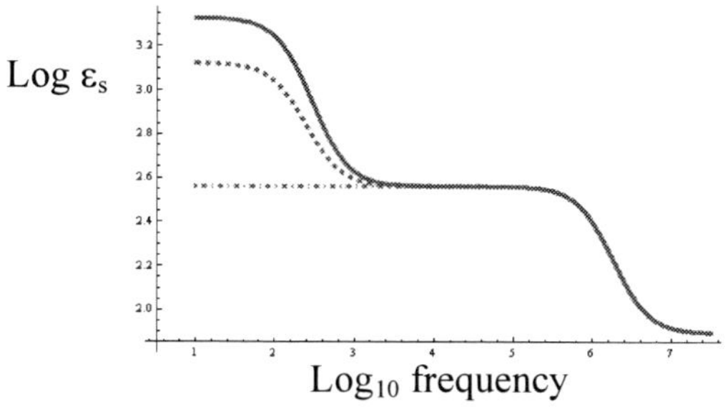

Figure 1A Permittivity spectra:$\Delta\varphi_0$=-150mV (solid), $\Delta\varphi_0$=-75mV(dashed), $\Delta\varphi_0$ ~0V(dotted)

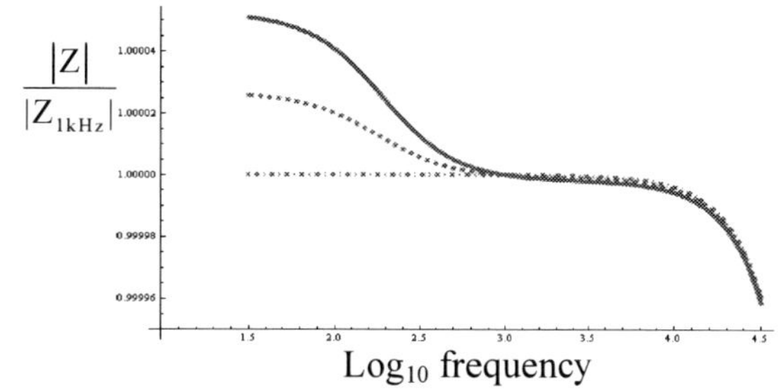

Figure 1B. Impedance magnitude spectra relative to the value at 1 kHz: $\Delta\varphi_0$=-150mV (solid), $\Delta\varphi_0$ =-75mV (dashed) & $\Delta\varphi_0$~0V (dotted)

When considering suspensions of larger cells (R_1 ~ 0.5 mm, as for the case of *Xenopus laevis* oocytes), impedance variation significantly enhances, but α dispersion occurs at lower frequencies (<10 mHz) which considerably increases measurement time and experimental constraints (due to electrode polarization, probe stability etc.). Figure 2 (A, B) reveals the dependency of α and β dispersions on membrane conductivity for all parameters as before, but R_1 ~ 0.5 mm ($\Delta\varphi_0$=-150mV).

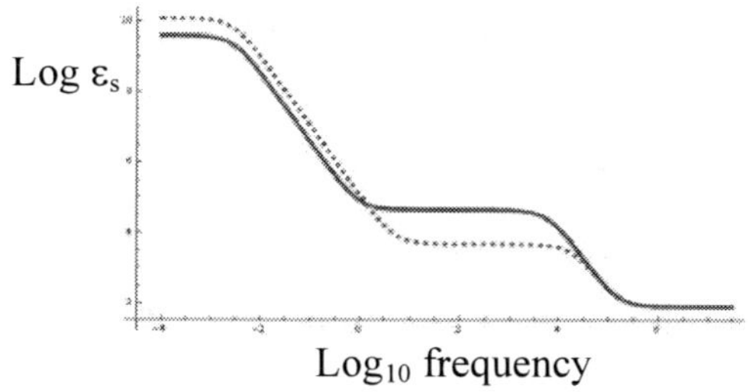

Figure 2.A Permittivity spectra of *oocytes,* σ_2=10^{-6}S m^{-1} (solid); σ_2=10^{-5}S m^{-1} (dashed)

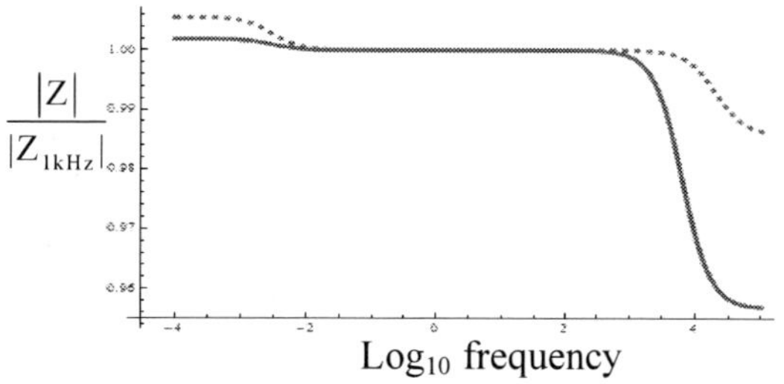

Figure 2.B Impedance magnitude spectra of *oocytes* relative to 1kHz, σ_2=10^{-6}S m^{-1} (solid), σ_2=10^{-5}S m^{-1} (dashed)

Figure 2A presents the permittivity spectra, whereas figure 2 B illustrates the spectra corresponding to impedance magnitude. Notably, impedance variation related to α dispersion relative to its value at 100 Hz (level prior to β dispersion), increases to ΔZr ~0.5%. Therefore, we stress on the possible biosensing applications related to assessment of pore formation effect of various stimuli by impedance spectroscopy when considering large cells, or (charged) vesicles.

3. Experiments

A dedicated front end enabling low frequency (≥10mHz) four point measurements using a 1260 Solartron impedance analyzer was developed in house, such as to control the low frequency polarization effects. The system was used to derive the impedance spectra of suspensions of living cells, *Xenopus laevis oocytes* (R_1 ~ 0.5mm, p ~ 0.3) and POPG liposomes (R_1 ~ 5μm, p ~ 0.1) as shown in figure 3, in both cases σ_3~ 0.03 S/m. The measurement time for each spectrum spanned ~ 3 minutes. Notably, for measuring α dispersion on *oocytes* the frequency spectrum should start at an even lower frequency ≤ 1mHz (as shown in figure 2B), as for POPG liposomes, the diffusion of the outer medium D_3 should be decreased towards ~10^{-13} m^2/s [11].

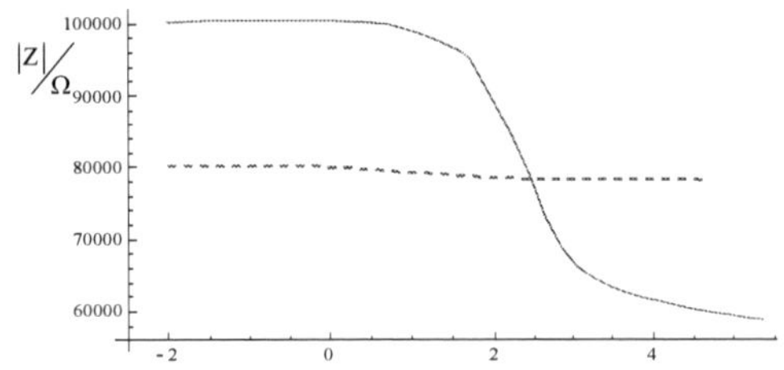

Figure 3.A Impedance magnitude spectrum *oocytes* (solid) and POPG liposomes (dashed)

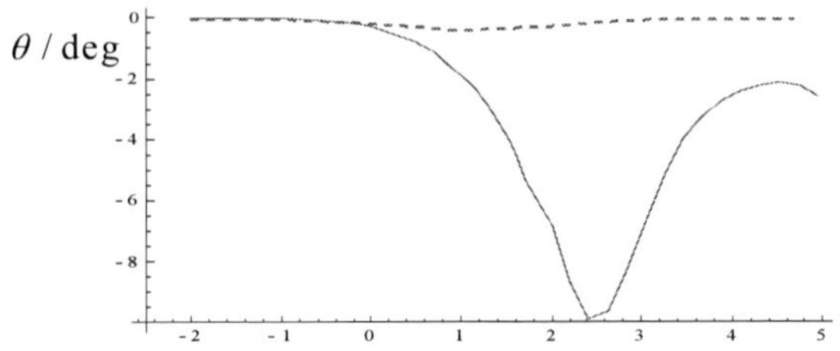

Figure 3.B Phase spectrum: *Xenopus laevis oocytes* (solid) and POPG liposomes (dashed)

4. Discussion

This study emphasises the dependency of α dispersion (related to displacement of counter-ions) on membrane potential and on several parameters pertaining to membrane and to suspending media, highlighting a set of experimental constraints for measuring impedance spectra exhibiting both α and β dispersions in suspensions of living cells or charged liposomes. The dependency on membrane potential is clearly shown by permittivity spectra, whereas impedance spectra reveal very tiny variations pertaining to α dispersion. We show that larger size of the cells/vesicles is accompanied by an increased variation of the impedance; however, the dispersion shifts towards lower frequencies, a domain more challenging experimentally (due to electrode polarization and specimen stability). A slight increase of membrane conductivity is accompanied by an increase of α dispersion in both permittivity and impedance spectra. Thus, possible biosensing application of bioimpedance assays are related to assessment of pore formation effect of various stimuli (e.g. toxins, membrane disrupting peptides etc.).

Advances of the experimental set-up and appropriate choice of both type of cells (or liposomes) and of parameters of the suspending medium could lead to detailed interrogation and improvement of the theoretical model, as well as open new applications in biomedicine and related fields.

5. **Acknowledgment:** we acknowledge support of grant BIOSCOPE (PN-II-ID-PCCE, C#11/2012).

References

[1] Schwan H P 1957 *In: Advances in Biological and Medical physics* vol 5 ed Lawrence J H and Tobias C A (New York Academic Press) p 147
[2] Grimnes S, Martinsen ØG 2010 *J. Physics: Conference Series* **224** 012073
[3] Schwarz G 1962 *J. Phys. Chem.* **66** 2636
[4] Vrinceanu D, Gheorghiu E 1996 *Bioelectrochem Bioenerg* **40** 167
[5] Gheorghiu E, Balut C, Gheorghiu M 2002 *Phys. Med. Biol.* **47** 341
[6] Sandu T, Vrinceanu D, Gheorghiu E 2010 *Phys. Rev. E* **81** 2 0219131
[7] Gheorghiu E 1993 *Phys. Med. Biol* **38** 979
[8] Gheorghiu E 1994 *J. Phys. A: Math. Gen.* **27** 3883
[9] Prodan E, Prodan C, Miller JH , 2008 *Biophys. J.* **95** 4174
[10] Di Biasio A, Ambrosone L, Cametti C 2010 *Biophys. J.* **99** 163
[11] Gheorghiu E, 2011 *J Electr Bioimp.* **2** 93

Investigation of a Bubble Detector based on Active Electrolocation of Weakly Electric Fish

M Mohan[1], K Mayekar[2], R Zhou[1], G von der Emde[2] and H Bousack[1,*]

[1] Forschungszentrum Jülich, Peter Grünberg Institute PGI-8, Jülich, Germany

[2] University of Bonn, Institute of Zoology, Bonn, Germany

E-mail: * h.bousack@fz-juelich.de

Abstract. Weakly electric fish employ active electrolocation for navigation and object detection. They emit an electric signal with their electric organ in the tail and sense the electric field with electroreceptors that are distributed over their skin. We adopted this principle to design a bubble detector that can detect gas bubbles in a fluid or, in principle, objects with different electric conductivity than the surrounding fluid. The evaluation of the influence of electrode diameter on detecting a given bubble size showed that the signal increases with electrode diameter. Therefore it appears that this detector will be more appropriate for large sized applications such as bubble columns than small sized applications such as bubble detectors in dialysis.

1. Introduction

The detection of gas bubbles in fluids is an important issue in areas such as medical technology or process engineering, for example during dialysis, infusions or two-phase flows. For these purposes, different measuring principles are used such as ultrasound, along with capacitive, thermal, optical or impedance sensors.

The sensor presented here is partly based on the principle of active electrolocation in the weakly electric fish, *Gnathonemus petersii,* and a simplified process tomography. Weakly electric fish orient in their environment and locate prey by means of an electrical field emitted from their tail as an electrical organ discharge. Objects with electrical properties other than those of the surrounding water distort the generated field. The fish perceives these distortions using about 3000 electroreceptor organs distributed over a large area of the its skin. Thus, an object within the range of the electrical field projects an electrical image onto the skin and the fish can determine object parameters, such as material, size, shape and distance [1]. Electrical impedance tomography (EIT) [2] uses a circular array of evenly spaced electrodes in a single plane of a vessel or several planes axially arranged. A pair of electrodes is connected with a sinusoidal current source and the resulting potential difference is measured between all electrode pairs. For the determination of the spatial distribution of the electrical conductivity above the plane surface based on the measured potential differences, an inverse problem must be solved. Hence, both *Gnathonemus petersii* and EIT use the spatial distribution of the electrical conductivity as an electrical image to detect and characterize objects.

2. Principle of the bubble detector

The method presented here uses two ring shaped electrodes arranged axially in a tube, see Fig. 1. Between the sending electrode (1 V, 1 KHz) and the receiving electrode an electrical field is

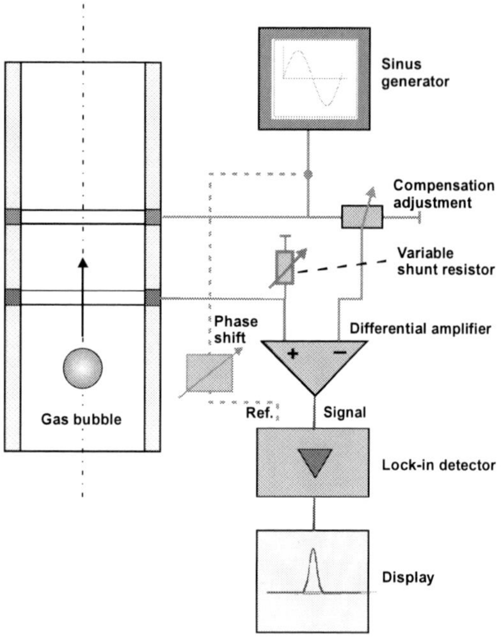

Figure 1. Principle of the measurements.

generated, which is disturbed when an object, e.g. a gas bubble, passes through the rings. Similar to the fish and in contrast to EIT, the potential difference measured at the receiving electrode is processed by a differential amplifier that eliminates the signal sent. A lock-in detector, used as a narrowband amplifier tuned to the sending frequency, extracts the signal from the noisy environment [3]. Thus, a time-dependent signal of the object is received, averaged over the cross section of the tube.

3. Experiments

For an easy modification of the electrode diameter and distance, the electrodes were manufactured from silver plated wires with a diameter of 0.7 mm. Both wire electrodes were attached to a holder that allows an adjustment of the electrode distance. To ensure a geometrical similarity in all electrode sets, the distance between the electrodes was kept equal to the electrode diameter. Based on this, electrode sets with diameters of 14, 18, 26 and 33 mm were used in the experiments. The upper curve in figure 2 shows a time series measurement of bubbles with an increasing baseline. This increase is due to ambient temperature changes that influence the conductivity of the fluid. In addition, peaks due to high frequency interference can occur. For the removal of a low frequency drift, the airPLS function in MATLAB was used as a baseline correction. The signal was further smoothened by using a 2nd order Butterworth low pass filter with a cutoff frequency of 0.17 Hz before the peak analysis of the signals was carried out. The filtered time series is shown in the lower diagram in figure 2. Based on the filtered time series, a histogram and a best-fitting normal distribution was calculated, which allows assigning a mean and a standard deviation for each measurement. Using this procedure, all electrode sets were measured and evaluated.

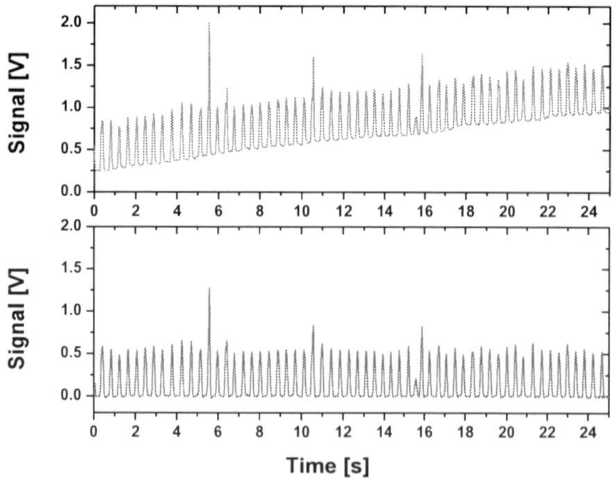

Figure 2. Time series of a bubble measurement. Above: raw data, below: filtered data. Parameters of the measurement: electrode diameter: 18 mm, electrode distance: 18 mm, air bubble diameter: 1mm, conductivity water: 80 μ S/cm.

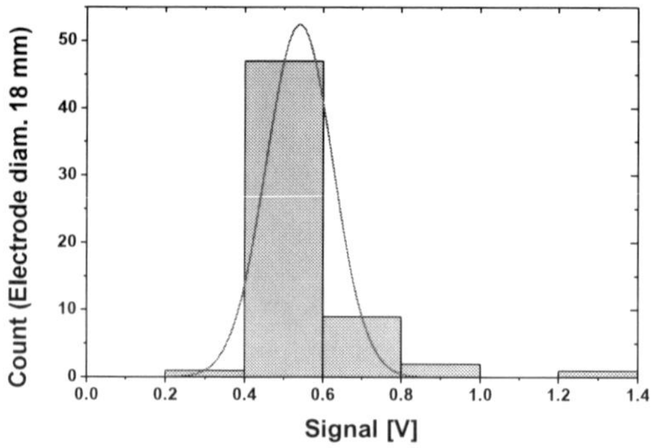

Figure 3. Histogram with superimposed best-fitting normal distribution for the time series shown in Fig. 2. The mean is 0.57 V, the median is 0.55 V and the standard deviation is 0.12 V. The diameter of the electrode is 18 mm and the diameter of the air bubble is 1 mm. Parameters of the measurement see Figure 2.

For control of reproducibility, the measurements were repeated several times on different days. It turned out that the reproducibility was not as good as expected. Several reasons are responsible for this: a) In general, the formation of a gas bubble detaching from a nozzle in a liquid is not stable, because the bubble surface tends to vibrate, and sometimes the bubble splits in two parts. b) During its rise, the bubble is flattening and spirals upward inducing turbulent water flow for the following bubbles. c) Due to the spiral path of the bubble and an unevenness of the electrical field as function of electrode radius, even two absolutely similar bubbles will generate slightly different signals. d) The mechanical stability of the fragile wire electrodes is insufficient, and the assembly of the same electrode set-up will result in a little different performance. The reasons a) to c) are system-inherent. Because of the real bubble behavior, different signals must be expected. Only the reason d) can be reduced by using more stable electrodes.

The reason a) was investigated with respect to the bubble formation. The tubing pump used (ISMATEC Reglo Digital MS-4/8, Idex Health & Science, Oak Harbor, USA) allows to produce a continuous stream of bubbles or a single bubble on demand, which may result in more homogenous bubbles. Figure 4 shows a comparison of the mean of the bubble signal with standard deviation as a function of the different electrode sets using the continuous bubble flow and a series of single bubbles. Except for the electrode diameter of 33 mm, the signals of the different measurements are more or less the same, although the differences of the standard deviations of the two measurements with continuous bubble flows are remarkable. Here, the series of single bubbles appears to yield smaller deviations. A phase shift adjustment (see Fig. 1) can compensate the phase shift between the receiving and the sending electrode. For this reason, the phase shift adjustment influences the measured signal quality. In fig. 5, the influence of the phase shift is shown using a linear fit with the standard deviation as weight. Obviously, the consideration of the phase shift resulted in reduced signals and in a significantly lower standard deviation.

In summary, it can be concluded that the signal of the bubble detector increases almost linearly with the electrode diameter. This result is surprising because during the experiments, the bubble size stayed constant for the different electrode diameters, which lead to a decreasing relationship of the bubble cross section and the electrode cross section. Due to the larger field distortions, smaller electrode diameters should yield larger signals.

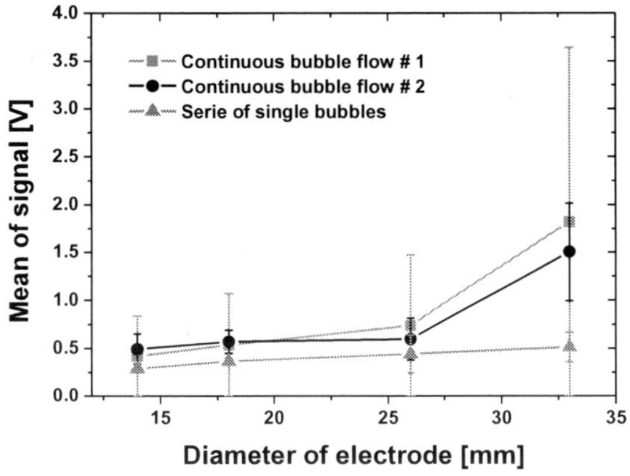

Figure 4. Comparison of the mean signals of continuous bubble flow and a series of single bubbles as function of the electrode diameter. The standard deviation is indicated. Parameters of the measurement: air bubble diameter: 1mm, conductivity water: 80 μ S/cm, phase shift φ = 0°

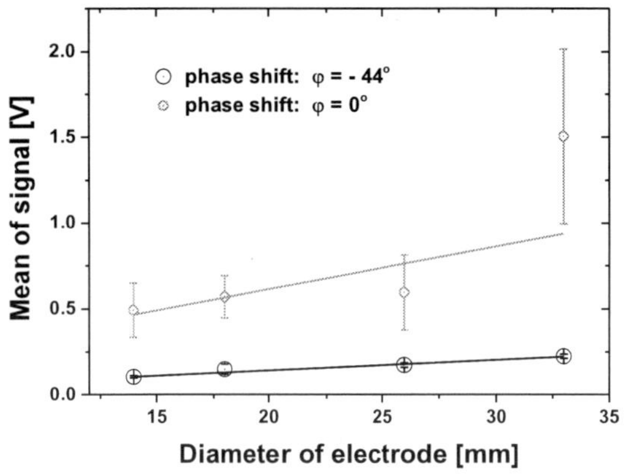

Figure 5. Comparison of the mean signals of continuous bubble flow with different phase shift. The standard deviation is indicated. Parameters of the measurement: see Fig. 4.

Journal of Physics: Conference Series **434** (2013) 012088 doi:10.1088/1742-6596/434/1/012088

4. Simulation

For the purpose of securing the principal linear course of the signal as a function of the electrode diameter, a finite element (FEM) analysis was carried out. Figure 6 shows the calculated field distribution of the electric potential with and without a bubble. It is obvious that for lower distances of the electrodes, the radial field distribution is inhomogeneous, which causes different signals depending on the radial position of the bubble. Furthermore it is evident that the field deformation is very small. Using this model, the diameters of the electrodes were varied between 15 mm and 40 mm, while all other parameters stayed constant. In figure 7, the calculation of the current in the shunt resistor (Fig. 1) resulted in a linear dependence of the current from the diameter of the electrodes. This confirmed the experimental result, which suggested a linear dependence.

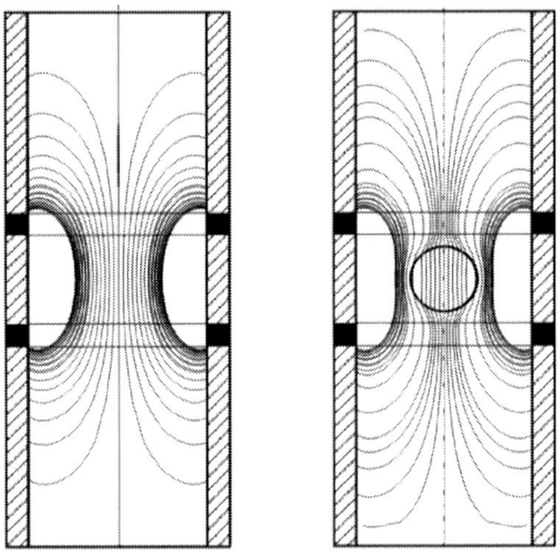

Figure 6. Field distribution of the electric potential without air bubble (left) and with air bubble (right). Parameters see Fig. 7

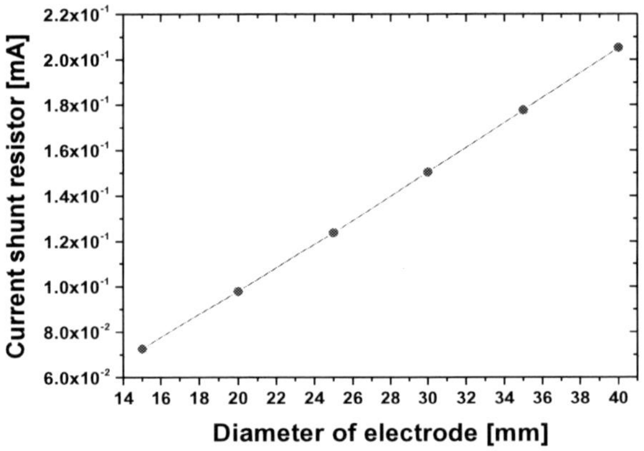

Figure 7. Simulation of the current in the shunt resistor as function of the electrode diameter. Parameters are: distance electrodes 10 mm, conductivity water 5 μS/cm, air bubble diameter 10 mm

References

[1] von der Emde G *et al* 2010 3-Dimensional Scene Perception during Active Electrolocation in a Weakly Electric Pulse Fish *Frontiers in behavioral neuroscience* **4** 26
[2] York T 2001 Status of electrical tomography in industrial applications *J Electron Imaging* **10** 608
[3] Bousack H *et al* 2012 patent application Blasendetektor Forschungszentrum Jülich and Universität Bonn

In situ characterization of fouling in reverse osmosis membranes using electrical impedance spectroscopy

Terry Chilcott[b], Alice Antony[a] Hans Coster[b], Greg Leslie[a]

[a]UNESCO Centre for Membrane Science and Technology, The University of New South Wales, Sydney 2052, Australia

[b]School of Chemical and Biomolecular Engineering, The University of Sydney, Sydney 2006, Australia

terry.chilcott@sydney.edu.au

Abstract. Analytical solutions of the Nernst-Planck, Poisson and continuity equations for a membrane undergoing reverse osmosis in a cross-flow system reveal that the flow of alternating ionic charge induced in the membrane during impedance measurements is actively assisted by the flow of water. The actively driven current manifested "inductive" responses in impedance measurements of a Filmtec BW30 reverse osmosis membrane mounted in an Inphaze flat-bed cross-flow module after 16 hours of filtering a mineral salt solution seeded with $CaCl_2$ and $NaHCO_3$ at pressure of 900 kPa. Fitted transfer functions resolved conduction and capacitive properties of four membrane layers, diffusion/concentration phenomenon and a pseudo "inductor" shunted by a conductor. A 10-fold decrease in the shunt conductance correlated with smaller increases in the conductance values for the filtrate and membranous layers, and the onset of fouling diagnosed by a rapid increase in flux decline.

1. Introduction

Membranes separate solutions of differing solute concentrations and thus are subject to differences in osmotic pressures arising from the membrane regulating solute and solvent flows that would otherwise allow the solutions to equilibrate. Extracellular "skeletons" in cellular systems and polymer composites in synthetic industrial systems elastically absorb osmotic as well as environmental and applied pressures that would otherwise rupture the membrane. So supported, membranes can sustain substantial pressure differences capable of reversing the direction of osmotically driven solvent flows whilst inhibiting solute flows. Such occurs in reverse osmosis membranes used in industry for desalination of seawater and recovery of water from industrial waste. Forward osmosis is an important function of ultrafiltration membranes used in hemodialysis treatments of the blood of patients suffering loss of kidney function.

A general deficiency of membrane systems is the accumulation of material and in some cases the eventual precipitation of soluble materials such as $CaSO_4$ or $CaCO_3$, on the membrane surface. If not detected and removed in time, this can cause irreversible fouling of the membrane.

Presented here is an *in situ* electrical impedance characterization of a reverse osmosis system that is consistent with a theoretical analysis demonstrating the detectability of membrane fouling.

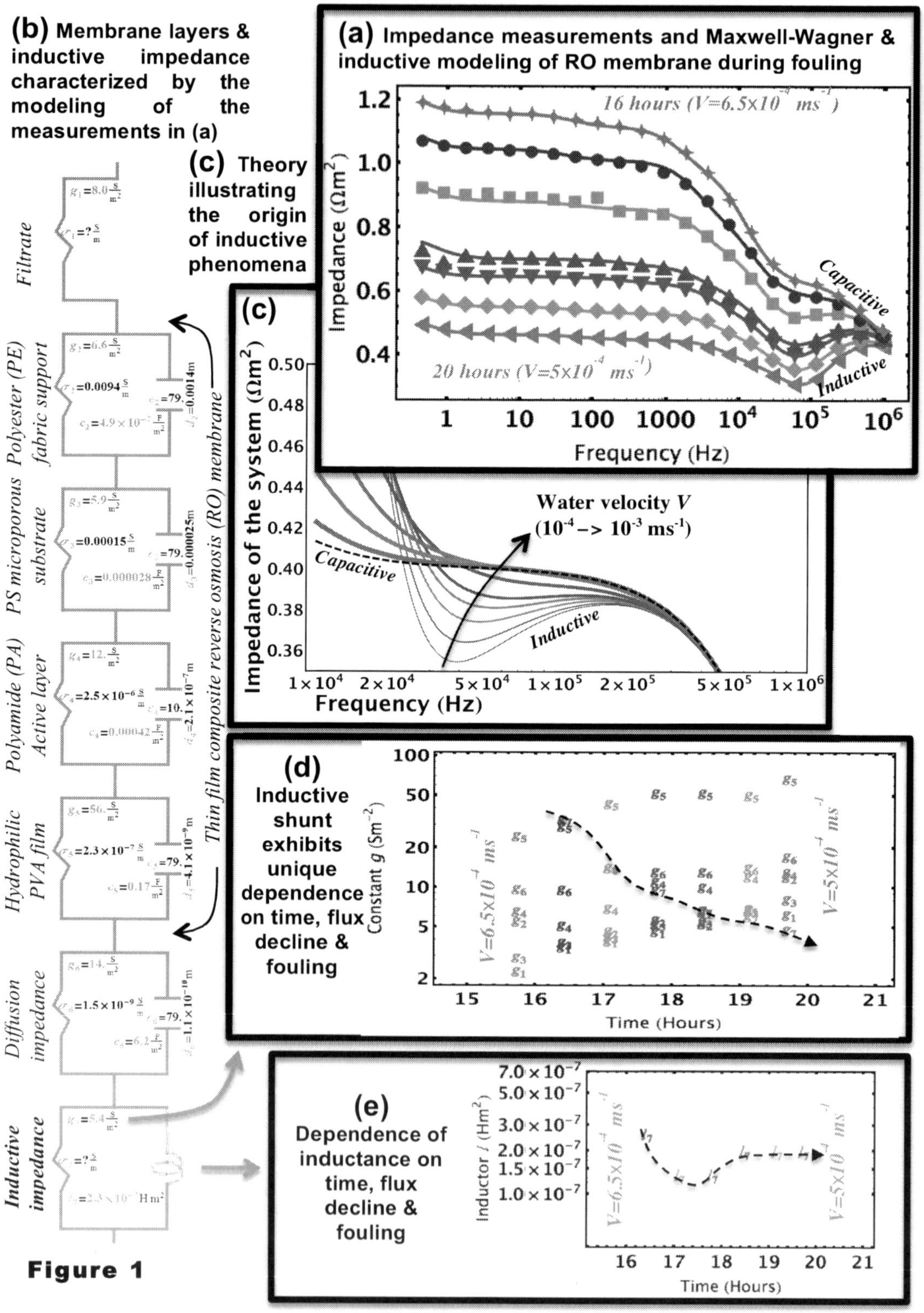
(b) Membrane layers & inductive impedance characterized by the modeling of the measurements in (a)
(a) Impedance measurements and Maxwell-Wagner & inductive modeling of RO membrane during fouling
16 hours (V=6.5x10-4 ms-1)
20 hours (V=5x10-4 ms-1)
Capacitive
Inductive
Impedance (Ωm²)
Frequency (Hz)
(c) Theory illustrating the origin of inductive phenomena
Impedance of the system (Ωm²)
Water velocity V (10-4 -> 10-3 ms-1)
Filtrate
Polyester (PE) fabric support
PS microporous substrate
Polyamide (PA) Active layer
Hydrophilic PVA film
Diffusion impedance
Inductive impedance
Thin film composite reverse osmosis (RO) membrane
(d) Inductive shunt exhibits unique dependence on time, flux decline & fouling
Constant g (Sm-2)
Time (Hours)
(e) Dependence of inductance on time, flux decline & fouling
Inductor l (Hm²)

Figure 1

2. Theory

Analytical solutions of the Nernst-Planck, Poisson and continuity equations [1] for a planar membrane of thickness λ, conductivity σ and dielectric permittivity ε, mounted in a cross-flow system separating a feed solution of conductivity σ_{feed} and a filtrate solution of surface conductivity $\sigma_{filtrate}$ yield the following expressions for the area specific impedance of the membrane in SI units of Ωm^2, i.e.;

$$z_{membrane} = \frac{\lambda}{\sigma + j\omega\varepsilon} + z_V \quad (1)$$

where ω is the angular frequency in units of s^{-1} and

$$z_V \equiv \frac{x_m\left[\frac{2x_m}{\lambda}\sinh\left(\frac{\lambda}{2x_m}\right)\right] - x_p\left[\frac{2x_p}{\lambda}\sinh\left(\frac{\lambda}{2x_p}\right)\right]}{\left(\frac{\left[\frac{2x_m}{\lambda}\sinh\left(\frac{\lambda}{2x_m}\right)\right] + \left[\frac{2x_p}{\lambda}\sinh\left(\frac{\lambda}{2x_p}\right)\right]}{1 - \frac{\sigma_{feed} + j\omega\varepsilon_{water}}{\sigma_{filtrate} + j\omega\varepsilon_{water}}} - \frac{x_m \exp\left(\frac{\lambda}{2x_m}\right) - x_p \exp\left(-\frac{\lambda}{2x_p}\right)}{\lambda}\right)(\sigma + j\omega\varepsilon)} \quad (2)$$

in which the Debye and characteristic lengths are defined in units of meters so;

$$\lambda_D \equiv \sqrt{\frac{\varepsilon k T}{2q^2 C}}; \qquad x_m \equiv \frac{2\lambda_D^2 \sigma}{\sqrt{V^2\varepsilon^2 + 2\lambda_D^2\sigma(\sigma + j\omega\varepsilon)} - V\varepsilon}; \qquad x_p \equiv \frac{2\lambda_D^2 \sigma}{\sqrt{V^2\varepsilon^2 + 2\lambda_D^2\sigma(\sigma + j\omega\varepsilon)} + V\varepsilon} \quad (3)$$

V is the velocity of water in units of ms^{-1} (as distinct from the cross-flow velocity), C the ionic concentration in the membrane (m^{-3}), k the Boltzmann constant (1.38×10^{-23} J/K), T the temperature (300 K) and q the magnitude of the electronic charge (1.602×10^{-19} C). The first term in Equation (1) derives exclusively from conductive, dielectric and geometrical properties of the membrane as described by a Maxwell-Wagner layer of conductance and capacitance per area given by, respectively;

$$g = \frac{\sigma}{\lambda} \qquad \text{and} \qquad c = \frac{\varepsilon}{\lambda} \quad (4)$$

The second term, z_V, derives from the flow of alternating ionic charge induced in the membrane during impedance measurements that will be actively driven by the flow of water. Equations (2) and (3) reveal the dependency of this term on the velocity V of the water and in particular on σ_{feed} and $\sigma_{filtrate}$ at the respective membrane surfaces. σ_{feed} is characteristically greater than $\sigma_{filtrate}$ and σ, which effectively eliminates the contribution of the feed as a Maxwell-Wagner layer from the impedance measurements. Thus the impedance of this simplified membrane system is essentially given by;

$$z_{system} \approx \frac{\lambda}{\sigma + j\omega\varepsilon} + z_V\left(\sigma_{feed}\right) + \frac{\lambda_{filtrate}}{\sigma_{filtrate} + j\omega\varepsilon_{water}} \quad (5)$$

where the dependency of z_V on σ_{feed} at the membrane surface, where ions will concentrate and promote precipitation, potentially provides a means of monitoring the degree of fouling in the system.

3. Materials and Methods

Filtration experiments were performed in an Inphaze stainless steel flat-bed cross-flow filtration module that facilitates simultaneous measurements of impedance and water velocity as described in detail by Antony et al [2]. The feed and filtration channels in the module are 310 mm long, 60 mm wide and 0.80 mm thick and separated by a supported membrane of operational area of 0.0124 m^2. The opposing faces of the channels contain stainless steel electrodes of comparable area to the membrane, which were used to inject a range of electrical current stimuli of differing frequency but uniform density through the feed, membrane and filtrate system. Insulated small voids located centrally in these electrodes, support voltage-sensing electrodes that monitored the voltage responses of the system to the stimuli, independently of the strongly frequency-dependent responses of the ionic double layers that form between the current-injecting electrodes and solutions. The amplitudes and phase differences of stimuli and responses were measured with an impedance spectrometer (Inphaze Pty Ltd Australia).

The Filmtec BW30 reverse osmosis membranes were preconditioned and compacted in deionised water (Millipore, Australia) for 24 hours until a stable permeate flux of deionized water was reached. Filtration, water flow and impedance measurements were performed simultaneously using a feed seeded with $CaCl_2$ and $NaHCO_3$ at, cross-flow velocity of 0.2 ms^{-1}, 20°C and 900 kPa.

4. Results and Discussion

The magnitudes and phases of impedances of frequencies from 1 to 10^6 Hz were acquired using an Inphaze spectrometer every 15 minutes after 15 hours of filtration. Transfer functions of the form;

$$z_{TF} = \frac{1}{g_1} + \sum_{k=2}^{5} \frac{1}{g_k + j\omega c_k} + \frac{1}{g_6 + j\omega c_6} + \frac{1}{g_7 + 1/j\omega l_7} \quad (6)$$

were fitted to the mean and standard deviation values of every three consecutive spectra. Reduced-χ^2 statistics [3] provided justification for the fits which are shown in Figure 1(a).

The transfer function is a general form of the Maxwell-Wagner model used to characterize spectra over the first 16 hours of filtration [2] (e.g. see 1st spectrum in Figure 1(a)). In common are the 1st term attributed principally to the filtrate, the next four terms (k=2...5) attributed to sub-structural layers of the membrane as assigned in Figure 1(b) and the 6th term attributed to concentration/diffusion polarization effects that exhibit a similar dependency on frequency to that of a Maxwell-Wagner layer [4]. The 7th term exhibits a different frequency dependency to that of a Maxwell-Wagner layer, which becomes discernable in spectra acquired after 16 hours in Figure 1(a) as positive slopes in the impedance dispersions with frequency in the range of 10^5-10^6 Hz. This type of dispersion has been reported in other membrane systems [2,5,6,7,8] and described as pseudo "inductive" phenomenon.

Figure 1(c) reveals similar "inductive" phenomena in theoretical dispersions of Equation (5) for a membrane of comparable bulk electrical properties covering ranges of frequency and water flux V encountered experimentally and a value for σ_{feed} reflecting fouling conditions at the membrane surface diagnosed by a by a rapid increase in flux decline. The generally consistency of theory with experiments suggests that the "inductive" behavior is attributable to the term z_V in Equation (5), which Equation (2) reveals exhibits different dependencies on frequency to that of a Maxwell-Wagner layer.

Whilst the capacitance values in Equation (6) did not vary significantly from the values shown in Figure 1(b), the time-courses of the conductance values shown in Figure 1(d) reveal that conductance values for the filtrate, membranous layers as well as the diffusion/polarization impedance generally increased with time as the flux declined. However, the most sensitive variation with increasing flux decline was the 10-fold decrease in values for g_7, the conductor shunting the inductor l_7.

5. Conclusions

Pseudo "inductive" phenomena in impedance measurements of membrane systems can arise from alternating ionic charge induced by these measurements that is actively driven by flows of water established by applied and osmotic pressures. Characterizations of the phenomena can quantify properties of the system that are potentially predicative of membrane fouling. Further detailed theoretical analyses of the origins of the phenomena can improve predictability.

We thank Water Quality Research Australia and Water Research Foundation for financial support as well as the Mark Wainwright Analytical Center for access to the Inphaze spectrometer.

References

[1] Chilcott TC, Antony A, Coster HGL and Leslie G 2013 Inductive responses induced in membrane systems by osmosis *ICEBI and EIT Conference Proceedings*
[2] Antony A, Chilcott TC, Coster HGL and Leslie G 2013 *J. Membr. Sci.* **425-426** 89-97
[3] Kenney JF and Keeping ES 1951 Mathematics and statistics 2nd edition "Princeton, NJ: Van Nostrand"
[4] Coster HGL and Chilcott, TC 1999 Surface Chemistry and Electrochemistry of Membranes "T.S. Sørensen (Ed.), Marcel Deker, Inc., New York" 749-792
[5] Alcaraz A, Holdik H, Ruffing T, Ramírez P and Mafé S 1998 *J. Membr. Sci.* **150** 43-56
[6] Zhao K, Yasuhiro M, Asaka K, Asami K and Hanai T 1991 *J. Membr. Sci.* **64** 163-172
[7] Göhr H, Mirnik M, Schiller C.A. 1984 *J. Electroanal. Chem.* **180** 273
[8] Sluyters-Rehbach M and Sluyters J.H. (1984) AC techniques, in: Comprehensive Treatise of Electrochemistry "Plenum Press, New York"

Dynamic assessment of Amyloid oligomers – cell membrane interaction by advanced impedance spectroscopy

M Gheorghiu, S David, C Polonschii, D Bratu and E Gheorghiu.

International Centre of Biodynamics, Intr. Portocalelor 1B, 060101 Bucharest, Romania, www.biodyn.ro

e-mail: mgheorghiu@biodyn.ro

Abstract. The amyloid β (Aβ) peptides are believed to be pivotal in Alzheimer's disease (AD) pathogenesis and onset of vascular dysfunction. Recent studies indicate that $A\beta_{1-42}$ treatment influences the expression of tight junction protein complexes, stress fibre formation, disruption and aggregation of actin filaments and cellular gap formation.
Aiming for functional characterization of model cells upon $A\beta_{1-42}$ treatment, we deployed an advanced Electric Cell-substrate Impedance Sensing for monitoring cell evolution. A precision Impedance Analyzer with a multiplexing module developed in house was used for recording individual electrode sets in the 40 Hz - 100 KHz frequency range. In a step forward from the classical ECIS assays, we report on a novel data analysis algorithm that enables access to cellular and paracellular electrical parameters and cell surface interaction with fully developed cell monolayers.
The evolution of the impedance at selected frequencies provides evidence for a dual effect of $A\beta_{42}$ exposure, at both paracellular permeability and cell adherence level, with intricate dynamics that open up new perspectives on $A\beta_{1-42}$ oligomers – cell membrane interaction. Validation of electrical impedance assays of the amyloid fibrils effect on cell membrane structure is achieved by both AFM analysis and Surface Plasmon Resonance studies.
The capabilities of this noninvasive, real time platform for cell analysis in a wider applicative context are outlined.

1. Introduction

The amyloid β (Aβ) peptides are believed to be pivotal in Alzheimer's disease (AD) pathogenesis and onset of vascular dysfunction. Recent studies [1],[2] indicate that Aβ treatment influences the expression of tight junction protein complexes, stress fiber formation, disruption and aggregation of actin filaments and cellular gap formation. Nevertheless, the exact mechanisms involved in these effects are still a matter of debate due to the variability in the investigation methods used, the type of cells analyzed and the nature of Aβ compound (type, oligomer/fibril form).

In a "reductionist approach" we focus on $A\beta_{1-42}$ effect on cellular barrier properties outside the complex microenvironment of the brain, on selected renal cells incorporated in an ECIS type device developed recently in ICB.

The ECIS devices [3] enable noninvasive, continuous assessment of cell attachment, spreading and proliferation and hence can be a powerful replacement for the traditional Transepithelial Electrical Resistance (TEER) determinations, a validated marker of barrier integrity in real time, for assessment of cytotoxic effects. They are based on AC impedance measurements using weak, noninvasive signals (up to 50 mV) in the frequency range: 1 Hz - 10 MHz. Attachment and spreading of cells on the electrode surface change the impedance enabling direct inference of the morphological information on the attached cells. Multifrequency impedance monitoring, as opposed to single frequency assays, is a prerequisite for discerning between the various processes involved in cell response and enable data deconvolution by modeling.

The platform enables dynamic assessment of amyloid oligomers – cell membrane interaction by advanced impedance spectroscopy and provides enhanced resolution on cell-surface and cell-cell interactions modulated by membrane related protein apparatus, applicable as well to other adherent cell types.

2. Experimental section

The measurement system is equipped with 4 pairs of planar, circular electrodes, manufactured by photolithography and metal deposition processes on glass slides, placed as bottom of the cell growth chambers.

For parallel control and test experiments we used a measurement chamber with two identical compartments [4]. An amount of 10^5 cells were seeded into each compartment of the measuring chamber and left to grow (standard growth conditions) into a subconfluent layer (2 days). Upon treatment, the culture medium was removed from both compartments and replaced with fresh, serum free medium (control) and 5 μM "aged" $A\beta_{42}$ in serum free medium (Aβ). The ageing process, 48 h at 37 °C, 5% CO_2, determines forced aggregation of $A\beta_{42}$ monomers into oligomers and (proto)fibrils.

A 4294 A Precision Impedance Analyzer, Agilent Technologies Ltd., Japan, with a multiplexing module developed in house was used for recording the four individual electrode sets. An AC potential was used with zero bias, 50mV pp amplitude, 40 Hz - 100 KHz frequency range (100 frequency points with logarithmic distribution).

Data analysis was performed based on a modified cell index V_N given by Eq. 1:

$$V_N(t) = \frac{V(frl)}{V(frh)}(t) - \frac{V(frl)}{V(frh)}(t_0) \qquad (1)$$

where fr_l and fr_h correspond to measurement points in the low and high frequency domain, respectively, and t(0) corresponds to a reference moment chosen after cell culture stabilization.

This normalization eliminates environmental differences (e.g. temperature, growth medium changes) while providing a broad sensitivity response.

3. Results

Based on Eq.1, the impedance spectra covering the frequency range 40Hz – 100 kHz are normalized, for each frequency point (by deriving the relative variation), versus real and imaginary impedance spectra at the beginning of the experiment or upon $A\beta_{42}$ addition. Figure 1 shows the distinct dynamics of the cell layer under $A\beta_{42}$ exposure and in controlled conditions (frl=100 Hz and frh=10 kHz).

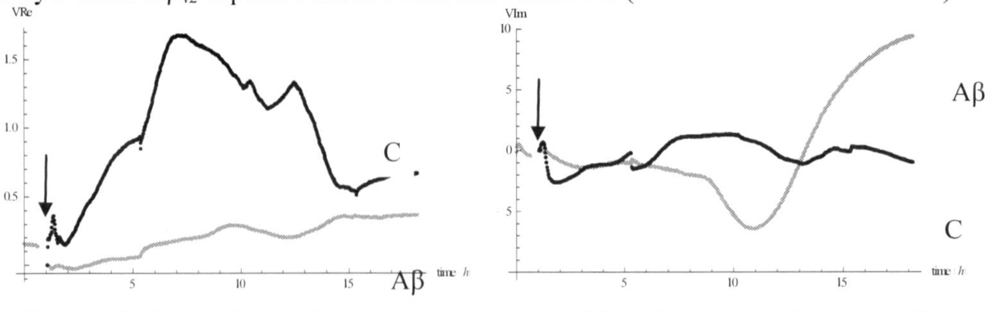

Figure 1. Time evolutions of Normalized impedance Real and Imaginary parts for control (black) and $A\beta_{42}$ exposed cells (red); arrow indicates Aβ addition or medium exchange

In the experimental time frame, there are no cell-cell junctional complexes formed, yet, distinct dynamics are evident for both cell-surface contacts and cell growth (as revealed in different regions of the frequency spectrum), figure 2.

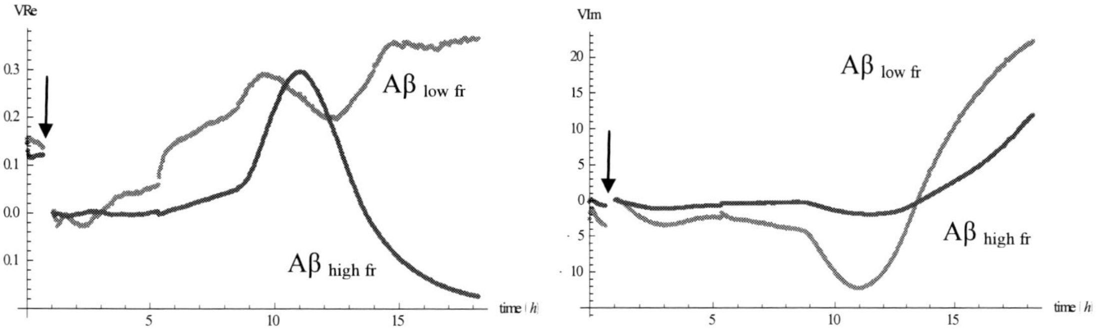

Figure 2. Effect of $A\beta_{42}$ exposure on MDCK cells as revealed by different frequency regions considered in the estimation of V_N (frl=100Hz and frh=20 kHz, red; frl=2.5 kHz and frh=100 kHz, blue); the arrow indicates Aβ addition

The multiphasic dynamics evidenced in Figure 2 upon $A\beta_{42}$ addition can be related to cell-surface tightening or protein mobilization toward the membrane (above 10 h) but discrimination between these processes cannot be made.

When coping with the whole frequency domain, the experimental data can be fitted with equivalent electrical circuits, that associate distinct functional entities in the biological probe (e.g. cell monolayer, paracellular space, cell – surface and cell-to-cell contacts) with abstract representations (e.g. constant phase elements, resistors and capacitors). Nevertheless, such a circuit has to be optimized for increased robustness and rendering capability to fit large time series of impedance data. Therefore, we considered a circuit depicted in figure 3, comprising a resistor for the subcellular and the bulk media, capacitor for the electrode system and a constant phase element, CPE for the cellular layer (characterized by 2 parameters: CPE_T and CPE_P) according to the following equation:

$$Z_{CPE} = \frac{1}{CPE_T \cdot (i \cdot \omega)^{CPE_P}}. \qquad (2)$$

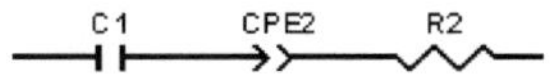

Figure 3. Equivalent circuit

The parameter CPE_P is such that if $CPE_P = 1$, the impedance of a CPE is that of an ideal capacitor, and when $CPE_P = 0$, the CPE is a pure resistor. The frequency dependence of this element has been ascribed to a wide variety of effects such as surface roughness and heterogeneity, nonuniform current distribution, and distribution of relaxation times. Given the possible $A\beta_{42}$ effect on the permeability of cell junctions, a change of R_2 is expected (reflecting changes in the adherence to neighbour cells) as usually evidenced by classical TEER. Nevertheless, a more general toxic effect (Aβ exposure induces generation of toxic reactive oxygen species [5]), should influence also cell adherence points to the substrate and overall cell growth. An increase of CPE_T should occur due to cells rounding up and displaying a larger surface area, as well as a decrease of CPE_P (since the heterogeneity of the surface decreases). Noteworthy, we have recently revealed a similar effect when reactive species are generated upon mechanical and chemical stimulation [4]. The evolution of fitted data (Figure 4) confirms this latter effect of $A\beta_{42}$ exposure cell adherence level: better cell-surface attachment is evident after 10 h of $A\beta_{42}$ exposure (see also figure 4 inset).

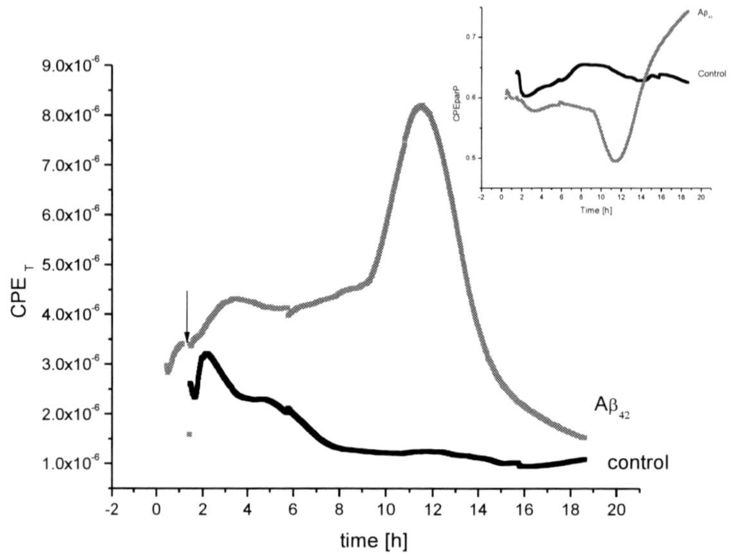

Figure 4. Time evolution of derived values for the CPE_T parameter corresponding to cell layer (Constant phase element - CPE) for control (black) and $A\beta_{42}$ exposed cell cultures (red); inset the evolution of CPE_P parameter of the cell layer.

Validation of electrical impedance assays of the amyloid fibrils effect on cell membrane structure can be achieved by both AFM analysis and Surface Plasmon Resonance studies [6].

4. Conclusions

Based on an improved ECIS approach, we reveal the complex, non-monotonous cell behavior in response to Aβ exposure, otherwise difficult to be assessed through end-point measurements. The proposed cellular platform is ideally positioned for more general, label free, cytotoxic evaluation of nonlethal effect of complex mixtures or unknown compounds as it noninvasively provides continuous assessment of: cell – surface and cell-cell interaction ,proliferation as well as intracellular changes.

5. Acknowledgment

We gratefully acknowledge the financial support of grant BIOSCOPE (PN-II-ID-PCCE, Contract No 11/2012).

References

[1] Marco S and Skaper S D 2006 *Neurosci Lett.* **401**(3), 219-224.
[2] Nagababu E, Usatyuk P V, Enika D, Natarajan V and Rifkind J M 2009 *J. Alzheimers Dis.* **17**(4), 845-854.
[3] Giaever I and Keese C R 1993 *Nature* **366**, 591–592
[4] Gaspar S, David S, Polonschii C, Marcu I, Gheorghiu M and Gheorghiu E 2012 *Anal. Chim. Acta* **713** 115-120
[5] Wang H, Ma J, Tan Y, Wang Z, Sheng C, Chen S and Ding J 2010 *J. Alzheimers Dis.* **21** 597-610
[6] Gheorghiu M, David S, Polonschii C, Olaru A, Bajenaru O, Gaspar S, Popescu B O and Gheorghiu E submitted 2013 *Biosens. Bioelectron.*

Label-free impedimetric biosensor for thrombin using the thrombin-binding aptamer as receptor

D Frense, S. Kang, K Schieke, P Reich, A Barthel, U Pliquett, T Nacke, C Brian, D Beckmann
Institut für Bioprozess- und Analysenmesstechnik e.V., Rosenhof
D-37308 Heilbad Heiligenstadt, Germany

E-mail: dieter.frense@iba-heiligenstadt.de

Abstract. This study presents the further establishment of impedimetric biosensors with aptamers as receptors. Aptamers are short single-stranded oligonucleotides which bind analytes with a specific region of their 3D structure. Electrical impedance spectroscopy is a sensitive method for analyzing changes on the electrode surface, e.g. caused by receptor-ligand-interactions. Fast and inexpensive prototyping of electrodes on the basis of commercially available compact discs having a 24 carat gold reflective layer was investigated. Electrode structures (CDtrodes [1]) in the range from few millimetres down to 100 microns were realized. The well-studied thrombin-binding aptamer (TBA) was used as receptor for characterizing these micro- and macro-electrodes. The impedance signal showed a linear correlation for concentrations of thrombin between 1.0 nM to 100 nM. This range corresponds well with most of the references and may be useful for the point-of-care testing (POCT).

1. Introduction

In this work, an aptamer-based biosensor for the detection of the clinical relevant protein thrombin was developed. This biosensor consists of electrodes (CDtrodes) prepared from CDs [1] as transducer and the thrombin-binding aptamer (TBA) [2] as receptor for the detection of thrombin.

Thrombin is a protein of importance which regulates many processes in inflammation and tissue repair at the blood vessel wall. Its concentration in blood is normally around 0.01 nM. TBA itself has a well-known structure and was investigated in several previous biosensor studies, such as electrochemical gold nanoparticles [3], microfabricated thin film electrodes [4], and electrochemiluminescence [5].

The demand for convenient methods of detecting and measuring specific proteins in biological and environmental samples has motivated the development of commercial biosensors. One of the main difficulties, however, is the instability of the biological recognition element. Aptamers are well-known for their excellent stability. This is why aptamers can contribute to producing stable biosensors.

Our work concentrates on the characterization of this biosensor using immobilized streptavidin and the biotinylated aptamer as sensitive layer on the electrodes. It also contributes to the development of impedimetric biosensors, which are portable, inexpensive, provide rapid response, and can be manufactured with ease.

2. Material and methods

Recordable compact discs were used as an inexpensive and abundant material for preparing electrodes. CDs typically consist of four layers of materials:

- polycarbonate base for providing mechanical stability
- polymer layer which will be sensitized during the recording process
- metallic reflective layer with thickness of about 100 nm
- polymeric lacquer film for protecting the underlying metal layer

In this work we used CDs from SONY having a 24 carat gold layer. For removing the lacquer film, the CDs were submerged in nitric acid (50%) for 5 min. After this procedure, the CD exposed a clean gold surface which does not require an abrasive cleaning process, such as polishing, or the piranha method typically used. The surface was rinsed with ethanol and then hydrophilized under UV. Afterwards the gold surface was sprayed with positive photoresist lacquer for photolithographic wet-etching of the various electrode patterns (etchant: 4 g KI and 1 g I_2 in 40 mL water for 5 min [1]).

We developed two electrode patterns for biosensor fabrication. The first pattern (Figure 1A) has a fixed measurement chamber glued on the electrodes. The samples are pipetted in and out between the measurements. Volumes of 50 µL to 400 µL can be used. The second pattern is the interdigitated finger electrode (IDE) (Figure 1B) which uses a flow-through channel chamber for the electrolyte (Figure 1C). Samples of 15 µL to 90 µL can be applied.

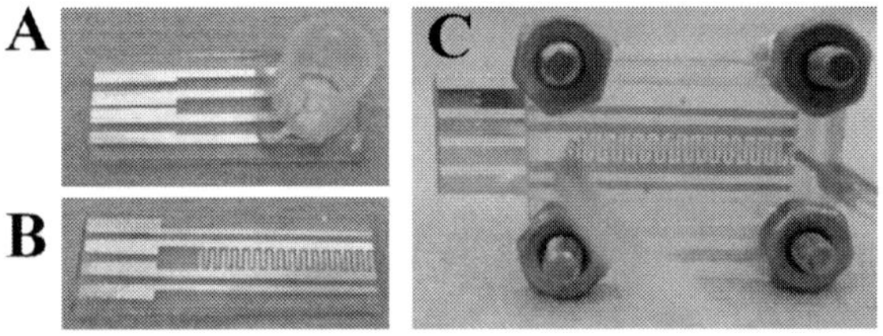

Figure 1. Photographs of electrode pattern with the glued measurement chamber (**A**) and the IDE pattern (**B**) which uses a self-made flow-through channel chamber in PDMS (**C**).

The surface of the clean gold electrode was treated with thiol for creating a self-assembled monolayer. We chose cysteamine (20 mM for 2h) for the initial layer and glutaraldehyde (300 mM for 2h) for the protein binding, according to the procedures described by Ferriera et al. [6]. Streptavidin, 1 µg/mL diluted in PBS-buffer, was then immobilized via the aldehyde groups (at 4°C overnight). After blocking the non-specific binding sites with bovine serum albumin (BSA, 1.0 mg/mL in PBS for 1 h) we immobilized the biotinylated aptamer TBA (1 µM diluted in TA-buffer) for 1 h. The TA-buffer consisted of 20mM Tris-HCl, 140 mM NaCl, 5 mM KCl, 1 mM $CaCl_2$, 1 mM $MgCl_2$. The pH-value was adjusted to 7.4. TBA, which we used in this work, was a generous gift from the working group of Mrs Strehlitz, UFZ, Leipzig [2].

The measurement procedure was as follows: A baseline was established using the TA-buffer solution. Thrombin diluted in the TA-buffer was applied for ten minutes, after which the unbound thrombin was washed with the TA-buffer. Upon reaching a stable line, regeneration solution (2 M sodium chloride in water) was applied for two minutes.

Electrochemical impedance spectroscopy (EIS) was performed using a BioLogic Science Instruments Model SP-150 in the frequency range from 100 mHz to 200 kHz. As electrolyte we used TA-buffer with 5 mM potassium hexacyanoferrate(II) and 5 mM potassium hexacyanoferrate(III). Frequency response of the impedance was expressed as Nyquist plot and modelled using a Randles circuit consisting of an active electrolyte resistance R_s in series with the parallel combination of the double-layer capacitance C_{dl} and the impedance of a Faradaic reaction. The impedance of the Faradaic reaction consisted of an active charge transfer resistance R_{ct} and a specific electrochemical element of diffusion W, which is the well-known Warburg element.

For some experiments the aptamer was heat treated according to Tombelli, et al. [7] before the immobilization step. This heat treatment consisted of heating the aptamers in TA-buffer to 90°C for one minute, and then cooling the solution on ice for 10 minutes.

3. Results and discussion

Preliminary thrombin-binding experiments resulted in Nyquist plots that showed noticeable changes in the real and imaginary parts of the impedance. Higher sensitivity was observed at lower frequencies, in agreement with our earlier observation of the Bode impedance curve. In order to record the thrombin binding kinetics with the biosensor we considered measuring the impedance at a fixed frequency. Figure 2 shows the time course during a thrombin binding measurement (111 nM) at four different frequencies (10 Hz to 100 kHz). The figure illustrates a typical binding and dissociation curve that is comparable to most of the biosensors. Discontinuities in the curves were a result of pipetting the solutions in and out of the chamber.

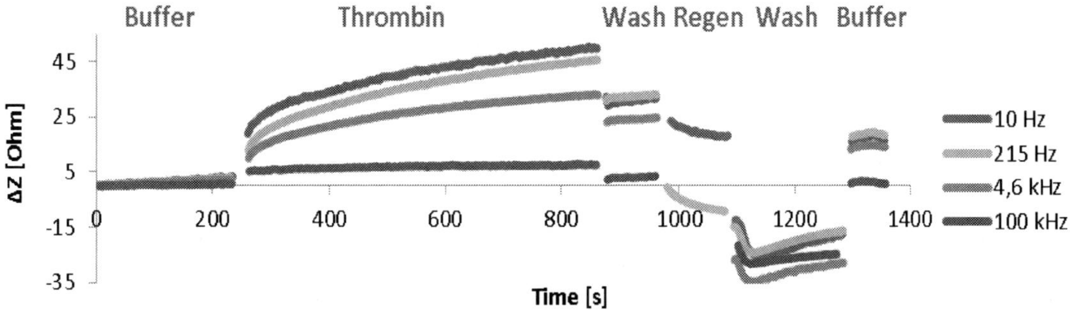

Figure 2. Thrombin binding measurement across a range of frequencies.

In order to determine the specificity of the biosensor we compared specific binding to thrombin and to 1.0 mg/mL BSA as the negative control at the fixed frequency of 10 Hz. The biosensor showed good specific binding to thrombin while the negative control produced a similar response to that of only TA-buffer. The sensitivity of TBA for thrombin was investigated in the range from 1.0 nM to 330 nM (Figure 3). There was a strong correlation between the concentrations of thrombin and the total change in impedance at 10 Hz after 10 minutes.

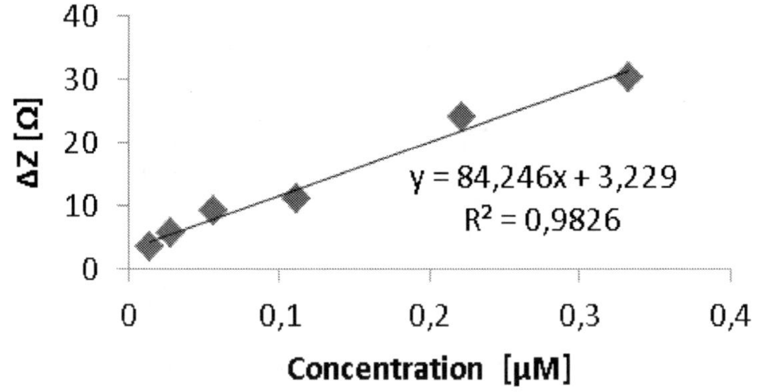

Figure 3. Correlations between the changes in the impedance and various concentrations of thrombin.

Tombelli, et al. [7] concluded that heat treatment of the aptamers influences intramolecular folding. The sDNA unfolds so that the biotin label and the recognition structure are available for interaction with streptavidin and the ligand. We performed the same heat treatment with our TBA and tested the sensitivity to thrombin and the stability of the biosensor. The aptamers which were not heat-treated showed higher sensitivity to thrombin. In contrast, the heat-treated aptamer were more stable (see Figure 4). The measurements consisted of thrombin binding (concentration: 111 nM) for ten minutes followed by the regeneration solution for one to two minutes followed by another thrombin binding test. Differences in impedance were measured at 10 Hz after 10 minutes binding. Detection cycle number 7 was treated as an outlier due to storage of the biosensor at 4°C overnight. The standard deviation of the other cycles was found to be around 15%.

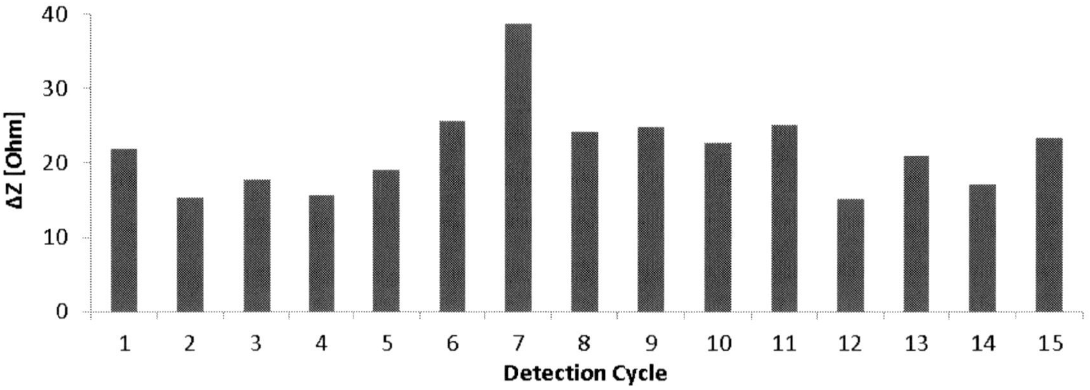

Figure 4. Regeneration of the heat-treated aptamers.

4. Discussions

In blood, plasma clots begin to form when thrombin concentrations reach a level of 5 nM, with the highest concentrations reaching 100 nM to 200 nM approximately ten minutes after clot formation. Thrombin is typically present in the blood in the 0.01 nM range. Low nanomolar concentrations of thrombin are generated early in haemostasis, and high range in patients suffering from diseases known to be associated with coagulation abnormalities. Therefore further studies are required to improve the sensitivity of this biosensor for thrombin-related diseases.

5. Conclusions

In this work we present inexpensive, stable, and easily producible biosensors for the purpose of thrombin detection. The electrodes showed good selectivity towards thrombin and were insensitive to BSA. The sensitivity of the electrodes to thrombin was not quite as good as some other results reported in literature earlier, but the detection limit was nevertheless in the clinical relevant range of 1 nM to 100 nM.

Regeneration of the biosensor was successful at least up to 15 times for electrodes with the heat-treated aptamers. This heat treatment, however, was found to deteriorate the sensitivity of the aptamer for thrombin detection. The efficiency of our label-free impedimetric biosensor was compared to measurements using the same aptamer and the quartz-crystal-microbalance (QCM) and the BIAffinity system, which is based on RIfS technique for detection of molecular interaction. These investigations achieved similar sensitivity and linear ranges (results not shown).

The advantage of the proposed biosensor is that the inexpensive electrode materials allow single-use applications. The fixed frequency of 10 Hz is in a range, for which inexpensive impedance systems can be realized with ease. This may overcome the low sensitivity, at least for POCT-applications. The deviations of our biosensor were about 15%. Similar values can be found for such simple systems. Nevertheless, more effort is needed to overcome this limitation before commercialization.

References

[1] Angnes L, Richter E M, Augelli MA and Kume G H 2000 *Anal. Chem.* **72** 5503.
[2] Strehlitz B, Nikolaus N and Stoltenburg R 2008 *Sensors* **8** 4296.
[3] Li L, Zhao H, Chen Z, Mu X and Guo L 2011 *Sens. Actuat. B* **157** 189.
[4] Cai H, Lee T M-H and Hsing I M 2006 *Sens. Actuat. B* **114** 433.
[5] Fang L Y, Lu Z Z, Wei H and Wang E 2008 *Anal. Chim. Acta* **628** 80.
[6] Ferreira C, Fugivara C, Barrazo S, Suegama P, Yamanaka H and Benedetti A 2009 *J. Electroanal. Chem.* **634** 111.
[7] Tombelli S, Mascini M and Turner A P 2002 *Biosens. Bioelectr.* **17** 929.

Time-domain based impedance measurement: strengths and drawbacks

Uwe Pliquett
Institut für Bioprozess- und Analysenmesstechnik, Heilbad Heiligenstadt, Germany

E-mail: uwe.pliquett@iba-heiligenstadt.de

ABSTRACT: While most applications of bio-impedance measurements for characterization of cells or tissue do not have any time requirements, newer developments like real time monitoring of moving cells or membrane recovery after high voltage application depend on fast measurement. This is not compatible with sweeping the frequency through the desired range while assessing magnitude and phase of the material under test (MUT). A high speed is only achievable by using broad-bandwidth excitation signals and monitoring the response in time-domain. Time-domain based methods can be distinguished by the excitation signals as well as by assessing the transmission or reflection behavior of the MUT.
Although there is good agreement regarding the advantages of fast measurements, time-domain measurements are often rejected because of low precision and noise sensitivity. This paper points not only the advantages of impedance measurements in time domain but shows also drawbacks together with possible solutions.

1. Introduction

The electrical characterization of biological material is well established but lacks applicability for a variety of potential use. Especially if the MUT is fast changing like after the application of electrical pulses for cell manipulation of if the material is fast moving along the electrodes like by monitoring cells in a sample stream in capillaries, established methods like frequency sweep for assessment of the impedance spectrum fail. Using broad bandwidth excitation signals and tracing the electrical response in time will greatly enhance the temporal resolution [1]. Moreover, due to the fast measurement, multichannel arrangements with reasonable time for testing all channels become affordable.
Biological objects composed of cells surrounded by insulating cell membranes show a β-dispersion but also α-dispersion at low frequency and γ-dispersion above the ß-dispersion [2]. Depending on the application, narrow bandwidth measurements at one frequency or broad bandwidth measurements spanning over 6 to 10 decades are required. A change in impedance or permittivity magnitude by several orders is likely for the majority of living objects.
Two characteristics of biological material challenge bio-impedance measurement: (1) broad bandwidth and (2) a high dynamic range in measurement amplitude. Although this can be solved in frequency domain as well, fast broad bandwidth measurements are not compatible with this approach.

2. Frequency domain vs. time domain

Although impedance measurement in time domain is well established in many areas, the majority of bio-impedance measurements are done either at a single frequency or by sweeping over the desired frequency range. This was mainly supported by the broad availability of gain-phase and network analyzers. Although time domain spectroscopy (TDS), mostly in reflection mode [3], was used since

the 70is of last century it was almost not recognized until about 10 years ago. There have been several arguments against it:
- Time domain spectroscopy is not precise and susceptible to noise;
- Using the 'classical' square wave approach, the energy at high frequency declines fast;
- Devices which have been developed for bio-impedance measurements are rarely marketed.

The most important advantage over frequency domain based measurements is the speed. However, with only a few exceptions like process instrumentation (i.e. meat classification) or monitoring of fast processes (i.e. recovery of membranes after high voltage manipulation) there was no real demand for fast measurements. For instance, time domain reflectometry [3] (Y.Feldman termed it later time domain dielectric spectroscopy, TDDS) used a sampling approach where in each period only one point was sampled and converted using a high resolution ADC. In this way and together with averaging a single measurement could take as long as getting the same information in frequency domain.

The interest in time domain approach rose during the last ten years with exploring new areas in process measurement and biomedical instrumentation. The intention was now to have fast and robust measurements using simple and therefore economic devices with low energy consumption. Even though specialized integrated circuits for frequency domain approach exist (Analog Devices, Texas Instruments), time domain based instrumentation can be even simpler, for instance consisting of only one microcontroller. However, a front end for matching the electrodes to the digital circuitry but also to the input of the ADC is always advisable for practical approach [4].

TDS in general is the use of a broad bandwidth signal for excitation and monitoring the response in time. While the most prominent excitation signals (Dirac and step function, square wave) are successfully employed since more than 40 years, today more sophisticated excitation signals like multi-sine, pseudo random noise (i.e. maximum length sequences) or chirp are used or under development [5]. Two pronounced development targets are the minimization of the crest factor and the confinement of the energy within the frequency range of interest. Moreover, efficient algorithms aiming in speed increase and pre-compression of high data volumes are under investigation by several groups. Further data processing is either done direct in time domain by fitting the response signal to an biophysical model in order to obtain time constants and amplitude factors or the excitation as well as the response signal are transformed into frequency domain (e.g. by Fourier transformation) for calculation of the impedance as function of the frequency. Other methods like correlation, convolution or other transformations (wavelet, Gabor) are used for several applications but they are not widely distributed. Even though non-uniform sampling for great reduction of the data stream together with discrete Fourier transformation with non-equally spaced time vector is known since decades [6], there are still only a few applications in bio-impedance measurement practice. This raises questions: Are the new techniques not good enough? Are their advantages not understood? OR: Is there no need for new techniques due to the overwhelming quality of existing ones?

The simple answer is: There is the need for new techniques. Their acceptance increases with advancing the knowledge, which can be seen especially in the field of bio-impedance measurement during the last years. However, further success of TDS depends on managing drawbacks of this technique which includes a pragmatic decision for forcing its application only where the advantages make either the measurement or the post-processing better or simpler.

3. Advantages of time domain spectroscopy

The clear advantage of TDS is the speed but also a significantly simpler hardware. Further advantages, especially for use at high frequency, are less known and should be briefly mentioned.

Correlated transients form switching events or from reflections along the connection circuitry are simply detected and can be compensated. Using a reflectometry arrangement without directional couplers, the same channel for voltage and current is used, which circumvents the common problem of never perfect match of these both channels. Although reflection methods in frequency domain exist as well, they use direction couplers and separate channels for the incident and reflected wave. Due to the low impedance magnitude above about 100 MHz, similar to this of the wave guide impedance (coaxial

cable, stripe line) time domain reflectometry becomes advantageous over transmission spectroscopy. The available frequency range extends currently up to 200 GHz.
The bandwidth of time domain spectroscopy is easily scalable, irrespectively of excitation signal used. The lower cutoff frequency is determined by the time duration of the excitation signal while the upper frequency depends on the fastest change of the signal with respect to time (i.e. rise time of a step function). The critical quantities at the sample side are the speed and the resolution of the ADC.

4. Drawbacks of time domain spectroscopy – what can we do?

Although the robustness of new algorithms has been shown in numerous studies even in noisy environment, some significant drawbacks are well known. However, there are solutions making this approach emerging during the last years.
Random noise, originating from resistors, amplifiers, electrodes or also high frequency noise due to the wiring of the setup is a negligible problem when measuring at a single frequency due to the use of selective amplifiers (i.e. lock-in-principle). For TDS, broad bandwidth amplifiers are used, not enabling any noise reduction in a single acquisition mode. Three prominent principles for noise reduction apply: (1) For slow signals where the amplitude change within one sampling interval is significantly less than the resolution of the ADC, oversampling with averaging around a single point in time decreases not only the noise but increases also the resolution of the analog-digital-conversion. The advantage of oversampling is a high speed, since it still works in single shoot mode. (2) Averaging over several periods does not require extremely fast ADC but comprises time resolution. (3) Using a biophysical model for the biological material enables the meaningful fit of a time function. In case of a step function for excitation, the response will be sum of exponentials which can be easily fitted. It should be noted that even smoothly appearing signals after fitting a noisy curve are not better because the significance of fitted values decreases fast with increasing noise. It is in any case better to obtain good results by optimizing the hardware.
Besides random noise, crosstalk from power lines can greatly distort the impedance signal even though the lower frequency of the excitation signal is above 50 Hz (60 Hz) (e.g. 1 kHz rectangular wave). Besides proper grounding and shielding of the setup, a simple approach is the synchronization of the excitation signal to the frequency of the power line. This can be done by Schmitt-triggering the 50 Hz (60 Hz) in order to obtain a noise free digital signal where the frequency can be multiplied in a PLL (phase lock loop) with proper division of the VCO (voltage controlled oscillator) output. Practically, a suppression of the power line noise of more than 80 dB can be achieved. A further suppression is possible by averaging over a full period of the power line frequency. Although this comprises measurement speed to 50 (60) measurements per second, it pays back by a much higher significance of the single measurement result.
Correlated noise like signal distortion from switching events in the chain of excitation signal generation or arising from reflections at mismatches along the measurements chain do not disappear by averaging. Applying a simple filter, analog or digital, will distort the entire signal, especially parts with fast changing voltage or current and would be therefore an additional error source. Using a step function for excitation results in a sum of exponentials for typical biological materials. Replacing the transient distortion by a fitted exponential does not disturb the spectrum but removes artifacts.
In general, impedance measurements should be performed at linear objects only. However, biological materials especially in contact with metal electrodes show slightly non-linear behavior. The overall rule in bioimpedance measurement, independent of frequency or time domain, is to use excitation signals as small as possible. In frequency domain measurement with sine wave for excitation together with using selective amplifiers the non-linearities create harmonics which are completely suppressed and thus they are not detected. However, by using broad bandwidth amplifiers, such harmonics influence the high frequency behavior. Suppression is possible by employing known behavior of the response signal. In particular, when using symmetric square waves for excitation, averaging of the positive half period with the inverted negative half period suppresses harmonic distortion at least to some extent. This is particularly useful when the time varying dynamic impedance at an operating

point with dc-offset is assessed (e.g. electrode bias, electrical manipulation, a membrane with resting potential.

An often underestimated characteristic of the measurement system is the transient response of the amplifier used. While this does not matter for sine wave signals, it may considerably disturb the high frequency region of time domain signals, especially if their deviation yields singularities as in case of Dirac or step functions. This requires carefully selected amplifiers (high slew rate, negligible overshot, fast settling) and a well-designed layout of the printed board.

The resolution of the ADC and the jitter of the sampling system is an important limitation of any system working in time domain [4]. The choice of the sampling system does not only influence the accuracy and speed of the entire measurement setup but determines considerably the prize. While at fast changing parts of the signal the jitter is an important quality factor, the resolution of the ADC becomes increasingly important for slow changing parts.

Low cost but precise systems use the principle of the sampling oscilloscope, where a periodic signal is sampled once in each period at a continuously shifting time point until the entire curve is sampled (stroboscopic under sampling). This requires as many periods as sample points which yields a rather long measurements time. The great advantage of this approach is use of slow ADC with high resolution (16 – 24 bit) but is very sensitive to jitter in the timing of the sampling points.

Better systems use a single shot approach where the jitter is low but this needs much faster ADC which in turn comprises ADC-resolution (usually 8 or 12 bit). High end systems use very fast ADC together with oversampling where each point is the average of several fast samples. This increases the overall resolution without the need of averaging over several periods.

A problem is the high swing of signals in time domain measurements. For confining the signal arising from the MUT, difference or ratio measurements with known equivalent circuits, mimicking for instance the electrode system, are used. Since only changes are amplified, higher amplification factors become possible which in turn increases the sensitivity of the system.

In order to reduce the data volume, adaptive sampling with fast sampling at fast changing parts of the signal and slow sampling when the changes over time are slow [7], is used. This overcomes also problems associated with the low energy of higher harmonics, particularly due to square wave excitation. With adaptive sampling not the energy distribution of the excitation signal but the product of signal/time gradient and sampling interval determines the useful frequency range. Although, this requires sophisticated data processing and is not compatible with the most popular fast Fourier transformation, solution in hard- and software already exist.

5. Conclusion

This paper could only give a brief introduction into the problems associated with impedance measurements in time domain. The practical experience shows that time domain measurements can be made as precise as frequency domain measurements but it needs more sophisticated data processing and more attention in terms of generation of high quality signals and noise suppression.

References

[1] Grimnes,S. & Martinsen,O.G., Bioimpedance and Bioelectricity Basics. Academic Press., 2008

[2] Foster,K.R., Schwan,H.P., Dielectric properties of tissues and biological materials : A critical review. *CRC Crit. Rev. Biomed. Eng.* 17, 25-104, 1989

[3] R.H.Cole, Time domain reflectometry. Annu.Rev.Biophys.Biomol.Struct., 28, pp.183-300, 1977

[4] Pliquett, U., Schönfeldt, M., Barthel, A., Frense, D., Beckmann, D., Front end with offset-free symmetrical current source optimized for time domain spectroscopy, Physiol. Meas.32, pp.927-944, 2011

[5] Min, M., Pliquett, U., Nacke, T., Barthel, A., Annus, P., Land, R., Broadband excitation for short-time impedance spectroscopy. *Physiol. Meas.* 29, 185-192., 2008

[6] Kremer, F., Schönhals, A, Broadband dielectric spectroscopy, Springer-Verlag, Berlin Heidelberg New York, ISBN: 3-540-43407-0, 2003

[7] Pliquett, U. Gersing, E., Pliquett, F., Evaluation of fast time domain based impedance measurements on biological tissue. Biomedizinische Technik, 45, 6-13, 2000

Assessment of suspension medium conductivity by means of micro electrodes

Maximilian Westenthanner[1], Andreas Barthel[1], Ping He[1,3], Dieter Beckmann[1], Arndt Steinke[2], Dr. Ingo Tobehn[2], Dr. Thomas Frank[2], Uwe Pliquett[1]

[1]Institut für Bioprozess- und Analysenmesstechnik, Heilbad Heiligenstadt, Germany
[2]CiS Forschungsinstituts für Mikrosensorik und Photovoltaik GmbH
[3]Southern Illinois University, Carbondale, IL, USA

Email: maximilian.westenthanner@iba-heiligenstadt.de

Abstract. Electrolytic conductivity of cell-based suspensions is an important parameter which can be easily and non-destructively measured as part of the electrical impedance. The low frequency conductivity of a cell suspension with very low cell density equals nearly the medium conductivity. However, a high cell density decreases the low frequency conductivity due to the insulating behaviour of the cell membranes.
Here we use miniaturized electrode structures, smaller than the size of typical cells for impedometric conductivity measurement which allows an impedance measurement independent of the electrical properties of suspended cells or particles.

1. Introduction

There are many applications of impedometric monitoring, where not the cell membranes are the primary structure of interest, but the conductivity of the medium surrounding the cells. Typical examples are the monitoring of cellular growth in batch reactors, dialysis control [1], or diagnosis of dairy cattle mastitis. The last application uses the fact that the conductivity of milk from infected tips shows an elevated conductivity [2]. In case of low cell density, sufficiently low that the polarization of the cells (ß-dispersion) does not influence the measured impedance, the medium conductivity can be easily measured even with complicate electrode geometry. This approach needs only calibration with an electrolyte of known conductivity. This results in the bulk conductivity, which equals at low concentration the conductivity of the suspension medium. Since there is a functional dependence of the low frequency conductivity of a cell suspension, the density of cells and their passive electrical properties, it is theoretically possible to derive the medium conductivity from the spectrum within the ß-dispersion by using proper modelling. However, this approach depends on multiple factors and is critical if for instance the cell density of the cell size is poorly determined or simply unknown.
Another critical point is the homogeneity of the suspension. If the impedance is for instance measured in the presence of gas bubbles or additional particles, the selectivity in terms of medium conductivity vanishes. A way out is centrifugation and subsequent conductivity measurement using well calibrated equipment. This however, is not an option for inline measurement like for the detection of mastitis or monitoring of bioreactors.
However, even without excessive preparation, a conductivity measurement which is unaffected by cells can be achieved, if the electrode structures are smaller than the dimensions of the cells. Therefore the electrical field extension is limited to the bulk between the electrodes, a space free of cells. Moreover, due to the small dimension, a fast equilibration of the ion content between this region and the bulk electrolytes would be expected.

Figure 1 shows the principle of a microelectrode, where the main electric field extends into the suspension less than the size of a cell.

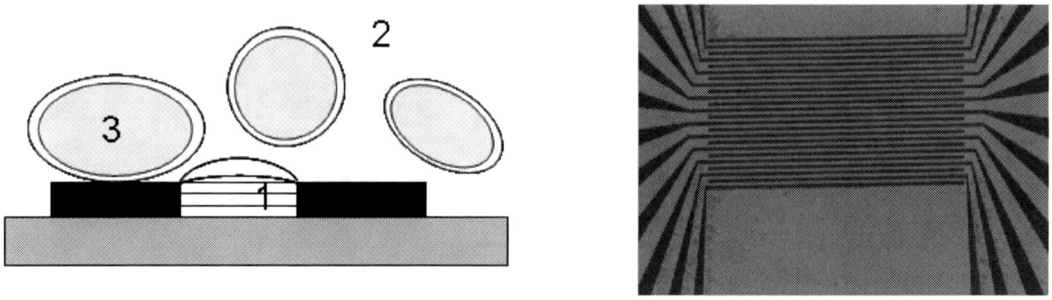

Figure 1A: scheme of a two-electrode microstructured electrode. The electrical field is concentrated between two electrodes (1) which results in a negligible sensitivity of the electrodes within the bulk electrolyte (2). This makes the electrodes essentially insensitive to cells (3). B: Microstructured gold-on-glass electrode chip with 24 individually contactable electrodes. The electrodes have a width and space of 5 μm.

The critical point of such a structure is a high sensitivity for the conductivity of electrolytes but a negligible sensitivity with respect to the presence of particles or cells.

1 Methods

Impedance spectra are recorded using different methods and devices. A phase sensitive multimeter (NumetriQ-PSM1735, Newtons4th Ltd, UK) in combination with a computer assessable multiplexer/frontend combination is used for frequency domain measurements. Other measurements are done in time domain by using a square wave generator (500 Hz, 2 ns rise time) for excitation and an oscilloscope (8GHz) for tracing current and voltage between the electrodes. An almost perfect match to the electrical behaviour of the electrodes was achieved using a custom built front end.

Unfortunately, the multiplexer made from solid state switches (ADG333) caused non-acceptable signal distortion. Since new multiplexer is under development, we undertook several measurements without multiplexing, by directly contacting the electrodes. This increased the sensitivity of the measurement but fortunately, did not change the overall outcome. To get the impedance spectra, the input (excitation) and the output (response) were transformed by the fast Fourier Transform algorithm (FFT) into frequency domain. The quotient FFT(U) / FFT(I) yielded the impedance spectrum [4].

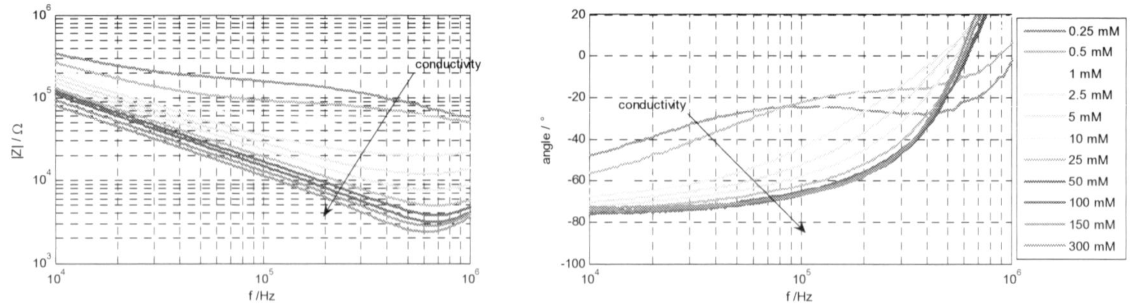

Fig. 2: Impedance spectra of a KCl dilution series, (left) magnitude, (right) phase angle

To evaluate the sensitivity of the system with respect to electrolytic conductivity, the smallest available electrode distance of 5 μm was selected to assess the impedance spectrum of varying concentrations of KCl from 0.25 mM to 300 mM.

The sensitivity for cell based suspensions is evaluated using model suspensions in PBS dilutions. Yeast (Saccharomyce Cervisiae) suspended in 20 mM PBS from 10^7 to 10^9 cells/ml and pig erythrocytes from $4x10^4$ to

$2.5x10^9$ cells/ml are used. The iso-osmolarity between the cytosol and the electrolyte was adjusted by adding sucrose.

2 Results

Two adjacent electrodes with a distance of 5 µm were directly contacted and the impedance spectra was taken using a time domain approach. As shown in Fig. 2, the impedance spectra of a miniaturized electrode pair is influenced by electrode polarization over the full accessible frequency spectrum.
The sensitivity of the electrode to the electrolytic conductivity is evident for the whole frequency range. As seen in Fig.3. the electrode polarization yields a non-linear characteristic, especially for the low frequency range.

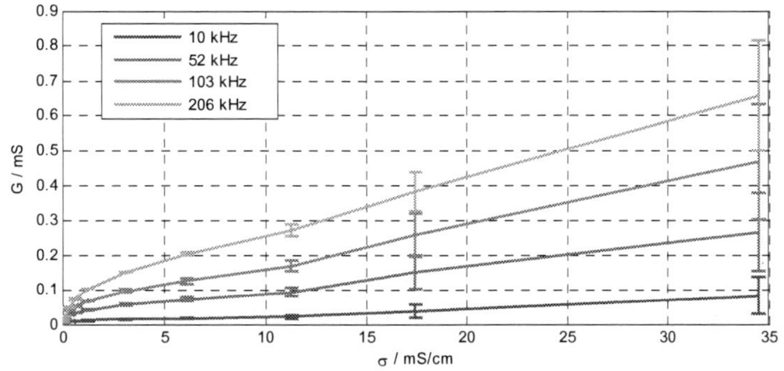

Fig. 3: Conductance with respect to the electrolytic conductivity at different frequencies revealing the sensitivity of the 5 µm structured micro electrodes. The measurements are corrected for the electrode resistance.

To test the hypothesis of dramatically decreased sensitivity to cells and particles, model suspensions of different density were used. Besides polysterene spheres (12 and 25 µm diameter) and glass spheres (5 – 150 µm), yeast and erythrocytes are used in 20 mM PBS and 140 mM PBS. Fig.4 shows the sensitivity of the 5 µm structured micro electrodes to yeast cells.

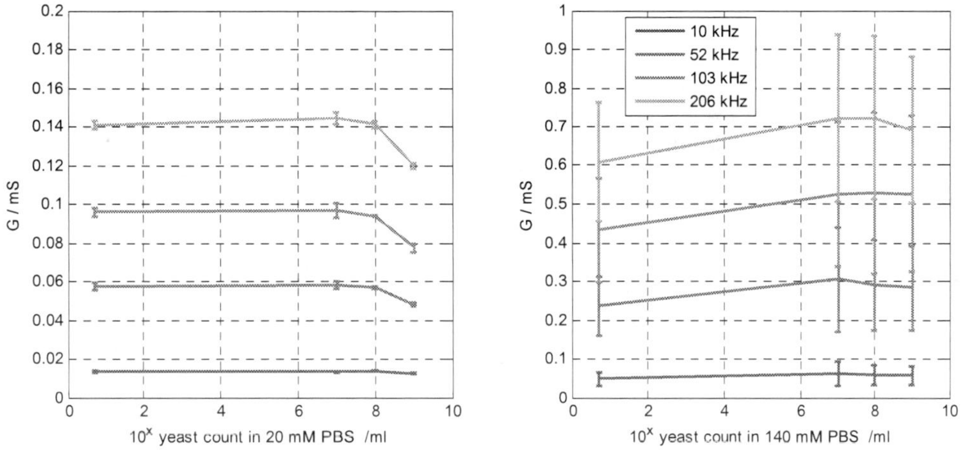

Figure 4: Conductance of cell suspensions with different cell density in PBS at different frequencies. A: 20 mM PBS; B: 140 mM PBS

By visual inspection, 140 mM are nearly insensitive to yeast cells, ranging between zero and 10^9 cells/ml. 20 mM PBS shows still some effect of the suspended cells on the conductance.
Using the same setup, pig erythrocytes in osmotically balanced PBS (20 mM and 140 mM) were investigated (Fig.5).

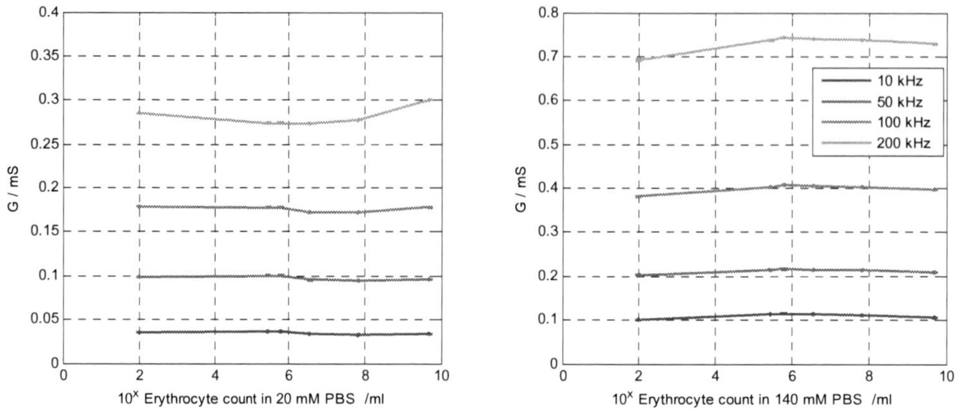

Figure 5: Conductance of erythrocytes suspension in PBS detected at different frequencies. A: 20 mM PBS ; B: 140 mM PBS

While cells in 140 mM PBS influence the impedance of the suspension only negligibly, a more pronounced effect was found for the cells suspended in 20 mM PBS. Although the medium conductivity was determined after centrifugation we cannot completely rule out the influence of dying cells during the measurement as well as the a kind of shadow effect arising from very small cells (< 5 µm) covering the electrode partially.

3 Results and Discussion

Impedance spectroscopy with electrodes having a dimension below this of the suspended cells or particles yields predominantly information about the medium conductivity. Here we showed the example of 5 µm electrodes in the presence of Yeast and Erythrocytes, both slightly bigger than the electrode structure.
These electrodes are sensitive for conductivity, even in a range where the electrode polarization dominates but they are essentially insensitive to the presence of cells and particles.

4 References

[1] Di Filippo S., Pozzoni P., Manzoni C., Andrulli S., Pontoriero G., Locatelli F., Relationship between urea clearance and ionic dialysance determined using a single-step conductivity profile; Kidney Int.; 68(5), 2389-95, 2005
[2] E. Norberg, Electrical conductivity of milk as a phenotypic and genetic indicator of bovine mastitis: A review, Livestock Production Science, 96(2-3):129-139, 2005
[3] R. D. Fomekong, U. Pliquett, F. Pliquett, Passive electrical properties of RBC suspensions: changes due to distribution of relaxation times in dependence on the cell volume fraction and medium conductivity, Bioelectrochemistry and Bioenergetics, 47-1, 81-88, 1998
[4] U. Pliquett, E. Gersing, F. Pliquett, Evaluation of Fast Time-domain Based Impedance Measurements on Biological Tissue, Biomed. Technik, 45(2000), 6-13, 2000

Determination of salt content in various depth of pork chop by electrical impedance spectroscopy

P Kaltenecker[1], D Szöllősi[1], L Friedrich[2] E Vozáry[1] ,
[1] Department of Physics and Control, [2]Department of Refrigeration and Livestock Products Technology, Corvinus University of Budapest,
Somlói st. 14-16. Budapest, H-1118 Hungary

E-mail: eszter.vozary@uni-corvinus.hu

Abstract. The salt concentration was determined inside of pork chop both by electrical impedance spectroscopy and by a conventional chemical method (according to Mohr). The pork chop in various depths (4 mm, 10 mm, 20 mm and 25 mm) was punctured with two stainless steel electrodes. The length of electrodes was 60 mm, and they were insulated along the length except 1 cm section on the end, so the measurement of impedance was realized in various depths. The magnitude and phase angle of impedance were measured with a HP 4284A and a HP 4285A LCR meters from 30 Hz up to 1 MHz and from 75 kHz up to 30 MHz frequency range, respectively at 1 V voltage. The distance between the electrodes was 1 cm. The impedance magnitude decreased as the salt concentration increased. The magnitude of open-short corrected impedance values at various frequencies (10 kHz, 100 kHz, 125 kHz, 1.1 MHz and 8 MHz) showed a good correlation with salt content determined by chemical procedure. The electrical impedance spectroscopy seems a prospective method for determination the salt concentration inside the meat in various depths during the curing procedure.

1. Introduction

Salting has mostly been used for meat preservation due to the achieved reduction in water activity and microbial growth effect inside meat products. Nowadays the healthy lifestyle and the healthy eating demand to minimize the salt intake. Therefore the salt concentration should be enough high (4.5-8 %) to avoid the microbial growth and should be enough low (not higher than 5-7 %) to provide the healthy eating

In practice the conventional curing processes of pork meat are realized by experience, the salt concentration inside of meat piece is measured usually only at the end of curing and only with destructive chemical methods. There is great importance to find a quick, non-destructive and low cost method for measurement of salt concentration inside of pork meat during various curing processes.

The electrical impedance of biological – both plant, and animal – tissue gives a characteristic spectrum in frequency range from 10 Hz up to 100 MHz [1]. The quality of meat during post-mortem ageing can be characterized by various impedance parameters [2]. Some parameters can give information about the quality of meat [2,3,4] and the integrity of cell membranes [2,4], other parameters can be used for predicting the sodium chloride content inside meat piece [5,6].

In this work the impedance spectrum of pork chop was determined daily over four days curing process. The spectra in four different depths were measured with a home made electrode. The salt content of the different part of pork chop was determined with a conventional chemical procedure. A mathematical relationship between salinity and impedance magnitude was found.

2. Materials and Methods

A pork chop (24 hours after rigor mortis, pH_{24}=5.8) was purchased on the local market. Four, each 9 cm length piece were cut from the chop transversally to the fibre direction and these pieces were put into sodium chloride solution of 16 %. Daily over the four days period a piece was taken out from curing solution and the impedance spectrum in four various depths – 4, 10, 20 and 25 mm - was measured in three various side of piece – on the two sides and on the top covered a thin fat layer perpendicular to the fibre. The diffusion coefficient of sodium chloride in meat is $2\text{-}5*10^{-10}$ m^2/s [7]. After impedance measurement the chop was sliced and the salt content on each side and in each depth of measurement was determined with a conventional chemical method according Mohr [7].

Two precision LCR meters – Hewlet-Packard 4824A and 4825A, were used for determination of impedance spectra in frequency range from 20 Hz up to 1 MHz, and from 75 kHz up to 30 MHz frequency range, respectively. The measured spectra were open-short corrected in order to eliminate the stray capacitance and inductance above 100 kHz. The level of measuring voltage was 1 volt. The length of electrodes was 60 mm, and they were insulated along the length except 1 cm section on the end, so the measurement of impedance was realized in various depths. The distance between the two electrodes was 1 cm, the electrical field strength is 10^2 V/m which is enough low not to cause physiological changes in membrane.

3. Results and Discussion

At the beginning of the experiment, on the 0th day (not shown) the impedance magnitude on each place and in each depth was relative high – about 100 ohm – at frequencies higher than 10 kHz. Over time of curing process the magnitude of impedance gradually decreased at frequencies higher 10 kHz, at first in the least depth, then in other depths (figure 1a). In lower frequency range (below 1 kHz) the high impedance magnitude is caused probably by electrode polarization [1]. The phase angle of the impedance in the function of frequencies also showed a gradually shift (figure 1b) during curing process, as the salt content increased inside the pork chop piece.

In the cured meat the salt concentration varied from 3 up to 9 m/m % (g NaCl/ 100g cured meat). The salt content increased inside of pork chop during the curing process (figure 2a). The salt concentration in 4 mm depth practically reached the saturation value during the first day. In other depths the salt content gradually increased during the curing process. Concentration data on figure 2a are the average of concentrations obtained from the both sides of pork chop. The salt concentrations determined from the upper part of meat piece (not shown) had lower values, because the thin fat layer slowed the salt diffusion. The change of salt content in various depths at various times can be described with diffusion of salt molecules.

The impedance magnitude in frequency range from 10 kHz up to 30 MHz can be used for characterization of salt content inside of meat (figure 1a). The change of impedance magnitude in the function of curing time and depth was determined at frequencies of 10 kHz, 100 kHz, 125 kHz, 1.1 MHz and 8 MHz. The functions were similar to each other. Typically the impedance magnitude decreased as the salt content increased (figure 2b) according to other experiments [5,6].

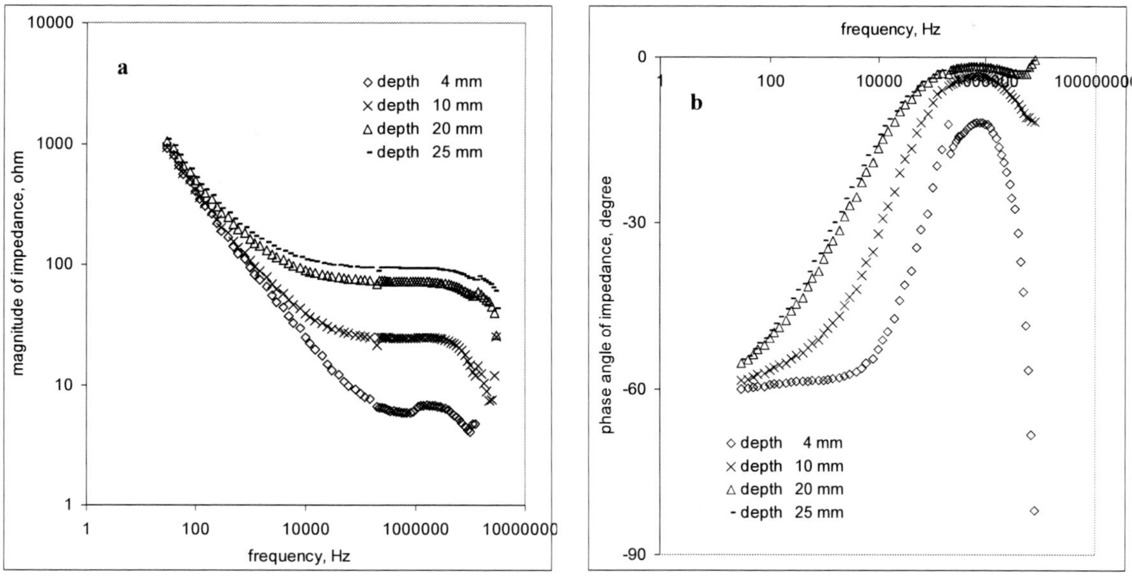

Figure 1. The impedance magnitude (a) and phase angle (b) as the function of frequency in a pork chop piece on the third day of curing process. The meat was punctured with electrodes on the side perpendicular to the fibres and the depth of measurement place was varied from 4 mm to 25 mm.

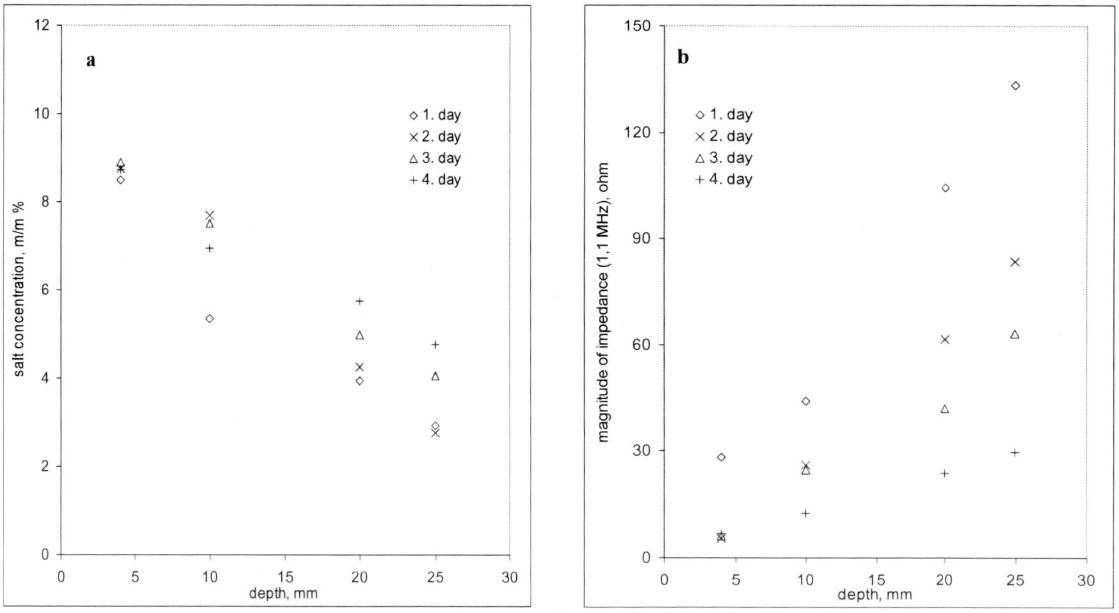

Figure 2. The salt concentration (a) and impedance magnitude (b) at 1.1 MHz frequency at various depths in pork chop side perpendicular to the fibres and at various time of curing process.

The change of impedance magnitude at 1.1 MHz in the function of depth was different for different days of measurement (figure 2b). In depth 4 mm – very near to the surface – the impedance magnitude decreased during the first two days of curing process to a value, which remained constant during the whole experiment. In other depths (10 mm, 20 mm and 25 mm) the decrease of impedance magnitude could be observed in all days of curing process.

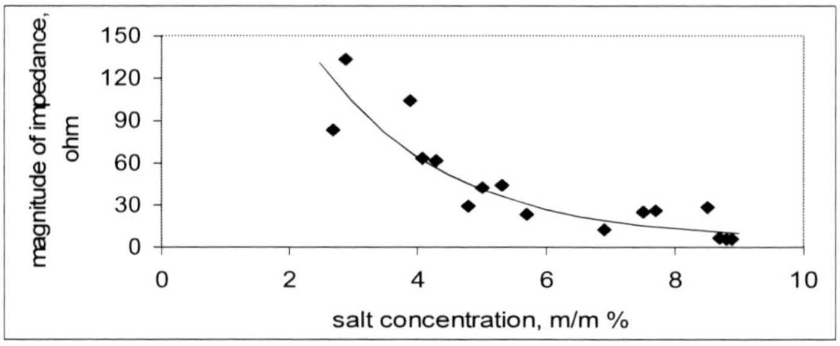

Figure 3. The impedance magnitude at 1.1 MHz in the function of NaCl content determined with a chemical method at the place of impedance measurement.

In the impedance magnitude decrease the higher rate at lower salt concentration (2-4 m/m %) and the lower rate at higher salt concentration (5-9 m/m %) [8] can be explained by hydration of proteins at lower salt concentration and by shrinkage of membranes at higher salt concentrations [7]. The various diffusion processes make difficult the fit of a general diffusion model to change of salt concentration. So an exponential connection between the impedance magnitude, Z, and the salt concentration, c, (figure 3) independently from the place of impedance measurement was found:

$$Z = 440\Omega * \mathrm{e}^{-0.5\mathrm{m/m}*c} + 5\Omega$$

In the future this mathematical relationship can be the basis of a quick and non-destructive quality tester of meat during the curing processes.

4. Conclusion

The magnitude of electrical impedance at 1.1 MHz frequency inside pork meat depends on the salt content. On the basis of this observation a quick, non-destructive impedance meter can be constructed for detection the salt content inside of meat during a curing process.

References

[1] Grimnes S and Martinsen O G 2000 *Impedance and Bioelectricity Basics* (London: Academic press) p 87

[2] Pliquett U, Altmann M, Pliquett F and Schöberlein L 2003 P_y – a parameter for meat quality *Meat Science* **65** 1429-1437

[3] Damez J-L., Clerjon S, Abouelkaram S and Lepetit J 2007 Dielectric behaviour of beef meat in the 1-1500 kHz range: Simulation with the Fricke/Cole-Cole model *Meat Science* **77** 512-519

[4] Oliver M Á, Gobantes I, Arnau J, Elvira J, Riu P, Grebol N and Monfort J M 2001 Evaluation of the electrical impedance spectroscopy (EIS) equipment for ham meat quality selection *Meat Science.* **58** 305-312

[5] Labrador R H, Masot R, Alcaniz M, Baigts D, Soto J, Martinez-Manez R, Garcia-Breijo E, Gil L and Barat J M, 2010 Prediction of NaCl, nitrate and nitrite contents in minced meat by using a voltametric electronic tongue and an impedimetric sensor *Food Chemistry* **122** 864-870

[6] Karásková P, Fuentes A, Fernández-Sgovia I, Alcaniy M, Masot R and Barat J M 2011 Development of a low-cost non-destructive system for measuring moisture and salt content in smoked fish products *Procedia Food Science* **1** 1195-1201

[7] Körmendy L 1991 Biochemical and chemical processe during curing *Practice and theory of processing of cured meat, II Meat Industry Training Days (in Hungarian)* **I** 78-84

[8] Masot R, Alcaniz M, Fuentes A, Schmidt F C, Barat J M, Gil L, Baigts D, Martinez-Manez R and Soto J 2010 Design of a low-cost non-destructive system for punctual measurements of salt level in food products using impedance spectroscopy *Sensors and Actuators A: Physical* **158** 217-223

Journal of Physics: Conference Series **434** (2013) 012095 doi:10.1088/1742-6596/434/1/012095

Bioimpedance Harmonic Analysis as a Diagnostic Tool to Assess Regional Circulation and Neural Activity

I S Mudraya , S V Revenko , L A Khodyreva, T G Markosyan, A A Dudareva, A R Ibragimov, V V Romich, and V I Kirpatovsky

Research Institute of Urology, National Institute of Cardiology, Moscow, Russia

E-mail: mis51ser@yahoo.com

Abstract. The novel technique based on harmonic analysis of bioimpedance microvariations with original hard- and software complex incorporating a high-resolution impedance converter was used to assess the neural activity and circulation in human urinary bladder and penis in patients with pelvic pain, erectile dysfunction, and overactive bladder. The therapeutic effects of shock wave therapy and Botulinum toxin detrusor injections were evaluated quantitatively according to the spectral peaks at low 0.1 Hz frequency (M for Mayer wave), respiratory (R) and cardiac (C) rhythms with their harmonics. Enhanced baseline regional neural activity identified according to M and R peaks was found to be presumably sympathetic in pelvic pain patients, and parasympathetic – in patients with overactive bladder. Total pulsatile activity and pulsatile resonances found in the bladder as well as in the penile spectrum characterised regional circulation and vascular tone. The abnormal spectral parameters characteristic of the patients with genitourinary diseases shifted to the norm in the cases of efficient therapy. Bioimpedance harmonic analysis seems to be a potent tool to assess regional peculiarities of circulatory and autonomic nervous activity in the course of patient treatment.

1. Introduction

The most diseases of genitourinary system are manifested by the disorders in blood flow and neural control. While specific methods are employed to individually evaluate any of these pathogenic factors, no method is available to assess simultaneously the changes in regional circulation and neural activity. Pelvic pain (PP) and overactive bladder (OAB) are complex syndromes involving sympathetic and parasympathetic disorders. However, there are no methods to obtain quantitative description of these painful and disturbing syndromes to objectively evaluate the treatment results.

In this study, we demonstrated the possibilities of the novel method based on the bioimpedance harmonic analysis to assess regional neural and vascular activity in human genitourinary organs during some genitourinary diseases and their treatment.

2. Study design

Regional circulatory and neural activities in the minor pelvic organs were assessed by bioimpedance harmonic analysis (BHA) in patients with pelvic pain (PP), overactive bladder (OAB) syndromes, and erectile dysfunction (ED) prior to and after the treatment.

2.1. Patient groups, examination, and treatment

All patients included in the study were divided into groups according to clinical diagnosis, therapeutic modality, and examined organ.

Patients with chronic PP (n=16), ED (n=6), and Peyronie disease (n=6) received the course of shockwave therapy (SWT). To assess regional circulatory and neural activity, two Ag-Ag/Cl electrodes 0.4 mm in diameter were applied to the patient's penis for measuring penile electric activity. To measure vesical impedance, the electrodes were applied to the front abdominal wall in projection of urinary bladder. The patients were examined prior to SWT as well as during and after the course of treatment (5-10 procedures).

In the patients with documented OAB (n=8), vesical impedance measurements were performed prior to and on day 1-4 after Botulinum toxin (BT) injections into vesical wall and sphincter.

Healthy human voluntaries composed two control groups examined with vesical (n=6) and penile (n=16) BHA.

2.2. *Method and calculations*

Two silver wire electrodes were covered with a narrow gauze bandage soaked in physiological saline, applied to penis or to the front abdominal wall in projection of urinary bladder, and connected to the impedance converter in bipolar electrode mode. The rhythmic variations of the vesical or penile impedance were recorded, processed, and automatically analyzed with fast Fourier transform implemented in an original software-hardware system ('Biola', Russia).

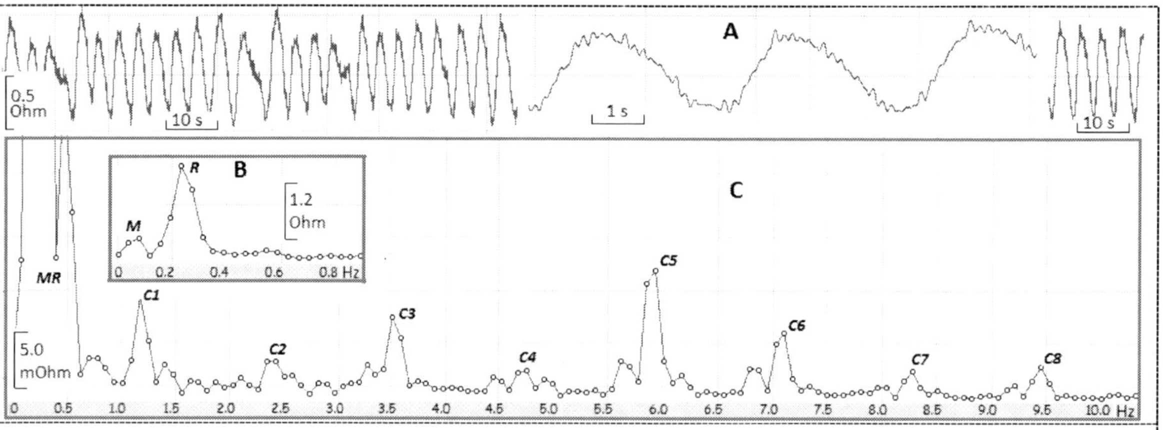

Figure 1. Examples of (A) real time recording of the bladder impedance, (B) bioimpedance harmonic analysis to reveal neural activity with epoch of 25.6 sec, and (C) BHA to assess pulsatile activity with epoch of 12.8 sec.

Calculations were performed as follows. A record of variable component (figure 1, A) was cut into 12.8 or 25.6-sec fragments containing 2048 or 4096 points. FFT was calculated with Hanning windows (figure 1, B, C). The bioimpedance variations were analyzed in the frequency band of 0.05-15.0 Hz. The periodicities in small variations of vesical and penile bioimpedance included the major spectrum peaks at the frequencies of heart beats (C), respiration (R), and Mayer (M) wave (0.1 Hz). The magnitude of selected peaks was calculated as 'efficient impedance' being equal to the square root of the sum of squared amplitudes of all individual spectrum components comprising these peaks. This procedure reflects the fact that any 'smooth' peak in digital Fourier transform consists of a number of several sinusoidal components. Total pulsatile activity (TPA) was calculated as effective impedance on the basis of cardiac C-peak and its harmonics with the epoch of 12.8 sec (2048 points) according to the formula:

$$TPA = \sqrt{C1^2+C2^2+C3^2+Cn^2 \ldots}$$

Neural activity (NA) was calculated as effective impedance corresponding to the frequency band encompassing M and R peaks measured with the epoch of 25.6 sec (4096 points). Mayer to respiratory peak ratio (M/R) was calculated to characterize the regional sympathetic-parasympathetic balance.

3. Results

Recordings of variable component of the urinary bladder impedance led from the abdominal wall made it possible to observe the bioimpedance oscillations of vascular and respiratory origin (figure 1, A). The low-frequency peak M was conventionally termed as 'Mayer wave'. Oscillations of arterial pressure at this frequency range are supposed to reflect the neural vasomotor sympathetic activity [1]. The respiration-related oscillations were usually presented with respiratory R-peak and its harmonics. The pulse-induced variations were reflected by one or several cardiac harmonics C1, C2, …, Cn. The harmonic spectra of vesical and penile bioimpedance were characterized with high M- and R- peaks, which were several times greater than the cardiac harmonics. This fact is interpreted as indication on the neurogenic nature of M and R rhythmic bioimpedance variations in these organs. Moreover, our experimental studies supported the idea that M-peak reflects sympathetic activity, while R-peak is an indicator of parasympathetic regional activity [2].

The parameter NA was elevated in vesical and penile spectra compared to the norm in patients with pelvic pain syndrome evidencing to enhanced neurogenic activity in minor pelvic organs. The increased neurogenic activity was presumably sympathetic, as revealed by elevated M/R ratio. SWT diminished the pain syndrome by decreasing NA in both organs, while M/R ratio decreased and approximated to the normal values thereby evidencing moderation of sympathetic activity.

Table 1. Parameters of bioimpedance analysis in patient groups.

Patient groups	NA (mOhm)	M/R	TPA (mOhm)
Bladder, pelvic pain	359.2±70.9	0.65±0.27	24.0±3.1
After SWT	106.4±18.6	0.21±0.04	14.2±2.4
Overactive bladder	238.9±120.6	0.16±0.05	16.6±4.4
After BT treatment	250.6±188.2	0.32±0.09	37.8±21.3
Normal bladder	126.6±43.0	0.21±0.06	13.9±1.2
Penis, pelvic pain	31.4±12.7	1.93±0.44	16.6±5.5
After SWT	12.7±5.0	0.99±0.37	4.2±1.7
Penis, ED	43.6±16.5	1.85±0.76	6.72±3.29
After SWT	28.5±10.6	1.18±0.07	6.08±1.20
Penis, ED+Peyronie	32.6±12.2	1.12±0.10	6.92±1.52
After SWT	29.4±7.1	1.41±0.51	5.96±0.88
Normal penis	16.0±0.8	0.61±0.11	4.6±0.8

The abnormally elevated NA parameter was documented in patients with OAB. In these patients, an enhanced neurogenic activity was probably determined by the parasympathetic component as evidenced by diminished M/R ratio. The neurogenic activity remained enhanced in BT-treated bladder with M/R ratio shifted in favor of sympathetic activity.

In patients with pelvic pain, TPA was significantly elevated in comparison with the normal value. After SWT, it diminished significantly in both examined organs. In these patients, the changes in TPA were paralleled with a decrease of sympathetic activity suggesting the potency of SWT to dramatically moderate the painful syndrome due to depression of the regional neural activity. Moreover, TPA diminished due to disappearance of resonant harmonics in successfully treated patients (figure 2, a).

In patients with OAB, TPA and sympathetic activity increased after BT treatment. The elevation of pulsatile activity resulted from a 2.5-fold increase of fundamental C1-harmonic (31.1 vs 12.2 mOhm), while the resonant harmonics decreased in these patients (figure 2, b) probably evidencing the changes in viscoelastic properties of the vesical arteries.

BHA helped to delineate the diagnostics of the pathologic features of ED, and to provide objective information on therapeutic efficacy. The elevated parameter NA in patients with ED decreased after SWT in the group of patients with ED only, but it did not change in the patients with ED complicated by Peyronie disease. The M/R ratio diminished slightly in patients with ED, although, it increased in the group "ED+Peyronie". Analysis of the changes in C harmonics in the above groups showed that the resonant harmonics successfully disappeared in the group "ED" (figure 2, c), while they remained pronouncedly elevated in the group "ED+Peyronie" (figure 2, d).

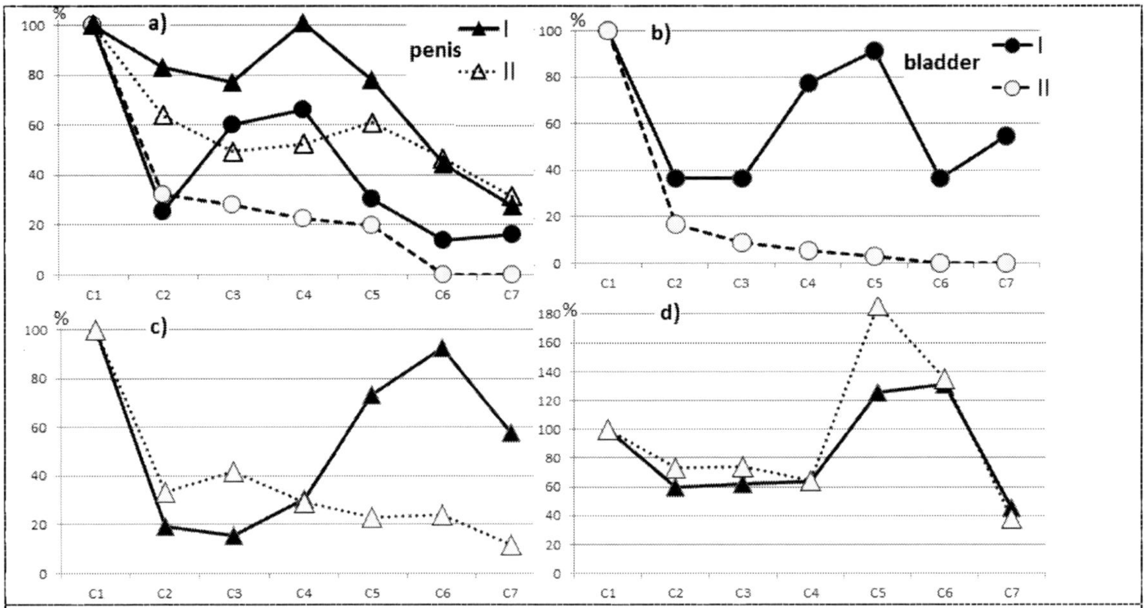

Figure 2. Percent changes in C-harmonics in the bladder (circles) and penis (triangles) impedance spectra before (**I**- solid lines) and after (**II**- dotted lines) treatment of the patients with **a)** pelvic pain; **b)** OAB syndrome; **c)** ED; and **d)** ED+Peyronie disease.

Thus, BHA can be routinely used prior to therapy for comprehensive diagnostics of the functional disorders on the basis of quantitative physiologic indices; moreover, it can assess the early and long-term effects of selected treatment modality.

4. Conclusions

The study showed that BHA can be an informative tool to assess regional neural and circulatory status in diverse pathologies. Moreover, this method gives the objective data on therapeutic efficacy.

The increased neurogenic activity in minor pelvic organs of the patients with pelvic pain originated mostly from enhanced sympathetic component, while in the patients with overactive bladder, it resulted from augmented parasympathetic component. In patients with pelvic pain and erectile dysfunction, circulation disorders were characterized by elevated total pulsatile activity in the bladder and penile impedance spectra. Large pulsatile resonances attested to the changes in viscoelastic properties of the arteries probably indicated abnormalities in vascular conditions.

Physiotherapy employing the focused shock waves seems to have a considerable advantage in treatment of the patients with chronic pelvic pain due to moderating abnormally enhanced regional sympathetic activity. In the patients with overactive bladder, Botulinum toxin treatment shifted regional sympathetic-parasympathetic balance in favor of sympathetic activity. A high fundamental cardiac harmonic with moderate pulsatile resonances indicate the healthy status of regional circulation in all examined patients and attest to efficacy of the corresponding treatment.

References

[1] Julien C 2006 *Cardiovasc. Res.* **70** 12
[2] Mudraya I S, Revenko S V, Nesterov A V, Gavrilov I Yu and Kirpatovsky V I 2011 *Physiol. Meas.* **32** 959

AUTHOR INDEX

AUTHOR INDEX

AUTHOR INDEX

AUTHOR INDEX